Operations Research for Military Organizations

Hakan Tozan
Istanbul Medipol University, Turkey

Mumtaz Karatas
National Defense University, Turkey

A volume in the Advances in Logistics, Operations, and Management Science (ALOMS) Book Series

IGI Global
DISSEMINATOR OF KNOWLEDGE

Published in the United States of America by
IGI Global
Information Science Reference (an imprint of IGI Global)
701 E. Chocolate Avenue
Hershey PA, USA 17033
Tel: 717-533-8845
Fax: 717-533-8661
E-mail: cust@igi-global.com
Web site: http://www.igi-global.com

Library of Congress Cataloging-in-Publication Data

Names: Tozan, Hakan. editor. | Karatas, Mumtaz, editor.
Title: Operations research for military organizations / Hakan Tozan and Mumtaz Karatas, Editors.
Description: Hershey, PA : Information Science Reference, IGI Global, 2018.
Identifiers: LCCN 2017045805| ISBN 9781522555131 (hardcover) | ISBN 9781522555148 (ebook)
Subjects: LCSH: Operational art (Military science) | Operations research--Methodology.
Classification: LCC U104 .O64 2018 | DDC 355.6/88--dc23 LC record available at https://lccn.loc.gov/2017045805

This book is published in the IGI Global book series Advances in Logistics, Operations, and Management Science (ALOMS) (ISSN: 2327-350X; eISSN: 2327-3518)

British Cataloguing in Publication Data
A Cataloguing in Publication record for this book is available from the British Library.

For electronic access to this publication, please contact: eresources@igi-global.com.

Advances in Logistics, Operations, and Management Science (ALOMS) Book Series

John Wang
Montclair State University, USA

ISSN:2327-350X
EISSN:2327-3518

Mission

Operations research and management science continue to influence business processes, administration, and management information systems, particularly in covering the application methods for decision-making processes. New case studies and applications on management science, operations management, social sciences, and other behavioral sciences have been incorporated into business and organizations real-world objectives.

The **Advances in Logistics, Operations, and Management Science** (ALOMS) Book Series provides a collection of reference publications on the current trends, applications, theories, and practices in the management science field. Providing relevant and current research, this series and its individual publications would be useful for academics, researchers, scholars, and practitioners interested in improving decision making models and business functions.

Coverage

- Finance
- Services management
- Decision analysis and decision support
- Risk Management
- Organizational behavior
- Marketing engineering
- Computing and information technologies
- Networks
- Production management
- Political science

IGI Global is currently accepting manuscripts for publication within this series. To submit a proposal for a volume in this series, please contact our Acquisition Editors at Acquisitions@igi-global.com or visit: http://www.igi-global.com/publish/.

The Advances in Logistics, Operations, and Management Science (ALOMS) Book Series (ISSN 2327-350X) is published by IGI Global, 701 E. Chocolate Avenue, Hershey, PA 17033-1240, USA, www.igi-global.com. This series is composed of titles available for purchase individually; each title is edited to be contextually exclusive from any other title within the series. For pricing and ordering information please visit http://www.igi-global.com/book-series/advances-logistics-operations-management-science/37170. Postmaster: Send all address changes to above address.

Titles in this Series

For a list of additional titles in this series, please visit: www.igi-global.com/book-series

Cases on Quality Initiatives for Organizational Longevity
Archana Krishnan (Institute of Innovation in Technology and Management, India)
Business Science Reference • copyright 2018 • 285pp • H/C (ISBN: 9781522552888) • US $195.00 (our price)

Enhancing Competitive Advantage With Dynamic Management and Engineering
Carolina Machado (University of Minho, Portugal) and J. Paulo Davim (University of Aveiro, Portugal)
Business Science Reference • copyright 2018 • 315pp • H/C (ISBN: 9781522553601) • US $195.00 (our price)

Information Technology as a Facilitator of Social Processes in Project Management and Collaborative Work
Timothy C. Bagwell (Walden University, USA) Robert A. Cropf (Saint Louis University, USA) and Sheryl L. Foster-Gadkari (Strategic Impact, USA)
Business Science Reference • copyright 2018 • 217pp • H/C (ISBN: 9781522534716) • US $185.00 (our price)

Handbook of Research on Cross-Cultural Business Education
Chandan Maheshkar (University of Indore, India) and Vinod Sharma (Christ Institute of Management, India)
Business Science Reference • copyright 2018 • 463pp • H/C (ISBN: 9781522537762) • US $235.00 (our price)

Handbook of Research on Supply Chain Management for Sustainable Development
Ulas Akkucuk (Bogazici University, Turkey)
Business Science Reference • copyright 2018 • 439pp • H/C (ISBN: 9781522557579) • US $265.00 (our price)

Globalization and Trade Integration in Developing Countries
Ahu Coşkun Özer (Marmara University, Turkey)
Business Science Reference • copyright 2018 • 320pp • H/C (ISBN: 9781522540328) • US $205.00 (our price)

Handbook of Research on Intrapreneurship and Organizational Sustainability in SMEs
Rafael Perez-Uribe (EAN University, Colombia) Carlos Salcedo-Perez (EAN University, Colombia) and David Ocampo-Guzman (EAN University, Colombia)
Business Science Reference • copyright 2018 • 475pp • H/C (ISBN: 9781522535430) • US $295.00 (our price)

Entrepreneurship, Collaboration, and Innovation in the Modern Business Era
Mehdi Khosrow-Pour, D.B.A. (Information Resources Management Association, USA)
Business Science Reference • copyright 2018 • 367pp • H/C (ISBN: 9781522550143) • US $215.00 (our price)

IGI Global
DISSEMINATOR OF KNOWLEDGE

701 East Chocolate Avenue, Hershey, PA 17033, USA
Tel: 717-533-8845 x100 • Fax: 717-533-8661
E-Mail: cust@igi-global.com • www.igi-global.com

Editorial Advisory Board

Table of Contents

Detailed Table of Contents

In the past decades, facility location problems have attracted much attention among researchers and practitioners from different disciplines. Among those problems, location models observed in military organizations have significant impact to the performance of the military organization since they require large amounts of money, resource, and people. Moreover, an efficient planning of military resources often leads to a good direction to victories. In this chapter, considering a number of selected papers, the authors give a brief survey of facility location models and solution techniques employed for military organizations. After providing the features of core location models, they analyze the military facility location models with respect to the context they are handled. After categorizing the articles with respect to the formulations and solution approaches employed, the authors highlight potential issues for further research.

In this chapter, locating and routing of a UAV fleet is discussed. Since the research in the location and routing of UAVs is very limited, the related problems are reviewed. A basic problem and its extended version found in the literature are redefined along with their MILP formulation. Both of the problems are characterized by a prize collecting objective function and a homogeneous fleet of fixed number of UAVs based on naval platforms. The aim is described as maximization of the collected importance value associated to interest points. A solution method that is based on ACO is discussed. The usage of this robust method for both the basic and the extended versions of the problem is explained. The improvement room in the present solution method and future research directions are also discussed.

As more and more Commercial Off The Shelf (COTS) parts are being used in sustainment-dominated systems where average product lifecycle is above 25 years, obsolescence management becomes a greater concern for program managers. The lack of management and poor planning for obsolescence cause companies, governments, and military organizations to spend progressively more to deal with aging systems. For a successful obsolescence management, program managers should consider both the cost-effectiveness and system availability issues simultaneously and should implement appropriate solution approaches. In this chapter, the authors first define the obsolescence management in sustainment-dominated systems and then give a brief summary of the related literature. They finally discuss and propose multiple criteria decision-making methodologies and evolutionary algorithms to tackle the management problem.

The U.S. Navy's at-sea replenishment system is a mobile supply line designed to support the deployed carrier task force (CTF)/cruiser/destroyer (CRUDES) surface action group (SAG) and forward deployed units while at sea. In the Pacific, the main component of the mobile supply line, the combat logistics force (CLF) ship, has become a possible target with the development of the anti-ship ballistic missile. With the ability to target and disable a CLF, an enemy can now disable a deployed CTF/CRUDES fleet by eliminating its required resources. With the goal of preserving the CLF's capabilities to perform its mission while avoiding ASBM threat, the authors consider the possibility of utilizing a "mini-CLF" to shuttle fuel between CLFs operating in a safe environment and warships operating in a threat zone. The authors perform two analyses: they (1) analyze the feasibility of using the Littoral combat ship/joint high-speed vessel, reconfigured as a shuttle to transport resources, and (2) analyze requirements for development of a new class of ships to support the CTF/CRUDES SAG while deployed in the Pacific.

The manual process for flight scheduling at Naval Air Station Lemoore accommodates the needs of 16 fighter resident squadrons as well as constraints imposed by limited military operating area availability. Given the complexity of this problem, attempting to additionally avoid periods of high activity and resultant congestion would challenge the manual process. However, congestion leads to long wait times for flight-line services. Refueling operations are particularly costly when operational time is lost, and resources are backlogged. The problem of avoiding inefficient periods of high demand is complicated by the two types of refueling available: hot refueling, which occurs when the aircraft's engine is running, and cold refueling, which occurs when the aircraft is shut down. Achieving a balance between the two refueling methods is key to maximizing operational effectiveness. The authors discuss an optimization model designed to determine the best daily flight schedules based on the squadrons' flying and training requirements, the refueling infrastructure, and range availability.

The success of military operations mainly relies on the proper flow of the logistical supplies such as water, food, ammunition, etc. from source to the operation theater on time. There are special types of transportation vessels regarding the feature of supply. However, when transporting special material like ammunition, most navies usually prefer utilizing their own transportation capabilities since they require special treatment. For this reason, such material is carried in special boxes, called containers. To minimize the transportation cost and time, an efficient container stowage plan is necessary in terms of loading and unloading these containers. This chapter aims to develop a solution methodology to the problem with the focus on military logistics planning. For this purpose, the author develops a mathematical model that attempts to minimize the transportation time by creating proper loading and unloading sequence of containers to military cargo ships.

The main goal of the bearing-only target motion analysis (BOTMA) is to determine the target's kinematic parameters such as position, course, and speed by only using the bearings reported by an onboard passive sensor (e.g., a sonar or an ESM [electronic support measures] device). This chapter provides a brief description of the BOTMA problem. Next, it discusses the implementation of the Kalman filtering technique to solve the problem. The authors then discuss the variations of the Kalman filtering (i.e., the extended and unscented Kalman filters). They also propose a genetic algorithm metaheuristic that incorporates a novel search space narrowing technique to solve the BOTMA problem and present numerical results for different noise conditions. They finally highlight the future research directions for modeling and solving the BOTMA problem.

Today, in the era of data and computing, fast and reliable retrieval of information has become of great importance for security and military applications, and continues to be such, as the amount available digital data increases every second. While the search and retrieval of text data has produced mature products and are today being used in search engines everyday by everyone, the retrieval of spoken content still remains a young research, especially for low resource languages where the available data is scarce to train reliable speech recognition systems. This chapter provides a thorough introduction of a speech retrieval task called "keyword search" and presents a novel similarity measure optimization-based approach. The

Foreword

During World War II, the need for a new approach for the solutions of hard military problems in addition to the improvements in military equipment and weapons ended up with the arise of a new scientific discipline called Operations research (OR). In time OR became a vital approach that provides advanced analytical methods and techniques to solve complex decision problems. And; in addition to military domain, OR techniques have been came into use in various fields such as business, industry, healthcare to make better strategic decisions.

This book is a new source of knowledge which artily gets inside military OR literature. With its sixteen chapters, the book presents a unique reference to military organizations in the field of OR applications.

The book is designed to advance scientific knowledge and to foster innovative engineering solutions for the military engineering community.

Chapter 1 presents a brief survey of facility location models and solution approaches employed for military organizations. A classification mechanism about military facility location models are provided.

The second chapter discusses the problem of locating and routing Unmanned Aerial Vehicles. Ant Colony Optimization is presented and discussed as a heuristic solution methodology with experimental results.

Chapter 3 overviews set covering problems which are generally employed for searching the optimal alternatives, planning, and allocating resources for public support and service systems. The author discusses the applicability of set covering models to military decision problems.

The fourth chapter proposes a methodology to allocate Search and Rescue resources optimally. The authors utilize simulation to test the performance of their proposed solution under stochastic demand. A case study conducted in the Aegean Sea is also presented.

The fifth chapter focuses on the concepts in Test and Evaluation for weapon systems in the navy and introduces the process for its implementation as part of a procurement decision.

The sixth chapter investigates the resource allocation problem in planning the military missions and its solution methodologies such as Artificial Neural Network, Genetic Algorithm, Simulated Annealing.

The seventh chapter focuses on the classical Lanchester equations, significant extensions of classical models and valuable applications in war modeling.

The eighth chapter aims at providing basic information about Multi-criteria Decision Making (MCDM) methods such as Electre, Topsis, Promethee for their possible application in military decision-making problems.

The ninth chapter introduces a fuzzy MCDM framework based on Hierarchical Fuzzy Technique for Order Preference by Similarity to Ideal Solution method. The study is aimed at solving the best missile system in a fuzzy environment.

The tenth chapter proposes a MCDM approach using evolutionary algorithms for obsolescence management in sustainment dominated military systems.

The eleventh chapter introduces the U.S. Navy's at-sea replenishment system. The authors discuss the feasibility of using the Littoral Combat Ship/Joint High-Speed Vessel, reconfigured as a shuttle to transport resources, and analyze requirements for development of a new class of ships to support the Carrier Task Force (CTF)/Cruiser/Destroyer (CRUDES) Surface Action Group (SAG) while deployed in the Pacific.

Chapter 12 proposes an optimization model in order to determine the best daily flight schedules based on the squadrons' flying and training requirements, the refueling infrastructure, and range availability.

The thirteenth chapter suggests an algorithm that yields efficient loading plan of containers via minimizing time spent in any port for the proper flow of logistical supplies.

The fourteenth chapter focuses on bearing-only Target Motion Analysis (BOTMA) concept. A new approach is presented. The authors employ evolutionary algorithms to solve the BOTMA problem.

Chapter 15 offers a speech retrieval task called 'keyword search' and presents a novel similarity measure optimization-based approach. The case study is experimented on telephone conversations in three different languages.

The sixteenth chapter focuses on the importance of clutter removal algorithms and analyzes the performance of clutter removal algorithms on the simulated dataset that is created by using the gprMax simulation software where it contains four different challenging scenarios.

Eventually, it is obvious that the development of OR methodologies and its applications in military organizations continue to flourish. Hence, this book is a good step in that direction. In fine; this book provides a novel single source to policy makers and managers of military organizations about operating their units in a more effective and efficient way.

Finally, I thank all the authors, editors and reviewers for the excellent work you put in to making this book a great success!

Marta Harničárová
Slovak University of Agriculture, Slovakia

Preface

The modern field of operations research (OR) arose during World War II in an effort to enhance the effectiveness of weapons and equipment used in the battlefield. Since then, OR techniques have been used to solve many sophisticated and complex defense-related problems not only limited to combat operations but also encompassing logistics, manpower planning, equipment procurement, training, infrastructure defense, and many other areas. Indeed, the global military community has fully embraced OR as a critical tool for ensuring efficiency and readiness; the existence of permanent OR branches in many military services and the growing number of research papers, books, and teaching and training activities attest to this fact.

OBJECTIVE OF THE BOOK

The goal of this book is to provide readers with detailed descriptions of military problems and solutions incorporating a broad array of contemporary OR techniques. In particular, this book addresses and features original research on (1) innovative applications of OR techniques and methodologies in defense and military-related problems, and (2) relevant theoretical frameworks and the latest empirical research findings in the area.

Operations Research for Military Organizations features spotlights on some important applications of location models, multi-criteria decision-making methodologies, heuristic & metaheuristic solution approaches, neural network learning, and different tools utilized for solving problems observed in the military context.

TARGET AUDIENCE

The target audience for this book includes researchers, consultants, academics, military and civilian personnel of defense other and government departments, students in military schools and universities, and practitioners involved in all related fields.

ORGANIZATION OF THE BOOK

The book is organized into 16 chapters. We now provide a brief overview of each chapter.

Chapter 1

The first chapter by Karatas et al. is a brief survey of location problems observed in the military field. Facility location models have received considerable attention from many researches in the past decades and several applications have been conducted in different fields. Among the problem types studied in location science, problems observed in military organizations are especially notable due to their significant practical impact. These organizations typically require large amounts of money, resources, and personnel in order to operate, and the impact of poor planning can be large, both in terms of dollars and lives. This survey aims to study the location problems observed in a military context and draw a general picture of the research in the field by delineating several sub-problem types, such as military logistics planning, infrastructure security and protection, sensor deployment, unmanned aerial vehicle (UAV) location, and search and rescue (SAR) operations planning. The authors also report a summary of their literature analysis with respect to the structure of models considered and the solution techniques available, such as heuristics, metaheuristics, and exact mathematical programming models. They finally highlight potential issues for further research.

Chapter 2

In the second chapter, Yakıcı et al. introduce a new variant of a location and routing problem for a fleet of ship-borne UAVs. These vehicles can be used in many military missions such as search and rescue, damage assessment, and surveillance. In all applications, stationing and routing are critical to performance. The introduced model takes as input a set of surveillance (demand) points, a UAV fleet, a naval fleet, and a set of candidate stations for the ships. While not all of the demand points are required to be visited, each demand point brings a certain benefit if it is visited by a UAV. The objective of the problem is to maximize the total benefit collected by the UAVs. The decision process creates an assignment of ships to a subset of stations, an assignment of UAVs to these ships, and a routing for each UAV. The authors also analyze an extension of the problem which incorporates time windows. Due to the complexity of these problems, the authors apply of a variant of ant colony optimization method. This chapter presents a novel variant of a well-studied problem, as well as an efficient heuristic method to solve it.

Chapter 3

In the third chapter, Argun presents a brief survey of set covering problems and their applications in the military context. After providing a formal description of the problem and its mathematical model, the author first gives a brief survey of studies which incorporate set cover models in real-world problems. Next, she provides a review of military applications which utilize set covering approach.

Chapter 4

In the fourth chapter, Karatas et al. consider the problem of determining the optimal locations of SAR resources. A SAR operation aims to eliminate or minimize injuries, deaths, and material damage by rendering aid to persons in distress or imminent danger. Considering the uncertainties and unfriendly environment of maritime zones, a SAR operation conducted at sea is a critical and risky process. Hence, locating SAR resources to appropriate bases geographically is as important as using correct SAR vehicles, materials, and methods. Allocating SAR vessels is a strategic-level process that must be carried out efficiently in order to react rapidly to emergencies at sea. Motivated by this fact, this chapter seeks to assess the performance of a SAR boat location and allocation plan using simulation. The proposed method first implements a mathematical model with the objective of determining an optimal SAR boat allocation scheme. At this phase of the solution, the authors adopt a previously developed decision support tool called the "Incident Based Boat Allocation Model (IB-BAM)". IB-BAM first ranks and assigns a weight to each incident type observed in the area of interest. Second, it generates aggregated weighted demand locations utilizing historical incident data. This is called as the Zonal Distribution Model. Finally, IB-BAM employs a multi-objective mixed integer program which seeks to determine locations and responsibility zones of each SAR boat. However, the mathematical model used in this phase does not incorporate the uncertainty of the SAR operations in terms of the number of distress calls, demand generation times, times between calls, etc. To answer this need, in the next phase the authors use simulation to test the performance of our analytical solution under stochastic demand. Using this methodology, the authors study the problem of locating SAR resources in the Aegean Sea coast of Turkey.

Chapter 5

In Chapter 5, Eriskin and Gunal review the methodology of test and evaluation (T&E) in weapon systems. T&E is an essential phase of the weapon systems acquisition process regardless of the system being procured or developed. All countries require T&E prior to deployment. Two broad types of T&E are specified: developmental T&E and operational T&E. Developmental T&E focuses on engineering-level tests which check whether the system meets its technical and performance specifications as documented. Operational T&E examines a system to determine whether it fulfills its operational-level mission requirements. This chapter provides details concerning both types of T&E, such as which entities will coordinate and perform the tests, as well as the overall process of T&E. Additionally, the chapter also discusses the role of modeling and simulation and the Simulation, Test, and Evaluation Process (STEP) methodology.

Chapter 6

The sixth chapter by Metin and Bulkan is a condensed survey of the resource allocation problems and their solution methods. In recent years, advancing technology has introduced a wide range of resources for military forces. However, these rapid improvements have also caused resource allocation management difficulties in combat missions. The resource allocation problem has a greater importance in military organizations since these organizations mostly deal with large amounts of personnel, equipment, and assets of high value. The authors consider the problem of determining an optimal sequence of resource allocations which is used for maximizing enemy kills or minimizing friendly casualties. In this chapter, the authors provide a brief literature survey and categorize the various resource allocation problem types, such as weapon deployment and weapon mix problems. They also give an overview of mathematical modeling and solution approaches for a number of small-sized examples in the military context, and they describe heuristic optimization models. Finally, they highlight potential directions for future research.

Chapter 7

The seventh chapter by Karacan et al. consists of a review of attrition models based on Lanchester's Laws and their applications. Attrition models are introduced to the literature purely for modeling the attrition of fighting units of opposing forces during a battle. The classical models are composed of two types of ordinary differential equations. Lanchester's Linear Law is applied to model the ancient war that includes duels of units with simple weapons such as swords, axes, and arrows. Lanchester's Square Law is used to explain the actions of firearms such as rifles, cannons, artilleries, and aircraft that are intended to inflict multiple casualties simultaneously. Both laws have been utilized to model many types of wars. Since the equations are restricted by various simplifying assumptions, many researchers have developed modified versions of the equations to handle more complicated real-life combat scenarios. Novel equations for reinforcements of units, the effects of intelligence, mixed models that can handle both aimed and unaimed fire, and multiple opponents instead of duels are common examples of contemporary approaches. The Battle of Trafalgar, Battle of Iwo Jima, and Battle of Atlantic are some of the well-known historic battles that are modeled and analyzed using Lanchester equations. In the chapter, the authors also give a number of real world problem examples that occur in different domains, such as modeling market shares, behavior of social animals, and network attacks.

Chapter 8

In Chapter 8, Temucin considers the multi-criteria nature of decision problems observed in the military context and provide insight on the use of Multi-Criteria Decision Making (MCDM) techniques. In a military context, a decision problem generally concerns the selection of the best alternative or a set of alternatives under certain conditions and assumptions. Decision-making techniques for such problems observed in military organizations should consider the complex structure, multi-criteria nature, and high uncertainties involved in military operations. These factors significantly complicate many real-life military decision problems. Therefore, MCDM methods are commonly used in this domain since they fit to the chaotic, complex, and ambiguous nature of military problems. In this context, this chapter provides an overview of the MCDM techniques used in the military domain and describes the basis of decision making and decision processes in military decision problems.

Chapter 9

In Chapter 9, Karadayi et al. present an application of MCDM techniques to a weapon system selection problem. Clearly, weapon selection is crucial due to the fact that the weapon systems used in a war have a significant impact on the outcome of the war. However, this process requires a detailed assessment of weapon systems with respect to multiple criteria. In this chapter, the authors adopt a fuzzy MCDM framework based on Hierarchical Fuzzy Technique for Order Preference by Similarity to Ideal Solution (HFTOPSIS). The proposed approach is able to incorporate both crisp data and fuzzy data and has the advantage of defining the complexity of the problem via its hierarchical structure. In particular, the authors study the problem of selecting a surface-to-air missile system to illustrate the application of the proposed MCDM approach. They assess the performance of those weapon systems in terms of 19 criteria under three categories including basic capabilities, operational capabilities, and cost & technical effects.

Chapter 10

The tenth chapter by Ozkan and Bulkan is a brief survey of obsolescence management problems observed in sustainment dominated systems. Ever since the end of the Cold War, the consumer electronics and integrated circuit markets became dominant and high market shares moved from military products to small-scale consumer products. Fast technological advances created new consumer markets in which the needs are constantly changed by either producers or consumers. Triggered by these technological advances, the life cycle of consumer products has shortened through the years. This fact created a new category for product acquisition: Commercial off-the-Shelf (COTS). Although there are several advantages of using COTS products, many disadvantages exist as well, including product volatility, obsolescence, and security. Obsolescence is a primary concern sustainment-dominated systems. Due to the high complexity of the environment in which these products are used, they have higher standards for re-qualification and re-certification in the event of a design change. In some cases, certain COTS products go obsolete even before production starts. Therefore, obsolescence is a major problem for systems with a long life cycle, and this problem has to be addressed by efficient obsolescence management plans. The lack of management and poor planning for obsolescence cause companies, governments, and military organizations spend progressively more to deal with aging systems. In this chapter of the book, authors present a brief summary of operations research models for obsolescence management in sustainment-dominated systems. They also propose a new approach for using multi-criteria decision-making techniques for this problem.

Chapter 11

In Chapter 11, Colburn and Craparo use OR techniques to explore a potential new concept of operations for wartime logistics. Due to advancing enemy missile technology, the current dependence on large combat logistics force (CLF) ships to conduct at-sea replenishment is becoming increasingly risky. The loss of a single CLF ship could have a devastating impact on operations. Colburn and Craparo consider the possibility of using a hypothetical fleet of "mini-CLF" ships designed to shuttle fuel from large CLF ships to warships, thus allowing the large CLF ships to remain outside enemy missile range. Such a significant change to current operations would be enormously costly and would entail significant modifications to current operational patterns. Thus, it is an ideal candidate for in-depth OR analysis prior to implementation. Colburn and Craparo perform the first step of such an analysis by developing a mixed-integer linear program that prescribes the movement and refueling schedules of a set of warships, CLF ships, and mini-CLF ships. They find that the "mini-CLF" concept has clear merit and allows large CLF ships to remain outside enemy missile range while increasing on-station time for the warships. However, the authors also find that the computational aspects of planning such operations are considerable. This highlights the need for future work to develop alternative models or heuristics before such a setup can be considered in practice.

Chapter 12

While Chapter 11 uses OR techniques to evaluate a completely novel concept of operations, in Chapter 12 Rosas and Craparo use similar techniques to explore more modest modifications to an existing process. Specifically, they model the refueling process at a particular naval air station in order to analyze its current level of efficiency and the station's ability to serve a large number of aircraft in the future. Their model, a mixed-integer linear program, prescribes refueling schedules to aircraft and assigns resources to support these schedules. In order to model the existing process of scheduling training flights, large modifications to the aircraft's requested flight times are not allowed. Using this model, the authors find that the current refueling infrastructure is able to accommodate additional aircraft with no major changes to the planning process. The authors also discover a new route to efficiency: by allowing additional flexibility in the process of assigning aircraft to hot refueling pits, the station could achieve more rapid refueling with less wasted fuel.

Chapter 13

Chapter 13 by Bilican is a brief survey of Container Stowage Plan (CSP) problems observed in the military field. Military logisticians often utilize containerized transportation in the battlefield. Rapid delivery of military goods is crucial for the success of the operations, and thus an efficient CSP is crucial. Over the past two decades, there has been a continuous increase in container transportation leading to deployment of larger container vessels. Along with the increase in vessel sizes, determining a proper and efficient loading sequence is of great importance both to shorten port visit times and to avoid unnecessary container movements. This is the primary goal of the CSP. Considering the close relationship between military transportation and the CSP, this chapter aims to develop a solution methodology to the problem with a focus on military logistics planning. For this purpose, the author first provides a formal description of the CSP and reports a summary of the related literature on the topic in terms of modeling techniques and solution methodologies. After setting forth the relation between CSP and military transportation problems, he develops a mathematical model which attempts to minimize the transportation time (the number of unnecessary container movements) by creating proper loading and unloading sequence of containers to military cargo ships. After presenting numerical results for problems arising from the transportation of military containers by a navy vessel, he finally highlights the potential studies reserved for future research.

Chapter 14

In Chapter 14, Aytun and Bulkan discuss the application of a genetic algorithm to solve the Bearing Only Target Motion Analysis (BOTMA) problem. Target Motion Analysis (TMA) mainly aims to extract the kinematic parameters (e.g. speed and position) of a platform of interest by gathering and using available information such as bearing, range, and frequency. TMA is mostly used in anti-submarine warfare (ASW) operations in which surface assets search for hidden enemy submarines. Hence, an accurate solution to the TMA problem has a significant impact on the success of the operation and potentially on the outcome of the war. For this reason, the TMA problem has attracted a considerable amount of attention and several solution methods have been proposed by many researchers from different disciplines. This chapter first provides brief descriptions and the critical observability requirements of the TMA and BOTMA problems. Next, it describes a commonly-used filtering technique, Kalman filtering, to solve the problem. Additionally, it discusses variations of the Kalman filtering technique, such as the Extended and Unscented Kalman filters are discussed. The authors then propose a genetic algorithm metaheuristic which incorporates a novel search space narrowing technique to solve the BOTMA problem, and they present numerical results for different noise conditions. They finally highlight the future research directions for modeling and solving the BOTMA problem.

Chapter 15

In the fifteenth chapter, Gundogdu and Saraclar propose a similarity optimization methodology, based on neural network learning of a new distance metric in a detection task. They investigate the issue in a case study, in which pertinent parts of speech on recordings of telephone conversations are sought. This task is called keyword search and it is very important in terms of both military applications and homeland security issues in the prevention or detection of terrorist activities. In the similarity optimization methodology proposed by the authors, the frame representations of the text queries are learned using labeled training data, along with a proper distance metric to be used with these representations. The text queries are artificially modeled in forms in which the audio had been converted and hence a similarity matching-based search is feasible to retrieve the parts of interest in the audio. The proposed distance metric learning methodology is carried out with asymmetric Siamese neural networks, such that the network takes two vectors as input and emits a scalar distance value between them. This distance measure is then used in the dynamic time warping based search between the query representation and the audio representation. The chapter shows that such a learning of representations and the distance metric performs significantly better than the known distance metrics.

Chapter 16

The sixteenth chapter by Kumlu and Erer presents a survey of the popular clutter removal methods and their comparisons for efficient landmine detection. This is still an open problem, and landmines pose a significant threat to civilians where developing nations and humanitarian aid organizations are actively struggling to clear them. It is estimated that there are more than 110 million buried landmines in approximately 70 countries, and every year nearly 26,000 victims die due to landmine explosions. According to the current demining rate, clearing of all existing mines could take approximately 450-500 years. Therefore, accurate and fast landmine detection methods are of great importance. A popular tool for landmine detection is ground-penetrating radar; however, its performance is greatly affected by clutter. This survey explains the reasons for clutter, such as ground-bounce, direct wave arrival, other candidate objects, etc. It then examines the popular clutter removal methods available in the literature by testing them using the simulated dataset which is constructed by gprMax simulation software tool. By using the constructed simulated dataset, quantitative analysis can be given in addition to visual results and thus the comparison becomes more objective. The authors finally highlight the advantages and disadvantages of the methods.

Finally, it is our pleasure to express our sincere thanks to our authors, referees and editorial board members for their cooperation. Without their assistance and encouragement, the task of putting this book together would have been a lot less pleasant. Many thanks to all of them.

We hope you find our book useful.

Hakan Tozan
Istanbul Medipol University, Turkey

Mumtaz Karatas
National Defense University, Turkey

Chapter 1
Military Facility Location Problems:
A Brief Survey

Mumtaz Karatas
National Defense University, Turkey

Ertan Yakıcı
National Defense University, Turkey

Nasuh Razi
Turkish Naval Forces, Turkey

ABSTRACT

In the past decades, facility location problems have attracted much attention among researchers and practitioners from different disciplines. Among those problems, location models observed in military organizations have significant impact to the performance of the military organization since they require large amounts of money, resource, and people. Moreover, an efficient planning of military resources often leads to a good direction to victories. In this chapter, considering a number of selected papers, the authors give a brief survey of facility location models and solution techniques employed for military organizations. After providing the features of core location models, they analyze the military facility location models with respect to the context they are handled. After categorizing the articles with respect to the formulations and solution approaches employed, the authors highlight potential issues for further research.

INTRODUCTION

In the past decades, location science has attracted many researchers and practitioners from different disciplines and fields. Being a very rich field, location science includes several types of location problems, solution techniques as well as considerable amount of theoretical modeling frameworks and solution techniques. Its connection and interaction with other disciplines such as mathematics, geography, logistics, economics is the main driving force behind its development (Laporte et al., 2015).

DOI: 10.4018/978-1-5225-5513-1.ch001

Lying at the core of this discipline, facility location problems are mainly concerned with optimally locating, relocating or expanding facilities such as hospitals, warehouses, schools, fire stations, bank branches, post stations, military installations, with respect to different objectives. High construction, setup and relocation costs of facilities make facility location projects long-term investments (Farahani and Hekmatfar, 2009). Moreover, deployment of a new construction requires to take a number of factors into consideration such as distances, travel times, customer types and locations, long-term relocation plans. It is of high importance to conduct an analytic work due to the remarkable cost of changing decisions on facility locations. Hence, these problems are critical components of strategic decision making for many private and public companies (Owen & Daskin, 1998).

There is a vast literature covering the subject of facility location problems in broad terms. Many models have been developed to help decision making in this area. The readers who are interested in learning about facility location models are referred to the early works of Francis and White (1974), Handler and Mirchandani (1979), Love, Morris, and Wesolowsky (1988), Francis, McGinnis, and White (1992), Mirchandani and Francis (1990), Daskin (1995, 1997), and Drezner (1995). More recent studies include Drezner and Hamacher (2002), Nickel and Puerto (2005), Church and Murray (2009).

There are also several facility location books. Among those, Mirchandani and Francis (1990) present methods and solution approaches for discrete location problems. Francis et al. (1992) presents both qualitative and quantitative methods utilized in facility location problems and provides a review of the field's literature. The book by Daskin (1995) considers the network-based location theory and besides providing theory on classical location problems, it also discusses real-life applications of those basic models. Drezner (1995) includes a review important contemporary approaches in locational theory. Drezner and Hamacher (2002)'s book reports some of the main-stream facility location topics observed in the industry and management science. In his book, Chan (2001) integrates practical and theoretical location studies and employs a number of contemporary quantitative models. The book also discusses case studies for locating facilities such as transportation terminals, nuclear power plants, military bases, emergency shelters, etc. Later, the book by Nickel and Puerto (2005) includes a through survey of both continuous and network-based facility location models. The book by Farahani and Hekmatfar (2009) is a comprehensive source of information on location theory. A more recent book by Laporte et al. (2015) provides the reader with the basic and advanced concepts of location science. It also includes a number of applications in telecommunications, healthcare, rapid transit networks, disaster events, etc. Eiselt and Marianov (2015) report some important real-world applications of location analysis. In another book, Karakitsiou, A. (2015) reviews and provides modeling approaches in discrete competitive location problems.

However, location studies concerning military organizations are relatively limited. Studies conducted in military context have a significant impact on organizations and supply chains, as the management and maintenance of military organizations require large amounts of money, resource, and people. Additionally, efficient location planning of military organizations leads to a good direction to victories (Yang et al., 2008). For this reason, we find it useful to examine this limited literature on military facility location problems and report the most common problem classes and solution approaches implemented in this domain.

The rest of the paper is organized as follows: Next section provides an overview of basic facility location classes and their properties as well as examples of their applications. Then, we present a review of facility location models in the field of the military. In particular, we group them with respect to the context of location problem studied. The following section reports a number of statistical results regarding the problem types and solution methodologies used in the field. We finally conclude our brief survey by pointing out the unexplored areas in the domain.

CORE LOCATION MODELS

Although our survey focuses on the military applications of facility location models, we find it useful to have a brief look at the history, properties, and examples of well-known facility location models implemented in the literature.

Location theory has emanated from Pierre de Fermat, Evagelistica Torricelli (a student of Galileo), and Battista Cavallieri who stated the basic Euclidean spatial median problem early in the seventeenth century; but the study of Alfred Weber in 1909 which was about locating a single warehouse with the objective of minimizing distances between it and customers is accepted as the origin of location science (Owen and Daskin 1998).

Classical facility location problems comprise of four main components: (1) demands/customers that are located at certain points, (2) facilities that will be installed or relocated, (3) a space (mostly corresponds to a geographical region) in which demands and facilities exist within, and (4) a metric which represents time or distance in system (ReVelle and Eiselt, 2005; Farahani and Hekmatfar 2009). As decision makers formulate their problem with these items in different forms, facility location problem has various types of sub-classes, of which security-military firms, safety organizations, and emergency response systems constitute a large amount.

Some of the latest reviews on facility location problems include Farahani et al. (2010), Farahani et al. (2012), Ahmadi-Javid et al. (2017), and Konak et al. (2017). The review by Farahani et al. (2010) studies multi-criteria location models and categorizes them as bi-objective, multi-objective and multi-attribute problems. Farahani et al. (2012) review the models, solution techniques and application areas of the covering problems. Ahmadi-Javid et al. (2017) report a survey on facility location problems in the field of healthcare. Different from previous review papers, Konak et al. (2017) presents a survey of competitive facility location problems.

Location problems are generally categorized as minisum, covering and minimax problems or a combination of the three. There is no doubt to say that distance is in inverse proportion to accessibility for a facility location. As traveled distance increases, accessibility and effectiveness of a facility decrease drastically. Thus, traveled distance is a significant factor to measure facility's effectiveness. Minisum problems, firstly introduced by Hakimi (1964, 1965), are also known as the p-median problem. The p-median problem seeks to determine the best locations of p facilities among n candidate locations such that the total weighted distance (or weighted travel time) between all demands and their nearest facilities is minimized (Tansel et al., 1983). It has been widely used in problems related to locating factories, distribution centers, warehouses, hospitals, etc.

Implementing the p-median approach may not always be the right choice criterion for siting public facilities. Minimizing average distance tends to favor customers who are located close to population centers while decreasing service quality to spatially dispersed ones significantly. Providing a quick or sufficient level of service to all demand nodes is critical especially in emergency service operations (police, fire brigades, ambulances). To answer this need, different from the p-median problem, the p-center problem targets the poorly serviced demand nodes by minimizing the maximum distance of any demand to its closest facility is minimized (Sylvester 1857). The *p*-center problem is also known as the minimax problem (Razi and Karatas, 2016).

Besides traveled distance, *coverage* is another common measure of effectiveness for a facility in a location problem (Karatas, 2017). In literature, coverage is first utilized as an objective by Toregas et al. (1971). A distinguished review by Farahani et al. (2010) state that decision makers define coverage factor as distance, population coverage and pre-determined time threshold. In other words, the demand is said to be covered if it can be served within a specified time or distance. Such problem types are studied in the name of covering problems (Laporte et al. 2015) and are categorized into two basic groups as the maximal covering location problems (MCLP) and set-covering problems (SCP). The MCLP formulation was first introduced by Church and ReVelle (1974) and White and Case (1974). It basically seeks to maximize the total number of demand nodes supplied (covered) with a limited number of facilities and designated budget (Balcik and Beamon 2008). The SCP (Beasley and Jørnsten 1992, Badri et al. 1998, Caprara et al. 2000), on the other hand, seeks to minimize the number or cost of resources required to ensure a pre-defined level of coverage. In their study, Karatas et al. (2016a) and Karatas et al. (2017a) compare the performances of the *p*-median problem, *p*-center problem, and MCLP with respect to a number of performance criteria under the requirement of backup coverage.

In most real-world problems, decision makers are likely to consider multiple objectives for a successful implementation of the available resources (Current et al., 1990). This requirement transforms the classic location problem to a multi-objective problem which incorporates conflicting objectives that demonstrate inherent tradeoffs. Farahani et al. (2010) report some of the commonly used objectives as:

- Minimizing total/fixed/setup/operating cost.
- Minimizing the longest distance from activated facilities.
- Minimizing average time/distance traveled.
- Minimizing maximum time/distance traveled.
- Minimizing total time/distance traveled.
- Minimizing total number of facilities activated.
- Maximizing service level/coverage.
- Maximizing responsiveness.

Therefore, depending on the nature of the problem, decision-makers should consider incorporating multiple (and possibly conflicting) objectives to their location problems. There exist several exact, heuristic and meta-heuristic algorithms in the literature for solving such problems. Among those, scalarization and Pareto approaches are the most common ones. In a recent study, Karatas and Yakıcı (2018) develop an iterative solution methodology (a combination of the branch & bound and goal programming techniques) for solving multi-objective facility location problems.

LOCATION PROBLEMS IN MILITARY

This section summarizes facility location problem applications in the military context that have been reported in the literature.

Logistics Planning

Location problems observed in the concept of military logistics mostly deal with the planning of transportation, distribution, design, storage, evacuation, disposition, and maintenance of military assets, personnel, medical service, etc.

One of the earlier works in this context is the study of Segall (2000). In this work, the author reports the location models and their applications that are used to optimally determine prior locations of military health care delivery systems over a planar region. The author discusses six different cases of medical facility placement problem within combat zones.

Military logistic problems also consider meeting the requirements of forces in the battlefield. As an example, Johnstone (2002) formulates a mixed integer model to determine the optimum pre-positioning strategy for guided munitions while minimizing response time which is computed by summing the loading, proceeding and offloading times to demand points. In a similar study, Bell (2003) considers the munitions prepositioning problem to determine storage locations and inventory levels during a possible conflict in future. In contrast with the traditional methods, the author formulates the location problem as a combinatorial optimization model which determines facility locations and storage inventories simultaneously. He employs a simulated annealing heuristic method to solve the problem. Gue (2003) develops a multi-period facility location and a multi-commodity flow model as a Mixed Integer Linear Program (MILP) with the objective of supporting the U.S. Marine Corps from a land-based distribution system. The model determines the locations of mobile support units during a battle over time with minimum inventory.

Considering possible regional/global conflict scenarios, Amouzegar et al. (2004a) attempt to determine the Forward Support Locations (FSLs) to meet War Reserve Material (WRM) requirements. They formulate a MILP model named "The Forward Support Location Site Selection and Transportation Model", that selects capacitated FSLs while minimizing costs of opening, transport, procurement, and penalties for not satisfying demands of Forward Operating Locations (FOL). Amouzegar et al. (2004b) study a more detailed version of the same problem by considering all requirements for material, equipment, and personnel.

In another study, Farahani and Asgari (2007) investigate the problem of locating distribution centers (DCs) for a military logistics system. They define a two-stage problem as determining (1) the best alternative locations for the DCs, and (1) the minimum number of DCs required to cover all supported centers (demand). For the first stage, they implement the TOPSIS methodology, a Multi-Attribute Decision-Making model, which incorporates decision-maker assessments with respect to multiple criteria. Next, they develop a multiple objective SCP formulation that seeks to minimize the number of DCs and maximize the quality of the selected facilities.

Stationing army forces is a common topic for researchers in OR. Dell et al. (2008) study this problem and present an Integer Linear Program (ILP) model named as Optimally Stationing Army Forces (OSAF). In their model, they seek optimal stationing plan of Army Forces for a planned period with two objectives: (1) minimizing installation, realignment and closure costs (existing dollars), (2) maximizing Army Value. They experiment OSAF for supporting decisions about 2011 in 2005 round of Base Realignment and Closure (BRAC) and provide 20 year-net saving of $7.6 billion. Yang et al. (2008) consider military facility location models which incorporate risks on transportation routes. The authors develop a MILP formulation for an armament maintenance support system with one service dispatcher and multiple demand nodes. The model seeks to minimize both costs of facility establishing and security and is solved by a Postorder Traversal Algorithm. Wang et al. (2008) propose a multi-criteria integrated military support facility location and allocation method which relies on the MCLP formulation. The method aims to maximize the satisfaction levels of supporting and supported military units and is solved by a Lagrangian relaxation based heuristic.

Rappold and Van Roo (2009) tackle the combinatorial problem of facility location and multi-echelon inventory optimization with stochastic demand. Motivated by the maintenance repair and aircraft engine and other subsystems deployment problem of USAF, they formulate an Integer Nonlinear Program (INLP) model. Next, they decompose it into two steps and use simulation to validate the accuracy of the solution approach. Morton et al. (2009) develop a stochastic mixed-integer model to determine military cargo deployment plan while considering a possible biological attack on shippings. They call their model as Stochastic Sealift Deployment Model (SSDM). In their model, the authors construct a two-stage stochastic model which incorporates the uncertainty of attack time. Ferrer (2010) studies the problem of storing two different types of F-16 aircraft spare engines (Pratt & Whitney and G&E). The author utilizes p-median approach to find best locations considering two types in one pool as open architecture. The model is solved with an Ardalan heuristic.

Toyoglu et al. (2011) develop a decision support tool to improve the ammunition distribution and storing performance in a battlefield using both fixed and mobile transfer points. They design a three-layer facility location and routing algorithm to efficiently distribute multiple types of products from multiple sources to demand nodes by using capacitated heterogeneous vehicles under time windows constraints. The authors formulate a static MILP model and solve multiple instances within reasonable CPU times. Considering military operations conducted with respect to a sea base, Qiu and Sharkey (2013) define the Integrated Dynamic Single-Facility Location and Inventory Planning (IDFLIP) problem. IDFLIP problem seeks to determine the location, order and inventory levels of the facility in order to meet the dynamic demands over a finite horizon. Defining two objectives as (1) minimizing the total logistical costs, and (2) minimizing the maximum cost of satisfying a demand, the authors solve the problem with dynamic programming algorithms.

Chen et al. (2013) study the location problem for military maintenance equipment during wartime. They model and solve the location problem by using fuzzy preference relationship model. In a more recent paper, Rettke et al. (2016) study the problem of dispatching aerial military medical evacuation vehicles to enhance the performance of military emergency medical units by increasing the survivability rates of casualties. The authors develop a Markov decision process model and solve it by approximate dynamic programming techniques. They demonstrate the effectiveness of their proposed approach on a case study based on contingency operations in northern Syria. Lang and McGarvey (2016) develop mathematical models to assist decision makers in determining the locations of forward supply locations (FSLs) which will be used for prepositioning rapid response material to support military operations around the globe. The authors first develop a MILP model to locate FSLs and allocate resources to opened FSLs with minimum cost. Next, they also extend their formulation to a reliability model which incorporates uncertainty in the network. Bastian et al. (2016) design a multiple-criteria analysis (MCA) framework that optimizes the military humanitarian assistance/disaster relief (HA/DR) aerial delivery supply chain network. In their work, they deal with costs, logistics and storage locations, supply and inventory levels by stochastic, mixed-integer, weighted-goal programming model. They experiment their model with the problem of determining optimal supply chain network of The U.S. Army Natick Soldier Research, Development and Engineering Center (NSRDEC) HA/DR aerial delivery capability.

Infrastructure Security and Protection

The protection of critical infrastructure is about achieving a certain level of preparedness to protect important facilities such as military facilities, financial centers, energy or transportation hubs, health centers, etc., from terrorist attacks, criminal activities, natural disasters or other malicious behaviors.

The U.S. Army National Guard utilizes combat vehicle training simulators, named mobile trainers, to train their units. Murty and Djang (1999) study the problem of determining bases and training stations for 21 combat vehicle trainer simulators. They seek optimal base and secondary training sites for mobile trainers while minimizing total distance covered by the trainers and trained units. They employ a heuristic decomposition method to solve the problem.

Eberlan (2004) utilizes three traditional methods, SCP, p-median, and p-center, to allocate strip alert sites where host air defense units for intercepting threats in the Contiguous United States (CONUS) immediately. The author designs four model settings that contain a variation on aircraft launch time and aircraft speed and evaluates the performance of each methodology.

In another paper, Church et al. (2004) develop a mixed-integer linear model, called r-interdiction model (RIM), that provides a determination of critical r in a total of p facilities, which have the highest impact on the network in case of a possible interdiction. The model presents the most critical r facilities that increase the total distance covered after interdiction. Then, Church and Scaparra (2007) formulate an integer linear model, called the Interdiction Median Problem with Fortification (IMF), to allocate q fortification resources for minimizing the impact of r-interdiction. Additionally, Scaparra and Church (2008) formulate a bilevel mixed-integer programming model, considering the p-median approach, to deploy protection sources on behalf of minimizing the effect of possible attacks on facilities. The authors solve it by an implicit enumeration (IE) algorithm. For the same problem, Liberatore et al. (2001) take the uncertainty of attacks in real-world into consideration and generate a Stochastic r-Interdiction Median Problem with Fortification (S-RIMF) model. In a similar study, Dong et al. (2009) consider the IMF problem for military facilities. The authors build upon this model by incorporating a time satisfaction function to protect critical facilities with the objective of maximizing system time satisfaction when a subset of facilities is eliminated. They use a combined genetic algorithm and simulated annealing approach to solve the problem.

Sathe and Miller-Hooks (2005) seek to locate a limited number of guarding forces such as military and police units for covering as many facilities as possible in a probabilistic network in which travel times and the demand data are uncertain. Operational support hubs are strategic locations where improve operability and effectiveness of a military logistics system. They formulate a multi-objective MILP which aims to minimize cost and maximize secondary coverage and solve it with a modified NSGA (nondominated sorting genetic algorithm) II metaheuristic.

Dawson et al. (2007) develop a hybrid technique which combines two traditional methods, p-median, and p-center, to locate security teams of US Air Force Intercontinental Ballistic Missile Systems (ICBM). The authors first solve the p-center problem and add the maximum distance obtained from this step as an additional constraint to the classic p-median formulation. They employ the Bisection method and the GRASP heuristic to solve the center and median problems, respectively. With this approach, they obtain a compromise solution between the two objectives. Similarly, Alkanat (2008) considers the problem of optimally locating Surface-to-Air Missile (SAM) sites to defend Turkish airspace. The author applies both SCP and MCLP models to determine the minimum number of SAM sites required to cover the area of interest, and to determine the maximum possible coverage for a limited number of SAM sites.

Arslan (2009) studies the problem of locating alert aircrafts to protect Turkish airspace under varying threat anticipation. The author first formulates a multi-objective location problem with backup coverage requirements which minimizes (1) the number of alert sites, (2) the maximum response time, (3) average response time for both primary and backup coverage requirements. Next, he develops a user-friendly tool capable of generating optimal aircraft locations for different threat anticipation levels. Sarikaya (2009) seeks the optimal allocation of orbit locations for Turkish Airborne Early Warning and Control (AEW&C) aircrafts to cover all responsibility regions. The author formulates a modified MCLP model and applies it to three different combat scenarios. In a similar study, Overholts et al. (2009) utilize a two-stage MCLP modeling approach which aims to improve scheduled maintenance activities of ICBM within the desired security level.

Ghanmi (2011) develops a discrete facility location model to determine hub location with optimizing aircraft routes for the Canadian Forces (CF). The author formulates the model with a Mixed Integer Non-linear Program (MINLP) which aims to maximize the average relative cost avoidance and experiments for the Canadian Forces with historical data. Bell et al. (2011) study the problem of aircraft alert site selection and develop a two-stage solution methodology. In the first stage, they formulate LSCP model to determine the minimum number of aircraft alert sites that cover critical locations in the maximum allowable distance. Then, they utilize the results of LSCP in the second stage which employs a p-median to minimize the total distance between alert sites and critical locations.

Sensor Deployment

Determining the locations of surveillance/detection sensors such as radars, sonars, in military applications is another important location problem that has been widely studied. In this section, we provide a selection of those studies.

In his thesis, Schick (1992) utilizes a modified MCLP and p-median formulation to locate imaging radars in Canada for imaging spaceborne objects. He maps the classical formulation to a network-flow problem and solves it with side constraints. Kierstead and DelBalzo (2003) attempt to design search paths against mobile targets in continuous space and time. They propose a Genetic Algorithm (GA) metaheuristic to solve the problem. The results show that the proposed approach outperforms the paths generated by standard Navy doctrines by 46% in terms of detection probability.

Considering the chemical attack threats to military bases, Gencer et al. (2008) formulate a model based on the MCLP and MEXCLP. In the formulation, they seek to find the best locations for detectors and alarms in a military region to protect the facilities from such asymmetric threats.

In recent years a number of researchers focused on the deployment of multistatic sonars for underwater target detection purposes. A multistatic sonar, different from the traditional monostatic sonar, includes geographically independent sources and receivers. The special case where only one source-receiver pair exists is called the bistatic sonar. In his technical report, Washburn (2010) develops back of the envelope type analysis of detection performance for multistatic sonar systems. In the following study, Washburn and Karatas (2015) study pattern optimization and develop an INLP formulation that maximizes area coverage. They also perform a cost-effectiveness analysis that compares multistatic sonars with monostatic. Using the geometrical concepts explained in Karatas (2013), Karatas and Akman (2015) tackle the problem of determining optimal locations of sensors in a bistatic setting. Later, using Monte Carlo simulation, Karatas and Craparo (2015) and Karatas et al. (2016b) validate some theoretical results presented in Washburn and Karatas (2015) and evaluate the performance of multistatic sonar fields with moving source.

Different from other studies concerning multistatic sonars, Craparo et al. (2017), Craparo and Karatas (2018), and Craparo et al. (2018) approach the point coverage problem where a point represents a critical facility or unit that must be protected from hostile underwater assets. The authors first formulate a nonlinear program (NLP) and an INLP which seek to locate sensors with the objective of maximizing total expected points covered. Next, assuming fixed receivers, they develop the Divide Best Sector (DiBS) algorithm which provides an optimal source position. As another point coverage application, Jourdan and de Weck (2004) consider determining the optimal locations of wireless sensors dropped from an aircraft for monitoring critical facilities located in a hostile region. The authors implement a multi-objective genetic algorithm metaheuristic to develop effective sensor location schemes.

Another location planning problem which seeks to determine the best locations of sensors emerges in the context of barrier coverage. A barrier is a line or belt-shaped region protected by a number of sensors to prevent the intrusion of hostile units into protected areas. Karatas and Onggo (2016) develop an INLP model which maximizes the total detection probability of intruders trying to penetrate a barrier line. Their model seeks to determine the best subset of locations among a larger set of candidate locations on the barrier line while considering for different types of sensor and intruders. They use Agent-Based Simulation (ABS) to validate the mathematical model. In a more recent study, Karatas (2018) consider a hybrid barrier and point coverage problem. The author develops NLP, INLP and GA models to solve the problem and report results on generic problem data sets. Studies by Chen et al. (2015), Wang et al. (2016), and Gong et al. (2016) consider the optimal positioning of bistatic radar sources and receivers to prevent illegal crossing along a barrier region.

UAV Location Planning

The development in UAV (Unmanned Aerial Vehicle) technology is expected to revolutionize military tactics and strategies in the future. The operational experiences until now show great promise and it is obvious that employment of UAVs will make military forces to become more efficient with lower cost and less risk (Glade 2000). This new technology brings along many questions which attract the interest of researchers. Although the area of command and control of UAVs receives the lion's share, several researchers addressed location analysis questions.

Chandrashekar et al. (2004) consider the problem of obtaining full connectivity for a disconnected military ad-hoc network (MANET). They suggest a heuristic method to find the minimum number of UAVs required to obtain full connectivity, and determine the locations of UAVs. Wang et al. (2014) also address a similar problem. However, they consider many issues in the MANET such as capacity, linkage, load balance and reliability, and they provide several location problem models to accommodate these issues. To master the computational burden, they use a quadratic optimization method and solve these problems with a Tabu search heuristic. Kress and Royset (2008) suggest a model to assists operators and planners to utilize unmanned aerial vehicles (UAVs) effectively in special operations. The objective is to maximize the detected number of targets with the given resources. The model determines optimal locations for the ground units and optimal time-phased search areas for the UAVs. This problem is closely related to location-routing problems in distribution field (Laporte 1988 and Min et al. 1998), where a facility location decision is made before the routing of vehicles. Kladis et al. (2008) provide methods for determining the UAV paths to minimize energy requirement considering the limited propulsion energy in a dynamically changing environment. They use location theory tools to incorporate additional constraints for intermediate refueling points.

Ayöperken and Ermiş (2011) implement an MLCP formulation with the objective of maximizing the area coverage of selected UAV bases with a limited budgetary. The study by Kurban and Tuncay (2016) seeks to determine the best locations of ground control units of mini UAVs that are used to obtain urgent intelligence, surveillance, and reconnaissance. These units have different coverage capabilities and the authors first employ a standard MCLP formulation which maximizes overall coverage. Next, they develop a Maximum Expected Covering Location Problem (MECLP) which incorporates the effects of environmental and weather conditions. The MCLP and MECLP models are tested for five and twelve different scenarios, respectively.

Yildirim (2016) studies the problem of determining UAV hubs for border monitoring by solving a hub maximal covering problem (HMCoP) with a hybrid genetic algorithm (GA). The model of HMCoP model has been proposed to and tightened by a preprocessing method. The hybrid GA has been applied to choose the optimal hub locations and allocate the demand points to nearest hub. An extended problem in a similar context is studied by Sarıçiçek and Akkuş (2015). They choose the hubs to position UAVs among a set of airports, assign the demand points to these hubs and determine optimal routes for each chosen hub. They solve the problem in two stages. In the first stage, the *p*-hub median problem is solved to select hubs, which employs a fitness parameter obtained from ELECTRE method. In the second stage, mathematical models are employed to find optimal routes for each selected hub.

Yakıcı (2016) discusses the problem of locating and routing of UAVs and develop an ILP which seeks to maximize total reward collected by UAVs from visited demand nodes in the area. The author solves the combinatorial problem by employing a customized ant colony optimization metaheuristic.

We also wanted to include some studies addressing UAV location problems which are not discussed in a military context but can be directly related to military concepts. As Chandrashekar et al. (2004) and Wang et al. (2014) do, Raissi-Dehkordi et al. (2004) address the problem of providing full connectivity in an ad-hoc network by placing UAVs as relay nodes in optimal places. A formulation is given where the connectivity problem is defined as a clustering problem. Then, a clustering algorithm is applied to find the minimum number of required UAVs and their locations. Godzdanker et al. (2011) consider the problem of strategically locating self-leveling landing platforms in the work field of the UAVs. They report initial results obtained by a genetic search algorithm. Lim et al. (2016) present a two-phase model for power network damage assessment using UAVs. In the first phase, to determine UAV locations, a two-stage stochastic integer programming optimization model is suggested considering the possibility of an extreme weather event. In the second phase, UAV paths are determined to minimize operating cost and completion time of the damage assessment mission. Mozaffari et al. (2016) investigate the optimal deployment of UAVs acting as flying base stations where the aim is to minimize the total required transmission power of UAVs while satisfying the rate requirements. The optimal locations of UAVs and the boundaries of their coverage areas are determined. The problem is decomposed to two sub-problems that are solved iteratively. In the first sub-problem, given the cell boundaries, the optimal locations of the UAVs are decided, while in the second sub-problem, cell boundaries are determined assuming that the UAV locations are fixed. Sharma et al. (2016) consider the problem of UAV placement to optimize the network delays. The problem is addressed using a minimax facility problem and an optimal placement algorithm. Golabi et al. (2017) study a combined mobile and immobile pre-earthquake facility location problem assuming that the demand is uniformly distributed along the network edges and considering some network edges may collapse and prevents accessibility. UAVs are utilized for the people who are located on collapsed edges. The authors suggest a mathematical model which minimizes the aggregate traveling time over a set of possible scenarios. They also provide metaheuristic algorithms to solve the proposed model.

Search and Rescue Operations Planning

Search and rescue (SAR) operations can be counted as another important and common application area for locating military facilities. In SAR applications the effectiveness of a location plan is generally measured in terms of response time to incidents and/or distress calls. There are a number of studies in the literature which consider locating SAR resources.

Basdemir (2000) studies the problem of locating Turkish Air Force (TUAF) SAR helicopters in the Aegean and the Western Mediterranean regions of Turkey. The author formulates the problem as a classical MCLP model which seeks to maximize the number of demand points covered in the area of interest and solves it for three scenarios each with different a helicopter coverage range. Abi-Zeid and Frost (2005) design a geographic decision support tool called the SARPlan for the Canadian Forces. The tool utilizes search theory, gradient search methods, and constraint satisfaction programming to determine the best locations of available SAR resources such that the probability of success of a mission is maximized.

Azofra et al. (2007) approach the SAR boat location problem from a more strategic viewpoint and formulize two gravitational models, Individual Distribution Model (IDM) and Zonal Distribution Model (ZDM). IDM evaluates each location individually with respect to several criteria and determines a co-efficient to each of them. In the ZDM, maritime incidents are clustered into a number of zones where each zone is later represented with a centroid called the "superaccident". Finally, the effectiveness of all possible allocation plans is computed with respect to the distance between each port and superaccident, and superaccident weights.

Radovilsky and Koermer (2007) develop an ILP model to allocate small SAR boats to USCG stations. They aim to minimize capacity deviations at the stations located at the USCG Pacific Area. Their model does not allow assigning a boat to multiple demand points and does not take operation capacities into consideration. Following Radovilsky and Koermer (2007)'s study, Wagner and Radovilsky (2012) and Radovilsky and Wagner (2014) develop a Microsoft Excel-based decision support tool called BAT (Boat Allocation Tool) which is used to optimally locate SAR boats to USCG stations. BAT employs a multi-objective MILP model with three objectives: minimize (1) deviation of supply from demand, (2) number of SAR vessel types at a station, and (3) total cost. They define 10 performance metrics for BAT model and compare its results with that of the original boat location plan. The authors claim that the proposed approach provides a considerable improvement on USCG fleet performance.

Chan et al. (2008) propose a variant of the MCLP formulation to locate signal-receiving stations to improve the efficiency of the SAR operations. For this purpose, they develop an ILP approximation of a nonlinear objective function which maximizes the number of expected lines-of-bearings and minimizes the excess coverage of frequencies. They map the problem to a two-stage network-flow formulation to obtain computationally efficient solutions. They also use a greedy heuristic to compare the solution quality of the network formulation with the original nonlinear model.

Afshartous et al. (2009) propose an optimization and simulation methodology to determine optimal locations of United States Coast Guard (USCG) air stations. They formulate an ILP model and solve the problem as a p-Uncapacitated Facility Location Problem (p-UFLP), a hybrid of the p-median and the UFLP. Then, they assess the results obtained from of optimization model with simulation. In the simulation model, they develop a statistical model which generates stochastic distress calls from an inhomogeneous Poisson process, using historical data of the year 2000. The authors assume that the demand for each client is equal and served with a single resource.

Similarly, Nelson et al. (2014) develop a MILP model to locate USCG aircrafts with the aim of improving fleet operational performance at minimum cost. As they seek the optimal allocation and aircraft deployment, they take SAR and scheduled missions into consideration together. In their study, Ai et al. (2015) develop an INLP model which deals with location, allocation, and configuration of maritime emergency response bases and vessels. The model seeks to minimize total cost which consists of the construction cost of bases, reserve cost of supplies, and configuration cost of salvage vessels. They employ a hybrid heuristic and a genetic algorithm to solve the problem. In another example, Acar (2015) applies SCP, MCLP and *p*-median models to deploy the TUAF SAR resources. He utilizes each model for different objectives: SCP to find the minimum number of SAR Deployment Points (DPs) required for covering all training areas, MCLP to cover as much area as possible with a given SAR DP number and *p*-median to allocate SAR DPs with the objective of minimizing aggregate or average response time.

Razi and Karatas (2016) consider a similar problem for the Aegean Sea responsibility area of the Turkish Coast Guard. Their problem structure accounts for varying demand sizes (time required to respond) for different incident types and allows cooperative coverage of demands from multiple SAR boats. The authors first employ the Analytical Hierarchy Process (AHP) technique to determine weights for each incident type. Next, they develop a multi-objective MILP formulation which seeks to minimize deviations from critical response time to incidents, allocated budget, and supply from the pre-specified annual capacity of boats. In more recent studies, Razı et al. (2016) and Karatas et al. (2017b) propose a hybrid optimization and simulation approach to determine locations of SAR helicopters. In the work by Karatas et al. (2017b), the authors first develop an ILP formulation which seeks to minimize response time to incidents in the area of interest at the first stage of their methodology. This model can be considered as a customized *p*-median problem which incorporates additional constraints to satisfy business rules or other technical issues. The second stage includes a discrete event simulation (DES) model which measures the performance of the solution obtained from the previous stage under stochastic data. Finally, they develop a rule-based algorithm called the alternative plan generation, which generates a series of other candidate solutions. They iteratively test the performance of those solutions in the DES to find superior location schemes of helicopters.

There also exists a number of studies conducted for optimizing SAR vessel locations for the Canadian Coast Guard (CCG). The first study by Pelot et al. (2015) tackles the problem optimally locating lifeboats in Atlantic Canada by using the MCLP, capacitated MCLP (Pirkul and Schilling 1991), and Maximum Expected Coverage Location Model (MEXCLP), the stochastic MCLP models.

Akbari et al. (2017a, 2017b) consider a similar problem for the CCG. Employing the *p*-median and MCLP models, Akbari et al. (2017a) perform a multi-criteria analysis on the performance of the CCG boat locations under five criteria. Akbari et al. (2017b), on the other hand, develop a multi-objective goal programming MILP model. The objectives include the minimization of deviations from targeted primary coverage, backup coverage and access time. Both studies use historical incident data distributions and use simulated future demand data to obtain more reliable solutions.

LITERATURE ANALYSIS

In this section, we report a summary of our analysis with respect to the solution techniques employed to solve the military facility location problems. We also categorize the formulations as single or multiple objective models and report the objectives adopted in each study of our review.

Table 1 reports solution technique used for each study and categorizes them as Tabu Search, Simulated Annealing, Genetic Algorithm, Ant Colony Optimization, Greedy, GRASP, Other heuristics, Decomposition, and Math Programming approaches. As can be seen in Figure 1, the results show that out of the 73 papers included in this review, 49 of them develop a mathematical model for the problem of interest while the rest of them employ heuristic or metaheuristic approaches.

Table 1. Solution techniques employed to solve location problems

#	Authors	Year	Tabu Search	Simulated Annealing	Genetic Algorithm	Ant Colony Optimization	Greedy	GRASP	Other Heuristics	Decomposition	Math Programming
1	Schick	1992									X
2	Murty and Djang	1999								X	
3	Murty and Djang	1999									X
4	Basdemir	2000									X
5	Johnstone	2002									X
6	Bell	2003		X							
7	Kierstead and DelBalzo	2003			X						
8	Gue	2003									X
9	Bell	2003									X
10	Jourdan and de Weck	2004			X						
11	Chandrashekar et al.	2004							X		
12	Raissi-Dehkordi et al.	2004							X		
13	Eberlan	2004									X
14	Church et al.	2004									X
15	Amouzegar et al.	2004a									X
16	Amouzegar et al.	2004b									X
17	Sathe and Miller-Hooks	2005			X						
18	Abi-Zeid and Frost	2005									X
19	Dawson et al.	2007						X			
20	Radovilsky and Koermer	2007									X
21	Farahani and Asgari	2007									X
22	Church and Scaparra	2007									X

continued on following page

Table 1. Continued

#	Authors	Year	Tabu Search	Simulated Annealing	Genetic Algorithm	Ant Colony Optimization	Greedy	GRASP	Other Heuristics	Decomposition	Math Programming
23	Chan et al.	2008					X				
24	Wang et al.	2008							X		
25	Yang et al.	2008							X		
26	Alkanat	2008									X
27	Gencer et al.	2008									X
28	Dell et al.	2008									X
29	Kress and Royset	2008									X
30	Scaparra and Church	2008									X
31	Kladis et al.	2008									X
32	Overholts et al.	2009							X		
33	Dong et al.	2009		X	X						
34	Rappold and Van Roo	2009								X	
35	Arslan	2009									X
36	Sarikaya	2009									X
37	Afshartous et al.	2009									X
38	Morton et al.	2009									X
39	Ferrer	2010							X		
40	Washburn	2010									X
41	Ferrer	2010									X
42	Liberatore et al.	2011							X		
43	Godzdanker et al.	2011			X						
44	Ghanmi	2011									X
45	Ayöperken and Ermiş	2011									X
46	Bell et al.	2011									X
47	Toyoglu et al.	2011									X
48	Liberatore et al.	2011									X
49	Wagner and Radovilsky	2012									X
50	Wang et al.	2014	X								
51	Radovilsky and Wagner	2014									X
52	Ai et al.	2015			X						
53	Washburn and Karatas	2015									X
54	Acar	2015									X
55	Sarıçiçek and Akkuş	2015									X
56	Yakıcı	2016				X					
57	Rettke et al.	2016							X		
58	Yildirim	2016			X						
59	Sharma et al.	2016							X		

continued on following page

Table 1. Continued

#	Authors	Year	Tabu Search	Simulated Annealing	Genetic Algorithm	Ant Colony Optimization	Greedy	GRASP	Other Heuristics	Decomposition	Math Programming
60	Kurban and Tuncay	2016									X
61	Razi and Karatas	2016									X
62	Razı et al.	2016									X
63	Karatas and Onggo	2016									X
64	Lang and McGarvey	2016									X
65	Lim et al.	2016									X
66	Mozaffari et al.	2016									X
67	Golabi et al.	2017		X	X				X		
68	Craparo et al.	2017									X
69	Akbari et al.	2017a									X
70	Karatas et al.	2017b									X
71	Akbari et al.	2017b									X
72	Karatas and Yakıcı	2018									X
73	Karatas	2018			X						X

Figure 1. Numbers of solution techniques employed to solve location problems

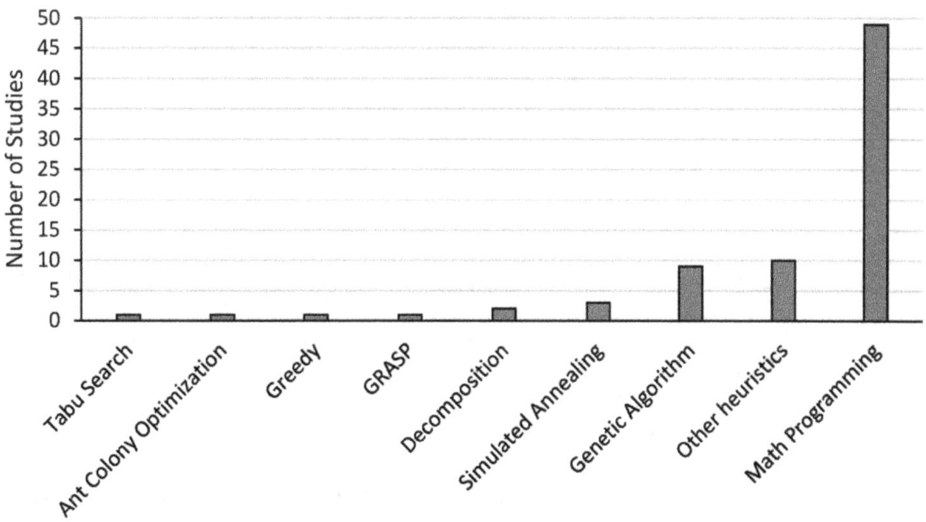

Table 2 reports the core location model (if implemented) and objectives incorporated to each study. Figure 2 shows the percentages of military location studies which incorporate single or multiple objectives. The results show that approximately a quarter of studies involve single objective whereas the majority involves multiple (and mostly conflicting) objectives.

Table 2. Objectives incorporated to military facility location models

#	Authors	Year	Core Location Model Type	Single/Multiple Objective	Objectives
1	Schick	1992	MCLP	Single	Maximize the number of satellite imagings
2	Murty and Djang	1999	*p*-median	Multiple	1. Minimizing distance covered 2. Maximing coverage
3	Basdemir	2000	MCLP	Single	Maximize the number of demand points covered
4	Johnstone	2002		Single	Minimize response time
5	Gue	2003		Single	Minimize total inventory
6	Bell	2003		Multiple	1. Minimize the total distribution costs 2. Minimize facility opening and closing costs
7	Kierstead and DelBalzo	2003		Single	Maximize detection probability
8	Jourdan and de Weck	2004		Multiple	1. Maximize coverage and survivability 2. Minimize number of sensors
9	Eberlan	2004	SCP, *p*-center, *p*-median	Single	Minimize the number of sites utilized
10	Church et al.	2004		Single	Maximize total distance traveled
11	Chandrashekar et al.	2004		Single	Minimize number of UAVs
12	Raissi-Dehkordi et al.	2004		Single	Minimize number of UAVs
13	Amouzegar et al.	2004a		Single	Minimize total cost
14	Amouzegar et al.	2004b		Single	Minimize the average duration time of an item in transit
15	Sathe and Miller-Hooks	2005		Multiple	1. Minimize cost 2. Maximize secondary coverage
16	Abi-Zeid and Frost	2005		Single	Maximize probability of SAR mission success
17	Dawson et al.	2007	*p*-median	Multiple	1. Minimize average distance 2. Minimize maximum distance
18 19	Radovilsky and Koermer	2007		Single	Minimize capacity deviations
20	Farahani and Asgari	2007	SCP	Multiple	1. Minimize the number of distribution centers 2. Maximize the quality of the selected facilities
21	Church and Scaparra	2007		Single	Minimizing total distance traveled
22	Alkanat	2008	SCP, MCLP	Single	Minimize the number of SAM sites required
23	Chan et al.	2008	MCLP	Multiple	1. Minimize the excess coverage of frequencies 2. Maximize the coverage within a predetermined
24	Gencer et al.	2008	MCLP, MEXCLP	Single	Maximize the total number of threat points covered
25	Dell et al.	2008		Multiple	1. Minimizing costs 2. Maximizing army value
26	Wang et al.	2008	MCLP	Single	Maximize satisfaction level of suppliers and demand nodes
27	Yang et al.	2008		Single	Minimize total cost
28	Kress and Royset	2008		Single	Minimize evasion probability

continued on following page

Table 2. Continued

#	Authors	Year	Core Location Model Type	Single/Multiple Objective	Objectives
29	Scaparra and Church	2008	*p*-median	Multiple	Minimize the maximum level of weighted distance or costs
30	Kladis et al.	2008		Single	Minimize energy requirement
31	Overholts et al.	2009	MCLP	Single	Maximize weighted sum of maintenance activities
32	Arslan	2009		Multiple	1. Minimize the number of alert sites 2. Minimize the maximum response time 3. Minimize average response time
33	Sarikaya	2009	MCLP	Single	Maximize covered demand points
34	Afshartous et al.	2009	*p*-UFLP	Single	Minimize average distance per call
35	Dong et al.	2009		Single	Maximize system time satisfaction
36	Rappold and Van Roo	2009		Single	Minimize total cost
37	Morton et al.	2009		Single	Minimize penalties of delivery failures
38	Washburn	2010			
39	Ferrer	2010	*p*-median	Single	Minimize total cost
40	Ghanmi	2011		Single	Maximize the average relative cost avoidance
41	Ayöperken and Ermiş	2011	MCLP	Single	Maximize the area coverage of selected UAV bases
42	Bell et al.	2011	LSCP, *p*-median	Multiple (Two-Stage)	1. Minimize the total number of activated aircraft alert site 2. Minimize facility opening and closing costs
43	Toyoglu et al.	2011		Single	Minimize total cost
44	Liberatore et al.	2011	*p*-median	Single	Minimize the expected cost of worst-case scenario
45	Godzdanker et al.	2011	MCLP	Single	Maximize the missions of the unmanned helicopters
46	Wagner and Radovilsky	2012		Multiple	1. Minimize deviation of supply from demand 2.Minimize number of SAR vessel types at a station, 3.Minimize cost
47	Qiu and Sharkey	2013		Multiple	1. Minimize total logistical costs 2. Minimize the maximum cost of satisfying a demand
48	Nelson et al.	2014		Single	Minimize total cost
49	Radovilsky and Wagner	2014		Multiple	1. Minimize deviation of supply from demand 2. Minimize number of SAR vessel types at a station 3. Minimize cost
50	Wang et al.	2014		Single	Minimize total distance
51	Washburn and Karatas	2015		Single	Maximize coverage
52	Acar	2015	SCP, MCLP, *p*-median	Single	Minimize the number of SAR Deployment Points
53	Ai et al.	2015		Single	Minimize cost
54	Pelot et al.	2015	MCLP, MEXCLP	Multiple	1. Minimize the average response distance 2. Maximize the coverage within a predetermined access time limit
55	Sarıçiçek and Akkuş	2015	p-median	Single	Maximize benefit
56	Kurban and Tuncay	2016	MCLP and MECLP	Single	Maximize coverage of UAV ground control units

continued on following page

Table 2. Continued

#	Authors	Year	Core Location Model Type	Single/Multiple Objective	Objectives
57	Yakıcı	2016		Single	Maximize total reward
58	Razi and Karatas	2016		Multiple	1. Minimize deviations from critical response time to incidents 2. Minimize deviations from budget, 3. Minimize average response time
59	Razı et al.	2016	*p*-median	Single	Minimize total response time to incidents
60	Karatas and Onggo	2016		Single	Maximize the total weighted-detection probability of intruders
61	Lang and McGarvey	2016		Single	Minimize cost
62	Rettke et al.	2016		Single	Maximize expected
63	Yildirim	2016	MCLP	Single	Maximize serviced flows
64	Lim et al.	2016		Single	Minimize cost
65	Mozaffari et al.	2016		Single	Minimize transmission power
66	Sharma et al.	2016	*p*-center	Single	Minimize the maximum distance
67	Craparo et al.	2017		Single	Maximize coverage
68	Golabi et al.	2017		Single	Minimize traveling time
69	Akbari et al.	2017a	*p*-median	Multiple	1. Minimize mean response time to incidents 2. Maximize number of demands covered within predetermined travel time
70	Karatas et al.	2017b	*p*-median	Single	Minimize total response time to incidents
71	Akbari et al.	2017b		Multiple	1. Minimize deviations from primary coverage 2. Minimize deviations from backup coverage 3. Minimize deviations from mean access time
72	Karatas	2018		Multiple	1. Maximize the overall detection probability of smugglers. 2. Maximize the weighted overall probability that critical facilities are protected against attacks
73	Karatas and Yakici	2018	*p*-median, MCLP, *p*-center	Multiple	1. Minimize average distance 2. Maximize covered demands 3. Minimize the maximum distance between a facility and its designated demand

CONCLUSION

In this chapter, we have reviewed a selection of papers considering important facility location problems for military organizations. We observed that the domain of military facility location problems is very active with several critical and interesting problems still being investigated, both from a formulation development and solution point of view. Including and analyzing only a selection of papers in this domain, we also noted that the literature on military facility location problems has been growing steadily. Approximately 75% of the papers were published in the past 10 years whereas roughly a third of them were published in the past five years. Our review also reveals that the topics SAR resource location/ allocation, barrier coverage and multistatic sensor location receive an increasing attention among re-

Figure 2. Percentages of studies incorporating single and multiple objectives

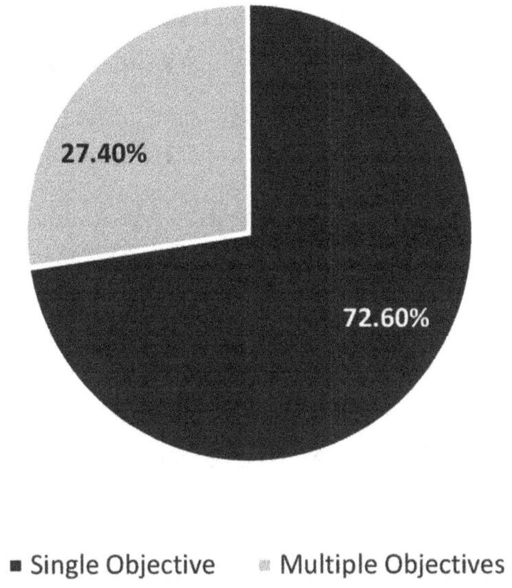

searchers in the last years. Finally, we observed that more than 25% of the articles attempt to incorporate multiple conflicting objectives to the location models while the remaining papers stick with the classical single objective models.

Regarding the recent developments in stochastic programming and robust optimization techniques, we expect future research to focus more on the stochastic and dynamic nature of the military operations to capture the complexities of realistic problem instances. Hence, we believe that there is still much room for the development of more realistic and flexible models as well as high-performance solution methodologies which can assist decision makers in military facility location planning.

NOTE

The views and conclusions contained herein are those of the authors and should not be interpreted as necessarily representing the official policies or endorsements, either expressed or implied, of any affiliated organization or government.

REFERENCES

Abi-Zeid, I., & Frost, J. R. (2005). SARPlan: A decision support system for Canadian Search and Rescue Operations. *European Journal of Operational Research*, *162*(3), 630–653. doi:10.1016/j.ejor.2003.10.029

Acar, M. (2015). *Optimization of Turkish Air Force SAR Units Forward Deployment Points for a Central Based SAR Force Structure (No. AFIT-ENS-MS-15-M-148)*. Air Force Institute of Technology Wright-Patterson AFB OH Graduate School Of Engineering And Management.

Afshartous, D., Guan, Y., & Mehrotra, A. (2009). US Coast Guard air station location with respect to distress calls: A spatial statistics and optimization based methodology. *European Journal of Operational Research, 196*(3), 1086–1096. doi:10.1016/j.ejor.2008.04.010

Ahmadi-Javid, A., Seyedi, P., & Syam, S. S. (2017). A survey of healthcare facility location. *Computers & Operations Research, 79*, 223–263. doi:10.1016/j.cor.2016.05.018

Ai, Y. F., Lu, J., & Zhang, L. L. (2015). The Optimization Model for the Location of Maritime Emergency Supplies Reserve Bases and the Configuration of Salvage Vessels. *Transportation Research Part E, Logistics and Transportation Review, 83*, 170–188. doi:10.1016/j.tre.2015.09.006

Akbari, A., Eiselt, H. A., & Pelot, R. (2017a). A maritime search and rescue location analysis considering multiple criteria, with simulated demand. *INFOR*, 1–23.

Akbari, A., Pelot, R., & Eiselt, H. A. (2017b). A modular capacitated multi-objective model for locating maritime search and rescue vessels. *Annals of Operations Research*, 1–26.

Alkanat, O. (2008). *Determining the Surface-to-Air Missile Requirement for Western and Southern Part of the Turkish Air Defense System. AFIT- Air Force Institute of Technology, AFIT/GOR/ENS/08-01.* Dayton, OH: Wright Patterson AFB.

Amouzegar, M. A., Tripp, R. S., & Galway, L. A. (2004b). Integrated logistics planning for the air and space expeditionary force. *The Journal of the Operational Research Society, 55*(4), 422–430. doi:10.1057/palgrave.jors.2601704

Amouzegar, M. A., Tripp, R. S., McGarvey, R. G., Chan, E. W., & Roll, C. R. Jr. (2004a). *Supporting air and space expeditionary forces: analysis of combat support basing options.* RAND Project Air Force.

Arslan, O. (2009). *Developing a Tool for the Location Optimization of the Alert Aircraft with Changing Threat Anticipation (No. AFIT/GOR/ENS/09-02).* Air Force Inst of Tech Wright-Patterson AFB OH Graduate School of Engineering and Management.

Ayöperken, E., & Ermiş, M. (2011). Modeling and optimizing the bases of unmanned aerial vehicles as a set-coverage problem. *Havacilik ve Uzay Teknolojileri Dergisi, 5*(1), 61–71.

Azofra, M., Pérez-Labajos, C. A., Blanco, B., & Achutegui, J. J. (2007). Optimum placement of sea rescue resources. *Safety Science, 45*(9), 941–951. doi:10.1016/j.ssci.2006.09.002

Badri, M. A., Mortagy, A. K., & Alsayed, C. A. (1998). A multi-objective model for locating fire stations. *European Journal of Operational Research, 110*(2), 243–260. doi:10.1016/S0377-2217(97)00247-6

Balcik, B., & Beamon, B. M. (2008). Facility location in humanitarian relief. *International Journal of Logistics, 11*(2), 101–121. doi:10.1080/13675560701561789

Basdemir, M. M. (2000). *Locating search and rescue stations in the Aegean and Western Mediterranean regions of Turkey (No. AFIT/GOR/ENS/00M-03). AFIT- Air Force Institute of Technology.* Dayton, OH: Wright Patterson AFB.

Beasley, J. E., & Jörnsten, K. (1992). Enhancing an algorithm for set covering problems. *European Journal of Operational Research, 58*(2), 293–300. doi:10.1016/0377-2217(92)90215-U

Bell, J. E. (2003). *A simulated annealing approach for the composite facility location and resource allocation problem: a study of strategic positioning of US Air Force munitions (No. C102-927)*. Auburn Univ.

Bell, J. E., Griffis, S. E., Cunningham, W. A. III, & Eberlan, J. A. (2011). Location optimization of strategic alert sites for homeland defense. *Omega, 39*(2), 151–158. doi:10.1016/j.omega.2010.05.004

Caprara, A., Toth, P., & Fischetti, M. (2000). Algorithms for the set covering problem. *Annals of Operations Research, 98*(1-4), 353–371. doi:10.1023/A:1019225027893

Chan, Y. (2001). *Location theory and decision analysis*. South-Western College Pub.

Chan, Y., Mahan, J. M., Chrissis, J. W., Drake, D. A., & Wang, D. (2008). Hierarchical maximal-coverage location–allocation: Case of generalized search-and-rescue. *Computers & Operations Research, 35*(6), 1886–1904. doi:10.1016/j.cor.2006.09.018

Chandrashekar, K., Dekhordi, M. R., & Baras, J. S. (2004, October). Providing full connectivity in large ad-hoc networks by dynamic placement of aerial platforms. In *Military Communications Conference, 2004. MILCOM 2004. 2004 IEEE* (Vol. 3, pp. 1429-1436). IEEE. 10.1109/MILCOM.2004.1495151

Chen, J., Wang, B., & Liu, W. (2015). Constructing perimeter barrier coverage with bistatic radar sensors. *Journal of Network and Computer Applications, 57*, 129–141. doi:10.1016/j.jnca.2015.07.015

Chen, X. W., Wang, X. Z., Ma, L., & Kong, L. (2013). Equipment Maintenance Position Location Uncertain Decision Method in Wartime. *Fire Control & Command Control, 4*, 21.

Church, R., & ReVelle, C. (1974). The maximal covering location problem. *Papers in Regional Science, 32*(1), 101–118. doi:10.1007/BF01942293

Church, R. L., & Murray, A. T. (2009). *Business site selection, location analysis and GIS*. New York: Wiley.

Church, R. L., & Scaparra, M. P. (2007). Protecting critical assets: The *r*-interdiction median problem with fortification. *Geographical Analysis, 39*(2), 129–146. doi:10.1111/j.1538-4632.2007.00698.x

Church, R. L., Scaparra, M. P., & Middleton, R. (2004). The r-interdiction median problem and the r-interdiction covering problem. *Annals of the Association of American Geographers, 94*, 491–502. doi:10.1111/j.1467-8306.2004.00410.x

Craparo, E. M., Fügenschuh, A., Hof, C., & Karatas, M. (2018). Optimizing Source and Receiver Placement in Multistatic Sonar Networks to Monitor Fixed Targets. *European Journal of Operational Research, 2018*. doi:10.1016/j.ejor.2018.02.006

Craparo, E. M., & Karatas, M. (2018). *A Method for Placing Sources in Multistatic Sonar Networks*. Technical Report NPS-OR-18-001. Naval Postgraduate School.

Craparo, E. M., Karatas, M., & Kuhn, T. U. (2017). Sensor placement in active multistatic sonar networks. *Naval Research Logistics, 64*(4), 287–304. doi:10.1002/nav.21746

Current, J., Min, H., & Schilling, D. (1990). Multiobjective analysis of facility location decisions. *European Journal of Operational Research, 49*(3), 295–307. doi:10.1016/0377-2217(90)90401-V

Daskin, M. (1997). Network and Discrete Location: Models, Algorithms and Applications. *The Journal of the Operational Research Society*, *48*(7), 763–763. doi:10.1057/palgrave.jors.2600828

Daskin, M. S. (1995). *Network and discrete location: Models, algorithms and applications*. New York: John Wiley and Sons. doi:10.1002/9781118032343

Dawson, M. C., Bell, J. E., & Weir, J. D. (2007). A Hybrid Location Method for Missile Security Team Positioning. *Journal of Business and Management*, *13*, 5–17.

Dell, R. F., Ewing, P. L., & Tarantino, W. J. (2008). Optimally stationing army forces. *Interfaces*, *38*(6), 421–435. doi:10.1287/inte.1080.0401

Dong, L., Xing-hua, K., & Xiang-tao, Y. (2009, October). A model for supply chain critical facility protection planning based on time satisfaction. In *Intelligent Computation Technology and Automation, ICICTA'09. Second International Conference on* (Vol. 3, pp. 903-906). IEEE.

Drezner, Z. (Ed.). (1995). *Facility location: A survey of applications and methods*. Berlin, Germany: Springer Verlag. doi:10.1007/978-1-4612-5355-6

Drezner, Z., & Hamacher, H. W. (Eds.). (2002). *Facility location: Applications and theory*. Berlin, Germany: Springer-Verlag. doi:10.1007/978-3-642-56082-8

Eberlan, J. A. (2004). *Location optimization of continental United States strip alert sites supporting homeland defense (No. AFIT/GLM/ENS/04-02)*. Air Force Inst of Tech Wright-Patterson AFB OH Dept of Operational Sciences.

Eiselt, H. A., & Marianov, V. (Eds.). (2015). *Applications of Location Analysis*. Springer International Publishing Switzerland; doi:10.1007/978-3-319-20282-2

Farahani, R. Z., & Asgari, N. (2007). Combination of MCDM and covering techniques in a hierarchical model for facility location: A case study. *European Journal of Operational Research*, *176*(3), 1839–1858. doi:10.1016/j.ejor.2005.10.039

Farahani, R. Z., Asgari, N., Heidari, N., Hosseininia, M., & Goh, M. (2012). Covering problems in facility location: A review. *Computers & Industrial Engineering*, *62*(1), 368–407. doi:10.1016/j.cie.2011.08.020

Farahani, R. Z., & Hekmatfar, M. (Eds.). (2009). *Facility location: Concepts, models, algorithms and case studies*. Heidelberg, Germany: Physica Verlag. doi:10.1007/978-3-7908-2151-2

Farahani, R. Z., Steadie Seifi, M., & Asgari, N. (2010). Multiple criteria facility location problems: A survey. *Applied Mathematical Modelling*, *34*(7), 1689–1709. doi:10.1016/j.apm.2009.10.005

Ferrer, G. (2010). Open architecture, inventory pooling and maintenance modules. *International Journal of Production Economics*, *128*(1), 393–403. doi:10.1016/j.ijpe.2010.07.034

Francis, R. L., McGinnis, L. F., & White, J. A. (1992). *Facility layout and location: An analytical approach* (2nd ed.). Englewood Cliffs, NJ: Prentice-Hall.

Francis, R. L., & White, J. A. (1974). *Facility layout and location an analytical approach* (1st ed.). Englewood Cliffs, NJ: Prentice-Hall.

Gencer, C., Aydogan, E. K., & Soydemır, A. (2008). Chemical agent detector placement methodology. *Applied Mathematics and Computation, 195*(2), 542–557. doi:10.1016/j.amc.2007.05.033

Ghanmi, A. (2011). Canadian forces global reach support hubs: Facility location and aircraft routing models. *The Journal of the Operational Research Society, 62*(4), 638–650. doi:10.1057/jors.2010.22

Glade, D. (2000). *Unmanned aerial vehicles: Implications for military operations.* Air Univ Press Maxwell AFB Al.

Godzdanker, R., Rutherford, M. J., & Valavanis, K. P. (2011, July). ISLANDS: a self-leveling landing platform for autonomous miniature UAVs. In *IEEE/ASME International Conference on Advanced Intelligent Mechatronics (AIM)*, 2011 (pp. 170-175). IEEE. 10.1109/AIM.2011.6027085

Golabi, M., Shavarani, S. M., & Izbirak, G. (2017). An edge-based stochastic facility location problem in UAV-supported humanitarian relief logistics: A case study of Tehran earthquake. *Natural Hazards*, 1–21.

Gong, X., Zhang, J., Cochran, D., & Xing, K. (2016). Optimal placement for barrier coverage in bistatic radar sensor networks. *IEEE/ACM Transactions on Networking, 24*(1), 259–271. doi:10.1109/TNET.2014.2360849

Gue, K. R. (2003). A dynamic distribution model for combat logistics. *Computers & Operations Research, 30*(3), 367–381. doi:10.1016/S0305-0548(01)00104-6

Hakimi, S. L. (1964). Optimum locations of switching centers and the absolute centers and medians of a graph. *Operations Research, 12*(3), 450–459. doi:10.1287/opre.12.3.450

Hakimi, S. L. (1965). Optimum distribution of switching centers in a communication network and some related graph theoretic problems. *Operations Research, 13*(3), 462–475. doi:10.1287/opre.13.3.462

Handler, G. Y., & Mirchandani, P. B. (1979). *Location on networks: Theory and algorithms.* Cambridge, MA: MIT Press.

Johnstone, D. P. (2002). *Modeling the pre-positioning of air force precision guided munitions (No. AFIT/GOR/ENS/02-09).* Air Force Inst of Tech Wright-Patterson AFB OH Dept of Operational Science.

Jourdan, D. B., & de Weck, O. L. (2004). Multi-objective genetic algorithm for the automated planning of a wireless sensor network to monitor a critical facility. In *Defense and Security* (pp. 565–575). International Society for Optics and Photonics. doi:10.1117/12.541685

Karakitsiou, A. (2015). *Modeling discrete competitive facility location.* Springer International Publishing. doi:10.1007/978-3-319-21341-5

Karatas, M. (2013). A multi foci closed curve: Cassini oval, its properties and applications. *Dogus University Journal, 14*(2), 233–250.

Karatas, M. (2017). A multi-objective facility location problem in the presence of variable gradual coverage performance and cooperative cover. *European Journal of Operational Research, 262*(3), 1040–1051. doi:10.1016/j.ejor.2017.04.001

Karatas, M. (2018). Optimal Deployment of Heterogeneous Sensor Networks for a Hybrid Point and Barrier Coverage Application. *Computer Networks, 132*(26), 129–144. doi:10.1016/j.comnet.2018.01.001

Karataş, M., & Akman, G. (2015). Bistatic sonobuoy deployment configuration for stationary targets. *Journal of Naval Science and Engineering*, *11*(2), 1–10.

Karatas, M., & Craparo, E. (2015). Evaluating the direct blast effect in multistatic sonar networks using Monte Carlo simulation. In *Proceedings of the 2015 Winter Simulation Conference* (pp. 1184-1194). IEEE Press. 10.1109/WSC.2015.7408244

Karatas, M., Gunal, M. M., & Craparo, E. M. (2016b). Performance Evaluation of Mobile Multistatic Search Operations via Simulation. In *Proceedings of the 2016 Spring Simulation Conference*. IEEE Press.

Karatas, M., & Onggo, B. S. (2016). Validating an Integer Non-Linear Program Optimization Model of a Wireless Sensor Network Using Agent-Based Simulation. In *Proceedings of the 2016 Winter Simulation Conference*. IEEE Press. 10.1109/WSC.2016.7822188

Karatas, M., Razi, N., & Gunal, M. M. (2017b). An ILP and simulation model to optimize search and rescue helicopter operations. *The Journal of the Operational Research Society*, *68*(11), 1335–1351. doi:10.105741274-016-0154-7

Karatas, M., Razi, N., & Tozan, H. (2016a). A Comparison of p-median and Maximal Coverage Location Models with Q–coverage Requirement. *Procedia Engineering*, *149*, 169–176. doi:10.1016/j.proeng.2016.06.652

Karatas, M., Razı, N., & Tozan, H. (2017a). A multi-criteria assessment of the p-median, maximal coverage and p-center location models. *Tehnicki Vjesnik-Technical Gazette*, *24*(Suppl. 2), 399–407. doi:10.17559/TV-20160508131905

Karatas, M., & Yakıcı, E. (2018). An iterative solution approach to a multi-objective facility location problem. *Applied Soft Computing*, *62*, 272–287. doi:10.1016/j.asoc.2017.10.035

Kierstead, D. P., & DelBalzo, D. R. (2003). A genetic algorithm applied to planning search paths in complicated environments. *Military Operations Research*, *8*(2), 45–59. doi:10.5711/morj.8.2.45

Kladis, G. P., Economou, J. T., Tsourdos, A., White, B. A., & Knowles, K. (2008, June). An emergency refuelling problem over a dynamically changing environment in the context of Unmanned Aerial Vehicles. In *Control and Automation, 2008 16th Mediterranean Conference on* (pp. 703-708). IEEE. 10.1109/MED.2008.4602182

Konak, A., Kulturel-Konak, S., & Snyder, L. (2017). A Multi-Objective Approach to the Competitive Facility Location Problem. *Procedia Computer Science*, *108*, 1434–1442. doi:10.1016/j.procs.2017.05.035

Kress, M., & Royset, J. O. (2008). Aerial search optimization model (ASOM) for UAVs in special operations. *Military Operations Research*, *13*(1), 23–33. doi:10.5711/morj.13.1.23

Kurban, Ö. F., & Tuncay, C. A. N. (2016). Allocation of mini unmanned aerial vehicles for urgent intelligence, surveillance and reconnaissance request. *Journal of Marmara University Social Sciences Institute/Öneri*, *12*(45).

Lang, T. E., & McGarvey, R. G. (2016). Determining Reliable Networks of Prepositioning Materiel Warehouses for Public-Sector Rapid Response Supplies. *Advances in Operations Research*.

Laporte, G., Nickel, S., & da Gama, F. S. (2015). Location science. In The p-Median Problem (p. 644). Berlin: Springer.

Liberatore, F., Scaparra, M. P., & Daskin, M. S. (2011). Analysis of facility protection strategies against an uncertain number of attacks: The stochastic R-interdiction median problem with fortification. *Computers & Operations Research*, *38*(1), 357–366. doi:10.1016/j.cor.2010.06.002

Lim, G. J., Kim, S., Cho, J., Gong, Y., & Khodaei, A. (2016). Multi-UAV Pre-positioning and Routing for Power Network Damage Assessment. *IEEE Transactions on Smart Grid*, 1. doi:10.1109/TSG.2016.2637408

Love, R., Morris, J., & Wesolowsky, G. O. (1988). *Facility location: Models and methods*. Amsterdam: North-Holland.

Mirchandani, P. B., & Francis, R. L. (Eds.). (1990). *Discrete location theory*. New York: Wiley-Interscience-Interscience.

Morton, D. P., Salmeron, J., & Wood, R. K. (2009). A stochastic program for optimizing military sealift subject to attack. *Military Operations Research*, *14*(2), 19–39.

Mozaffari, M., Saad, W., Bennis, M., & Debbah, M. (2016, May). Optimal transport theory for power-efficient deployment of unmanned aerial vehicles. In *Communications (ICC), 2016 IEEE International Conference on* (pp. 1-6). IEEE.

Murty, K. G., & Djang, P. A. (1999). The US Army national guard's mobile training simulators location and routing problem. *Operations Research*, *47*(2), 175–182. doi:10.1287/opre.47.2.175

Nelson, C., Boros, E., Roberts, F., Rubio-Herrero, J., Kantor, P., McGinity, C., . . . Hanson, K. (2014). ACCAM Global Optimization Model for the USCG Aviation Air Stations. In *Proceedings of IIE Annual Conference*. Institute of Industrial Engineers-Publisher.

Nickel, S., & Puerto, J. (2005). *Location theory: A unified approach*. Berlin, Germany: Springer Verlag.

Overholts, D. L. II, Bell, J. E., & Arostegui, M. A. (2009). A location analysis approach for military maintenance scheduling with geographically dispersed service areas. *Omega*, *37*(4), 838–852. doi:10.1016/j.omega.2008.05.003

Owen, S. H., & Daskin, M. S. (1998). Strategic facility location: A review. *European Journal of Operational Research*, *111*(3), 423–447. doi:10.1016/S0377-2217(98)00186-6

Pelot, R., Akbari, A., & Li, L. (2015). Vessel Location Modeling for Maritime Search and Rescue. In H. A. Eiselt & V. Marianov (Eds.), *Applications of Location Analysis* (pp. 369–402). Springer International Publishing. doi:10.1007/978-3-319-20282-2_16

Qiu, J., & Sharkey, T. C. (2013). Integrated dynamic single-facility location and inventory planning problems. *IIE Transactions*, *45*(8), 883–895. doi:10.1080/0740817X.2013.770184

Radovilsky, Z., & Koermer, T. (2007). Allocation of US Coast Guard boats utilizing integer programming. *Journal of the Academy of Business and Economics*, *7*(2), 130–135.

Radovilsky, Z., & Wagner, M. R. (2014). Optimal Allocation of Resources at US Coast Guard Boat Stations. *Journal of Supply Chain and Operations Management*, *12*(1), 50.

Raissi-Dehkordi, M., Chandrashekar, K., & Baras, J. S. (2004). *UAV placement for enhanced connectivity in wireless ad-hoc networks*. Technical Research Report.

Rappold, J. A., & Van Roo, B. D. (2009). Designing multi-echelon service parts networks with finite repair capacity. *European Journal of Operational Research, 199*(3), 781–792. doi:10.1016/j.ejor.2008.08.006

Razi, N., & Karatas, M. (2016). A multi-objective model for locating search and rescue boats. *European Journal of Operational Research, 254*(1), 279–293. doi:10.1016/j.ejor.2016.03.026

Razi, N., Karatas, M., & Gunal, M. M. (2016). A combined optimization and simulation based methodology for locating search and rescue helicopters. In *Proceedings of the 49th Annual Simulation Symposium* (p. 5). Society for Computer Simulation International.

Rettke, A. J., Robbins, M. J., & Lunday, B. J. (2016). Approximate dynamic programming for the dispatch of military medical evacuation assets. *European Journal of Operational Research, 254*(3), 824–839. doi:10.1016/j.ejor.2016.04.017

ReVelle, C. S., & Eiselt, H. A.ReVelle. (2005). Location Analysis: A Synthesis and Survey. *European Journal of Operational Research, 165*(1), 1–19. doi:10.1016/j.ejor.2003.11.032

Sarıçiçek, İ., & Akkuş, Y. (2015). Unmanned Aerial Vehicle hub-location and routing for monitoring geographic borders. *Applied Mathematical Modelling, 39*(14), 3939–3953. doi:10.1016/j.apm.2014.12.010

Sarikaya, N. (2009). *Determining the Orbit Locations of Turkish Airborne Early Warning and Control Aircraft over the Turkish Airspace. AFIT-Air Force Institute of Technology, AFIT/GOR/ENS/09-14*. Dayton, OH: Wright Patterson AFB.

Scaparra, M. P., & Church, R. L. (2008). A bilevel mixed-integer program for critical infrastructure protection planning. *Computers & Operations Research, 35*(6), 1905–1923. doi:10.1016/j.cor.2006.09.019

Schick, W. G. (1992). *Location an Imaging Radar in Canada for Identifying Spaceborne Objects*. Air Force Inst. of Technology, Wright Patterson AFB. Retrieved from http://handle.dtic.mil/100.2/ADA258969

Segall, R. S. (2000). Some quantitative methods for determining capacities and locations of military emergency medical facilities. *Applied Mathematical Modelling, 24*(5), 365–389. doi:10.1016/S0307-904X(99)00045-1

Sharma, V., Sabatini, R., & Ramasamy, S. (2016). UAVs assisted delay optimization in heterogeneous wireless networks. *IEEE Communications Letters, 20*(12), 2526–2529. doi:10.1109/LCOMM.2016.2609900

Sylvester, J. J. (1857). A question in the geometry of situation. *Quarterly Journal of Pure and Applied Mathematics*, 1.

Tansel, B. C., Francis, R. L., & Lowe, T. J. (1983). State of the art—location on networks: a survey. Part I: the p-center and p-median problems. *Management Science, 29*(4), 482–497. doi:10.1287/mnsc.29.4.482

Toregas, C., Swain, R., ReVelle, C., & Bergman, L. (1971). The Location of Emergency Service Facilities. *Operations Research, 19*(6), 1363–1373. doi:10.1287/opre.19.6.1363

Toyoglu, H., Karasan, O. E., & Kara, B. Y. (2011). Distribution network design on the battlefield. *Naval Research Logistics, 58*(3), 188–209. doi:10.1002/nav.20402

Wagner, M. R., & Radovilsky, Z. (2012). Optimizing Boat Resources at the US Coast Guard: Deterministic and Stochastic Models. *Operations Research*, *60*(5), 1035–1049. doi:10.1287/opre.1120.1085

Wang, B., Chen, J., Liu, W., & Yang, L. T. (2016). Minimum cost placement of bistatic radar sensors for belt barrier coverage. *IEEE Transactions on Computers*, *65*(2), 577–588. doi:10.1109/TC.2015.2423679

Wang, H., Huo, D., & Alidaee, B. (2014). Position unmanned aerial vehicles in the mobile ad hoc network. *Journal of Intelligent & Robotic Systems*, *74*(1-2), 455–464. doi:10.100710846-013-9939-y

Wang, W. F., Liu, X. L., & Guo, B. (2008). Research on the multi-criteria integrated military support facility location-allocation method. *Systems Engineering-Theory & Practice*, *5*, 20.

Washburn, A., & Karatas, M. (2015). Multistatic search theory. *Military Operations Research*, *20*(1), 21–38.

Washburn, A. R. (2010). A Multistatic Sonobuoy Theory. Naval Postgraduate School report NPS-OR-10-005. doi:10.21236/ADA526686

Weber, A. (1929). *Theory of the Location of Industries* (C. J. Friedrich, Trans.). Academic Press.

White, J., & Case, K. (1974). On covering problems and the central facility location problem. *Geographical Analysis*, *281*, 1974.

Yakıcı, E. (2016). Solving location and routing problem for UAVs. *Computers & Industrial Engineering*, *102*, 294–301. doi:10.1016/j.cie.2016.10.029

Yang, M., Cao, J., Song, J., & Guo, J. (2008, December). A Postorder Traversal Location Model Taking Security on Routes into Consideration. In *Intelligent Information Technology Application Workshops, IITAW'08. International Symposium on* (pp. 535-538). IEEE.

Yildirim, M. F. (2016). *Optimization of hub locations of UAVs for border monitoring as coverage problem* (Doctoral dissertation). State University of New York at Binghamton.

Chapter 2
The Problem of Locating and Routing Unmanned Aerial Vehicles

Ertan Yakıcı
National Defense University, Turkey

Mumtaz Karatas
National Defense University, Turkey

Oktay Yılmaz
National Defense University, Turkey

ABSTRACT

In this chapter, locating and routing of a UAV fleet is discussed. Since the research in the location and routing of UAVs is very limited, the related problems are reviewed. A basic problem and its extended version found in the literature are redefined along with their MILP formulation. Both of the problems are characterized by a prize collecting objective function and a homogeneous fleet of fixed number of UAVs based on naval platforms. The aim is described as maximization of the collected importance value associated to interest points. A solution method that is based on ACO is discussed. The usage of this robust method for both the basic and the extended versions of the problem is explained. The improvement room in the present solution method and future research directions are also discussed.

INTRODUCTION

An unmanned system can be described as an electro-mechanical system without operator aboard (Vargas 2012). Currently, these systems include ground vehicles, aerial vehicles, underwater vehicles, surface vehicles, unattended munitions and sensors. Among all of those, UAV is the most famous type. This is also clear from the fact that U.S., the leader country in unmanned system technology, has allocated 95% of the resources used for unmanned systems for UAVs (US DoD UAV Roadmap 2011). The beginning of the use of UAV technology for human benefit goes to early 20th century. The concerns about safety, cost

DOI: 10.4018/978-1-5225-5513-1.ch002

and ease in usage makes it more attractive to use UAVs instead of classical manned aircrafts, whenever it is possible. UAVs can be employed easily in cases where human life is in danger, such as military operations executed in bad weather, threatening environmental conditions or contaminated regions. Scientific research activities such as data collection, commercial and security duties like border watching, monitoring vehicle traffic, controlling protected areas like energy transmission lines are other cases where UAVs are employed frequently. Organizations have already spent billions of dollars for this technology, and apparently this expenditure will continue to rise with the increasing need in both governmental and civilan organizations. With development of flight capabilities like range, observation, endurance and altitude, UAVs seem to be indispensable factor for increasing military power (Glade 2000). In armed forces, UAVs can be used in executing a number of missions such as transportation, search and rescue, electronic warfare and signals intelligence, battle damage assessment, surveillance and reconnaissance and attacking an enemy target. In addition to all these capabilities of UAVs, especially vertical takeoff and landing is one of the most important capabilities for a navy. With this capability, small UAVs can be deployed to small platforms. Therefore, even small patrol ships can be utilized as stations to operate surveillance or similar operations. Utilizing these privileged systems efficiently is as important as having them as a force multiplier. One of the most important aspects related to the efficiency of their use is stationing and routing them in an optimal manner. In this book chapter, a decision problem is discussed to seek an answer to the combined question of how to locate and route a fleet of UAVs optimally which can station on ships. Note that, this problem can be adapted easily to land-based operations. In this problem, a set of demand points, fleets of UAVs and ships, and candidate stations are assumed. Not all demand points are required to be visited, however, each visited demand point has a certain benefit, while the objective is to maximize the total operation benefit collected by visiting these points. The selection of allowed stations, assignment of ships to these selected stations and assignment of UAVs to the ships are executed simultaneously. It is assumed that the number of ships is not a constraint and any number of ships can be assigned to any selected station. Moreover, the authors also deal with the extension of this basic problem which allows a demand point to have a time window for being visited by a UAV and also allows a UAV to take off and land on ships assigned at different stations. Note that location, routing and prize (benefit) collection are main aspects of the problems described above. Location and routing are well studied subjects, while there is relatively less attention in prize collection. Separately, location problem and routing problem are combinatorial optimization problems which have attracted attention of many researchers because they have several application fields such as logistics, telecommunications, military operations etc. These two problems are closely related to each other. Location problems focus on the decisions which provide an infrastructure for routing problems. Combination of these two problems creates another well-known problem type known as location and routing problem (LRP). A specific variant of this problem, the UAV location and routing problem can be defined as prize maximization LRP with additional constraints reflecting operational limitations.

In the following sections of this chapter, after reviewing the relevant literature, the authors present mixed integer linear programming (MILP) formulation for both types of UAV location and routing problem which are basic and extended versions. Since the problem is hard and, considering the required time to obtain a good solution solving a realistic problem with a commercial optimization software is not reasonable, a heuristic solution method, ant colony optimization, is discussed for obtaining a good solution along with the experimental results.

LITERATURE REVIEW

Since the UAV location and routing problem is defined as LRP with the objective of prize maximization, the authors review the relevant literatures addressing UAV involved routing, location and routing, and team orienteering which is characterized by collection of prizes by a team of players.

Routing of UAVs

With the development in the UAV technology, the need for using UAVs efficiently becomes more important matter. Users need to plan the activities effectively and share the information timely, while the improvements in the communication and information systems help practitioners find fast and effective solutions.

It is observed that UAVs are employed increasingly for surveillance tasks. Since the maximum flight time of a UAV is constrained, it is crucial to have an optimal route plan in order to cover more points, lines or regions of interest (Sundar and Rathinam 2014). Moreover, allowed operation time may also be tighter than flight time. Operation of multiple UAVs in a surveillance task leads the researchers to the routing problem which is a seminal field of operations research known as the Vehicle Routing Problem (VRP), introduced by Dantzig and Ramser (1959).

Vehicle Routing Problem (VRP) is generally used in the context of transportation operations with a distribution network of goods between a warehouse and customers. All of the routes to be used by vehicles are determined to minimize the transportation cost or total travel distance subject to certain constraints. VRP is an NP-hard problem, therefore, since it is assumed that P is not equal to NP, the computational power required for solution increases at a greater rate than polynomial as the problem size increases. The exact solution techniques such as mathematical programming approaches or dynamic programming are not practical for UAV routing problem. For large instances of the VRP, heuristic or metaheuristic approaches have been used in an attempt to overcome the problems associated with exact approaches. The survey of meta-heuristics shows that the proposed methods can find close-to-optimal and sometimes optimal solutions to large instances with a few hundred customers (Shetty 2008).

In recent years, UAV involved routing problems has become an attracting research topic. In the following paragraphs, a summary of relevant works in the UAV related literature is presented.

Ryan et al. (1998) suggest a Tabu Search heuristic within a simulation to solve routing problem for multiple UAVs which attempts to maximize expected target coverage. Weather and UAV survival are taken as probabilistic components in the decision process. Jun et al. (2002) propose an algorithm for path planning using a given map of the probability of threats. Kress and Royset (2008) suggest an optimization model to deploy and employ unmanned aerial vehicles (UAVs) in special operations missions effectively. They also incorporate intelligence regarding the targets in their model. The aim is to find optimal deployment locations for the ground units and optimal search areas for the UAVs to detect maximum number of targets. Royset et al. (2009) formulate and solve a routing problem encountered in aircraft mission planning in the presence of ground threats. They discretize the airspace into a grid of connected vertices. The chosen objective is minimization of risk from threats subject to constraints of fuel capacity and/or maximum flight time.

Shetty (2008) introduces a problem of strategic routing of a fleet of UAVs to attack targets which are prioritized. Lower and upper bounds of munitions for each target, payload capacities for each UAV in the fleet and the range based on fuel capacity and payload are considered in the model. A decomposition scheme is employed to address the target assignment problem and a vehicle routing problem. A Tabu Search Heuristic is employed in solving the problem. Mufalli et al. (2012) constructs a model which conducts sensor selection and routing simultaneously to maximize collection of intelligence by UAVs. Heuristic methods based on column generation are proposed for solving the problem. Avellar et al. (2015) introduces a problem addressing minimum time coverage of areas with a team of UAVs equipped with sensors. The problem is decomposed into two parts: the task assignment to UAVs and routing of UAVs over the area. Luo et al. (2017) introduces a model for the problem of routing the ground vehicle and its carried UAV, where the carrier travels on a road network and UAV travels to visit a number of targets unreached by the carrier. Two different heuristics are suggested to solve the problem.

Tian et al. (2006) solve the routing problem for multiple UAVs to conduct reconnaissance on targets with time window at minimum cost. Constraints satisfy the reconnaissance resolution demands of the targets, and the requirement of not violating the maximum travel time. A GA based approach is proposed to solve the problem. Liu et al. (2013) consider the problem of routing UAVs tasked with traffic surveillance for a road network where road segments are represented as nodes. The nodes are clustered by a clustering algorithm and then a traveling salesman problem is solved using simulated annealing algorithm for each cluster. Liu et al. (2012) and Liu et al. (2014) introduce two multiobjective problems involving UAV route planning for road traffic information collection. In the first problem, the number of the UAVs and UAV maximum cruise distance are considered as constraints and the model attempt to minimize the total cruise distance and maximize the number of detected road sections. The second one aims to minimize the total cruise distance and the number of UAVs used, subject to UAV maximum cruise distance, the number of UAVs available and time windows of targets. Multi-objective evolutionary algorithms are employed in solving both of the problems. Liu et al. (2014) consider the length of road segment also for a multi-objective UAV routing. Zhenhua et al. (2008) propose Multiobjective Ant Colony System (MACS) algorithm for solving a two-objective UAVs route planning problem based on VORONOI diagram. The objective functions are minimizing the route length and danger exposure. Mersheeva and Friedrich (2015) suggest a tailored method based on Variable Neighborhood Search metaheuristic to obtain and update the image of a disaster area to be used in the rescue mission conducted by a fleet of UAVs. Sundar and Rathinam (2014) consider a routing problem for a single UAV constrained by limited fuel and attempting to obtain minimum total travel cost. They assume multiple refueling depots. A mixed integer linear programming formulation and a heuristic approach are proposed to solve the problem. Coelho et al. (2017) introduce a routing problem for a heterogeneous UAV fleet which collect and deliver packages. They assume multiple charging stations and operational requirements. The model with seven objective functions is solved by a matheuristic algorithm. Zhang et al. (2015) presents a linear integer programming formulation to minimize the delay in detection and cost of operation with constraint on flying routes in the context of traffic monitoring system with fixed and mobile sensors. They propose a solution framework which decomposes the problem into sub-problems.

Lim et al. (2016) introduces a two-phase framework for the solution of a power network damage assessment problem. The first phase solves a two-stage stochastic optimization model in which the first stage determines the UAV locations considering an extreme weather event, and the second stage updates the UAV locations according to updated information about weather condition. In the second phase, UAV routes are determined to minimize both operating cost and damage assessment completion time.

Sarıçiçek and Akkuş (2015) consider a hub-location problem and a UAV routing problem separately in the context of border security. The first problem is the selection of hubs optimally among the airports and the assignment of demand points to hubs with respect to several criteria, utilizing a multi-criteria decision making tool. Second problem is finding optimal routes for each hub considering cost and monitoring frequency.

Savuran and Karakaya (2015) introduce a route optimization method for a UAV launched from carrier. In order to minimize the tour length they have designed a Genetic Algorithm (GA).

Karakaya and Sevinç (2017) propose a Genetic Algorithm as a solution method to a UAV routing problem in order to minimize the number of used UAVs and maximize the number of customers who are serviced in their time window requirement.

Up to now, a few researchers focused on the review of UAV routing. In their book chapter, Rathinam and Sengupta (2007) consider a class of routing problems and review the results and developments in this field. Enright et al. (2015) present a survey of algorithms on task assignment and scheduling for UAVs in an environment where targets arrive randomly. Finally, it is better to note that, a "Handbook of Unmanned Aerial Vehicles" is published recently (Valavanis & Vachtsevanos, 2014) which may serve as a reference text, covering many aspects from design to logistics and ethical issues.

Location Routing Problem (LRP)

A recent survey on LRP (Drexl & Schneider, 2015) defines the problem as a mathematical optimization problem in which at least two decisions are made simultaneously. These decisions are the selection of facilities out of a set of potential ones and selection of customer clusters visited by each vehicle from a given fleet along with their visit order. The research in this area has attracted attention with the advances in processor speed of computers. The motivation behind combining two problems (facility location problem and vehicle routing problem) and solving them simultaneously stems from inferior results when they are optimized separately (Salhi & Rand, 1989). It must be noted that including two NP-hard problems, LRP itself is also an NP-hard problem.

Basic LRP is described by the requirement of at most one trip for each vehicle which begins and ends at the same location (depot) and capacity limit on depots and vehicles, while its objective is minimization of the fixed cost and the routing cost.

Exact solution methods are studied by several researchers. Akca et al. (2009) propose a formulation based on set partitioning for LRP with vehicle and facility capacity limits. A branch-and-price approach is introduced for solution of the problem. Baldacci et al. (2011) also proposed a formulation based on set partitioning. A decomposition algorithm is utilized in solving the capacitated LRP. Another exact solution approach employing a Branch-and-Cut algorithm is presented by Belenguer et al. (2011). The authors strengten the formulation of the capacitated LRP with valid inequalities. Contardo et al. (2011) proposes an exact algorithm which includes both column and cut generation for solving a capacitated LRP formulated as a set-partitioning problem.

Some researchers focused on problem specific heuristics. Barreto et al. (2007) integrates several clustering methods to solve the capacitated LRP. Boudahri et al. (2013) utilizes a clustering-based location-routing model in a sequential manner in the context of planning a supply chain for poultry products.

Although many researchers offered metaheuristic techniques (see Drexl & Schneider, 2015; Prodhon & Prins, 2014), population based metaheuristics are not frequently applied in LRP solution. Metaheuristics based on memetic, GRASP and ant colony optimization is employed to solve capacitated LRP (CLRP)

(Prins et al. 2006, Duhamel et al. 2010, Ting and Chen 2013). Prins et al. (2006) have solved the problem in two phases. In the first phase which is described as a learning process for choosing depots, a GRASP based on extended and randomized Clarke and Wright algorithm is employed. In the second phase, solutions are generated using path relinking approach. Duhamel et al. (2010) describe a greedy randomized adaptive search procedure (GRASP) involving an evolutionary local search for solving the capacitated location-routing problem (CLRP). Ting and Chen (2013) propose an ant colony optimization algorithm having three levels. The levels determine open depots, assignment of customers to depots and visiting order of assigned customers for each open depot. Pheromone matrix is updated with construction and evaluation of each solution.

As in the problems that will be discussed in the next section, some researchers have interests in LRPs with unusual optimization criteria. Those criteria are mostly encountered in emergency logistics or material disposal problems (see Drexl and Schneider, 2015, Prodhon and Prins, 2014).

Ahn et al. (2008) has described a relevant problem. The problem is first discussed in the context of planetary surface exploration missions in the paper of Ahn et al. (2008). The detail about the solution method is presented by Ahn et al. (2012). The problem is introduced as the generalized location routing problem with profits (GLRPP), where the depots and the customers correspond to potential bases and exploration sites with profit values, respectively. The simultaneous decisions to be made in the problem are base selection, selection of mission strategy for each base, selection of sites to be visited along with visit order, and routing tactic selection. The problem is described mathematically and two solution methods are provided.

The readers who need more information on LRP literature are recommended to review the seminal survey papers published by Drexl and Schneider (2015) and by Prodhon and Prins (2014).

Team Orienteering Problem

The other relevant problem is team orienteering problem (TOP), characterized by a set of vertices, where a certain price is assigned to each element, and a set of players who belong to a team. The aim is to determine the paths of the players to maximize the total prices collected in a limited time. The single-player form of the problem was first introduced by Golden et al. (1987), while the generalized form allowing multiple paths is introduced by Chao et al. (1996). From the combinatorial optimization point of view, the single-player version, namely orienteering problem (OP), can be considered as the combination of Knapsack Problem (KP) and Travelling Salesperson Problem (TSP), where the initial goal is to maximize the total collected prices as in the KP, and the latter one is to minimize the total time or distance as in the TSP (Gunawan et al. 2016). The problem is known to be NP-Hard, and this makes heuristic methods necessary. Therefore, many of the researchers have chosen to solve the problem using heuristics and metaheuristics. In the following paragraphs, the solution approaches for TOP and TOP with time windows (TOPTW) are briefly reviewed.

Butt and Ryan (1999) utilize column generation in an exact algorithm to solve the TOP. Poggi et al. (2010) formulate the problem in three different ways. An exact approach, a branch-and-cut-and-price algorithm, is proposed. An exact method using branch-and-cut algorithm is proposed by Dang et al. (2013a). The authors also present new valid inequalities. Another exact method which employs a branch-and-price algorithm is proposed by Keshtkaran et al. (2016).

Several metaheuristic adaptations are proposed by the researchers. A Tabu search heuristic in an Adaptive Memory Procedure (AMP) is applied by Tang and Miller-Hooks (2005). Paths are combined to form an initial solution for the tabu search which improves them. Archetti et al. (2007) develop four different metaheuristic methods based on Tabu search and Variable Neighbourhood Search for TOP. Ant Colony Optimisation (ACO) technique is implemented by Ke et al. (2008). A local search procedure is also used to improve the solution. Vansteenwegen et al. (2009a,b) applied heuristic approaches in their research. The methods are based on Guided Local Search (GLS) (Vansteenwegen et al., 2009a) and Skewed Variable Neighbourhood Search (SVNS) (Vansteenwegen et al., 2009b). It is observed that the SVNS algorithm outperforms the GLS algorithm in both solution quality and solution time. An algorithm which is a combination of GA and local search techniques is proposed by Bouly et al. (2010). A metaheuristic, Discrete PSO (DPSO), is proposed by Muthuswamy and Lam (2011). Another approach based on PSO is proposed by Dang et al. (2011). The authors extend the algorithm presented by Bouly et al. (2010). Dang et al. (2013b) extend more by proposing a new PSO based method, namely PSO-inspired Algorithm (PSOiA), which can explore the feasible space effectively. Lin (2013) proposes a Multi-start Simulated Annealing (MSA) by combining SA and the multi-start hill climbing technique. The multi-start hill climbing reduces the risk of being trapped in a local optimum.

Boussier et al. (2007) made the problem richer by adding limits on vehicle capacity and travel distance. The authors utilize a branch-and-price algorithm originally designed for the TOP for solving the proposed new problem, namely the Selective Vehicle Routing Problem with Time Windows. Note that in the problem the authors deal with, visiting a vertex requires a service time. Therefore, differently, the time limit affects not only total travel time, but also total visiting time of a path.

TOPTW has also attracted attention of the researchers. Montemanni and Gambardella (2009) apply an ant colony optimization approach. Tricoire et al. (2010) have chosen to solve Multi-Period Orienteering Problem with Multiple Time Windows, which can be converted to TOPTW by setting certain parameters. A Variable Neighbourhood Search (VNS) method embedded with an exact algorithm is employed in solving the problem. Vansteenwegen et al. (2009c) proposed a fast Iterated Local Search (ILS) approach to handle TOPTW instances. Labadie et al. (2011) introduces a hybrid approach where Greedy Randomized Adaptive Search Procedure (GRASP) and an Evolutionary Local Search (ELS) algorithm is combined. The same authors propose a Variable Neighborhood Search to solve both OPTW and TOPTW (Labadie et al. 2012). The approach is tailored to deal with granularity for OPTW and TOPTW. An hybridization of GRASP and ILS is presented by Souffriau et al. (2013) for solving the Multi-Constraint Team Orienteering Problem with Multiple Time Windows (MC-TOPMTW) where some nodes may have more than one time window. Hu and Lim (2014) propose an iterative method with three components: two local search procedures and a simulated annealing procedure. The local search procedures explore the feasible space to find a collection of routes. Then, SA component derives the solutions with high quality. Cura (2014) proposes an Artificial Bee Colony (ABC) approach. A new food source acceptance criterion and a new scout bee search behavior are introduced to search for improvement in the solution quality.

The result of researchers' increasing attention to OP, TOP and their variants in recent years are surveyed in the papers authored by Vansteenwegen et al. (2011) and Gunawan et al. (2016). Note that, these review papers also covers several variants of the OP and TOP which are not discussed here.

MODELING AND SOLVING UAV LOCATION AND ROUTING PROBLEM

In this chapter, as already expressed before, two versions of UAV location and routing problem reported in previous studies of Yakıcı (2016a,b) are discussed.

We name the basic and extended problems as "Prize Collecting Location and Routing Problem (PCLRP)" and "Prize Collecting Location and Routing Problem with Time Windows (PCLRPTW)", respectively.

In the following subsections, definition of the problems are given along with their mathematical (Mixed Integer Linear Programming (MILP)) formulations, then solution approach, a metaheuristic solution approach based on Ant Colony Optimization is described and finally the results of numerical experiment, where it is applied to many different instances, is discussed.

Prize Collecting Location and Routing Problem (PCLRP)

In PCLRP, given sets are interest points, stations and UAVs. The aim of the problem is to maximize the total benefit collected by visiting interest points. The selection of (from a set of candidate points), assignment of UAVs to stations and path construction of each UAV are executed simultaneously under the assumption of a homogeneous UAV fleet. Each UAV has to complete its sortie (path) and land on its origin station before the maximum mission time is elapsed. An illustration of a feasible solution to PCLRP is presented in Figure 1.

The assumptions of PCLRP can be listed as follows:

- UAV fleet is composed of identical vehicles,
- Available number of UAVs is known,
- Each UAV flies single sortie.
- A UAV can take off and land on naval platform(s) which is(are) assigned at the same station where UAV is assigned.

The indices, sets, parameters, variables and the formulation are as follows:

Sets and Indices

$u \in U$: set of UAVs.
$i, j \in I$: set of interest points.
$i, j \in S$: set of stations.

Parameters

p_i: importance of interest point i.
t_i: expected time on interest point i.
d_{ij}: expected time in flight between i and j.
y_{max}: maximum number of active stations allowed.
t_{max}: maximum time between takeoff and landing for UAV.

Figure 1. An illustration of a feasible solution to PCLRP

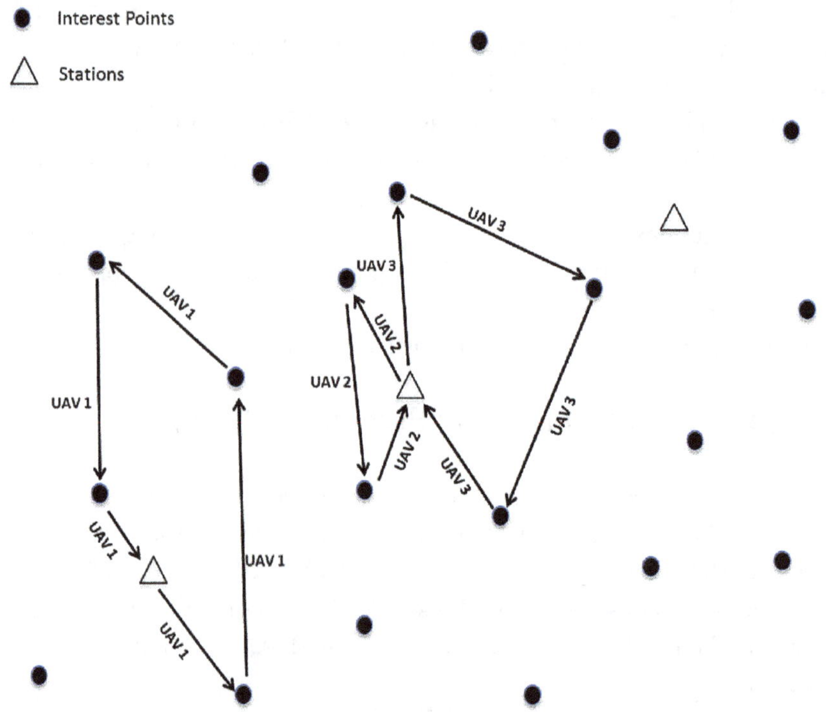

Decision Variables

X_{iju}: binary variable indicating if UAV u has a leg from point i to point j, or not.
Y_i: binary variable indicating if station i is activated, or not.
F_{iju}: a continuous variable.

Formulation of the Problem

$$\max z = \sum_{j \in I \cup S} \sum_{i \in I} \sum_{u \in U} p_i X_{jiu} \tag{1}$$

subject to

$$\sum_{i \in S} Y_i \leq y_{\max} \tag{2}$$

$$\sum_{j \in I} \sum_{u \in U} X_{iju} \leq Y_i \left| I \right|; \forall i \in S \tag{3}$$

$$\sum_{i\in S}\sum_{j\in I} X_{iju} \leq 1; \forall u \in U \tag{4}$$

$$\sum_{j\in I} X_{iju} = \sum_{j\in I} X_{jiu}; \forall i \in S, u \in U \tag{5}$$

$$\sum_{i\in I\cup S}\sum_{j\in I\cup S} X_{iju}(d_{ij} + t_j) \leq 1; \forall i \in S, j \in I, u \in U \tag{6}$$

$$\sum_{i\in I\cup S} X_{iju} = \sum_{i\in I\cup S} X_{jiu}; \forall j \in I, u \in U \tag{7}$$

$$\sum_{i\in I\cup S}\sum_{u\in U} X_{jiu} \leq 1; \forall j \in I \tag{8}$$

$$\sum_{i\in I\cup S}\sum_{u\in U} F_{jiu} - \sum_{i\in I\cup S}\sum_{u\in U} F_{iju} \leq \varepsilon * \sum_{i\in I\cup S}\sum_{u\in U} X_{iju}; \forall j \in I \tag{9}$$

$$F_{iju} \leq X_{iju}; \forall i \in I \cup S, j \in I \cup S, u \in U \tag{10}$$

$$X_{iju} \in \{0,1\}; \forall i \in I \cup S, j \in I \cup S, u \in U \tag{11}$$

$$Y_i \in \{0,1\}; \forall i \in S \tag{12}$$

$$F_{iju} \geq 0; \forall i \in I \cup S, j \in I \cup S, u \in U \tag{13}$$

The objective function 1 represents total importance value collected from interest points. Constraint 2 limits the number of stations that can be activated. Constraints 3 and 4 force each UAV to start its route from only one station to which it is assigned, while Constraint 5 force each UAV to return back to its departure point. Constraint 6 limits flight time for each UAV with a certain duration. Constraint 7 serves as well-known flow conservation requirement of network flow problems. Constraint 8 limits the number of departures of each UAV from an interest point at most to one. Constraints 9 and 10 prevent infeasible tours by eliminating subtours, where ε is a small positive real number. Constraints 11-13 identify the sets for decision variables. The objective function and these constraints numbered from 2 to 13 collectively define PCLRP.

Prize Collecting Location and Routing Problem With Time Windows (PCLRPTW)

PCLRP is extended by implementing two new features. One of them is allowing each UAV to land on any one of the active stations and the other is adding time windows for the task can be started on interest points. Removal of constraints 5 and 6, and employment of the constraints numbered from 14 to 22 converts the basic problem PCLRP to the extended problem PCLRPTW.

Note that, to obtain PCLRPTW, the assumption about taking off and landing on at the same station is removed by allowing UAVs to land any open station whether or not it is assigned initially to there. Moreover, each demand point has given a time window, such that no UAV can start operating on this demand point out of its time window. An illustration of a feasible solution to PCLRPTW is presented in Figure 2.

Additional parameters and constraints required to construct the mathematical formulation of PCLRPTW are given as follows:

Figure 2. An illustration of a feasible solution to PCLRPTW

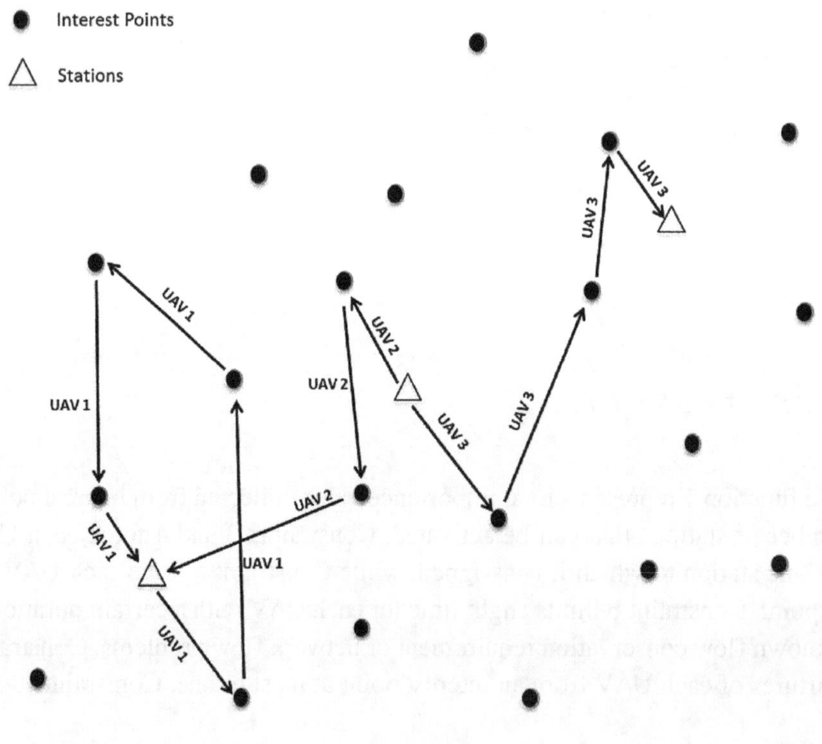

Parameters

b_i: beginning time for time window of interest point i.
e_i: ending time for time window of interest point i.
A_{ui}: arrival time of UAV u to interest point i.

Constraints

$$\sum_{j\in I}\sum_{u\in U} X_{jiu} \leq Y_i \left|I\right|; \forall i \in S \tag{14}$$

$$\sum_{j\in I}\sum_{u\in U} X_{jiu} \leq \sum_{j\in I}\sum_{u\in U} X_{iju} \left|U\right|; \forall i \in S \tag{15}$$

$$\sum_{i\in S}\sum_{j\in I} X_{jiu} \leq 1; \forall u \in U \tag{16}$$

$$\sum_{i\in S}\sum_{j\in I} X_{iju} = \sum_{i\in S}\sum_{j\in I} X_{jiu}; \forall u \in U \tag{17}$$

$$A_{uj} \leq \left(1 - \sum_{i\in I\cup S} X_{iju}\right) M + E_j; \forall j \in I, u \in U \tag{18}$$

$$A_{uj} \geq \left(\sum_{i\in I\cup S} X_{iju} - 1\right) M + B_j; \forall j \in I, u \in U \tag{19}$$

$$A_{uj} \geq A_{ui} + X_{iju} t_i + d_{ij} + \left(X_{iju} - 1\right) M; \forall i \in I \cup S, j \in I, u \in U \tag{20}$$

$$A_{uj} + t_i + X_{jiu} d_{ij} + \left(X_{jiu} - 1\right) M \leq t_{\max}; \forall i \in S, j \in I, u \in U \tag{21}$$

$$A_{uj} \geq 0; \forall j \in I, u \in U \tag{22}$$

Constraint 14 ensures that if station is not active, UAV cannot land on that station. Constraint 15 restricts that landing on a station can occur if a takeoff is realized (which means that the station is active) at that station. Constraint 16 forces each UAV to land on at most one station. Constraint 17 ensures that if a UAV takes off, it must land. Constraints from 18-20 provide the satisfaction of time window restrictions. Constraint 21 ensures that all UAVs should be landed on a station before a certain given time. Constraint 22 declares the domain for variable A_{uj}. The constant M, used in Constraints 18 to 21, represents a positive real number greater than t_{max}.

The objective function and the constraints numbered from except 5 and 6 construct the extended problem PCLRPTW.

Ant Colony Optimization as a Solution Method

Ant Colony Optimization (ACO) is a heuristic solution method inspired from ant behavior observed in the nature. Ants use pheromones to communicate each other and to mark their path in order to find the shortest path between a food source and their nest. Intensity of the pheromone trail helps them to make their selection between alternative ways. The pheromone trail evaporates over time. However, if many ants select a certain path and lay down pheromones, the intensity of trail on this path increases which causes this path to be used by more ants (Stützle and Hoos 1997).

The first example of an algorithm which is based on the behavior of ants is Ant System (AS) (Dorigo 1992). This initiation stimulated the research on algorithm and on applications of different problems such as TSP, scheduling, assembly line balancing, etc. In general, the ants implement a randomized construction heuristic utilizing pheromone trails and heuristic information based on the the problem to be handled. Algorithms which follow this framework, namely ACO metaheuristic, are called ACO algorithms (Dorigo and Stützle 2010). These algorithms have been used to solve several combinatorial optimization problems which are proven to be NP-Hard. Motivated with the success, a variant of ACO is tailored to solve the problems discussed in this study (Yakıcı 2016a,b).

Tailoring Ant Colony Optimization to PCLRP and PCLRPTW

Here, the algorithm tailored for PCLRP is explained, then the required alterations for the algorithm to be used in solving PCLRPTW is noted.

A famous ACO approach which is called MMAS (MAX-MIN Ant System) (Stützle and Hoos 1997, 2000) has inspired the solution approach implemented. Each UAV corresponds to an ant and all of the routes of ants construct one complete solution. Iterative construction process and update of pheromone trails are the main subroutines of the implementation. Two probability distributions which build the solution construction mechanism include visibility and learned knowledge components of ACO approach.

Before explaining the algorithm, the notation used in describing ACO implementation is given as follows:

- η_{ij}: A measure of visibility of point $j \in I$ from point $i \in I \cup S$ which corresponds to greedy component of the ACO approach.
- ζ_{ij}: Importance of interest point j per unit time elapsed both in transition from $i \in I \cup S$ to $j \in I$ and the time elapsed for executing the task at j

- $\tau^{gl}_{i,\,ni,\,j,\,k}$ and $\tau^{st}_{i,\,ni,\,j,\,k}$: A measure of contribution of solution component to the value of prior solutions (learned knowledge components or pheromone trails). The superscripts used in the notation, "*gl*" and "*st*", differentiates two pheromone trails as global and stationary, respectively.
- τ_{max} and τ_{min}: Minimum and maximum values for pheromone trails to prevent early search stagnation.

A basic component of a solution can be defined by a tuple of indices i, n_i, j and k, where $i \in S$ is the station of origin platform, $n_i \leq |U|$ is the number of UAVs based on the origin platform stationed at i, $j \in (I \cup S)$ is the point where UAV currently operates and $k \in I$ is the point which is considered to be visited next. A feasible collection of solution components consisting of UAV routes form a solution.

Employment of multiple pheromone trails ($\tau^{gl}_{i,\,ni,\,j,\,k}$ and $\tau^{st}_{i,\,ni,\,j,\,k}$) is preferred as encountered in a few of the previous works in the literature (Chircop and Buckingham 2014, Ngenkaew et al. 2008). By this strategy, the authors reflects the frequency of component (i, n_i, j, k)'s use in both of the previous global and stationary reference solutions (which will be defined later in this section).

The visibility component η_{ij} is given in Equation (23).

$$\eta_{i',j'} = \tau_{\min} + \frac{\zeta_{i',j'} - \min\limits_{i,j} \zeta_{i,j}}{\max\limits_{i,j} \zeta_{i,j} - \min\limits_{i,j} \zeta_{i,j}} (\tau_{\max} - \tau_{\min}) \tag{23}$$

where ζ_{ij} is equal to $p_j / (d_{ij}+t_j)$. Note that η_{ij} is normalized between the boundaries s_{max} and s_{min}.

For assigning UAVs to stations and for routing them afterwards, separate probability distributions are constructed using the same pheromone trails. Before describing the probability distributions, following functions need to be defined:

$$\gamma^{gl}_{i,j} = \sum\nolimits_{c|c>n^c_i} \tau^{gl}_{i,c,i,j} \tag{24}$$

$$\gamma^{gl}_{i,j} = \sum\nolimits_{c|c>n^c_i} \tau^{st}_{i,c,i,j} \tag{25}$$

where n^c_i is the current number of already assigned UAVs to station i. Note that, the functions reflect the sum of pheromone trails on the arc from station i to interest point j when there are greater number of UAVs than the current number of UAVs at station i.

Equation (26) defines the probability distribution that is used in the assignment of UAVs to stations. Any UAV which is not assigned to any station yet has a probability for leaving any station i' for going to any interest point j'. Therefore, if this leaving event is realized, the UAV is implicitly assigned to the station where it leaves. If the elapsed time in a tour starting from i', visiting j' and ending at i' is greater than t_{max}, then the corresponding probability is set to zero and UAV is not allowed to leave i' to visit j'.

$$p_{i',j'}^{assignment} = \frac{\left(\gamma_{i',j'}^{gl}\right)^{\alpha^{gl}} \left(\gamma_{i',j'}^{st}\right)^{\alpha^{st}} \left(\eta_{i',j'}\right)^{\beta^{a}}}{\sum_{i,j} \left(\gamma_{i,j}^{gl}\right)^{\alpha^{gl}} \left(\gamma_{i,j}^{st}\right)^{\alpha^{st}} \left(\eta_{i,j}\right)^{\beta^{a}}} \quad (26)$$

Note that these arcs are only temporarily taken by assigned UAVs for assignment phase. After assignment of a UAV its leaving from assigned station is erased, so that each one of these arcs leaving from stations is kept as a non-routed arc that is allowed to be taken by any unassigned UAV. The parameters used as exponents determine the relative effects of corresponding components. The superscript of parameter β^a stands for the "assignment" phase. Note here that, assignment of a UAV does not mean that it has to be used in routing.

After assignment of all of the UAVs, routing phase is started. The probability of a UAV leaving from its current point $j' \in I \cup S$ and flies directly to $k' \in I$ (given it is based on station i' and n_i' UAVs assigned to station i') is expressed as follows:

$$p_{i',n_{i'},j',k'}^{routing} = \frac{\left(\tau_{i',n_{i'},j',k'}^{gl}\right)^{\alpha^{gl}} \left(\tau_{i',n_{i'},j',k'}^{st}\right)^{\alpha^{st}} \left(\eta_{j',k'}\right)^{\beta^{r}}}{\sum_{i,n_{i'},j,k} \left(\tau_{i,n_{i'},j,k}^{gl}\right)^{\alpha^{gl}} \left(\tau_{i,n_{i'},j,k}^{st}\right)^{\alpha^{st}} \left(\eta_{j,k}\right)^{\beta^{r}}} \quad (27)$$

This probability is set to zero when any one of the problem constraints is not satisfied for the candidate solution component defined by the tuple (i', n_i', j', k'). As discussed in the explanation of assignment phase, the exponents determine the relative effects of corresponding components and the superscript of parameter β^r represents "routing" phase.

In the beginning, all of the pheromone values are initialized at τ_{max}. During the heuristic run, whenever a pheromone value is outside the allowed range which is between τ_{min} and τ_{max}, it is set to the closest value in the allowed range. The pheromone trails are updated at every m and n iterations according to the Expressions (28) and (29):

$$\tau_{i,n_i,j,k}^{gl} \leftarrow \rho\tau_{i,n_i,j,k}^{gl} + \Delta\tau_{i,n_i,j,k}^{gl} \quad (28)$$

$$\tau_{i,n_i,j,k}^{st} \leftarrow \rho\tau_{i,n_i,j,k}^{st} + \Delta\tau_{i,n_i,j,k}^{st} \quad (29)$$

where $\Delta\tau^{gl}$ $i,ni,j,k = z^R / \widehat{z^{gl}}$ and $\Delta\tau^{st}$ $i.ni,j,k = z^R / \widehat{z_i^{st}}$ if the solution component described by the tuple (i, n_i, j, k) is employed in the referenced solution (R), otherwise they both are set to zero. The evaporation rate ρ is taken between 0 and 1. R is either the best solution found, or the group best solution which is the best in last n iterations. The value of solution R is denoted by z^R. The values denoted by $\widehat{z^{gl}}$ and $\widehat{z_i^{st}}$ represent the best solution value and best contribution of station (i) to any solution.

A solution is evaluated in two different ways: global evaluation and stationary evaluation. In the first one, the total objective function value is taken. While the solution is evaluated with respect to a station i, only the importance values of points visited by the paths leaving station i are summed up. Therefore, there are four kinds of values utilized in update of global and stationary pheromones: best global solution, B^{gl}, the solution with best stationary contribution of station i, B^{st}_{i}, group best global solution, G^{gl}, and the solution with best stationary contribution of station i (within the group of last n iterations), G^{st}_{i}.

At the end of every m iterations, updates are executed with respect to best global and best stationary contributions, B^{gl} and B^{st}_{i} (for each station i). On the other hand, at the end of every n iterations, updates are executed with respect to group best contributions, G^{gl} and G^{st}_{i}, only if the referenced solution is within π percent of corresponding best solution B^{gl} and B^{st}_{i}. Otherwise, $\Delta\tau$ value is set to zero.

CONSTRUCTSOLUTION and UPDATEPHEROMONE procedures, called at each iteration, are described in Figures 2 and 3 (Yakıcı 2016a), respectively. In pseudocode, corresponding iteration is indexed with "*iter*".

As applied in MMAS implementations (Dorigo & Stützle, 2010), at every certain number of iterations additional pheromone trail may be laid on all of the solution components to support exploration.

```
procedure CONSTRUCTSOLUTION(iter)
        while any u ∈ U is not assigned to station i ∈ S do
            for ∀i ∈ S do
                if ( Σ_{i|i∈A} 1 < ymax ) ∨ (i ∈ A) then        ▷ A | A ⊆ S is set of assigned stations
                        Calculate γ^{gl}_{i,j} and γ^{st}_{i,j} ∀i ∈ S, j ∈ I
                end if
            end for
            for ∀i ∈ S, j ∈ I do
                    Calculate p^{assigment}_{i,j}
        end for
                    Choose the assigment component (i, j) randomly
      end while
    Σ_{i,n_i,j,k} (τ^{gl}_{i,n_{i'},j,k})^{α^{gl}} (τ^{st}_{i,n_{i'},j,k})^{α^{st}} (η_{j,k})^{β^τ} ← 1 ▷ sum is set to a positive number arbitrarily
    while Σ_{i,n_i,j,k} (τ^{gl}_{i,n_{i'},j,k})^{α^{gl}} (τ^{st}_{i,n_{i'},j,k})^{α^{st}} (η_{j,k})^{β^τ} > 0 do
            for ∀i ∈ S, j ∈ {i} ∪ I, k ∈ I do
                    Calculate p^{routing}_{i,n_i,j,k}
                end for
                Choose the routing component (i, n_i j, k) randomly
            end while
  end procedure
procedure UPDATEPHEROMONE(iter)
    if iter is a multiple of n then
        Update G^{gl}
    R ← G^{gl}
    for ∀(i, n_{i'}, j, k) do
```

if $z^{\hat{g}l}(1-\pi) \le z^R \wedge (i,n_{i'},j,k) \in R$ **then** $\tau^{gl}_{i,n_{i'},j,k} \leftarrow \rho\tau^{gl}_{i,n_{i'},j,k} + z^R / z^{\hat{g}l}$
else $\tau^{gl}_{i,n_{i'},j,k} \leftarrow \rho\tau^{gl}_{i,n_{i'},j,k}$
end if
end for
for $i \in S$ **do**
if i is assigned in last n iterations **then**
Update G^{st}_i
$R \leftarrow G^{st}_i$
for $\forall(i,n_{i'},j,k)$ **do**
if $z^{\hat{s}t}(1-\pi) \le z^R \wedge (i,n_{i'},j,k) \in R$ **then** $\tau^{st}_{i,n_{i'},j,k} \leftarrow \rho\tau^{st}_{i,n_{i'},j,k} + z^R / z^{\hat{s}t}$
else $\tau^{st}_{i,n_{i'},j,k} \leftarrow \rho\tau^{st}_{i,n_{i'},j,k}$
end if
end for
end if
end for
end if
if *iter* is a multiple of m **then**
Update B^{gl}
$R \leftarrow B^{gl}$
for $\forall(i,n_{i'},j,k)$ **do**
if $(i,n_{i'},j,k) \in R$ **then** $\rho\tau^{gl}_{i,n_{i'},j,k} \leftarrow \rho\tau^{gl}_{i,n_{i'},j,k} + z^R / z^{\hat{g}l}$
else $\tau^{gl}_{i,n_{i'},j,k} \leftarrow \rho\tau^{gl}_{i,n_{i'},j,k}$
end if
end for
for $i \in S$ **do**
Update B^{st}_i
$R \leftarrow B^{st}_i$
for $\forall(i,n_{i'},j,k)$ **do**
if $(i,n_{i'},j,k) \in R$ **then** $\tau^{st}_{i,n_{i'},j,k} \leftarrow \rho\tau^{st}_{i,n_{i'},j,k} + z^R / z^{\hat{s}t}_i$
else $\tau^{st}_{i,n_{i'},j,k} \leftarrow \rho\tau^{st}_{i,n_{i'},j,k}$
end if
end for
end for
end if
end procedure

To make the algorithm usable for *PCLRPTW*, a parameter $\theta_{j'k'}$, is defined and given in Equation 30, where ω represents the current time for corresponding UAV. It is utilized to decrease the probability of a UAV to arrive too early at a point before the start of its time window. The probability is decreased proportional to the waiting time until the start of time window.

$$\theta_{ij} = 1 - \frac{\max\left(0, b_i - \omega - d_{ij}\right)}{b_i} \tag{30}$$

As it is discussed before, in PCLRPTW, the constraint which forces each UAV to land on the same station where it has taken off is relaxed. This change can be handled by approval of a solution component as feasible without checking whether UAV has time to turn back to the station where it has taken off. Therefore, with these small changes in the algorithm, it is easily adapted to the extended problem PCLRPTW.

Numerical Experiments and Results

Here, the experiments and results reported in the studies Yakici (2016a and b) will be discussed. For the sake of conciseness, the details of the instances are not given in this section. They are available in the mentioned papers.

For the first problem PCLRP, 30 test instance types are created which differ in number of potential stations (10, 15), number of interest points (40, 60, 80, 100, 120, 150), allowed number of stations (2, 3, 4 and 5), flight time limit (200, 300, 600) and number of UAVs (6, 9, 12). For each instance type two instances are generated using random numbers to locate the nodes of the network. These instances have up to 329,000 constraints, 653,000 variables and 326,000 discrete variables. CPLEX (12.6.2.0) is used as an ILP solver on a computer with 1.9 GHz processor and 4 GB RAM. For heuristic implementation, JAVA (1.8.0_65) is employed.

Heuristic parameters are determined by parametric experiments conducted before the numerical tests. Each test instance is solved both by the proposed heuristic and ILP solver CPLEX 12.6.2.0 with the formulations given in Section 3.

For each test instance, the gap between the best heuristic solution and the best CPLEX solution which is found in one hour is compared. The gaps are presented in Figure 3. It is observed that the heuristic method obtains 55% (on average) better solutions for 56 test instances, while CPLEX obtains 5% (on average) better solutions than the heuristic for only 4 instances. For these problems, heuristic runs have taken from 2 minutes (for the smallest size instances) to 29 minutes (for the largest size instances).

For the second problem, PCLRPTW, the experiment settings, hardware/software tools and comparison criterion are all kept same as in the experiment of PCLRP. The algorithm is tested on 9 of the 30 instance types discussed previously. A number of interest points are chosen randomly and visits to these points are restricted with time windows.

In all of the experienced instances, CPLEX has performed very poor, while the tailored heuristic method has provided significantly better solutions in very short periods (between 2 minutes and 8 minutes). The gaps between the best solutions of CPLEX and heuristic method are given in Figure 4.

Figure 3. Gap between the best solution of CPLEX and heuristic method
Source: Yakici (2016a)

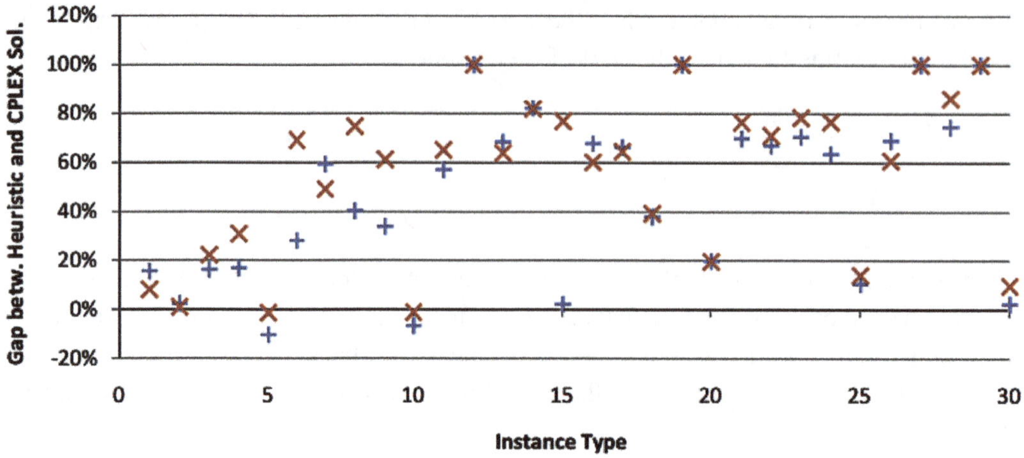

Figure 4. Gap between the best solution of CPLEX and heuristic method
Source: Yakici (2016b)

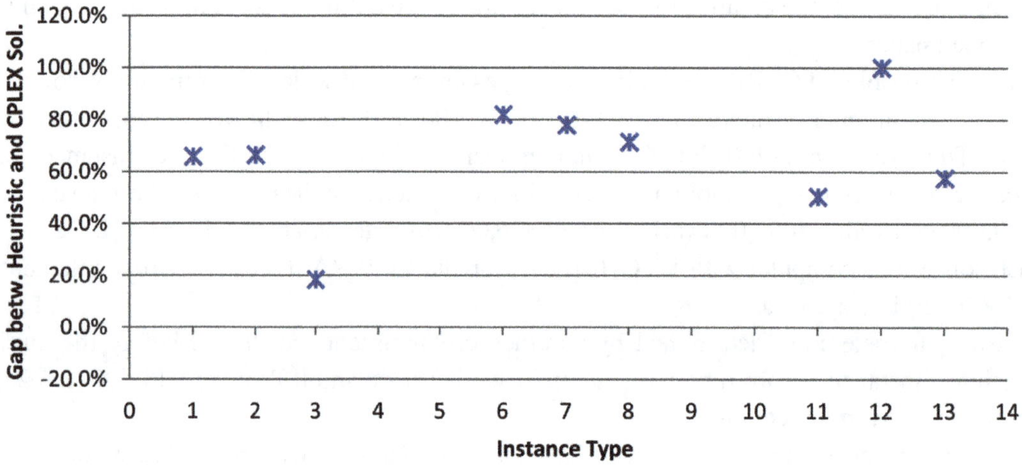

CONCLUSION

In this chapter, the authors discuss locating and routing a homogenous UAV fleet launched from ships. The routing problems in the context of UAV and the related problems are reviewed first. A basic problem and its extended version found in the literature are re-defined along with their MILP formulation. Both of the problems are characterized by a prize collecting objective function and a homogeneous fleet of fixed number of UAVs based on naval platforms. The aim is described as maximization of the collected importance value associated to interest points. Solution method which is based on ACO and proposed by Yakıcı (2016a) is discussed. The usage of this robust method for both of the basic and the extended versions of the problem is explained. The experiment results reported in the original study is briefly discussed.

As a future research, a more general problem which can be applied to many cases with less restrictive assumptions must be considered. Researchers may allow a number of heterogeneous UAVs on a station at any time at most as many as the capacity of that station which is equal to the total capacity of ships assigned to that station (conforming to the system requirements of landing UAV type). A UAV of a certain type can land on any station which has slack capacity of that certain type just before its fuel is exhausted and can depart again after refueling, if necessary. A path of a certain UAV has to start from any one of the open stations having a ship that can launch that certain type and end at any one of them. Each UAV has to finish its all sorties and land on a station before exceeding the allowed mission time. Another assumption which may be relaxed can be about the behavior of ships. Specifically, these launching platforms can be allowed to move on a network. Although this is a realistic assumption, this relaxation may increase the problem size very large and make the problem more complex to solve optimally (or near-optimally).

From the point of the solution approach presented in this chapter, an improvement room that a post-processing application may provide can be observed. After obtaining solutions from heuristic method, a few chosen good solutions can be improved by local search. More specifically, the flight time of a route (or routes) can be decreased by optimizing the visit sequence which may lead to the opportunity to visit an unvisited point before landing due to flight time limitation. Also, swapping visited points between routes can be considered. Of course, many promising solution methods applied in combinatorial problems can be considered again to implement separately of in a combined way in the solving this hard problem.

REFERENCES

Ahn, J., de Weck, O., Geng, Y., & Klabjan, D. (2012). Column generation based heuristics for a generalized location routing problem with profits arising in space exploration. *European Journal of Operational Research, 223*(1), 47–59. doi:10.1016/j.ejor.2012.06.018

Ahn, J., De Weck, O., & Hoffman, J. (2008). An optimization framework for global planetary surface exploration campaigns. *JBIS. Journal of the British Interplanetary Society, 61*(12), 487.

Akca, Z., Berger, R. T., & Ralphs, T. K. (2009). A branch-and-price algorithm for combined location and routing problems under capacity restrictions. In *Operations research and cyber-infrastructure* (pp. 309–330). Boston, MA: Springer. doi:10.1007/978-0-387-88843-9_16

Archetti, C., Hertz, A., & Speranza, M. G. (2007). Metaheuristics for the team orienteering problem. *Journal of Heuristics, 13*(1), 49–76. doi:10.100710732-006-9004-0

Avellar, G. S., Pereira, G. A., Pimenta, L. C., & Iscold, P. (2015). Multi-uav routing for area coverage and remote sensing with minimum time. *Sensors (Basel), 15*(11), 27783–27803. doi:10.3390151127783 PMID:26540055

Baldacci, R., Mingozzi, A., & Wolfler Calvo, R. (2011). An exact method for the capacitated location-routing problem. *Operations Research, 59*(5), 1284–1296. doi:10.1287/opre.1110.0989

Barreto, S., Ferreira, C., Paixao, J., & Santos, B. S. (2007). Using clustering analysis in a capacitated location-routing problem. *European Journal of Operational Research, 179*(3), 968–977. doi:10.1016/j.ejor.2005.06.074

Belenguer, J. M., Benavent, E., Prins, C., Prodhon, C., & Calvo, R. W. (2011). A branch-and-cut method for the capacitated location-routing problem. *Computers & Operations Research, 38*(6), 931–941. doi:10.1016/j.cor.2010.09.019

Boudahri, F., Aggoune-Mtalaa, W., Bennekrouf, M., & Sari, Z. (2013). Application of a clustering based location-routing model to a real agri-food supply chain redesign. In *Advanced methods for computational collective intelligence* (pp. 323–331). Berlin, Heidelberg: Springer. doi:10.1007/978-3-642-34300-1_31

Bouly, H., Dang, D. C., & Moukrim, A. (2010). *A memetic algorithm for the team orienteering problem.* 4OR: A Quarterly Journal of Operations Research, 8(1), 49-70.

Boussier, S., Feillet, D., & Gendreau, M. (2007). *An exact algorithm for team orienteering problems.* 4OR: A Quarterly Journal of Operations Research, 5(3), 211-230.

Butt, S. E., & Ryan, D. M. (1999). An optimal solution procedure for the multiple tour maximum collection problem using column generation. *Computers & Operations Research, 26*(4), 427–441. doi:10.1016/S0305-0548(98)00071-9

Chao, I. M., Golden, B. L., & Wasil, E. A. (1996). A fast and effective heuristic for the orienteering problem. *European Journal of Operational Research, 88*(3), 475–489. doi:10.1016/0377-2217(95)00035-6

Chircop, J., & Buckingham, C. D. (2014). A multiple pheromone ant clustering algorithm. In Nature Inspired Cooperative Strategies for Optimization (NICSO 2013) (pp. 13-27). Springer International Publishing. doi:10.1007/978-3-319-01692-4_2

Coelho, B. N., Coelho, V. N., Coelho, I. M., Ochi, L. S., Zuidema, D., Lima, M. S., & da Costa, A. R. (2017). A multi-objective green UAV routing problem. *Computers & Operations Research, 88*, 306–315. doi:10.1016/j.cor.2017.04.011

Contardo, C., Cordeau, J. F., & Gendron, B. (2011). *A branch-and-cut-and-price algorithm for the capacitated location-routing problem.* CIRRELT.

Cura, T. (2014). An artificial bee colony algorithm approach for the team orienteering problem with time windows. *Computers & Industrial Engineering, 74*, 270–290. doi:10.1016/j.cie.2014.06.004

Dang, D. C., El-Hajj, R., & Moukrim, A. (2013a, May). A branch-and-cut algorithm for solving the team orienteering problem. In *International Conference on AI and OR Techniques in Constriant Programming for Combinatorial Optimization Problems* (pp. 332-339). Springer, Berlin, Heidelberg. 10.1007/978-3-642-38171-3_23

Dang, D. C., Guibadj, R. N., & Moukrim, A. (2011, April). *A pso-based memetic algorithm for the team orienteering problem.* In *European Conference on the Applications of Evolutionary Computation* (pp. 471-480). Springer, Berlin, Heidelberg. 10.1007/978-3-642-20520-0_48

Dang, D. C., Guibadj, R. N., & Moukrim, A. (2013b). An effective PSO-inspired algorithm for the team orienteering problem. *European Journal of Operational Research*, *229*(2), 332–344. doi:10.1016/j.ejor.2013.02.049

Dantzig, G. B., & Ramser, J. H. (1959). The truck dispatching problem. *Management Science*, *6*(1), 80–91. doi:10.1287/mnsc.6.1.80

Department of Defense, *Unmanned Systems Integrated Roadmap FY2011-2036* (Washington, DC: Government Printing Office, October 2010), 89.

Dorigo, M. (1992). *Optimization, learning and natural algorithms*. Ph. D. Thesis, Politecnico di Milano, Italy.

Dorigo, M., & Stützle, T. (2010). Ant colony optimization: overview and recent advances. In Handbook of metaheuristics (pp. 227-263). Springer US. doi:10.1007/978-1-4419-1665-5_8

Drexl, M., & Schneider, M. (2015). A survey of variants and extensions of the location-routing problem. *European Journal of Operational Research*, *241*(2), 283–308. doi:10.1016/j.ejor.2014.08.030

Duhamel, C., Lacomme, P., Prins, C., & Prodhon, C. (2010). A GRASP× ELS approach for the capacitated location-routing problem. *Computers & Operations Research*, *37*(11), 1912–1923. doi:10.1016/j.cor.2009.07.004

Enright, J. J., Frazzoli, E., Pavone, M., & Savla, K. (2015). UAV routing and coordination in stochastic, dynamic environments. In *Handbook of Unmanned Aerial Vehicles* (pp. 2079–2109). Springer Netherlands. doi:10.1007/978-90-481-9707-1_28

Glade, D. (2000). *Unmanned aerial vehicles: Implications for military operations*. Air Univ Press Maxwell Afb Al.

Golden, B. L., Levy, L., & Vohra, R. (1987). The orienteering problem. *Naval Research Logistics*, *34*(3), 307–318. doi:10.1002/1520-6750(198706)34:3<307::AID-NAV3220340302>3.0.CO;2-D

Gunawan, A., Lau, H. C., & Vansteenwegen, P. (2016). Orienteering problem: *A survey of recent variants, solution approaches and applications. European Journal of Operational Research*, *255*(2), 315–332. doi:10.1016/j.ejor.2016.04.059

Hu, Q., & Lim, A. (2014). An iterative three-component heuristic for the team orienteering problem with time windows. *European Journal of Operational Research*, *232*(2), 276–286. doi:10.1016/j.ejor.2013.06.011

Jun, M., & D'Andrea, R. (2003). Path planning for unmanned aerial vehicles in uncertain and adversarial environments. In Cooperative control: models, applications and algorithms (pp. 95-110). Springer US. doi:10.1007/978-1-4757-3758-5_6

Karakaya, M., & Sevinç, E. (2017). An Efficient Genetic Algorithm for Routing Multiple UAVs under Flight Range and Service Time Window Constraints. *International Journal of InformaticsTechnologies*, *10*(1).

Ke, L., Archetti, C., & Feng, Z. (2008). Ants can solve the team orienteering problem. *Computers & Industrial Engineering*, *54*(3), 648–665. doi:10.1016/j.cie.2007.10.001

Keshtkaran, M., Ziarati, K., Bettinelli, A., & Vigo, D. (2016). Enhanced exact solution methods for the team orienteering problem. *International Journal of Production Research*, *54*(2), 591–601. doi:10.1080/00207543.2015.1058982

Kress, M., & Royset, J. O. (2008). Aerial search optimization model (ASOM) for UAVs in special operations. *Military Operations Research*, *13*(1), 23–33. doi:10.5711/morj.13.1.23

Labadie, N., Mansini, R., Melechovský, J., & Calvo, R. W. (2012). The team orienteering problem with time windows: An lp-based granular variable neighborhood search. *European Journal of Operational Research*, *220*(1), 15–27. doi:10.1016/j.ejor.2012.01.030

Labadie, N., Melechovský, J., & Calvo, R. W. (2011). Hybridized evolutionary local search algorithm for the team orienteering problem with time windows. *Journal of Heuristics*, *17*(6), 729–753. doi:10.100710732-010-9153-z

Lim, G. J., Kim, S., Cho, J., Gong, Y., & Khodaei, A. (2016). Multi-UAV Pre-positioning and Routing for Power Network Damage Assessment. *IEEE Transactions on Smart Grid*, 1. doi:10.1109/TSG.2016.2637408

Lin, S. W. (2013). Solving the team orienteering problem using effective multi-start simulated annealing. *Applied Soft Computing*, *13*(2), 1064–1073. doi:10.1016/j.asoc.2012.09.022

Liu, X. F., Gao, L. M., Guang, Z. W., & Song, Y. Q. (2013). A UAV allocation method for traffic surveillance in sparse road network. [English Edition]. *Journal of Highway and Transportation Research and Development*, *7*(2), 81–87.

Liu, X. F., Guan, Z. W., Song, Y. Q., & Chen, D. S. (2014). An optimization model of UAV route planning for road segment surveillance. *Journal of Central South University*, *21*(6), 2501–2510. doi:10.100711771-014-2205-z

Liu, X. F., Peng, Z. R., Chang, Y. T., & Zhang, L. Y. (2012). Multi-objective evolutionary approach for UAV cruise route planning to collect traffic information. *Journal of Central South University*, *19*(12), 3614–3621. doi:10.100711771-012-1449-8

Luo, Z., Liu, Z., & Shi, J. (2017). A Two-Echelon Cooperated Routing Problem for a Ground Vehicle and Its Carried Unmanned Aerial Vehicle. *Sensors (Basel)*, *17*(5), 1144. doi:10.339017051144 PMID:28513552

Mersheeva, V., & Friedrich, G. (2015, April). Multi-UAV Monitoring with Priorities and Limited Energy Resources. In ICAPS (pp. 347-356).

Montemanni, R., & Gambardella, L. M. (2009). An ant colony system for team orienteering problems with time windows. *Foundation Of Computing And Decision Sciences*, *34*(4), 287.

Mufalli, F., Batta, R., & Nagi, R. (2012). Simultaneous sensor selection and routing of unmanned aerial vehicles for complex mission plans. *Computers & Operations Research*, *39*(11), 2787–2799. doi:10.1016/j.cor.2012.02.010

Muthuswamy, S., & Lam, S. S. (2011). Discrete particle swarm optimization for the team orienteering problem. *Memetic Computing*, *3*(4), 287–303. doi:10.100712293-011-0071-x

Ngenkaew, W., Ono, S., & Nakayama, S. (2008). The deposition of multiple pheromones in ant-based clustering. *International Journal of Innovative Computing, Information, & Control*, *4*(7), 1349–4198.

Poggi, M., Viana, H., & Uchoa, E. (2010). The team orienteering problem: Formulations and branch-cut and price. In *OASIcs-OpenAccess Series in Informatics* (Vol. 14). Schloss Dagstuhl-Leibniz-Zentrum fuer Informatik.

Prins, C., Prodhon, C., & Calvo, R. W. (2006). *Solving the capacitated location-routing problem by a GRASP complemented by a learning process and a path relinking*. FOR: A Quarterly Journal of Operations Research, 4(3), 221-238.

Prodhon, C., & Prins, C. (2014). A survey of recent research on location-routing problems. *European Journal of Operational Research*, *238*(1), 1–17. doi:10.1016/j.ejor.2014.01.005

Rathinam, S., & Sengupta, R. (2007). Algorithms for routing problems involving uavs. In *Innovations in Intelligent Machines-1* (pp. 147–172). Springer Berlin Heidelberg. doi:10.1007/978-3-540-72696-8_6

Royset, J. O., Carlyle, W. M., & Wood, R. K. (2009). Routing Military Aircraft With A Constrained Shortest Path Algorithm. *Military Operations Research*, *14*(3), 31–52. doi:10.5711/morj.14.3.31

Ryan, J. L., Bailey, T. G., Moore, J. T., & Carlton, W. B. (1998, December). Reactive tabu search in unmanned aerial reconnaissance simulations. In *Proceedings of the 30th conference on Winter simulation* (pp. 873-880). IEEE Computer Society Press. 10.1109/WSC.1998.745084

Salhi, S., & Rand, G. K. (1989). The effect of ignoring routes when locating depots. *European Journal of Operational Research*, *39*(2), 150–156. doi:10.1016/0377-2217(89)90188-4

Sarıçiçek, İ., & Akkuş, Y. (2015). Unmanned Aerial Vehicle hub-location and routing for monitoring geographic borders. *Applied Mathematical Modelling*, *39*(14), 3939–3953. doi:10.1016/j.apm.2014.12.010

Savuran, H., & Karakaya, M. (2015). Route optimization method for unmanned air vehicle launched from a carrier. *Lecture Notes on Software Engineering*, *3*(4), 279–284. doi:10.7763/LNSE.2015.V3.204

Shetty, V. K., Sudit, M., & Nagi, R. (2008). Priority-based assignment and routing of a fleet of unmanned combat aerial vehicles. *Computers & Operations Research*, *35*(6), 1813–1828. doi:10.1016/j.cor.2006.09.013

Souffriau, W., Vansteenwegen, P., Vanden Berghe, G., & Van Oudheusden, D. (2013). The multiconstraint team orienteering problem with multiple time windows. *Transportation Science*, *47*(1), 53–63. doi:10.1287/trsc.1110.0377

Stützle, T., & Hoos, H. (1997). MAX-MIN ant system and local search for the traveling salesman problem. In *IEEE International Conference on Evolutionary Computation (ICEC'97)*. 10.1109/ICEC.1997.592327

Stützle, T., & Hoos, H. H. (2000). MAX–MIN ant system. *Future Generation Computer Systems*, *16*(8), 889–914. doi:10.1016/S0167-739X(00)00043-1

Sundar, K., & Rathinam, S. (2014). Algorithms for routing an unmanned aerial vehicle in the presence of refueling depots. *IEEE Transactions on Automation Science and Engineering*, *11*(1), 287–294. doi:10.1109/TASE.2013.2279544

Tang, H., & Miller-Hooks, E. (2005). A tabu search heuristic for the team orienteering problem. *Computers & Operations Research*, *32*(6), 1379–1407. doi:10.1016/j.cor.2003.11.008

Tian, J., Shen, L., & Zheng, Y. (2006). Genetic algorithm based approach for multi-UAV cooperative reconnaissance mission planning problem. *Lecture Notes in Computer Science*, *4203*, 101–110. doi:10.1007/11875604_13

Ting, C. J., & Chen, C. H. (2013). A multiple ant colony optimization algorithm for the capacitated location routing problem. *International Journal of Production Economics*, *141*(1), 34–44. doi:10.1016/j.ijpe.2012.06.011

Tricoire, F., Romauch, M., Doerner, K. F., & Hartl, R. F. (2010). Heuristics for the multi-period orienteering problem with multiple time windows. *Computers & Operations Research*, *37*(2), 351–367. doi:10.1016/j.cor.2009.05.012

Valavanis, K. P., & Vachtsevanos, G. J. (2014). *Handbook of unmanned aerial vehicles*. Springer Publishing Company, Incorporated.

Vansteenwegen, P., Souffriau, W., Berghe, G. V., & Van Oudheusden, D. (2009a). A guided local search metaheuristic for the team orienteering problem. *European Journal of Operational Research*, *196*(1), 118–127. doi:10.1016/j.ejor.2008.02.037

Vansteenwegen, P., Souffriau, W., Berghe, G. V., & Van Oudheusden, D. (2009b). Metaheuristics for tourist trip planning. In *Metaheuristics in the service industry* (pp. 15–31). Springer Berlin Heidelberg. doi:10.1007/978-3-642-00939-6_2

Vansteenwegen, P., Souffriau, W., Berghe, G. V., & Van Oudheusden, D. (2009c). Iterated local search for the team orienteering problem with time windows. *Computers & Operations Research*, *36*(12), 3281–3290. doi:10.1016/j.cor.2009.03.008

Vansteenwegen, P., Souffriau, W., & Van Oudheusden, D. (2011). The orienteering problem: A survey. *European Journal of Operational Research*, *209*(1), 1–10. doi:10.1016/j.ejor.2010.03.045

Vargas, R. A. (2012). *Unmanned Systems: Operational Considerations for the 21st Century Joint Task Force Commander and Staff*. ARMY COMMAND AND GENERAL STAFF COLL FORT LEAVENWORTH KS.

Xiaofeng, L. I. U., Zhongren, P. E. N. G., Zhang, L., & Li, L. I. (2012). Unmanned aerial vehicle route planning for traffic information collection. *Journal of Transportation Systems Engineering and Information Technology*, *12*(1), 91–97. doi:10.1016/S1570-6672(11)60186-4

Yakıcı, E. (2016a). Solving location and routing problem for UAVs. *Computers & Industrial Engineering*, *102*, 294–301. doi:10.1016/j.cie.2016.10.029

Yakıcı, E. (2016b). Generalization of a UAV location and routing problem by time windows. *Journal of Naval Science and Engineering*, *12*(2), 67–78.

Zhang, J., Jia, L., Niu, S., Zhang, F., Tong, L., & Zhou, X. (2015). A space-time network-based modeling framework for dynamic unmanned aerial vehicle routing in traffic incident monitoring applications. *Sensors (Basel)*, *15*(6), 13874–13898. doi:10.3390150613874 PMID:26076404

Zhenhua, W., Weiguo, Z., Jingping, S., & Ying, H. (2008, September). *UAV route planning using multiobjective ant colony system.* In Cybernetics and Intelligent Systems, 2008 IEEE Conference on (pp. 797-800). IEEE.

Zhu, N., Liu, Y., Ma, S., & He, Z. (2014). Mobile traffic sensor routing in dynamic transportation systems. *IEEE Transactions on Intelligent Transportation Systems*, *15*(5), 2273–2285. doi:10.1109/TITS.2014.2314732

Chapter 3
An Overview on Set Covering Problems With a Focus on Military Applications

Irem Duzdar Argun
Duzce University, Turkey

ABSTRACT

This chapter deals with the set covering problem, a specific type of a discrete location and representative combinatorial optimization problem, which occupies a huge space in military operations research with its various applications. After presenting a brief literature survey of set covering problems, the author provides a formal description and the mathematical model of the problem. Finally, the key position of those problem types in the military domain and with real-world applications is discussed.

INTRODUCTION

Location problems deal with the problem of determining the best locations of facilities to attain the highest possible level of service related to those facilities. Being an important element of strategic planning problems observed both in the public and private sectors, location problems assist decision-makers by considering many and possibly conflicting criteria such as cost, distance, level of service, etc. Interested reader can refer to studies such as Francis et al. (1992), Daskin (1995), Drezner (1995), Drezner and Hamacher (2002), Church and Murray (2009) and Farahani et al. (2012). Among several problem types, set covering is one of the commonly-used models and have been applied to many decision-making problems. Given a universe U consisting of n elements, assume that a number of subsets of U say $S = \{S_1, S_2, \ldots, S_m\}$ exists and each subset S_i is associated with a cost. The goal of the set covering model is to determine a minimum cost sub-collection of S that covers all elements of U.

Set covering models have been attracting many researchers from different disciplines due to their natural applicability to real-life problems, which generally deal with the location of service and emergency facilities. In some cases, the objective of the set covering problem can be satisfied when a service facility meets a demand in the determined critical range, even if it is not the nearest service point. The demands

DOI: 10.4018/978-1-5225-5513-1.ch003

are usually met depending on the distance between the facility and the customer. This implies that the customer can be served by any facility whose distance is less than or at least equal to a predetermined distance away (Farahani et al., 2012). If the problem is to decide the location and the number of service facilities (e.g., fire stations, waste disposal plants, emergency facilities, public buildings, public schools, hospitals, post offices, military bases, radar stations, banks etc.), one can easily employ a set covering model to solve the problem. In the literature, set covering problems are generally applied to decide the placement and/or number of certain facilities in urban areas. There are very few studies on the location of facilities in rural areas.

The main objective of this chapter is to analyse the applicability of set covering models to different real-world problem types observed in various contexts. Previous studies, in general, have dealt with the search for optimal alternatives in the planning and allocation of resources for public support and service systems. This chapter, on the other hand, gives an overview of the set covering models with a focus on military applications. In the next section we first give the mathematical model for the set covering model. Next, we describe the criteria used to filter the papers included in our survey. Then, we report the related literature on the set covering problem applications and describe the military related problems.

METHODOLOGY OF SET COVERING OPTIMISATION

The set covering problem (SCP) covers the rows of an m-row, n-column zero-one matrix (a_{ij}) by a subset of the columns at minimum cost. The problem is defined as $x_j = 1$ if column j (with cost $c_j > 0$) is in the solution and $x_j = 0$ (Beasley and Chu, 1996).

The SCP is formulated with the following indices, parameters, objective function, and constraints described as below:

Sets:

i: the index of demand nodes,
j: the index of candidate facility locations,

Parameters:

S: the maximum service distance of a facility,
c_j: the cost of locating a facility at location j,
a_{ij}: a binary parameter that is 1 if the distance from candidate location j to a demand i is less than S. The customer can be served by any facility whose distance is less than or at least equal to a pre-defined distance (Farahani et al., 2012).

Decision Variable:

x_j: a binary variable indicating whether the facility is sited at candidate location j or not,

Mathematical Model:

$$\min \sum_{j=1}^{n} c_j x_j$$

s.t.

$$\sum_{j=1}^{n} a_{ij} x_j \geq 1, i = 1,...,m$$

$$x_j \in \{0,1\}, j = 1,...,n$$

The objective function (1) minimises the total cost of the installed facilities. Constraint set (2) ensures that each demand node *i* is covered by at least one facility. Finally, the constraint set (3) defines the domain of decision variables.

The SCP aims to determine the set of facilities with minimum cost among a finite set of candidate facilities so that all demand nodes are covered by at least one facility. In a dual approach to SCP proposed by Church and ReVelle (1974), the maximal covering location problem (MCLP) is taken as the base. In their proposal, by employing a certain number of service points, the maximisation of the total weighted demand is achieved by using a linear programming model.

REVIEW CRITERIA

By searching the *Web of Science* to find published studies on set covering optimisation under the words "*Set*", "*Covering*", and "*Optimisation*", 3,606 studies were found, 68% of which were published in journal article format. The work done in these studies indicated the most common area to be "Set Covering Optimisation". Table 1 and Figure 1 show the number of annual publications. However, since 2005 in particular, a regular increase can be seen in the number of studies, thus, this review exhibits that the set covering model is a significantly attractive method to optimize these types of problems.

In addition, the search performed on the *Web of Science* using the word "*military*" revealed 14 studies (Table 2) and most of these studies were conducted after 2010. The application of set covering models to military decision-making problems implies that this method can be employed in a variety of new areas.

Table 1. Number of annual studies

Years	1980	1983	1988	1990	1991	1992	1993	1994	1995	1996	1997	1998	1999	2000	2001	
# of Studies	1	1	1	1	11	16	23	17	28	29	45	57	55	59	63	
Years	**2002**	**2003**	**2004**	**2005**	**2006**	**2007**	**2008**	**2009**	**2010**	**2011**	**2012**	**2013**	**2014**	**2015**	**2016**	**2017**
# of Studies	79	66	91	120	152	143	167	183	167	197	218	231	295	360	415	302

Figure 1. Distribution of annual studies

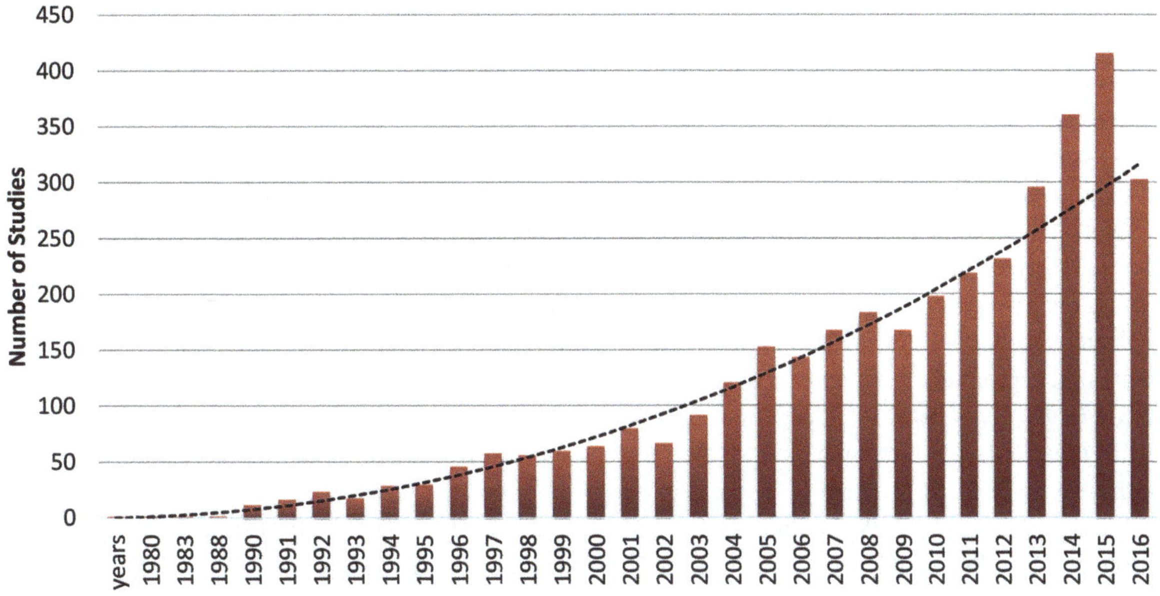

Table 2. Number of studies in the military field

Years	1992	2003	2005	2006	2007	2011	2013	2014	2016
# of studies	1	1	1	1	2	2	3	1	2

LITERATURE APPLICATIONS OF SET COVERING OPTIMISATION

The modelling and especially the optimisation of location problems is a technique that has been used for a long time with success. The objective of the set covering may be minimizing the number of service bases to satisfy all of the service needs (*set covering*), or maximizing the number of demands met by using a predetermined number of service facilities (*maximum covering*) (Farahani et al., 2014). Covering location problems can be identified in order to decide the minimum number of service points which will assure that each customer gets the required service at a predetermined distance or time, or sometimes both (Vemuganti, 1998). Toregas et al. (1971) proposed the SCP, which is the most elementary deterministic model. Its aim is to satisfy all of the service requirements in a defined area by employing the minimum number of service bases. Here, in satisfying all of the demands with the minimum number of service points, the number of demands are not considered. The covering model was used to solve problems concerning locations of Turkish bank branches by Başar et al. (2015).

Church and ReVelle (1974) introduced a new version of the coverage problem called the MCLP, which attempts to maximise the number of service requirements in a predefined area of service facilities. The Modified Maximal Covering Location Problem (*MMCLP*), introduced by Daskin and Stern (1981), is another application of the MCLP and has a second goal to maximise the service requirements met more than one time. The Maximal Expected Covering Location Problem (*MECLP*) created by Daskin (1983) involves an equally engaged expectation allocated to all service bases.

In the literature, one can find a great number of successful applications of this covering structure in different areas. A series of feasible schedules required for formulating a crew scheduling problem that is another type of set covering, which is the same as the employee scheduling problem. Rubin (1973) proposed a method to solve an airline crew scheduling problem by employing a massive set covering problem. Vemuganti (1998) applied set covering, set packing, and set partitioning models in a survey.

Bjorklund et al. (2003) improved set covering formulations used to find the solutions to problems. The results of the numerical applications proved that the column generation method is very successful in solving the linear programming (LP) relaxations of set covering problems. In addition, these bring close limitations for the integer problems. In another application of this model, many parts are produced in a plant and every unique part requires some operation on a special-purpose machine. These machines are placed in groups of cells and there is the same type of machine in each of the various cells. The objective of the problem is to decide how to minimise the number of cells to satisfy the need for processing on these machines. This model is frequently used to find a solution for minimising the number of fire stations (Bahri, 2016). ReVelle and Hogan (1989) introduce the Probabilistic Location Set Covering Problem (PLSCP) to minimise the service bases needed to serve an entire city. Their model has some similarities to the Maximum Availability Location Problem II (MALPII), which utilises close probabilities of busy servers. Subsequently, Marianov and ReVelle (1994) improved the PLSCP by applying a close probabilities hypothesis to the MALPII to introduce the Queuing Probabilistic Location Set Covering Problem (Q-PLSCP). In their model, every unique service base is seen as a multi-server loss system which first tries to find the busy neighbourhood probabilities, and then employs them as inputs to the system.

The studies of Karasakal and Karasakal (2004) and Karatas (2017) are among those which incorporated partial coverage in location problems. The partial coverage concept has several real-world applications in location science and the military domain such as underwater sensor location (Craparo et al., 2017; Craparo et al., 2018), border protection and surveillance problems (Karatas and Onggo, 2016; Karatas, 2018). The Conventional Set Covering (CSC) problem has been employed for placement of emergency medical intervention bases. However, there are some restrictions in computing and operating. The Hierarchical Objective Set Covering problem (HOSC) arises with the idea of evaluating the interrelations between the different regions of a city. In HOSC, the objective is to minimise the number of emergency service vehicles, while maximising coverage of all regions (Daskin, 1981). Moreover, three basic types of covering models were proposed by Daskin: the set covering model, the maximal covering model, and the p-centre model. These models can be solved by employing mixed integer programming methods. All of these models can be grouped under the non-deterministic polynomial-time hard (NP-hard) topic (Daskin, 2008). Karatas et al. (2016) and Karatas et al. (2017a) studied the three common location problems, i.e. the p-median, maximal covering model, and p-centre, and compared them with respect to multiple criteria under the assumption of the q-coverage requirement. In another study, Karatas and Yakıcı (2018) developed an iterative solution procedure for solving problems which incorporated the objectives of the p-median, maximal covering model, and p-centre problems at the same time. Razi and Karatas (2016), Razi et al. (2016) and Karatas et al. (2017b) used modified versions of the set covering and p-median problems for locating search and rescue facilities such as coast guard boats or helicopters. These applications of covering models are commonly used for civilian life support services. These models are likely to be employed for solutions to military problems requiring important tactical decisions.

SET COVERING IN MILITARY APPLICATIONS

There have been 14 studies published on the *Web of Science* about military applications of set covering optimisation (Table 3). These publications consist of eight articles, five proceeding papers and one review.

Some of the preliminary studies found in *Web of Science* can be applied to a set covering model for military decision-making problems as well. The set covering model has been utilised in many different fields. A study by Toregas et al. (1971) noted that the most common application of this model was to locate fire stations, but that it can be applied successfully for locating other urban services like public schools, libraries, hospitals, etc.

There is a special issue covering problem studies which included planning military resources, scheduling of staff, network prohibitions, and shipping and fleet routing (Boginski et al., 2015). Another study discussed the optimisation of military applications to improve resource assignment for military emergency services. The allocation of military emergency air aid facilities in Afghanistan was demonstrated in order to verify their concept. The methodology is easily applicable to other military allocation and operational problems (Bates, 2015).

The Maximal Covering Location Problem and the Location Set Covering Problem similarly deal with service requirements and service bases scattered on the field. However, the objective of the Location Set Covering Problem is to meet the maximum number of demands by using one or more of a restricted number of service bases at predetermined distances (ReVelle and Rosing, 2000). A military force arrangement problem was constructed by ReVelle and Rosing using the Set Covering and Maximal Set Covering Models. The goal of the Set Covering Deployment Problem is to defend each area by one or more military forces, called Field Armies (FA's), or by the help of only one FA to reach there from two FA areas in one movement. The object of the Movement Covering Deployment Problem (*MCDP*) is to defend the maximum number of areas with only one movement by using the minimum number of FA's settled in the field.

Konur et al. published another study in 2016, exploring the use of an SoS (system of system) architectural problem to express a military duty problem via a multi-objective mixed-integer-linear optimisation model, without considering its flexibility (Konur et al, 2016). In the case of all systems being flexible, with no requirement for connection of the systems, the SoS architectural problem (in the form of a set covering problem) and its subsets may be identified in order to provide capacities for inflexible systems.

The Digital Evaluation Model (*DEM*) was proposed by Shi and Xue as a new approach (Shi and Xue, 2016). It proposes to determine automatically the minimum number of watch points for achieving the maximum coverage of the digital area information. The results of this model can be used in various decision-making applications such as optimising and managing telecommunication infrastructures, militaristic tactics and operations and wildfire observation towers and their distribution, all of which depend on the operational area and its physical structure.

The most important factor limiting the design of wireless sensor networks is the energy consumption required to increase the lifetime of these network systems. Because of this, most studies done in the past were investigating wireless sensor networks. The problem called the Network Coverage Optimisation for Wireless Sensor Network aims to find the sensor joints satisfying the expectations. This problem is also called the Problem of Minimum Connected Covering Node Set. In a wireless sensor network generally used by the military or other organisations, many small sensors having the ability to sense the

Table 3. Set covering optimisation: Military applications

Publication Title	Publication Year	Authors	Publication Type	Application Area	Model	Cited
Deriving a minimum set of viewpoints for maximum coverage over any given digital elevation model data	2016	Shi, Xuan; Xue, Bowei	Article	Digital elevation model (DEM) data	Set covering	1
Coverage Optimization for Wireless Sensor Networks by Evolutionary Algorithm	2016	Li, Kangshun; Wen, Zhichao; Li, Shen	Proceedings Paper	Wireless Sensor Network	Set covering	0
A multi-objective evolutionary algorithm for facility dispersion under conditions of spatial uncertainty	2014	Wei, Ran; Murray, Alan T.	Article	Wireless Sensor Network	Set covering	1
Design Optimisation of a Reconfigurable Perching Element for Vertical Take-Off and Landing Unmanned Aerial Vehicles	2013	Erbil, Mehmet Ali; Prior, Stephen D.; Keane, Andrew J.	Article	small Unmanned Aerial Vehicles (sUAVs)	Anti-covering location problem	2
Balanced Energy and Coverage Guaranteed Protocol for Wireless Sensor Networks	2013	Le, Nam Tuan; Saha, Nirzhar; Mondal, Ratan Kumar; et al.	Proceedings Paper	Wireless Sensor Network	Set covering	0
Optimization of Disposition for Terminal Air Defence System Based on Set-Covering Model	2013	Han Yunjun; Wang Jinhua; Liao Xiaojing	Proceedings Paper	Terminal air defence system	Set covering	0
An optimization approach to strategic sourcing: A case study of the United States Air Force	2011	Apte, Aruna U.; Rendon, Rene G.; Salmeron, Javier	Article	Military strategic sourcing process	Set covering	3
Location optimization of strategic alert sites for homeland defence	2011	Bell, John E.; Griffis, Stanley E.; Cunningham, William A., III; et al.	Article	Aircraft in a defence emergency	Location Set covering	21
Combination of MCDM and covering techniques in a hierarchical model for facility location: A case study	2007	Farahani, Reza Zanjirani; Asgari, Nasrin	Article	distribution centres (DCs) in military logistics	Set covering	37
Coverage optimization based on improved NSGA-II in wireless sensor network	2007	Jia, Jie; Chen, Jian; Chang, Guiran; et al.	Proceedings Paper	Wireless Sensor Network	Set covering	5
Optimal area covering by sensors for planning a tracks collection	2006	Dambreville, Frederic	Proceedings Paper	Sensor Management	Set covering	0
Radio resource management across multiple protocol layers in satellite networks: A tutorial overview	2005	Barsocchi, P; Celandroni, N; Davoli, F; et al.	Review	Telecommunication	Set covering	12
Flight test data cycle map optimisation	2003	Panton, D; John, M; Lucas, S; et al.	Article	Flight tests	Set covering	0
THE MULTICOVERING PROBLEM	1992	HALL, NG; HOCHBAUM, DS	Article	Military	Set covering	11

controlling area by employing self-organised wireless connections (Li et al., 2015). Wide area detections are generally used for military purposes, and the main objectives are the coverage of the network, while the length of its life depends on the energy consumption of the system. Le proposed widening the covered area while balancing the energy consumption of whole sensing system (Le et al., 2013). Wei and Murray used an anti-covering location problem for a dispersion model (Wei and Murray, 2014). The study of Dambreville explored methods of obtaining the required information via optimisation of the coverage of a range of sensors, in addition to collecting information for directing and controlling military operations (Dambreville, 2006).

The topic of another study was telecommunication. Established modern communication technologies such as telephone, radio/television broadcasting, internet networks, sea/land navigation and military defence operations are carried out via satellite. In some cases, this can be the only useful way to communicate. The importance of satellite communication via cable or wireless is increasing daily as a result of the growing demands of navigation, video delivery, mobility, and interactive multimedia services. Consequently, researchers need to re-examine satellite transmission applications in future studies and propose new solutions (Barsocchi et al., 2005).

Another research work dealt with the optimal covering of requested information via sensors. This paper discussed the collection of information for military operations at the command level (Jia et al., 2007). The aim was to maximise the coverage by employing the least possible number of sensor nodes for a determined field.

The set covering model has been employed for air defence and logistic site selection problems in addition to military problem solving. Han et al. proposed the employment of the Terminal Air Defence System (TADS) problem together with the set covering model (Han et al., 2013). They investigated two different covering notions (the *range* and *angle of firing*) in order to find the optimum disposition and establish the correlative integer programming models. It can be concluded that the firing angle concept was more advantageous for the disposition solution than the firing range.

The study of Apte et al. proposed another optimisation model that used for the U.S. Air Force Strategic Sourcing Process to find a series of suggestions from multiple bidders in order to achieve the required operations using multiple armaments (Apte et al., 2011). A military application study on defending important regions is presented by the U.S. Department of Defence employs a Location Analysis Process to identify the U.S. Air Force warning bases (Bell et al., 2011). This approach requires two levels. In the first, one tries to find the minimum number of bases by employing the location set covering problem. A great increase was detected in the required number of warning bases with respect to the military forecasts.

Another military application is exploring the placement of warehouses which will be used as distribution bases in the military logistic system. A simple Set Partitioning Model (SPM) was used to assign each consuming point only to a single distribution base (Farahani and Asgari, 2007).

An algorithm was introduced by Panton et al. to create a Data Cycle Map (DCM) combining theoretic outcomes with a set covering model (Panton et al., 2003). The DCM was used for allocating huge data collections into blocks of test flight data for a new aircraft by examining design concepts, problems, insufficiencies, defects and faults. A study conducted by Hall and Hochbaum proposed an algorithm for solving large multicovering problems (Hall and Hochbaum, 1992). Dissimilarities in industrial, commercial, and military problems and aims make multicovering problems greater. Consequently, the aim of the study was to construct an algorithm to solve these kinds of large problems.

CONCLUSION

Because of the complexity of NP-hard covering problems, intuitive approaches are used to find solutions within an acceptable period. There are few studies in the literature that deal with analysing of real-world problems in a military context. Most of the studies seen in the literature covered optimisation conducted by researchers from the view of various disciplines. However, the covering model can be applied to military problems as well as to a wide variety of other fields

Various types of military covering applications can be found in the literature. The most frequently seen military covering studies can be divided into two main types: the location set covering problem (LSCP) and the maximal covering location problem (MCLP). The MCLP, proposed by Church and DeVelle (1974), tries to maximise the satisfied service demand by using predetermined service bases within a limited distance. There is no requirement to meet all of the service demands, but the objective is to maximise the meeting of demands. The objective of LSCP, formulated by Toregas et al. (1971), is to minimise the number of service bases needed to meet all of the service demands. The resources allocated to the service bases may be excessive in LSCP because of its aim to satisfy all service demands without considering the number of demands, distances, or population.

The studies reviewed in this paper demonstrate that there has been a great amount of research done on set covering method applications for solving critical military problems, and that the importance of its employment for military operational problems has been recognised. These applications cover problems of military personnel scheduling, resource planning, vehicle/aircraft routing and network exclusion. The number of studies on military applications of set covering optimisation is increasing and this trend appears to be making headway into the future. This compilation of applications was prepared with the expectation that it may provide an informative and stimulating base for future research on the subject.

REFERENCES

Apte, A. U., Rendon, R. G., & Salmeron, J. (2011). An optimization approach to strategic sourcing: A case study of the United States Air Force. *Journal of Purchasing and Supply Management, 17*(4), 222–230. doi:10.1016/j.pursup.2011.03.002

Bahri, S. (2016, April). 0-1 integer linear programming model for location selection of fire station: A case study in Indonesia. In AIP Conference Proceedings (Vol. 1723, No. 1, p. 030004). AIP Publishing.

Barsocchi, P., Celandroni, N., Davoli, F., Ferro, E., Giambene, G., Castano, F. J. G., ... Todorova, P. (2005). Radio resource management across multiple protocol layers in satellite networks: A tutorial overview. *International Journal of Satellite Communications and Networking, 23*(5), 265–305. doi:10.1002at.820

Başar, A., Kabak, Ö., Topçu, Y. İ., & Bozkaya, B. (2015). Location analysis in banking: a new methodology and application for a Turkish bank. In *Applications of Location Analysis* (pp. 25–54). Springer International Publishing. doi:10.1007/978-3-319-20282-2_2

Bates, A., Bell, Z., Mountford, A., & Evangelista, P. (2015). Military Resource Allocation as a Set Covering Problem. *Industrial and Systems Engineering Review, 3*(1), 1–6.

Beasley, J. E., & Chu, P. C. (1996). A genetic algorithm for the set covering problem. *European Journal of Operational Research, 94*(2), 392–404. doi:10.1016/0377-2217(95)00159-X

Bell, J. E., Griffis, S. E., Cunningham, W. A. III, & Eberlan, J. A. (2011). Location optimization of strategic alert sites for homeland defence. *Omega, 39*(2), 151–158. doi:10.1016/j.omega.2010.05.004

Berman, O., & Krass, D. (2002). The generalized maximal covering location problem. *Computers & Operations Research, 29*(6), 563–581. doi:10.1016/S0305-0548(01)00079-X

Bjorklund, P., Varbrand, P., & Yuan, D. (2003, March). Resource optimization of spatial TDMA in ad hoc radio networks: A column generation approach. In *INFOCOM 2003. Twenty-Second Annual Joint Conference of the IEEE Computer and Communications. IEEE Societies* (Vol. 2, pp. 818-824). IEEE. 10.1109/INFCOM.2003.1208919

Boginski, V., Pasiliao, E. L., & Shen, S. (2015). Special issue on optimization in military applications. *Optimization Letters, 9*(8), 1475-1476.

Church, R., & Velle, C. R. (1974). The maximal covering location problem. *Papers in Regional Science, 32*(1), 101–118. doi:10.1007/BF01942293

Church, R. L., & Murray, A. T. (2009). *Business site section, location analysis and GIS*. New York: Wiley.

Craparo, E. M., Fügenschuh, A., Hof, C., & Karatas, M. (2018). Optimizing Source and Receiver Placement in Multistatic Sonar Networks to Monitor Fixed Targets. *European Journal of Operational Research*. doi:10.1016/j.ejor.2018.02.006

Craparo, E. M., Karatas, M., & Kuhn, T. U. (2017). Sensor placement in active multistatic sonar networks. *Naval Research Logistics, 64*(4), 287–304. doi:10.1002/nav.21746

Dambreville, F. (2006, July). Optimal area covering by sensors for planning a Tracks Collection. In *Information Fusion, 2006 9th International Conference on* (pp. 1-7). IEEE. 10.1109/ICIF.2006.301821

Daskin, M. S. (1983). A maximum expected covering location model: Formulation, properties and heuristic solution. *Transportation Science, 17*(1), 48–70. doi:10.1287/trsc.17.1.48

Daskin, M. S. (1995). *Network and discrete location: Models, algorithms and applications*. New York: John Wiley and Sons. doi:10.1002/9781118032343

Daskin, M. S. (2008). What you should know about location modelling. *Naval Research Logistics, 55*(4), 283–294. doi:10.1002/nav.20284

Daskin, M. S., & Stern, E. H. (1981). A hierarchical objective set covering model for emergency medical service vehicle deployment. *Transportation Science, 15*(2), 137–152. doi:10.1287/trsc.15.2.137

Drezner, Z. (Ed.). (1995). *Facility location: A survey of applications and methods*. Berlin, Germany: Springer Verlag. doi:10.1007/978-1-4612-5355-6

Drezner, Z., & Hamacher, H. W. (Eds.). (2002). *Facility location: Applications and Theory*. Berlin, Germany: Springer-Verlag. doi:10.1007/978-3-642-56082-8

Farahani, R. Z., & Asgari, N. (2007). Combination of MCDM and covering techniques in a hierarchical model for facility location: A case study. *European Journal of Operational Research*, *176*(3), 1839–1858. doi:10.1016/j.ejor.2005.10.039

Farahani, R. Z., Asgari, N., Heidari, N., Hosseininia, M., & Goh, M. (2012). Covering problems in facility location: A review. *Computers & Industrial Engineering*, *62*(1), 368–407. doi:10.1016/j.cie.2011.08.020

Farahani, R. Z., Hekmatfar, M., Fahimnia, B., & Kazemzadeh, N. (2014). Hierarchical facility location problem: Models, classifications, techniques, and applications. *Computers & Industrial Engineering*, *68*, 104–117. doi:10.1016/j.cie.2013.12.005

Francis, R. L., McGinnis, L. F., & White, J. A. (1992). *Facility layout and location: An analytical approach* (2nd ed.). Englewood Cliffs, NJ: Prentice-Hall.

Hall, N. G., & Hochbaum, D. S. (1992). The multicovering problem. *European Journal of Operational Research*, *62*(3), 323–339. doi:10.1016/0377-2217(92)90122-P

Han, Y., Wang, J., & Liao, X. (2013, July). Optimization of disposition for terminal air defence system based on set-covering model. In *Control Conference (CCC), 2013 32nd Chinese* (pp. 8461-8464). IEEE.

Jia, J., Chen, J., Chang, G., Li, J., & Jia, Y. (2007, March). Coverage optimization based on improved NSGA-II in wireless sensor network. In *Integration Technology, 2007. ICIT'07. IEEE International Conference on* (pp. 614-618). IEEE. 10.1109/ICITECHNOLOGY.2007.4290391

Karasakal, O., & Karasakal, E. K. (2004). A maximal covering location model in the presence of partial coverage. *Computers & Operations Research*, *31*(9), 1515–1526. doi:10.1016/S0305-0548(03)00105-9

Karatas, M. (2017). A multi-objective facility location problem in the presence of variable gradual coverage performance and cooperative cover. *European Journal of Operational Research*, *262*(3), 1040–1051. doi:10.1016/j.ejor.2017.04.001

Karatas, M. (2018). Optimal Deployment of Heterogeneous Sensor Networks for a Hybrid Point and Barrier Coverage Application. *Computer Networks*, *132*(26), 129–144. doi:10.1016/j.comnet.2018.01.001

Karatas, M., & Onggo, B. S. (2016). Validating an integer non-linear program optimization model of a wireless sensor network using agent-based simulation. In *Winter Simulation Conference (WSC)*, 2016 (pp. 1340-1351). IEEE. 10.1109/WSC.2016.7822188

Karatas, M., Razi, N., & Gunal, M. M. (2017b). An ILP and simulation model to optimize search and rescue helicopter operations. *The Journal of the Operational Research Society*, *68*(11), 1335–1351. doi:10.105741274-016-0154-7

Karatas, M., Razi, N., & Tozan, H. (2016). A Comparison of p-median and Maximal Coverage Location Models with Q–coverage Requirement. *Procedia Engineering*, *149*, 169–176. doi:10.1016/j.proeng.2016.06.652

Karatas, M., Razi, N., & Tozan, H. (2017a). A multi-criteria assessment of the p-median, maximal coverage and p-centre location models. *Technical Gazette*, *24*(2), 399–407.

Karatas, M., & Yakıcı, E. (2018). An iterative solution approach to a multi-objective facility location problem. *Applied Soft Computing*, *62*, 272–287. doi:10.1016/j.asoc.2017.10.035

Konur, D., Farhangi, H., & Dagli, C. H. (2016). A multi-objective military system of systems architecting problem with inflexible and flexible systems: Formulation and solution methods. *OR-Spektrum*, *38*(4), 967–1006. doi:10.100700291-016-0434-2

Le, N. T., Saha, N., Mondal, R. K., Chae, S., & Jang, Y. M. (2014, February). Balanced energy and coverage guaranteed protocol for wireless sensor networks. In *Electrical Information and Communication Technology (EICT), 2013 International Conference on* (pp. 1-6). IEEE. 10.1109/EICT.2014.6777881

Li, K., Wen, Z., & Li, S. (2015, November). Coverage Optimization for Wireless Sensor Networks by Evolutionary Algorithm. In *International Symposium on Intelligence Computation and Applications* (pp. 52-63). Springer.

Marianov, V., & Revelle, C. (1994). The queuing probabilistic location set covering problem and some extensions. *Socio-Economic Planning Sciences*, *28*(3), 167–178. doi:10.1016/0038-0121(94)90003-5

Panton, D., John, M., Lucas, S., & Mason, A. (2003). Flight test data cycle map optimisation. *European Journal of Operational Research*, *146*(3), 486–497. doi:10.1016/S0377-2217(02)00232-1

Razi, N., & Karatas, M. (2016). A multi-objective model for locating search and rescue boats. *European Journal of Operational Research*, *254*(1), 279–293. doi:10.1016/j.ejor.2016.03.026

Razi, N., Karatas, M., & Gunal, M. M. (2016). A combined optimization and simulation based methodology for locating search and rescue helicopters. In *Proceedings of the 2016 Spring Simulation Conference*. IEEE Press.

Revelle, C., & Hogan, K. (1989). The maximum reliability location problem and α-reliablep-centre problem: Derivatives of the probabilistic location set covering problem. *Annals of Operations Research*, *18*(1), 155–173. doi:10.1007/BF02097801

ReVelle, C. S., & Rosing, K. E. (2000). Defendens imperium romanum: A classical problem in military strategy. *The American Mathematical Monthly*, *107*(7), 585–594. doi:10.1080/00029890.2000.12005243

Rubin, J. (1973). A technique for the solution of massive set covering problems, with application to airline crew scheduling. *Transportation Science*, *7*(1), 34–48. doi:10.1287/trsc.7.1.34

Shi, X., & Xue, B. (2016). Deriving a minimum set of viewpoints for maximum coverage over any given digital elevation model data. *International Journal of Digital Earth*, *9*(12), 1153–1167. doi:10.1080/17538947.2016.1207718

Toregas, C., Swain, R., ReVelle, C., & Bergman, L. (1971). The location of emergency service facilities. *Operations Research*, *19*(6), 1363–1373. doi:10.1287/opre.19.6.1363

Vemuganti, R. R. (1998). Applications of set covering, set packing and set partitioning models: A survey. In Handbook of combinatorial optimization (pp. 573-746). Springer US. doi:10.1007/978-1-4613-0303-9_9

Wei, R., & Murray, A. T. (2014). A multi-objective evolutionary algorithm for facility dispersion under conditions of spatial uncertainty. *The Journal of the Operational Research Society*, *65*(7), 1133–1142. doi:10.1057/jors.2013.58

Chapter 4
Assessing the Performance of a SAR Boat Location-Allocation Plan via Simulation

Mumtaz Karatas
National Defense University, Turkey

Nasuh Razi
Turkish Naval Forces, Turkey

Hakan Tozan
Istanbul Medipol University, Turkey

ABSTRACT

Maritime search and rescue (SAR) operation is a critical process that aims to minimize the loss of life, injury, and material damage by rendering aid to persons in distress or imminent danger at sea. Optimal allocation of SAR vessels is a strategic level process that is to be carried out with a plan to react rapidly. This chapter seeks to evaluate the performance of a SAR boat location plan using simulation. The proposed methodology in this chapter works in two stages: First, an optimal allocation scheme of SAR resources is determined via a multi-objective mathematical model. Next, simulation is used to test the performance of the analytical solution under stochastic demand. With the heaviest traffic and maritime risk, the methodology is applied to a case study in the Aegean Sea.

INTRODUCTION

Each year the Turkish Coast Guard (TurCG) receives hundreds of calls and distress signals from the vessels in danger. As the time difference between life and death can sometimes be measured in minutes, quick response to these signals plays a vital role in reducing fatalities and physical damage (Razi and Karatas, 2016). However, an average SAR operation requires substantial amounts of time, effort and money as well. Hence, emergency response operations should be planned with care and foresight. SAR operations planning can be counted as an emergency system planning. As discussed by Green and Colesar (2004),

DOI: 10.4018/978-1-5225-5513-1.ch004

despite many challenges, operations research and management science applications play an important role in decreasing the negative outcomes of emergencies. This argument is particularly applicable to SAR operations conducted by the TurCG.

TurCG conducts SAR operations in the Turkish Maritime SAR Zone, which is divided into four sub-responsibility areas as follows: the Black Sea, the Sea of Marmara and Adjacent Straits, the Aegean Sea and the Mediterranean Sea. Of all those sub-areas, the Aegean Sea has the heaviest marine traffic due to maritime transportation from and to the Black Sea, shipping, cruise tours, yachting, windsurfing and enormous illegal-border crossing activities.

As a consequence of the heavy maritime traffic, the Aegean Sea has an increased level of risk in maritime safety, which is as much high as the maritime incident rate. Together with (UNHCR Global Appeal, 2015) and (Giuliani, 2015)'s works, Razi and Karatas (2016)'s study show that the number of incidents tend to increase in the Aegean Sea responsibility area each year due to a number of reasons such as:

- Presence of narrow waters and dangerous routes among 3000 islands of various sizes
- Increased vessel traffic passing through the region as an outcome of the turmoil in Syria and Iraq
- Being the main route for immigrants to illegally cross EU borders
- Increased number of vessels carrying hazardous cargo
- Lack of designated shipping lanes

In 2014 alone, TurCG conducted 842 SAR operations, and 716 of them (85% of all operations) were in the Aegean Sea. The data provided by TurCG reveals that during those operations, 12,901 victims were saved, 190 boats were recovered undamaged and 154 lives were lost, while 62% of all operations were related to illegal-border crossing activities.

Considering the abovementioned risks related to the increasing traffic density, Razi and Karatas (2016) detailed the problem of establishing well-planned SAR organizations for the TurCG and they developed a decision support tool named as the "Incident Based-Boat Allocation Model (IB-BAM)". IB-BAM is a three step methodology designed to allocate search and rescue resources. The methodology first ranks and assigns a weight to each incident type observed in the region. Next, utilizing deterministic historical incident data, a Zonal Distribution Model (ZDM) generates aggregated weighted demand locations. Finally, it employs a multi-objective mixed integer program (mo-MIP) model to determine locations and responsibility zones of each SAR boat. Using IB-BAM, the authors attain a more efficient utilization of boats considering multiple objectives.

Although IB-BAM is a powerful and flexible tool in the sense that using real-world data and subjective decision maker assessments on the severity of incident types, it generates effective boat allocation plans, it also has some drawbacks. Currently, IB-BAM does not consider variability in the demand, that is, it generates allocation plans for deterministic incident data. In real-world the nature of the demand is stochastic, hence the performance of IB-BAM should also be evaluated for stochastic incident data. Secondly, IB-BAM aggregates the incidents to concentrated demand points, called *superincidents*, by applying a weighted k-clustering algorithm. Such aggregation techniques bring out some potential errors which may affect the accuracy of the results.

For those reasons, in this study we develop a discrete event simulation (DES) to (1) measure the performance of boat allocation plans given by IB-BAM under stochastic demand and (2) measure the error introduced by aggregating demands to superincident locations. The simulation model basically implements the output (allocation plan) of IB-BAM and tests its performance with respect to a number of performance metrics under stochastic incident type, demand, location and time. We also investigate the impacts of different scenarios and business rules when assigning boats to incidents.

The paper is organized as follows. Next section reviews related literature on DES and SAR applications. The following chapter describes the implemented optimization and simulation methodology. Next, we explain our simulation model and present our numerical results for the TurCG boat allocation plan. Finally, we discuss our results and conclude.

RELATED WORK

Simulation is an essential way to evaluate the performance of real-world systems and also assist to design systems for meeting future concerns (Banks, 1999, p.7-13). Shawki et al. (2015, p.533-540) categorize simulation models mainly as mathematical and physical while defining computer simulation as a sub-type of mathematical modeling application. Computer simulation is a decisive method for evaluating complex systems which requires great budget or enormous effort to construct physically in real. Besides, it is more practical than analytic approaches to deal with problems which consider stochastic aspects. Through the history, computer simulation has been applied in a number of private and public firms (Jahangirian et al., 2010, p.1-13). For example, a distinguished review by Aboueljinane et al. (2013, p.734-750) provides a good perspective about computer simulation models applications which are constructed for improving emergency medical services.

DES is a widely used simulation method to analyze and study systems that have state changes at discrete time intervals (Goldsman and Goldsman, 2015, p.103-109). Discrete event-systems have two major features: (1) randomness, (2) changes at discrete points in time. These systems are applicable to the traditional OR fields which are about queuing and inventory theory such as manufacturing, security, network and airport operations (Fu, 1994, p.199-247). In his survey, Smith (2003, p.157-171) displays studies of DES application about manufacturing, published between 1969 and 2002. Williams and Celik (1998, p.915-920) utilize discrete-event simulation method to evaluate the conveyor systems in automotive final assembly line. In their model, they consider a number of issues such as meeting production goal, flow time, determining bottlenecks and improving system performance. For another example of manufacturing application, Byrne and Heavey (2006, p.420-437) formulate a DES model for analyzing the performance of supply chain with multiple customer, distributors and product families. In military sector, Parsons and Krause (1999, p.1174-1178) use discrete-event simulation method to evaluate logistic issues of U.S. Marine Corps while Burke et al. (2000) develop a discrete-event simulation model, namely Transportation System Capability (TRANSCAP) model, for dealing with the problem of deployment of forces from bases. Similarly, Yıldırım et al. (2009, p.597-611) study military deployment planning problem and formulate simulation model with event graph methodology for analyzing efficiency of deployment plans. They experiment their approach with a given scenario which consider deployment of four battalions from north-western Turkey to southeast border. In another example, Varol and Gunal (2015, p.2037-2049) conduct a hybrid DES and Agent-Based Simulation (ABS) model to imitate and evaluate counter-piracy operations on the Gulf of Aden. In their model, they study the problem with behaviors of three main actors such as pirates, transporters and naval forces, in a piracy activity. Consequently, they state that helicopter is the most important vehicle in counter-piracy operations. Similarly, Onggo and Karatas (2015, p.254-265) and Onggo and Karatas (2016) develop a simulation model to analyze maritime search operations such as SAR and patrol. In their study, they try to generate an alternative approach to improve search operations with a simulation model that considers behaviors of searcher and target.

In addition to these areas, researchers confront with other application fields that healthcare and emergency medical services (EMS) constitute a large amount. Günal and Pidd (2010, p.42-51) state that there is a notable increase in the number of DES applications in healthcare over ten years and gather recent DES applications about healthcare systems favorably. They concern each study in literature in terms of application area and categorize studies into six groups such as accident and emergencies (A&Es), inpatient facilities, outpatient clinics, other hospital units, whole hospital simulation and other relevant issues. Ingolfsson et al. (2013, p.736-746) develop a DES to analyze the effects of changing ambulances shift organizations from multiple-start system (MS) to single-start system (SS) in which all ambulances start and end their shifts at the same location. They use historical data of 2000 summer and apply model for north-west and central Edmonton, Western Canada. Zhen et al. (2014, p.12-23) apply DES optimization method to analyze the operational efficiency of ambulance deployment and relocation plan in stochastic scene. Furthermore, responding to distress call as quickly as possible is the main goal for an EMS. However, several issues such as heavy traffic, unexpected failures and breakdowns, and lack of idle resources cause delay and affect the success of operations. Wu et al. (2009, p.1359-1366) study the problem of finding optimal deployment strategies and critical level of demand that requires the ambulance fleet size expanded. They propose a DES model to evaluate EMS of Tainan City while keeping a pre-defined service level as intercepting 90% of calls within 9 minutes. Aboueljinane et al. (2012, p.84) try to improve response time of the French Emergency Medical service (SAMU) in Val-de-Marne department. They develop a DES model to reduce three factors as waiting time, travel time and processing time which determine response time to a distress call. In their study, they experiment their approach with seven scenarios which consider expanding resource fleet, relocation of existing teams and reducing process time. In a different study, Aboueljinane et al. (2014, p.46-59) use DES in five different strategies to advance same system, SAMU, with the objective of intercepting patients' calls within 20 minutes. Wei et al. (2014, p.207-216) utilize DES to improve ambulance response times in Singapore EMS. They follow three main strategies which consist of relocation ambulances, adding private ambulances and modified dispatch policy and provide a significant improvement for the system. In another study, Nogueira et al. (2014, p.1-12) concentrate on minimizing response times of EMS with optimization and simulation approach. They formulate a multi-objective optimization model to determine deployment plan of ambulances. Then, they define the deployment plan obtained from optimization model as initial plan for DES model. They apply their methodology to EMS in Belo Horizonte, Brazil with six different scenarios which consider dynamic changes in number of ambulances, bases and hospitals.

Maritime SAR operations can be counted as another application field for DES to improve response time to incidents. However, there are few DES studies about maritime SAR systems in the literature. Afshartous et al. (2009, p.1086-1096) propose an optimization and simulation combination methodology to determine locations of the US Coast Guard air stations while minimizing total travel cost (distance). In optimization model, they build the model as p-Uncapacitated Facility Location Problem (p-UFLP), combination of p-MP and UFLP. Then, they apply simulation method to evaluate the solutions of optimization model. In simulation model, they formulate a statistical model which generates stochastic distress calls from an inhomogeneous Poisson process, using historical data of the year 2000, in order to utilize in simulation model. As another simulation application in maritime SAR, Goerlandt et al. (2013, p.1-6) identify the maritime search and rescue operation as a multi-server, multi-customer queuing system that incidents/incident locations are customers and boats at ports are servers. They develop a DES model to provide a decision-support tool for The Finnish Lifeboat Institution in determining total number and type of boats at each station for SAR organization. For model inputs, they utilize historical incident data from the year 2007 to 2010 in Gulf of Finland and categorize incidents into 21 different types. In addition, they import wave conditions at incident locations into model for imitating real-world state which is an important aspect in determining which type of boat should be assigned. Under these circumstances, they try to locate 29 boats to 11 different stations along Gulf of Finland.

In literature, we confront with few studies addressing locating maritime SAR resources. Even though each study tries to solve the problem from a different perspective, there are a number of common approaches applied in formulation. For example, authors focus mainly on cost minimization and coverage. In contrast with private services, user accessibility and response time should be the main concern in public and emergency services (Balcik and Beamon, 2008, p.101-121). Most studies use aggregated demand and try to meet that demand with the total capacity of boats. However, this assumption ignores actual locations of the demand and affects the solution quality due to the aggregation error. In real-world scenarios, sometimes technical limitations on resources and strategic business rules can be relaxed depending on the severity of the emergency. Hence, the actual outcome of an analytic solution may significantly vary in real world.

In the recent studies, Razi et al. (2016) and Karatas et al. (2017) consider locating SAR helicopters for the TurCG and they propose a combined optimization and simulation approach. Similar to our approach in this chapter, Razi et al. (2016) also employs simulation to measure the performance of SAR helicopter location plans. Karatas et al. (2017), on the other hand, propose a novel three step approach to locate helicopters. At the first stage, they employ a mathematical model which uses deterministic data to generate a good location plan. Next, they implement a heuristic called the alternative plan generation, which generates a number of candidate location schemes. Finally, the authors use simulation to test the performance of each alternative under stochastic data. In the following section, we describe our simulation methodology which addresses these issues.

METHODOLOGY

When planning the locations of SAR resources, decision makers aim to respond to all incidents at sea as rapidly as possible in order to increase the rate of survivals and decrease that of material damages. Moreover, a number of key factors such as real-world incident data, incident severity levels, resource constraints and limitations, geographical limitations, available ports and their capacities, and business rules of the organization should be taken into account when construction allocation models.

Razi and Karatas (2016)'s proposed methodology, IB-BAM, takes into account all above factors and generates effective boat allocation plans in three steps (see Figure 1). They first determine weights for each incident type using a widely used Multi Criteria Decision Making (MCDM) technique named Analytical Hierarchy Process (AHP). Next, the authors develop a ZDM to cluster incidents and generate superincident locations. Finally, they employ a mo-MIP model to generate boat allocation plans.

In this paper we first summarize the three steps of IB-BAM and extend it by implementing a DES model as shown in Figure 1. The DES uses the allocation plan determined by IB-BAM as an input and measures its performance under stochastic data. Next, it evaluates the impact of alternative business rules (rules of assigning a boat to an incident) as well as data aggregation.

Figure 1. IB-BAM and DES model methodology

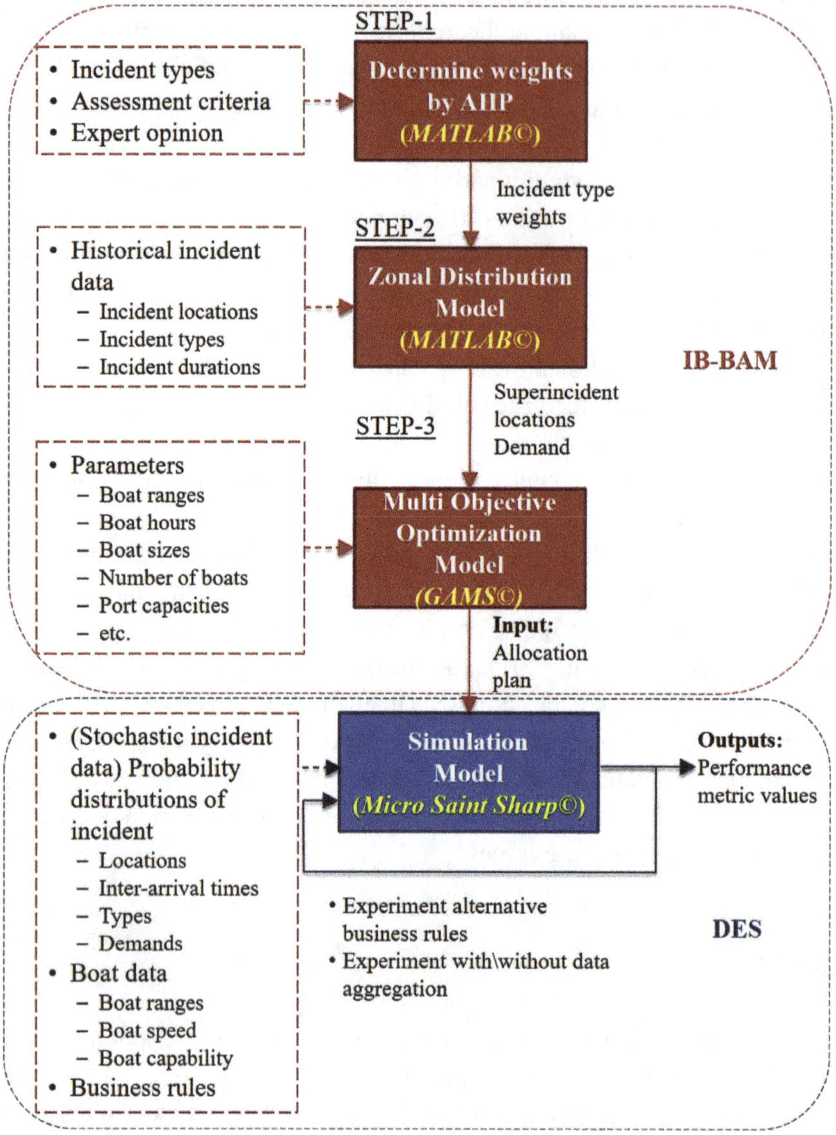

We now give an overview of the IB-BAM methodology proposed by Razi and Karatas (2016) and then describe the key aspects of our simulation approach in detail.

Incident Based-Boat Allocation Model (IB-BAM)

IB-BAM takes into account all above factors and generates effective boat allocation plans in three steps as explained below.

Weight Determination

Using AHP, IB-BAM first determines incident type weights. AHP allows a decision maker to determine relative importance of alternatives by conducting a series of pairwise comparisons with regard to a number of decision criteria.

A comprehensive review of Vaidya and Kumar (2006, p.1-29) analyzes the applications of the AHP in literature with a number of themes such as selection, assessment, cost-benefit analysis, allocations, planning and development, priority and ranking, and decision-making. In addition, they also categorize all studies in terms of the application area, such as manufacturing, engineering, personal, social, political, education, industry, government, and others. Korpela and Tuominen (1996, p.169-180) develop an approach to select warehouse sites by the AHP while both tangible and intangible criteria are considered. Similarly, Lirn et al. (2004, p.70-91) utilize the AHP technique to the problem of transshipment port selection. Amer and Daim (2011, p.420-435) overcome the problem of selecting renewable energy option for Pakistan with the AHP. They evaluate four alternative renewable energy options such as wind energy, solar photovoltaic, solar thermal and biomass energy with five criteria and 22 sub-criteria. Karacan (2015) develops a hybrid decision support tool that considers the AHP with other MCDM techniques such as TOPSIS/Fuzzy TOPSIS, VIKOR/Fuzzy VIKOR, and Goal Programming. The author tests the tool in selection of obesity surgery selection among three alternatives. Işıklar and Büyükozan (2007, p.265-274) study the problem of mobile phone selection by utilizing AHP methodology. Sivilevičius and Maskeliūnaite (2010, p.368-381) try to increase the quality of railway transportation and for that purpose, they get passengers evaluated a number of criteria which are categorized mainly into four groups and conduct the AHP technique. Recent studies of Ho (2008, p.211-228), Dağdeviren et al. (2009, p.8143-8151), Amiri (2010, p.6218-6224), Ishizaka et al. (2011, p.1801-1812), Stein (2013, p.157-171), Herva and Roca (2013, p.77-84), Ren and Sovacool (2014, p.589-597), Luthra et al. (2015, p.762-776) provide further AHP applications for interested researchers.

Razi and Karatas (2016) determine the criteria to assess the importance of an incident type is determined as fatality, material damage, response arduousness and environmental impact. The incident types observed in the Aegean Sea are drifting, capsizing, grounding, fire, collision, flooding, missing, and medical assistance. Denoting the set of incident types by $i \in I$, AHP technique provides the decision makers with a weight of each incident type *i*, w_i.

Zonal Distribution Model (ZDM)

Given a set of deterministic incident locations and incident type weights (from step-1), in this step

IB-BAM determines superincident locations with a ZDM. The ZDM uses Kerdprasop et al. (2005)'s "*weighted k-means clustering algorithm*" (*wk*-MCA) to generate superincident locations. The *wk*-MCA seeks to minimize the weighted sum of squared distances between each point and its *weighted*-centroid. Actual locations and superincident locations are shown in Figures 3(a) and 3(b), respectively.

Let $z \in Z$ denote the set of superincident zones. The number of zones is determined by *the rule of thumb method* (Kodinariya and Makwana,2013), and computed by the formula $\|Z\| \approx \sqrt{K/2}$, where *K* denotes the number of incidents.

In a facility location problem, locations constitute the majority of inputs in a model. However, as the amount of location data increases, it makes the optimization model so complex to solve in a decent time. In addition, using large-sized data brings out a number of issues by which the decision-maker must confront such as data collection cost, modeling cost, computing cost, confidentiality concerns and statistical data uncertainty (Francis et al., 2009, p.171-208). Thus, data aggregation that Fredrikson et al. (1999) state also as "*a single point that represents or summarizes the group of data points*" is a practicable method to deal with these hardships.

In literature, researchers encounter with a number of clustering techniques. With a more than 50 years history, *k*-means clustering algorithm (*k*-MCA) is a prevalent method for data clustering due to its simplicity, efficiency, and implementation with ease (Jain, 2010, p.651-666). The algorithm provides *k* centers, called "*centroids*", on given *n* data while minimizing the squared error between the center of the cluster and each point in it. In the application process, *k*-MCA follows three main steps (Jain and Dubes, 1988):

1. Initial partition with *k*-cluster,
2. Assigning new patterns to nearest cluster for generating a new partition,
3. Determining new centroids. Re-apply steps 2 and 3 until obtaining stable clusters.

The main parameter of *k*-MCA, *k*-number of clusters to be generated, can be determined in various ways. While Tibshirani et al. (2001, p.411-423) develop a heuristic method, the gap statistic, Kodinariya and Makwana (2013, p.90-95) mention six approaches: rule of thumb, elbow method, information criterion approach, information theoretic approach, choosing *k* using the Silhouette and cross-validation in their review. The algorithm runs separately for any *k* value no matter which method is used for determining "*k*" (Jain, 2010, p.651-666).

Data aggregation has been applied for a number of purposes such as grouping, defining patterns and determining similarity of data. There are a number of papers that contain data aggregation for different purposes in facility location models. Nemes et al. (2004, p.15-20) study the problem of evaluating the performance of AIDS care clinics where deliver antiretroviral treatment in Brasil. They conduct a questionnaire at 27 sites to determine service quality and utilize *k*-means clustering to gather health services in four groups as "Best", "Medium", "Poor" and "Very Poor" in terms of questionnaire scores. Liao and Guo (2008, p.323-339) generate a clustering-based method to deal with capacitated facility location problem. At first, they determine the allocation plan with minimum cost. Then, they utilize *k*-means clustering-based iterative optimization method to allocate facilities while minimizing demand-weighted distance between demand and associated facility. They also evaluate their methodology with Genetic Algorithm (GA) and state that clustering-based method provides better solutions and lower cost of time. In their study, Li et al. (2012, p.1103-1117) develop a clustering-based scatter search method to determine locations of transshipment points (TPs) which allow multi-commodity flow. They use *k*-means clustering approach to obtain initial solution for their method. In another application, Esnaf and Küçükdeniz (2009, p.259-265) propose a fuzzy clustering-based hybrid method for the multi-facility location problem. The problem is about to deploy multiple facilities for serving demand points. In their method, authors utilize clustering approach to group demand points with the aim of degrading problem from multiple to single facility location problem. They partition demand points into clusters and try to locate a single facility in each cluster. Similarly, Sahraeian and Kazemi (2011, p.1098-1102) study the problem of locating multiple facilities among a number of sites whereas all demand points are covered at least one facility. In their three stage method, firstly they determine the minimum number of facilities needed, in other words, k, with fuzzy set covering approach. Then, they implement k-means clustering approach to group demand points into k. Finally, they seek the optimal locations of k facilities in each group (cluster).

Optimization Model

As the third step, IB-BAM generates effective allocation plans using a *mo*-MIP model. The model considers incident types, candidate ports, superincident zones, boat types, boat features and capacities, and some organizational rules. The sets, indices, parameters, decision variables, objective function, and constraints used in the formulation are explained below.

Sets and Indices

$i \in I$: Set of incident types.
$p \in P$: Set of candidate ports.
$z \in Z$: Set of (superincident) zones.
$b \in B$: Set of boat types.
$B_i \subset B$: Set of boat types that are appropriate for incident type $i \in I$.
$B_p \subset B$: Set of boat types that are allowed at port $p \in P$.
$B_{pz} \subset B$: Set of boat types that can respond to incidents in zone $z \in Z$ from port $p \in P$.
$g \in G$: Set of goals.

Parameters

cf_{bp} = Yearly fixed cost of utilizing one type b boat at port p (\$/year).
cv_b = Variable cost of operating one type b boat for one hour (\$/hr).
U_b = Total available number of type b boats.
$u_b^{\min}$ = Minimum allowable number of type b boats at a port if allocated.
$u_b^{\max}$ = Maximum number of type b boats that can be allocated to a port.
S_b = Annual Operation Capacity (AOC) of a single type b boat (hrs).
r_b = Maximum allowable percentage of AOC that can be exceeded for type b boat.
γ_b = Multiplier to provide minimum allowable yearly total task time for boats of type b assigned from a port to a zone for a single incident type (hrs).
w_i = Weight of incident type i.
D_{zi} = Aggregated yearly demand in zone z for incident type i (hrs).
n_{zi} = Number of type i incidents aggregated to zone z.
rt_{bpz} = Response time of a type b boat from port p to zone z (hrs).
crt_i = Critical response time threshold for incident type i (hrs).
L_p = Physical capacity of port p (mts).

ℓ_b = Length of type b boat (mts).
FB = Annual fleet operating budget (M$).
w_g = Weight of goal g (assigned by the decision maker)
θ_g = Normalization factor for goal g (assigned by the decision maker)
M = Large number

Decision Variables

x_{bp} = Number of type b boats allocated to port p (integer).
y_{bp} = Variable indicating whether or not type b boats are utilized in port p (binary).
s_{bpzi} = Total number of yearly hours (supply) of type b boats assigned from port p to zone z for incidents type i (hrs).
λ_{bpzi} = Variable indicating whether or not type b boats are utilized in port p to zone z for incidents type i (binary).
$dev_{bpzi}^{crt+}, dev_{bpzi}^{crt-}$ = Positive and negative deviations associated with critical response time threshold for each individual incident type i in zone z responded from port p by type b boat (hrs).
dev^{fb+}, dev^{fb-} = Positive and negative deviations associated with the annual fleet operating budget ($).
$dev_{bp}^{s+}, dev_{bp}^{s-}$ = Positive and negative deviations associated with the AOC for type b boats in port p (hrs).

Objective Function

Utilizing incident type weights from step-1 and superincident locations from step-2, it provides an optimal allocation plans while minimizing three terms in the objective function (1):

- The total weighted deviations resulting from exceeding critical response times to incidents.
- The deviation resulting from exceeding the allocated fleet budget.
- The total deviation of supply from capacity (annual operation capacity of SAR boats).

As an output the model provides the number of each boat type allocated to each port.

$$\min z = \omega_1\theta_1 \sum_{b\in B}\sum_{p\in P}\sum_{z\in Z}\sum_{i\in I} w_i n_{zi} dev_{bpzi}^{crt+} + \omega_2\theta_2 dev^{fb+} + \omega_3\theta_3 \sum_{b\in B}\sum_{p\in P}\left(dev_{bp}^{s+} + dev_{bp}^{s-}\right) \tag{1}$$

Although there are a number of solution techniques for multi-objective problems, we choose weighted sum method to solve our model. A brief discussion for solving multi objective location models also exist in Karatas (2017, 2018) and Karatas and Yakici (2018). However, each section of the objective function has a different scale and this difference could affect the correctness of results. Thus, the normalization of the objective function component is necessary.

Regarding the type of our optimization model, it is proper to utilize the notion of Pareto optimality which provides no other solutions that improves at least one part of the objective function without degradation in another (Zitzler and Thiele, 1999, p.257-271). At this point we use the normalization technique described in (Grodzevich and Romanko, 2006, p.89-101). In this technique, two special points are calculated for each section of the objective function individually, namely Nadir and Utopia points. Utopia point, z^U, represents the lower bound of mentioned section and could be obtained by minimizing each section i as follows:

$$z^* = \arg\min_x \left\{ f_i\left(x\right) : x \in \Omega \right\} \tag{2}$$

Nadir Point, z^N, represents the upper bound of each objective section i and could be defined as:

$$z_i^N = \max_{1 \le j \le k}\left(f_i\left(x^{[j]} \right) \right), \forall i = 1,\ldots,k \tag{3}$$

After calculations of these points, normalization coefficient value for each section, θ_i, could be determined as in equation (4):

$$\theta_i = \frac{1}{z_i^N - z_i^U} \tag{4}$$

Each of these coefficient values, θ_i, normalize scales and make the magnitude of each section equalized. Furthermore, it is clear that utilization of normalization technique bounds the objective function value as:

$$0 \le \frac{f_i(x) - z_i^U}{z_i^N - z_i^U} \le 1 \tag{5}$$

The θ_g values are determined by computing the Nadir and Utopia values as described in Grodzevich and Romanko (2006). These coefficients are used to scale objective function terms which have different magnitudes.

Constraints

$$\sum_{p \in P} x_{bp} \le U_b, \forall b \in B \tag{6}$$

$$y_{bp} u_b^{\min} \le x_{bp} \le y_{bp} u_b^{\max}, \forall b \in B, p \in P \tag{7}$$

$$\sum_{p \in P} \sum_{b \in B} s_{bpzi} = D_{zi}, \forall z \in Z, i \in I \tag{8}$$

$$\lambda_{bpzi} \gamma_b \leq s_{bpzi} \leq \lambda_{bpzi} M, \forall b \in B, p \in P, z \in Z, i \in I \tag{9}$$

$$\sum_{z \in Z} \sum_{i \in I} s_{bpzi} \leq x_{bp} M, \forall b \in B, p \in P \tag{10}$$

$$\sum_{b \in B} x_{bp} \ell_b \leq L_p, \forall p \in P \tag{11}$$

$$\sum_{b | b \in B \setminus B_p} x_{bp} = 0, \forall p \in P \tag{12}$$

$$\sum_{b | b \in B \setminus B_i} \sum_{p \in P} \sum_{z \in Z} s_{bpzi} = 0, \forall i \in I \tag{13}$$

$$\sum_{b | b \in B \setminus B_{pz}} \sum_{i \in I} s_{bpzi} = 0, \forall p \in P, z \in Z \tag{14}$$

$$rt_{bpz} \lambda_{bpzi} + dev_{bpzi}^{crt-} - dev_{bpzi}^{crt+} = crt_i, \forall b \in B, p \in P, z \in Z, i \in I \tag{15}$$

$$\sum_{z \in Z} \sum_{i \in I} \left(s_{bpzi} + 2 n_{zi} rt_{bpz} \lambda_{bpzi} \right) + dev_{bp}^{s-} - dev_{bp}^{s+} = S_b x_{bp}, \forall b \in B, p \in P \tag{16}$$

$$\left(\sum_{b \in B} \sum_{p \in P} cf_{bp} x_{bp} + \sum_{b \in B} \sum_{p \in P} \sum_{z \in Z} \sum_{i \in I} cv_b s_{bpzi} \right) + dev^{fb-} - dev^{fb+} = FB \tag{17}$$

$$dev_{bp}^{s+} \leq S_b x_{bp} r_b, \forall b \in B, \forall p \in P \tag{18}$$

$$x_{bp} \in \{0,1,2,...\}, \forall b \in B, p \in P \tag{19}$$

$$y_{bp} \in \{0,1\}, \forall b \in B, p \in P \tag{20}$$

$$\lambda_{bpzi} \in \{0,1\}, \forall b \in B, p \in P, z \in Z, i \in I \tag{21}$$

$$s_{bpzi}, dev_{bpzi}^{crt+}, dev_{bpzi}^{crt-} \geq 0, \forall b \in B, p \in P, z \in Z, i \in I \tag{22}$$

$$dev^{fb+}, dev^{fb-} \geq 0 \tag{23}$$

$$dev_{bp}^{s+}, dev_{bp}^{s-} \geq 0, \forall b \in B, p \in P \tag{24}$$

Constraint set (6) ensures that the number of boats allocated does not exceed the total number of boats available. Constraint set (7) implements the business rule for the minimum and maximum number of boats requirements, and ensures that for each boat, there should be at least $u_b^{\min}$ and at most $u_b^{\max}$ of it respectively. Considering scheduled events such as maintenance or overhauling, or the unexpected situations such as breakdowns or crew shortages, this constraint provides backup supply to ports. Constraint set (8) satisfies the supply-demand balance for each zone and incident type. Constraint set (9) determines the minimum allowable s_{bpzi} hours. Constraint set (10) ensures that supply from a boat type *b* in port *p* can be positive only if x_{bp} is positive. Constraint set (11) restricts the number of assigned boats for a port. According to business rules, certain boat types are not allowed at certain ports. These rules are implemented by constraint set (12). Constraint set (13), on the other hand, dictates appropriate boat type-incident type assignments. Constraint set (14) makes sure boats operate within their maximum ranges. Constraint sets (15), (16) and (17) represent the goals for response time, boat capacity and budget. Constraint set (18) ensures that, for each boat type in a port, the positive deviation associated with the AOC does not exceed the maximum allowable time.

Simulation Model

In this study, we identify Maritime SAR process as server-to-customer queuing system which incidents/demands are customers and boats/resources are servers and perform simulation methodology to evaluate the system. Simulation optimization technique requires advanced computers and takes respectable time to conduct. Thus, we apply DES model to test the performance of optimization model.

To evaluate the optimization model properly and to measure aggregation error, we demonstrated two distinctive simulation models: model with grids (SimModel-1), and model with data aggregation (SimModel-2). These models will be discussed in the following sections. In both models, we followed four different business rules about supply capacity policy. Firstly, we ran the model with the business rule that does not allow decision-makers to use boats having already consumed all of the yearly supply capacity (AOC). It was followed by three alternative business rules that allow 10%, 20% and unlimited exceedance on the AOC. Experiments were conducted in Micro Saint Sharp© version 3.7 on a computer with Intel Core i5-3330S CPU@2.70 Ghz processor and 6.00 GB memory. Our simulation model is constructed with three stages as (1) Inputs, (2) Process Flow, and (3) Outputs. We now briefly explain these stages.

Input

In our simulation model, we utilize three inputs: (1) Incidents (demand), (2) Ports, (3) SAR boats (resources). In stage of determining incidents, we use historical incident data for year 2014 provided by TurCG. However, utilizing all incidents individually makes the model more complex as in optimization model. Thus, we divide the responsibility area into rectangular grids. Each grid represents a sub-area with its individual incident patterns. It is clear that generating smaller grids approximates the system to the real case. However, geographical features of the area could not let it to do easily. More importantly, generating a smaller grid structure enables to represent risky sub-regions. Then, we analyze real-world data to generate the below-mentioned probability distributions for each grid:

- Inter-arrival time distributions for each incident type,
- Incident location distributions,
- Incident duration (demand).

For other inputs, we take the data of activated ports and assigned boats from IB-BAM optimization model's outputs. We implement the allocation plan provided by the optimization model as an input to the simulation model.

Process Flow

A maritime SAR operation usually starts with a distress call from the vessel in danger. Decision-makers in an operation center assess the situation and pass all related information about incident to the nearest port which hosts at least one idle capable boat. When all information is recorded, port dispatches a capable boat with respect to incident type, e.g. decision-makers must assign a faster boat to medical assistance incident for transferring patient to nearest medical services as quickly as possible. Then, boat conducts the operation unless it is in need of refueling otherwise it turns back to port for refueling. After that, it goes on duty if requires, otherwise, stands in port for next missions.

In our DES model structure (see Figure 2), a SAR process begins with a distress call to the operation center. Then, the model calculates distances from incident position to each port and sorts them. Port-zone distances are calculated by an online tool called the Marine vessel traffic ©. Next, the operation center checks ports in ascending order (with respect to distance) for an idle capable boat. If the operation center determines port and boat type to assign, designated boat dispatches for conducting the SAR mission. However, in some cases, there could be more than one capable and idle boat for a mission. Thus, we implement two different selection criteria for determining which boat to assign among capable ones: (1) remaining capacity and (2) speed.

Remaining Capacity

The main goal of this criterion is to maintain balanced workload among boats. Hence, operation center selects the capable one with higher remaining capacity, even if there is a faster idle boat. Furthermore, we try to represent the third section of the objective function in optimization model with this criterion. After selection, the boat conducts the duty. However, in case of no idle boat for a duty, incident waits for an idle boat in the queue (FIFO).

Speed

With this criterion, the model follows a greedy approach by assigning the available fastest boat for each incident. When an incident occurs and there is more than one idle boat, the model (operation center) selects the fastest idle boat among capable ones. It is clear that this approach could help to reduce response times. However, giving priority to faster ones could also cause consuming faster boats toward the end of the simulation model. After selection, the boat conducts the duty. However, in case of no idle boat for a duty, incident waits for an idle boat in the queue (FIFO). Model executes a process flow for 24x365=8760 hours which imitate a year.

Output

We use our simulation model to evaluate the performance of the optimization model in terms of 12 measure of effectiveness (MoE) values summarized in Table 1.

NUMERICAL APPLICATION

Input

In our simulation model, we consider locating 9 types of SAR boats to 25 candidate ports. For confidentiality purposes we name boats as $B1, B2,\ldots, B9$ and ports as $P1, P2,\ldots, P25$. The historical data for year 2014 includes 716 incidents of 8 different types in the Aegean Sea. Key parameters with respect to incidents (incident type weights, number of incidents observed and average demand per incident) and boat types are presented in Tables 2 and 3, respectively.

As the allocation plan input, we used the plan obtained from IB-BAM strategy #12 in Razi and Karatas (2016)'s paper. In that strategy, the optimization model activates 13 ports and utilizes 8 different boat types. TurCG defines the threshold of the critical response time to an incident as 0.5 hours for all types of incidents.

To generate incident input data, we divided the area of interest, The Aegean Sea sub-responsible region, into 0.25x0.25 degree-sized grids with respect to incident density. Figure 3 displays (a) the locations of the 716 incidents (each represented by a diamond) having occurred in the Aegean Sea sub-responsible region throughout, (b) the 18 superincidents derived by the IB-BAM ZDM and all candidate ports, and (c) generated grid structure of the area of interest.

Figure 2. Flow chart of simulation model

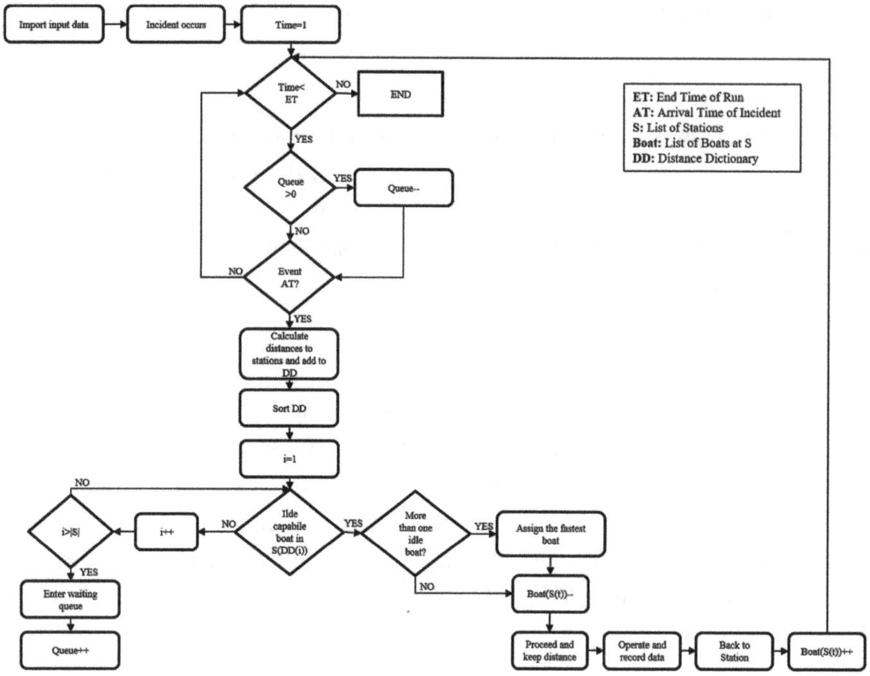

In grid view structure, we confronted with 71 regions where at least one incident was observed. It is assumed that incident creation is a Poisson process. After an incident type breakdown, the inter-incident time distribution for each incident type in a grid is computed as an exponential distribution with a mean of (1 year)/(number of incidents of that type). We further assume that incidents are distributed in sub-areas random uniformly. To determine the demand (required operation time) distribution, we analyzed the demand size of all incidents (see Figure 4 for a histogram of demand hours). Our analysis showed that distribution of operation times fit in the geometric distribution with parameter $p = 0.00147$/min. Consequently, we utilized the allocation plan obtained from strategy #12 to supply stochastic demands that were generated in a grid for evaluating the performance of optimization model.

Simulation Model

The structure of SimModel-1 is shown in Figure 5. Each grid is represented by 71 task modules on the left side of the figure. Incident entity that is generated in a grid takes a step forward and enters into the "Center" task module. In the "Center", the model tries to determine port and boat assignments and sends incident entity to the determined port task. However, in case of no idle boat, incident entity lines up in the queue (FIFO) in "Waiting" task module to wait for an idle capable boat. Port modules represent the proceeding process to incident locations. Then, incident entity follows the path to "Operation#" task where SAR operation is conducted. Finally, task groups on the right side of the figure release boats which complete the duty and collects all related data.

Table 1. Model outputs (AOC: annual operation capacity of a boat)

	Performance Metric	Definition
MoE$_1$	Total delay (hrs)	$\triangleq$ Sum of positive deviations associated with the critical reaction time threshold
MoE$_2$	Total weighted delay (hrs)	$\triangleq$ Sum of weighted positive deviations associated with the critical reaction time threshold
MoE$_3$	Shortage on the budget (M$)	$\triangleq$ Positive deviation associated with the annual budget
MoE$_4$	Ratio of budget used	$\triangleq \dfrac{\text{used fleet budget}}{\text{planned fleet budget}}$
MoE$_5$	Total excess AOC (hrs)	$\triangleq$ Sum of negative deviations associated with the AOCs
MoE$_6$	Total shortage on the AOC (hrs)	$\triangleq$ Sum of positive deviations associated with the AOCs
MoE$_7$	Total deviation on the AOC (hrs)	$\triangleq$ Sum of (positive and negative) deviations associated with the AOCs
MoE$_8$	Ratio of utilized fleet size	$\triangleq \dfrac{\text{total number of boats utilized}}{\text{total number of boats available}}$
MoE$_9$	Ratio of incidents intercepted	$\triangleq \dfrac{\text{total number of incidents intercepted}}{\text{total number of incidents occured}}$
MoE$_{10}$	Ratio of demands satisfied	$\triangleq \dfrac{\text{total number of demands satisfied}}{\text{total number of demands generated}}$
MoE$_{11}$	Total reaction time (hrs)	$\triangleq$ Sum of reaction times to incidents intercepted
MoE$_{12}$	Reaction time per incident (hrs)	$\triangleq \dfrac{\text{total reaction time}}{\text{total number of incidents intercepted}}$

Table 2. Key parameters related to incident types

Incident Type	Weight	Number of Incidents Observed	Demand per Incident (hrs)
Drifting	0.044	474	8.506
Capsizing	0.249	30	74.533
Grounding	0.060	11	7.182
Fire	0.188	4	11.500
Collision	0.139	2	3.000
Flooding	0.085	34	10.618
Missing	0.119	43	24.930
Medical Asst.	0.116	118	2.381

Table 3. Key parameters related to boat types (For confidentiality purposes, some parameters are not presented in this table)

Boat Type	Range (nm)	Speed (kts)	Variable Cost ($M/hr)
*B*1	300	22	0.00417
*B*2	110	27	0.00080
*B*3	100	27	0.00071
*B*4	95	27	0.00081
*B*5	90	18	0.00060
*B*6	75	45	0.00052
*B*7	65	47	0.00070
*B*8	55	62	0.00051
*B*9	20	54	0.00041

Figure 3. (a) Incident locations occurred in 2014. Solid line represents the TurCG responsibility zone border. (b) Superincident (*) and port (•) locations (c) Grid structure of sub-responsible region

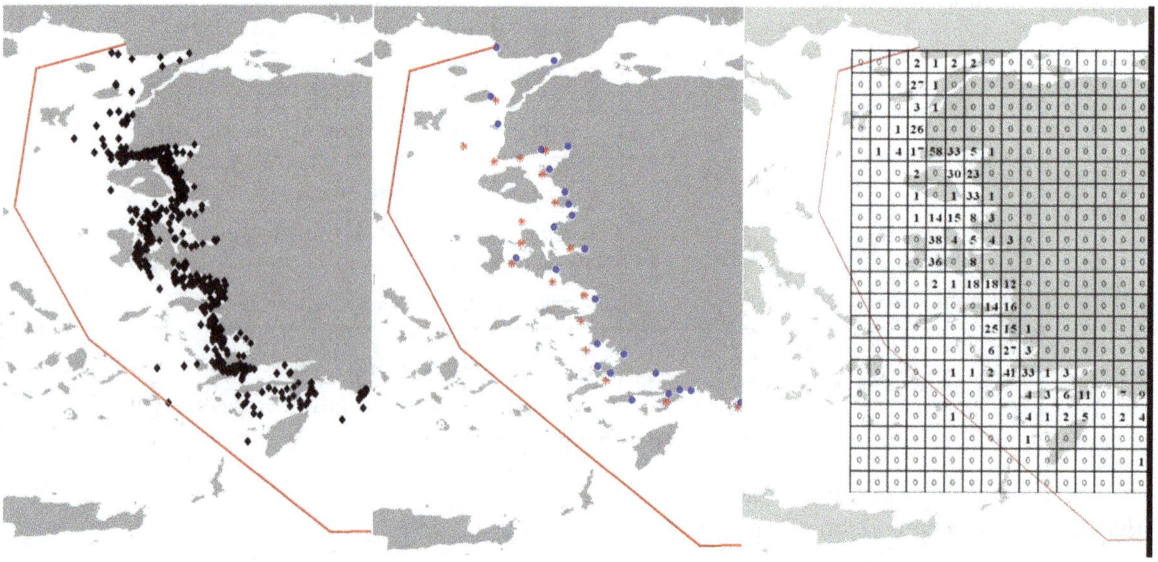

In SimModel-2, we utilized 18 zones obtained from data aggregation (ZDM) to represent regions in which incidents occur. Except this difference, we formulated SimModel-2 with the same assumptions and business rules implemented in SimModel-1. The structure of SimModel-2 is shown in Figure 6:

Results of Simulation Model

Simulation models were performed 2000 times for each business rule and selection criteria. Table 4 displays results of both optimization model and simulation model with grids (SimModel-1). Best values for the simulation model results are highlighted in bold.

Figure 4. Demand distribution

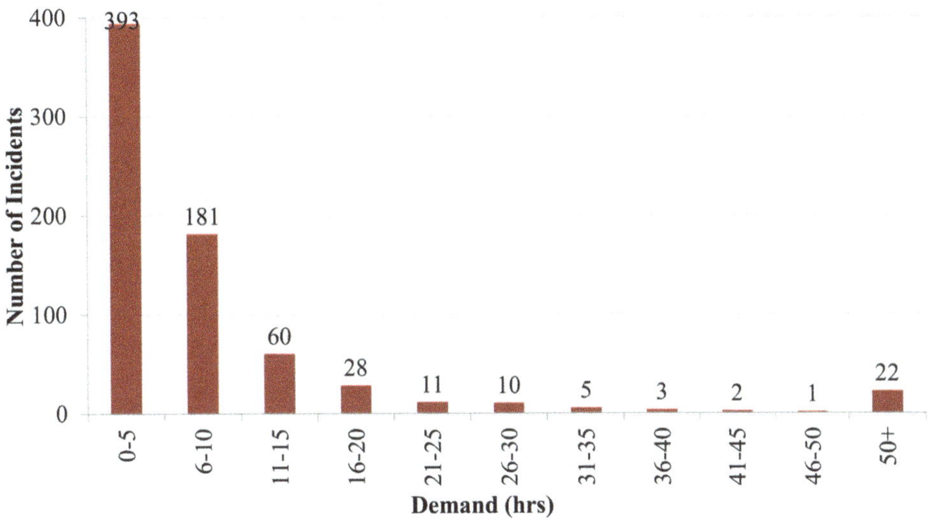

Figure 5. Simulation model structure of SimModel-1 in Micro Saint Sharp©

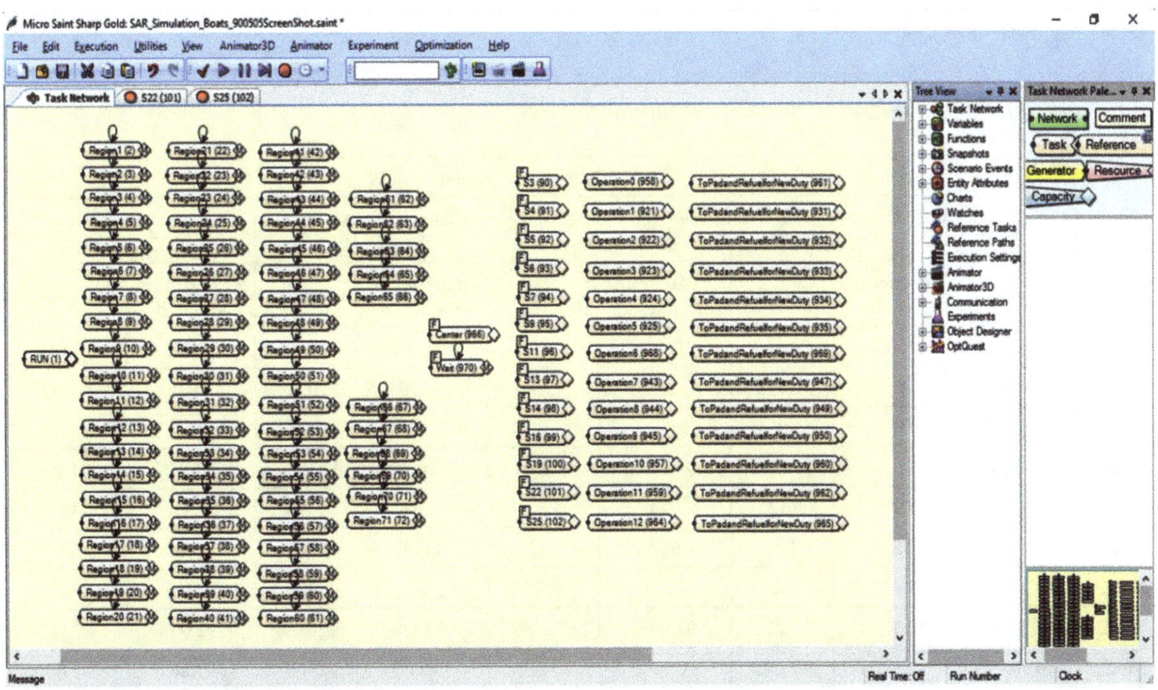

Firstly, results show that both optimization model and SimModel-1 provide no risk in budget. As we apply the allocation plan of analytic approach in simulation models, ratio of utilized fleet (MoE_8) values and surely fixed costs are equal for each experiment. SimModel-1 uses 94.83% of the budget at most whereas assigned boats exceed total of 1118 hours in the AOC. Secondly, in the case where AOC exceedance is not allowed (optimization model assumption), there is difference between analytic and

Figure 6. Structure of the SimModel-2 in Micro Saint Sharp©

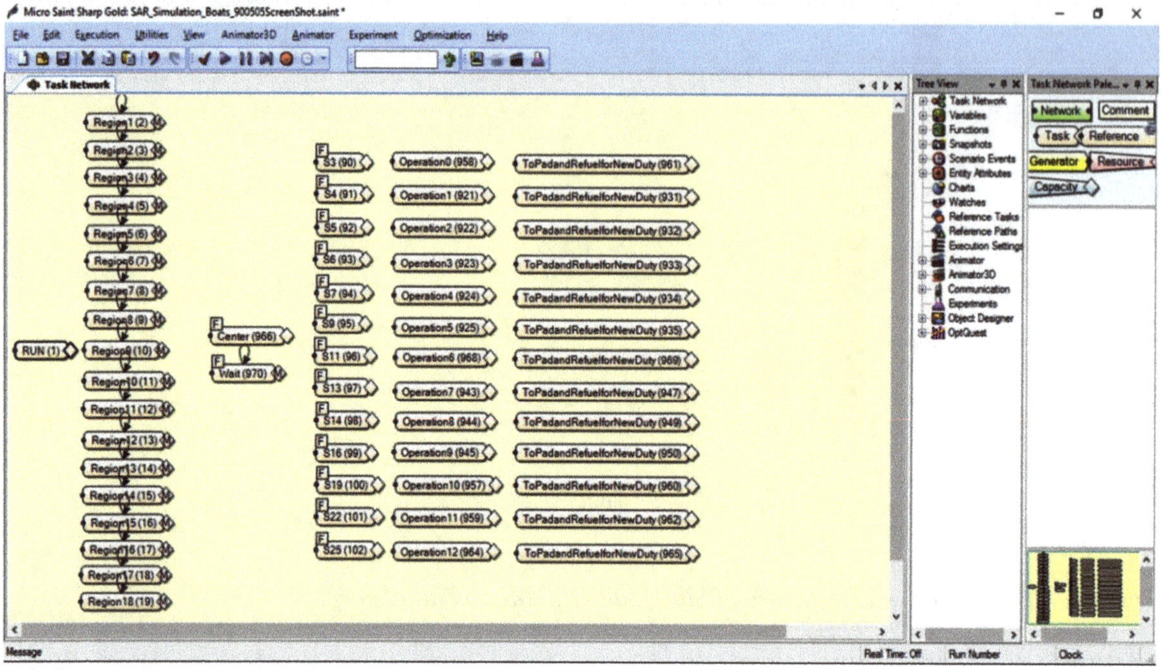

Table 4. Results of optimization model and SimModel-1

MoE#	IB-BAM Result	Simulation Model Results							
		Case-I: Shortage on the AOC 0%		Case-II: Shortage on the AOC 10%		Case-III: Shortage on the AOC 20%		Case-IV: AOC Unlimited	
		Remaining Capacity	Speed	Remaining Capacity	Speed	Remaining Capacity	Speed	Remaining Capacity	Speed
MoE_1	2.14	193.69	224.15	170.06	189.20	141.44	140.13	79.39	**42.69**
MoE_2	0.38	14.60	16.05	12.75	13.93	10.72	10.50	6.16	**3.47**
MoE_3	0	0	0	0	0	0	0	0	0
MoE_4	94.00%	93.07%	93.30%	94.09%	94.17%	94.83%	94.46%	94.78%	**91.99%**
MoE_5	224	693	**612**	914	983	1231	1598	2331	4680
MoE_6	39	**154**	183	624	692	1118	1311	2082	4016
MoE_7	263	847	**795**	1538	1675	2349	2909	4413	8696
MoE_8	92.00%	92.00%	92.00%	92.00%	92.00%	92.00%	92.00%	92.00%	92.00%
MoE_9	100%	92.38%	93.35%	95.84%	96.54%	98.89%	98.89%	**99.86%**	**99.86%**
MoE_{10}	100%	92.36%	93.43%	95.96%	96.72%	98.98%	98.95%	**99.94%**	**99.94%**
MoE_{11}	302.532	438.89	470.442	422.259	440.418	397.722	391.159	321.743	**256.247**
MoE_{12}	0.422	0.658	0.698	0.610	0.632	0.557	0.548	0.446	**0.355**

SimModel-1 in terms of total delay (MoE_1) gets its highest value. The optimization model provides 2.143 hours of total delay whereas SimModel-1 results in 193.69 and 224.146 hours of total delay. Besides, reaction time per incident (MoE_{12}) increases by 55.7% even if 7.62% of incidents are not responded (MoE_9). Particularly, 0.658 hours of reaction time per incident (MoE_{12}) indicates that SAR boats are getting late to each incident about 9 minutes in SimModel-1.

With the other three alternative business rules, SimModel-1 provides better solutions in terms of MoE_1, MoE_9, MoE_{10}, MoE_{11} and MoE_{12}. However, ratio of the budget used (MoE_4) worsens especially when "Remaining Capacity" criterion is implemented. As stated in Table 3, larger-capacitated units have a higher variable cost than smaller-faster ones. Thus, SimModel-1 is to bear a higher cost when the "Remaining Capacity" criterion is applied. In the case where 10% exceedance on the AOC is allowed, total delay (MoE_1) decreases by 15.6% and boats respond to each incident almost 4 minutes earlier (MoE_{12}). Besides, exceedance on supply capacity provides 3.5% more incidents to be intercepted. In the case that permits 20% exceedance on the AOC, boats respond to almost %98.9 of stochastically generated incidents while 2349 and 2909 hours of deviation on the AOC occurs. The model provides 25.9% improvement on total delay (MoE_1) when the "Speed" criterion is implemented. Besides, reaction time per incident (MoE_{12}) decreases to 33 minutes.

In the last case that allows full relaxation on capacity policy, SimModel-1 with "Speed" criterion dominates the model with "Remaining Capacity" in all MoE values except MoE_5, MoE_6 and MoE_7. The model provides this dominance along with a %3 lower budget. Outputs of this experiment indicate that we can have at least 42.69 hours of total delay (MoE_1), even though we utilize uncapacitated boats in the model. However, the model provides 0.3554 hours of reaction time per incident (MoE_{12}), which means responding to each incident in almost 21 minutes.

To sum up, the first three business rules (cases) in SimModel-1 yields higher values for the reaction time per incident (MoE_{12}) when compared with the optimization model. As we relax the capacity policy, total delay (MoE_1), ratio of incidents intercepted (MoE_9) and response time per incident (MoE_{12}) values decrease drastically and align with the analytic solution. For the last business rule (unlimited capacity), SimModel-1 provides the best values for all MoEs with the maximum deviation on the AOC. In order to summarize the scene, we displayed response time of each incident in Figures 7 and 8.

In Table 5, we presented the results of SimModel-2 which is demonstrated with data aggregation (ZDM) technique. Best values for the simulation model results are highlighted in bold.

Before we discuss results of all cases individually, there are a number of general points that should be mentioned about simulation model outputs. Firstly, "Speed" criterion in Case-IV is the dominant experiment over all tests in terms of all MoEs (except MoE_7). However, for other cases, the model with "Speed" criterion provides worse values in total delay (MoE_1) and total deviation on the AOC (MoE_7). Secondly, the model provides reaction time per incident (MoE_{12}) value under the critical time threshold (0.5 hour) in all cases (except the first case with "Speed" criterion). Finally, as in SimModel-1, implemented budget is enough to conduct each business rule and the model spends 0.41% more funds than the optimization model does at most.

In the first case, SimModel-2 is far worse than the optimization model in terms of total delay (MoE_1), even though there is not much difference in the other MoEs. While the total delay in optimization model is 2.143 hours, boats respond to incidents with 97.45 and 128.278 hours of total delay (MoE_1) in Sim-

Figure 7. Response time of each incident with "Remaining Capacity" (SimModel-1)

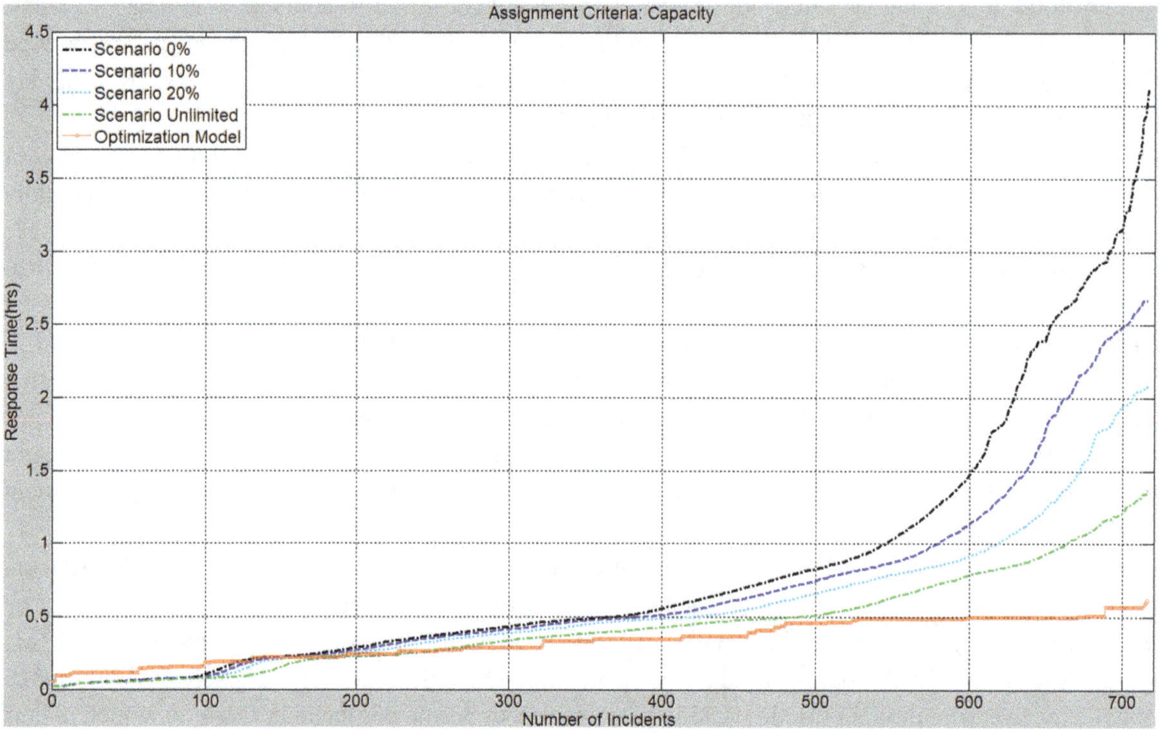

Figure 8. Response time of each incident with "Speed" (SimModel-1)

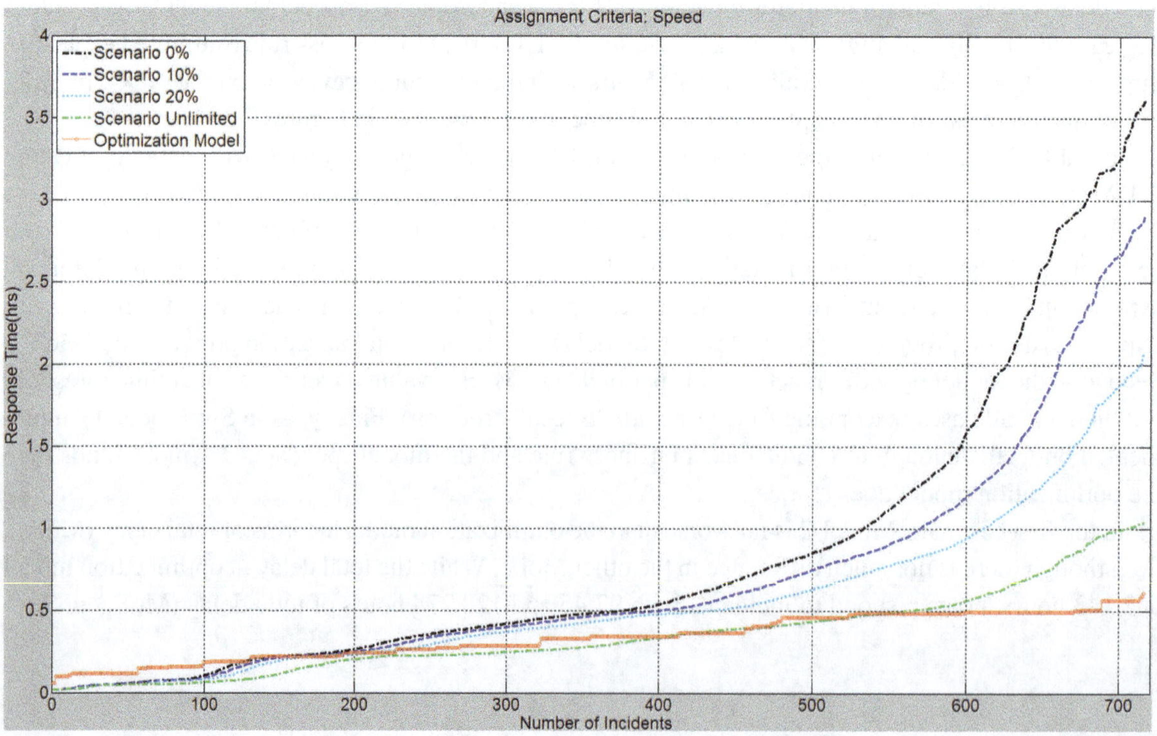

Table 5. Results of optimization model and SimModel-2

MoE#	IB-BAM Result	Simulation Model Results							
		Case-I: Shortage on the AOC 0%		Case-II: Shortage on the AOC 10%		Case-III: Shortage on the AOC 20%		Case-IV: AOC Unlimited	
		Remaining Capacity	Speed	Remaining Capacity	Speed	Remaining Capacity	Speed	Remaining Capacity	Speed
MoE_1	2.14	97.45	128.28	58.99	89.46	53.48	58.88	26.01	**9.12**
MoE_2	0.38	6.33	8.39	4.16	6.15	3.78	4.18	1.99	**0.89**
MoE_3	0	0	0	0	0	0	0	0	0
MoE_4	94.00%	93.07%	93.12%	94.41%	93.80%	94.30%	93.89%	94.23%	**91.50%**
MoE_5	224	**891**	967	1203	1506	1245	1910	1412	4157
MoE_6	39	**158**	204	680	695	718	1056	900	3201
MoE_7	263	**1049**	1171	1883	2201	1963	2966	2312	7358
MoE_8	92.00%	92.00%	92.00%	92.00%	92.00%	92.00%	92.00%	92.00%	92.00%
MoE_9	100%	94.18%	94.46%	98.61%	97.51%	98.89%	98.48%	**100.00%**	**100.00%**
MoE_{10}	100%	93.96%	94.38%	98.41%	97.33%	98.82%	98.44%	**99.86%**	**99.86%**
MoE_{11}	302.532	325.73	353.666	295.84	317.424	289.287	284.53	260.39	**208.853**
MoE_{12}	0.422	0.479	0.519	0.415	0.451	0.405	0.400	0.361	**0.289**

Model-2. SimModel-2 also provides five times higher exceedance on the AOC (MoE_6) than the optimization model does. Even if 158 and 204 hours of exceedance occur on the AOC, approximately 5.5% of generated incidents cannot be responded by any capable boat. For intercepted incidents, mean time to be responded is about 28.75 minutes (Remaining Capacity) and 31.11 minutes (Speed). In addition, SimModel-2 utilizes 1% lower budget than the analytic approach does.

SimModel-2 improves the delay time about 40% (Remaining Capacity) and 30% (Speed) when 10% exceedance is allowed on the AOC. With this relaxation, total exceedance (MoE_6) jumps from 158 and 204 hours to 680 and 695 hours. Moreover, ratio of incidents intercepted (MoE_9) increases by 4% and reaction time per incident (MoE_{12}) decreases by 13% to 24.9 and 27 minutes. Along with these improvements, ratio of the budget used (MoE_4) worsens about 1%. When relaxation on the AOC is expanded to 20% (Case-III), we confront with a different improvement ratio for each selection criterion. Total delay (MoE_1) decreases by 35% with the "Speed" criterion while it is by 10% with that of "Remaining Capacity". This difference also appears in the reaction time per incident (MoE_{12}). While boats respond to incidents 0.6 minutes earlier with the "Remaining Capacity" criterion, reaction time per incident (MoE_{12}) decreases to 24 minutes in the other criterion.

In the last case, the most important point is to mention the ratio of incidents intercepted (MoE_9). While SimModel-2 cannot provide full coverage in other cases, it responds to all generated incidents in Case-IV. Total delay (MoE_1) also gets the lowest value of 9.119 hours with the "Speed" criterion whereas reaction time per incident decreases to 17.3 minutes. Furthermore, SimModel-2 utilizes 2.5% lower budget than the optimization model itself, with that, boats exploit the highest exceedance on the AOC with 3201 hours.We plotted the response time of each incident in Figures 9 and 10 in order to visualize the dominance of any cases in terms of that factor.

Figure 9. Response time of each incident with "Remaining Capacity" (SimModel-2)

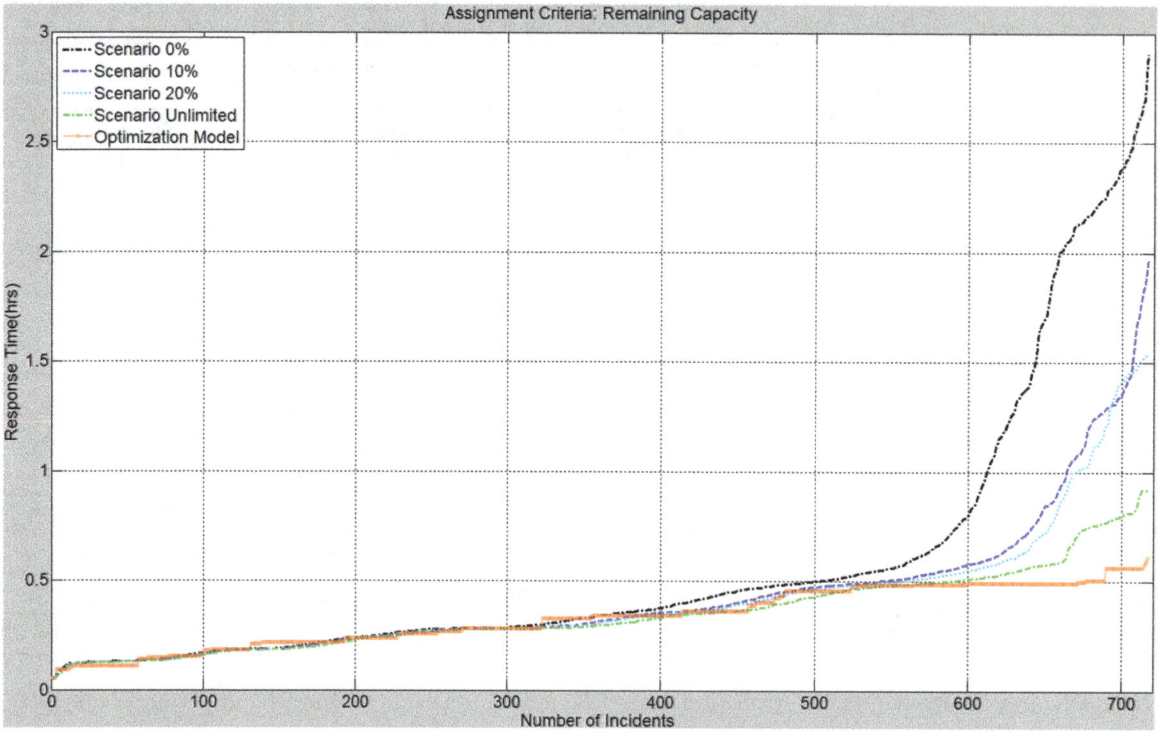

Figure 10. Response time of each incident with "Speed" (SimModel-2)

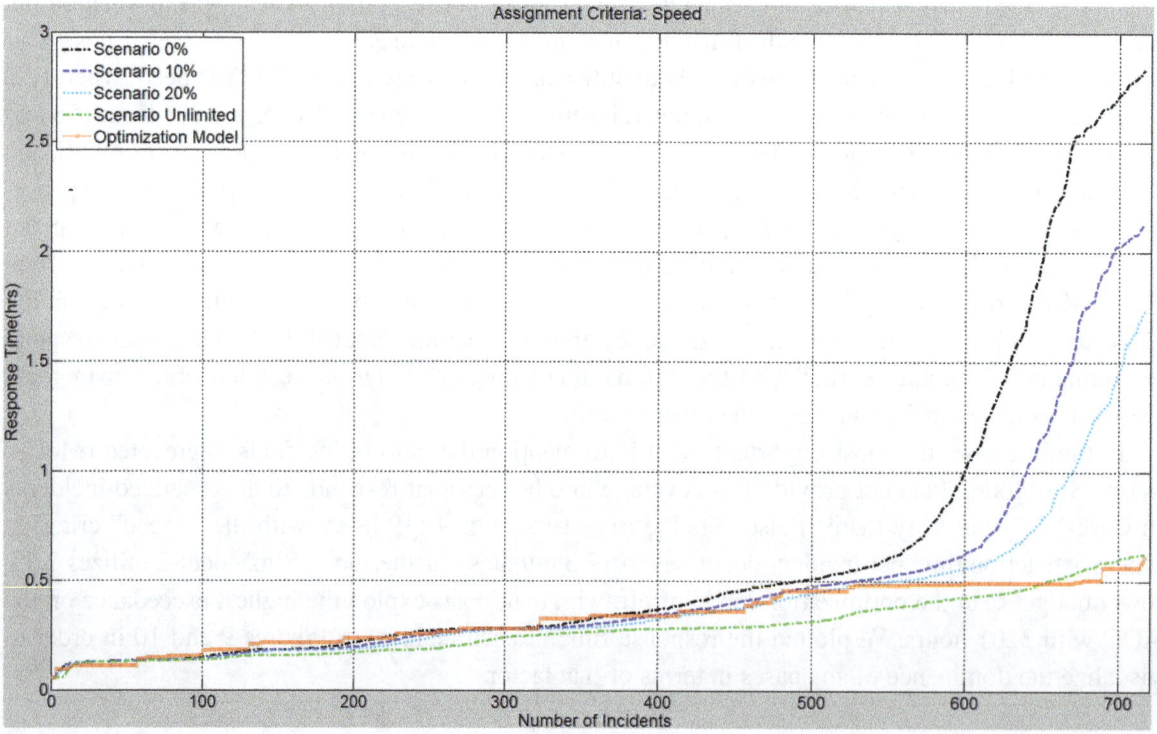

DISCUSSION

When we compare results of simulation models with IB-BAM results, we notice that there is a significant difference in terms of total delay (MoE_1). However, with the capacity relaxation (Case-II, Case-III and Case-IV), MoE_1, MoE_{11} and MoE_{12} improve and the value of total delay (MoE_1) starts getting closer to output of optimization model. Making use of the "Remaining Capacity" selection criterion, the operation center assigns the boat with more available capacity in case of an incident and maintains balanced workload at a port. As high-capacity units are slower, the operation center begins operations with slower units and utilizes them till their capacity gets equal with the faster ones. This approach provides better solutions than other selection criterion when limited capacity policy is conducted (Case-I and Case-II). However, results of this criterion worsen when boat capacities are unbounded.

SimModel-1 gives the best solution with "Speed" selection criterion in Case-IV that allows using boats unrestrictedly. Total delay (MoE_1) has the value of 42.69 hours at least whereas the optimization model provides 2.143 hours. The model also improves reaction time per incident (MoE_{12}) for four minutes. In this criterion, as boat capacities are unbounded, operation center assigns the fastest available boat for each incident that occurs in range. Thus, incidents near to the shore are responded in a short time, which decreases the total reaction time (MoE_{11}) and reaction time per incident (MoE_{12}) for sure. As expected, bigger units have long range but low speed. Therefore, operation center assigns slower boats when the incident occurs far from the port (long distance), which undoubtedly causes delay.

Despite the improvement in reaction time per incident (MoE_{12}), a number of incidents are responded late even though the model allows using faster boats unrestrictedly. It is clear that the main reason of this problem is the data aggregation in optimization model. The optimization model utilizes each incident on the superincident of its zone whereas the SimModel-1 applies grid structure that generates the incident in nearly its real location and provides longer distances than data aggregation. In some cases, the boat spends more time in proceeding to incident than the operation time on scene. Therefore, proceeding to far distances consumes a considerable amount of supply from annual capacity and makes the boat out of order earlier than expected in terms of capacity limitation. In the beginning, the model responds to incidents within 30 minutes since all boat types are operable. However, reaction time to incident increases especially when faster boats start to get out of order. It can also be seen in Figures 9 and 10 that response time to incidents rises up to 2-4 hours with capacity limitation.

To reveal the effects of data aggregation clearly, we generate SimModel-2 which utilizes 18 zones directly obtained from the ZDM. Although there is a significant difference between optimization model and SimModel-2 in terms of total delay (MoE_1) in Case-I (optimization model assumption), SimModel-2 yields 97.45 and 128.278 hours of total delay that are approximately %50 lower than those of SimModel-1. The benefit of data aggregation in SimModel-2 rises to 60%-65% with higher percentages of relaxed capacity (Case-II and Case-III). However, SimModel-2 provides the best results when boats are utilized unrestrictedly with the "Speed" criterion (Case-IV). In this case, boats respond to all generated incidents with a total delay of 9.119 hours and reaction time per incident decreases to 17.3 minutes. According to this result, using grid structure to generate demand undoubtedly results in five times worse result than the model with data aggregation. However, it can be assumed that 9.119 hours of total delay in 716 incidents is negligible and SimModel-2 gets close to the optimization model in terms of total delay (MoE_1). Along with data aggregation, scheduling is another reason for the delays in the system. The optimization model locates SAR resources according to supply-demand relation and does not consider scheduling which is also an important factor that deteriorates performance of the optimization model. In real world, more than one incident may occur simultaneously and there could not be enough idle resources to respond them all. In that case, the incident waits for an idle boat and delay happens. As we develop two DES models, scheduling factor is taken into consideration and causes delays in a number of incidents.

The other factor that causes delay in the system is the stochastic demand data. We believe that solving simulation models by using the "exact" locations of the incidents as in optimization model would be a repetition of the past in terms of model input data. Instead, we utilize stochastic demand that relies on historical data. We believe that this approach would enable models to represent the uncertainty. It also helps the model to be tested for different demand sizes. On the other hand, this demand variability affects operation matters and causes difference in solutions.

Simulation models (SimModel-1 and SimModel-2) help to test the performance of the optimization model in realistic settings where scheduling and stochastic demand data are allowed. Comparison of both models reveals that data aggregation elicits an error in the model regarding the delay to incidents. In other respects, simulation models perform close to the optimization model with regard to total delay and reaction time per incident in the following cases:

- SimModel-1 with the "Speed" criterion in Case-IV,
- SimModel-2 with both criteria in Case-IV.

In real life, we would expect that the operation center would assign the fastest boat without considering the remaining capacity at a given time. Thus, we believe that the analytic model is a good representation of the real world SAR operation process, and decision makers can apply it to manage their SAR organizations.

CONCLUSION

We have developed a model (IB-BAM) for allocating SAR resources in an area of interest. The model consists of three steps: weighting incident types with AHP technique, data aggregation (ZDM) to generate demand data and mo-MIP optimization model that provides optimal allocation plan. It aims at minimizing the following three objectives: total weighted delay over critical time threshold, exceedance on budget and deviation on the annual operation capacity of boats. To observe its performance, we utilized our model to locate SAR resources into ports along the Aegean Sea. The model was performed for 15 different weighting strategies. Results show that the optimization model helps decision makers respond to 716 incidents with only a three-minute delay in total. The sensitivity analysis shows that respond to incidents within a pre-defined time threshold is only achievable with the help of analytic approach which locates SAR resources optimally.

Along with this result, we developed simulation models (SimModel-1 and SimModel-2) to test results of the optimization model in real world settings. Scheduling and stochastic demand are considered and three different capacity policies are performed. Outputs of both models indicate that simulation models provide similar results with the optimization model under fully relaxed capacity policy. Scheduling and stochastic demand are considered and three different capacity policies are performed. Outputs of both models indicate that simulation models provide similar results with the optimization model under fully relaxed capacity policy.

In this thesis, we applied deterministic inputs into the optimization model. As a future study, stochastic and fuzzy theory can be used to determine the demands in the problem. It enables forecasting future incident rates and also makes the model more flexible to deal with different possible scenarios. As another future study, a combined model utilizing both aircraft and surface units together may be developed to organize joint SAR operations.

Disclaimer—The views and conclusions contained herein are those of the authors and should not be interpreted as necessarily representing the official policies or endorsements, either expressed or implied, of any affiliated organization or government.

REFERENCES

Aboueljinane, L., Jemai, Z., & Sahin, E. (2012, December). Reducing ambulance response time using simulation: The case of Val-de-Marne department Emergency Medical Service. *Simulation Conference (WSC) Proceedings of the*, *2012*(Winter), 1–12.

Aboueljinane, L., Sahin, E., & Jemai, Z. (2013). A review on simulation models applied to emergency medical service operations. *Computers & Industrial Engineering*, *66*(4), 734–750.

Aboueljinane, L., Sahin, E., Jemai, Z., & Marty, J. (2014). A simulation study to improve the performance of an emergency medical service: Application to the French Val-de-Marne department. *Simulation Modelling Practice and Theory*, *47*, 46–59.

Afshartous, D., Guan, Y., & Mehrotra, A. (2009). US Coast Guard air station location with respect to distress calls: A spatial statistics and optimization based methodology. *European Journal of Operational Research*, *196*(3), 1086–1096.

Amer, M., & Daim, T. U. (2011). Selection of renewable energy technologies for a developing county: A case of Pakistan. *Energy for Sustainable Development*, *15*(4), 420–435.

Amiri, M. P. (2010). Project selection for oil-fields development by using the AHP and fuzzy TOPSIS methods. *Expert Systems with Applications*, *37*(9), 6218–6224.

Balcik, B., & Beamon, B. M. (2008). Facility location in humanitarian relief. *International Journal of Logistics*, *11*(2), 101–121.

Banks, J. (1999). Introduction to simulation. In *Simulation Conference Proceedings*, 1999 Winter (Vol. 1, pp. 7-13). IEEE.

Burke, J. F., Love, R. J., Macal, C. M., Howard, D. L., & Jackson, J. (2000). *Modeling force deployment from army installations using the transportation system capability (TRANSCAP) model.* Argonne National Laboratory.

Byrne, P. J., & Heavey, C. (2006). The impact of information sharing and forecasting in capacitated industrial supply chains: A case study. *International Journal of Production Economics*, *103*(1), 420–437.

Dağdeviren, M., Yavuz, S., & Kılınç, N. (2009). Weapon selection using the AHP and TOPSIS methods under fuzzy environment. *Expert Systems with Applications*, *36*(4), 8143–8151.

Esnaf, Ş., & Küçükdeniz, T. (2009). A fuzzy clustering-based hybrid method for a multi-facility location problem. *Journal of Intelligent Manufacturing, 20*(2), 259–265.

Francis, R. L., Lowe, T. J., Rayco, M. B., & Tamir, A. (2009). Aggregation error for location models: Survey and analysis. *Annals of Operations Research, 167*(1), 171–208.

Fredrikson, A., North, C., Plaisant, C., & Shneiderman, B. (1999, November). Temporal, geographical and categorical aggregations viewed through coordinated displays: a case study with highway incident data. In *Proceedings of the 1999 workshop on new paradigms in information visualization and manipulation in conjunction with the eighth ACM internation conference on Information and knowledge management* (pp. 26-34). ACM.

Fu, M. C. (1994). Optimization via simulation: A review. *Annals of Operations Research, 53*(1), 199–247.

Giuliani, J. (2015, April 14). *The Challenge of Illegal Immigration in the Mediterranean*. Retrieved from http://www.robert-schuman.eu/en/doc/questions-d-europe/qe-352-en.pdf

Goerlandt, F., Torabihaghighi, F., & Kujala, P. (2013). A model for evaluating performance and reliability of the voluntary maritime rescue system in the Gulf of Finland. In *11th International Probabilistic Safety Assessment and Management Annual European Safety and Reliability Conference* (pp. 1-6). Academic Press.

Goldsman, D., & Goldsman, P. (2015). *Modeling and Simulation in the Systems Engineering Life Cycle*. London: Springer.

Green, L. V., & Kolesar, P. J. (2004). Anniversary article: Improving emergency responsiveness with management science. *Management Science, 50*(8), 1001–1014.

Grodzevich, O., & Romanko, O. (2006). Normalization and other topics in multi-objective optimization. In *Proceedings of the Fields–MITACS Industrial Problems Workshop* (pp.89-101.), Toronto, Canada: Academic Press.

Günal, M. M., & Pidd, M. (2010). Discrete event simulation for performance modelling in health care: A review of the literature. *Journal of Simulation, 4*(1), 42–51.

Herva, M., & Roca, E. (2013). Ranking municipal solid waste treatment alternatives based on ecological footprint and multi-criteria analysis. *Ecological Indicators, 25*, 77–84.

Ho, W. (2008). Integrated analytic hierarchy process and its applications–A literature review. *European Journal of Operational Research, 186*(1), 211–228.

Ingolfsson, A., Erkut, E., & Budge, S. (2003). Simulation of single start station for Edmonton EMS. *The Journal of the Operational Research Society, 54*(7), 736–746.

Ishizaka, A., Balkenborg, D., & Kaplan, T. (2011). Does AHP help us make a choice? An experimental evaluation. *The Journal of the Operational Research Society, 62*(10), 1801–1812.

Işıklar, G., & Büyüközkan, G. (2007). Using a multi-criteria decision making approach to evaluate mobile phone alternatives. *Computer Standards & Interfaces, 29*(2), 265–274.

Jahangirian, M., Eldabi, T., Naseer, A., Stergioulas, L. K., & Young, T. (2010). Simulation in manufacturing and business: A review. *European Journal of Operational Research, 203*(1), 1–13.

Jain, A. K. (2010). Data clustering: 50 years beyond K-means. *Pattern Recognition Letters, 31*(8), 651–666.

Jain, A. K., & Dubes, R. C. (1988). *Algorithms for clustering data*. Prentice-Hall, Inc.

Karacan, İ. (2015). *A New Hybrid Decision Support Tool and an Application to Health Technology Selection* (Doctoral dissertation). Naval Science and Engineering Institute, Department of Operations Research, Turkish Naval Academy.

Karatas, M. (2017). A multi-objective facility location problem in the presence of variable gradual coverage performance and cooperative cover. *European Journal of Operational Research, 262*(3), 1040–1051.

Karatas, M. (2018). Optimal Deployment of Heterogeneous Sensor Networks for a Hybrid Point and Barrier Coverage Application. *Computer Networks, 132*(26), 129–144. doi:10.1016/j.comnet.2018.01.001

Karatas, M., Razı, N., & Gunal, M. (2017). An ILP and Simulation Model to Optimize Search and Rescue Helicopter Operations. *The Journal of the Operational Research Society*, 1–17. doi:10.105741274-016-0154-7

Karatas, M., & Yakıcı, E. (2018). An iterative solution approach to a multi-objective facility location problem. *Applied Soft Computing, 62*, 272–287.

Kerdprasop, K., Kerdprasop, N., & Sattayatham, P. (2005). Weighted k-means for density-biased clustering. *Data Warehousing and Knowledge Discovery*, 488-497.

Kodinariya, T. M., & Makwana, P. R. (2013). Review on determining number of Cluster in K-Means Clustering. *International Journal (Toronto, Ont.), 1*(6).

Korpela, J., & Tuominen, M. (1996). A Decision Aid in Warehouse Site Selection. *International Journal of Production Economics, 45*(1), 169–180.

Li, J., Prins, C., & Chu, F. (2012). A Scatter Search for A Multi-Type Transshipment Point Location Problem with Multicommodity Flow. *Journal of Intelligent Manufacturing, 23*(4), 1103–1117.

Liao, K., & Guo, D. (2008). A Clustering-Based Approach to the Capacitated Facility Location Problem. *Transactions in GIS, 12*(3), 323–339.

Lirn, T. C., Thanopoulou, H. A., Beynon, M. J., & Beresford, A. K. C. (2004). An application of AHP on transhipment port selection: A global perspective. *Maritime Economics & Logistics, 6*(1), 70–91.

Luthra, S., Kumar, S., Garg, D., & Haleem, A. (2015). Barriers to renewable/sustainable energy technologies adoption: Indian perspective. *Renewable & Sustainable Energy Reviews, 41*, 762–776.

Nemes, M. I., Carvalho, H. B., & Souza, M. F. (2004). Antiretroviral therapy adherence in Brazil. *AIDS (London, England), 18*, S15–S20. PMID:15322479

Nogueira, L. C., Pinto, L. R., & Silva, P. M. S. (2016). Reducing Emergency Medical Service response time via the reallocation of ambulance bases. *Health Care Management Science, 19*(1), 31–42. PMID:24744263

Onggo, B. S., & Karatas, M. (2015, December). Agent-based model of maritime search operations: A validation using test-driven simulation modelling. In *Winter Simulation Conference (WSC)*, 2015 (pp. 254-265). IEEE.

Onggo, B. S., & Karatas, M. (2016). Test-driven simulation modelling: A case study using agent-based maritime search-operation simulation. *European Journal of Operational Research*, *254*(2), 517–531.

Parsons, D. J., & Krause, L. C. (1999). Tactical logistics and distribution systems (TLoaDS) simulation. In *Simulation Conference Proceedings*, 1999 Winter (Vol. 2, pp. 1174-1178). IEEE.

Razi, N., & Karatas, M. (2016). A multi-objective model for locating search and rescue boats. *European Journal of Operational Research*, *254*(1), 279–293.

Razi, N., Karatas, M., & Gunal, M. M. (2016). A Combined Optimization and Simulation Based Methodology for Locating Search and Rescue Helicopters. In *Proceedings of the 2016 Spring Simulation Conference*. IEEE Press.

Ren, J., & Sovacool, B. K. (2014). Enhancing China's energy security: Determining influential factors and effective strategic measures. *Energy Conversion and Management*, *88*, 589–597.

Sahraeian, R., & Kazemi, M. S. (2011, December). A fuzzy set covering-clustering algorithm for facility location problem. In *Industrial Engineering and Engineering Management (IEEM), 2011 IEEE International Conference on* (pp. 1098-1102). IEEE.

Shawki, K. M., Kilani, K., & Gomaa, M. A. (2015). Analysis of earth-moving systems using discrete-event simulation. *Alexandria Engineering Journal*, *54*(3), 533–540.

Sivilevičius, H., & Maskeliūnaite, L. (2010). The criteria for identifying the quality of passengers' transportation by railway and their ranking using AHP method. *Transport*, *25*(4), 368–381.

Smith, J. S. (2003). Survey on the use of simulation for manufacturing system design and operation. *Journal of Manufacturing Systems*, *22*(2), 157–171.

Stein, E. W. (2013). A comprehensive multi-criteria model to rank electric energy production technologies. *Renewable & Sustainable Energy Reviews*, *22*, 640–654.

Tibshirani, R., Walther, G., & Hastie, T. (2001). Estimating the number of clusters in a data set via the gap statistic. *Journal of the Royal Statistical Society. Series B, Statistical Methodology*, *63*(2), 411–423.

UNHCR. (n.d.). *Global Appeal 2015 Update*. Retrieved from http://www.unhcr.org/5461e5f80.html

Vaidya, O. S., & Kumar, S. (2006). Analytic hierarchy process: An overview of applications. *European Journal of Operational Research*, *169*(1), 1–29.

Varol, A. E., & Gunal, M. M. (2015). Simulating prevention operations at sea against maritime piracy. *The Journal of the Operational Research Society*, *66*(12), 2037–2049.

Wei Lam, S. S., Zhang, Z. C., Oh, H. C., Ng, Y. Y., Wah, W., & Hock Ong, M. E. (2014). Reducing ambulance response times using discrete event simulation. *Prehospital Emergency Care*, *18*(2), 207–216. PMID:24134647

Williams, E. J., & Çelik, H. (1998, December). Analysis of conveyor systems within automotive final assembly. In *Proceedings of the 30th conference on Winter simulation* (pp. 915-920). IEEE Computer Society Press.

Wu, C. H., & Hwang, K. P. (2009). Using a Discrete-event Simulation to Balance Ambulance Availability and Demand in Static Deployment Systems. *Academic Emergency Medicine*, *16*(12), 1359–1366. PMID:20053259

Yıldırım, U. Z., Tansel, B. Ç., & Sabuncuoğlu, İ. (2009). A multi-modal discrete-event simulation model for military deployment. *Simulation Modelling Practice and Theory*, *17*(4), 597–611.

Zhen, L., Wang, K., Hu, H., & Chang, D. (2014). A simulation optimization framework for ambulance deployment and relocation problems. *Computers & Industrial Engineering*, *72*, 12–23.

Zitzler, E., & Thiele, L. (1999). Multiobjective evolutionary algorithms: A comparative case study and the strength Pareto approach. *IEEE Transactions on Evolutionary Computation*, *3*(4), 257–271.

Chapter 5
Test and Evaluation for Weapon Systems:
Concepts and Processes

Levent Eriskin
Turkish Naval Command, Turkey

Murat M. Gunal
Naval Science and Engineering Institute, Turkey

ABSTRACT

Defense systems are complex and expensive and decision makers acquiring these systems aspire better value for money. As part of a procurement decision, test and evaluation (T&E) is conducted to assess if the system is meeting desired requirements. This chapter reviews the concepts in T&E for weapon systems in the navy and presents the process for its implementation. The authors point out the role of modeling and simulation and how to use them to support T&E. The review reveals that developmental and operational T&E can examine weapon systems requirements before their deployment, T&E process can effectively be implemented in six steps, and mission/capacity requirements must be decomposed into predefined evaluation criteria.

INTRODUCTION

Decisions made for weapon systems acquisition are hard in nature due to high technology and costs. Number of decision variables in weapon systems increases the complexity in acquisition since holistic evaluation is not possible with human brain. One way of taming the complexity dragon is to divide and conquer: portioning the acquisition problem into manageable pieces and solving each piece of problem using analytical techniques. Operations Research/Management Science (OR/MS) provide many analytical methods and techniques to solve complex decision problems.

DOI: 10.4018/978-1-5225-5513-1.ch005

In today's world where countries tend to cut defense budgets, weapon system acquisition decisions turn out to be more compelling since one need to meet defense capability requirements with minimum cost. Hence, among various alternatives having different costs, determining which particular system meets capability requirements requires thorough evaluation of system alternatives.

There is more than one way of procuring a weapon system. One weapon system may be developed by a country itself while another system might be procured as Commercial Off-The Shelf (COTS). Without loss of generality, we can classify types of weapon system acquisition as;

- Non-Developmental Item (NDI) or COTS (Developed or produced by another country)
- NDI-COTS with some custom modifications
- Co-developing with another country
- Another country's development and co-production
- Country's full development and production

Regardless of the way of procurement, the main questions to be answered remain to be the same: "How much a particular system alternative meets mission or capability requirements?" and "Which weapon system is superior in terms of meeting capability requirements?".

Acquisition of NDI or COTS weapon systems involves some risks thereby needs rigorous evaluation. Most of the NDI/COTS systems are developed based on developing country's needs, threat evaluation and capability requirements. These systems might be designed for different doctrine, user skills and combat environment. Moreover, they may not be compatible and interoperable with buyer's systems that are already in inventory. Even though custom modifications might be considered to overcome aforementioned problems, these modifications may turn out to be infeasible and unaffordable.

Developing and producing a weapon system without another country's involvement poses different challenges. At each phase of the development, the system needs to be evaluated to verify that it meets designated system specifications. Even though a holistic evaluation after all development process completed seems to be plausible and practical, this might generate greater problems since it may not be possible to solve or fix design problems, which are caused by early design flaws, after the system becomes a COTS.

The need for evaluating weapon systems is not only limited to acquisition process. For instance, in order to provide input to new concept development process, we may desire to measure performance of a weapon system that is already in inventory for some particular operational environment. We may also wish to verify weapon system performance after some years of service. This evaluation may help us to decide whether to put particular system out of service or to continue using it. In another case, it might be required to compare two or multiple in-service weapon systems' performances for operational planning. This evaluation provides insight for weapon system allocation and new operational tactic development.

All these aforementioned needs for observing, verifying or measuring weapon system performance requires extensive test and evaluation planning and implementation. A test can be defined as a procedure to obtain, verify, or provide data for determining performance, capability, suitability, vulnerability, and lethality of systems in terms of meeting stated objectives. Evaluation corresponds to assembling, analyzing and comparing qualitative or quantitative data obtained from the test process to provide support to decision makers. In the context of weapon system analysis, the combined process of test and evaluation is a process where a system or components are tested and related data are evaluated in order to compare them against requirements and specifications or alternative system solutions.

There are several test and evaluation types defined in the literature, however, all test and evaluation procedures can be classified in two main categories;

1. Developmental Test and Evaluation (DTE)
2. Operational Test and Evaluation (OTE)

DTE is conducted as a support to engineering process in order to verify that the developed system meets its technical and performance specifications. As the name itself reveals, this process is implemented for own development or co-develop weapon systems, in general. OTE, on the other hand, aims to test and evaluate a weapon system under realistic conditions, where several factors are considered but only some of them are controlled. Even though a combination of DTE and OTE, where DTE provides inputs to OTE for a better test and evaluation planning, is plausible and should be conducted whenever possible, different types of acquisition and operational conditions rarely makes it feasible to perform accordingly.

Test and evaluation of a weapon system has specific properties and requirements that are different from other generic systems. This usually stems from the fact that weapon systems are high technology and high cost systems, moreover, the terminology used in for these systems require different methodological definitions. In this respect, there are numerous text books, guides and research papers about test and evaluation procedures, and experimental design, however, only some focus on weapon systems and military experimentation. Claxton et al. (2005) explain United States military acquisition process and regulations for test and evaluation. They make required definitions and provide a taxonomy for test and evaluation process, as well as present Department of Defense (DoD) methodology including responsibilities of organizations and documentation for test and evaluation. The topic is also discussed in other research areas such as Systems of Systems (SoS) concept (Dahmann & Baldwin, 2008). SoS is a terminology where a system's definition is done by providing its sub-systems' definitions and their interactions. These definitions are rather abstract and conceptual, however, for test and evaluation SoS is quite helpful to breakdown a system's constituent elements.

In the application side, Fox et al. (2004) mention test and evaluation of aircraft and weapon systems while primarily focusing on cost aspect of the process. They also discuss trends in test and evaluation. Kass (2006) explains how military experiments should be designed and shows that robust experimentation methods from the sciences can be adapted and applied to military experimentation and will provide the foundation for continual advancement in military effectiveness. McShea (2010) also focuses on test and evaluation of aircraft avionics and weapon systems. He explains test and evaluation procedures for avionics and weapon systems in detail and relates these procedures with generic test and evaluation methodology.

Other than the organization of test and evaluation efforts, there are statistical and analytical aspects such as hypothesis testing, interval estimation of performance metrics, and decision modelling. National Defense Council (1998) in the US published a reference book on statistics, testing, and defense acquisition. Furthermore, Buede and Bresnick (1992) reviews decision analysis models that can be used for military decision making, including decision made in acquisition phase. This phase includes test and evaluation efforts. In their study, they conducted decision analysis for the concept selection phase of the systems acquisition process for US Marine Corps. In statistics context, Mathew and Webb (2005) used statistical test to investigate gun tube accuracy of an M1 Series tank. Their study extends ANOVA test in which a unified approach to solve problems is presented using the concepts of generalized p values and generalized confidence intervals. Greiner et al (2003) presents a hybrid decision support methodology for use in the screening of weapon systems development projects. In their methodology, an Analytical Hierarchy Process (AHP) with 0-1 integer portfolio optimization model is used and applied in Air Force projects.

As the review of literature reveals, there is no "cookbook" for test and evaluation methodology for weapon systems and therefore a generic approach to create a holistic solution for better decision-making lacks. This approach starts from conceptualizing acquisition problem on hand in order to reveal the needs of test and evaluation. Additionally, none of the guides touch on analytical or statistical methods for evaluating data obtained from the test phase. After all, this important application area needs to bring data analytics, methodology and experience from the field together to plan, build and implement a realistic test and evaluation for weapon systems. In this respect, the main aim of this chapter is to summarize concepts and provide the processes to implement tests and evaluation.

This chapter is organized as follows: In the next section test and evaluation methodology is presented including the two types; developmental and operational. The process for test and evaluation is also explained in this section which presents a hierarchical breakdown of evaluation criteria. In section 3, modelling and simulation is discussed from test and evaluation point of view. We sum up our discussion in the conclusion section.

TEST AND EVALUATION METHODOLOGY

The primary purpose of test and evaluation is to provide decision support to decision makers for verifying that system under evaluation meets capability requirements/mission needs and technical specifications, and operationally efficient in terms of suitability, survivability, compatibility, interoperability, reliability and maintainability. Considering test and evaluation of a domestic development weapon system, the process evolves from testing attainment of required system specifications and technical performance parameters to verifying if the system is operationally efficient for the intended use. In harmony with this evolvement, test and evaluation process can be partitioned into two main categories: DTE and OTE. Both planning and evaluation processes of DTE and OTE present some differences. Due to specific attributes of these two test and evaluation categories, we will discuss them respectively.

Types of Test and Evaluation

Developmental Test and Evaluation

DTE is conducted usually by the contractor to determine that the developed system meets technical specifications. Additionally, it is the primary means for demonstrating that the engineering design and development is complete (Claxton, Cavoli, & Johnson, 2005). Output of DTE helps developing agency to reduce risks while procuring country to verify that the system is ready for OTE. DTE is conducted iteratively as needed. Hence, all respective processes of designing, building and producing the system are kept under control while related risks are minimized. Consequently a DTE program aims to (Claxton, Cavoli, & Johnson, 2005);

- Verify that system meets criteria described in the mission need or capability requirements document.
- Verify that designed system is ready for the next development step.

- Show that system functions as intended.
- Assess reliability, compatibility, interoperability, maintainability.
- Provide data for cost.
- Show that system is ready for OTE.

DTE is performed in a controlled and repeatable environment. All factors having effect on the measure of interest are controlled and varied according to test plan. In this respect DTE evaluation is rather technical. Most of the time quantitative data is collected and statistical analysis is used for evaluation. For this reason, performance metrics should be measured meticulously. DTE usually contains instrumented tests, hence, measurements are performed by technicians and engineers accompanied by matter experts.

Operational Test and Evaluation

The main purpose of OTE is to measure and evaluate to what extent a weapon system accomplishes a mission. For instance, the primary focus of OTE is to assess capability of a frigate to conduct surface warfare, rather than measuring its surface to surface missile's probability of kill against a target type. Therefore, OTE tests are conducted under realistic conditions where various threats are present. Factors having effect on the performance of weapon system are not controlled intentionally. Regular user military personnel operates the systems and adequacy of organization, doctrine, tactics, training, and logistics are also assessed.

OTE explains system performance in terms of operational effectiveness and operational suitability. These terms can be explained as follow:

- **Operational Effectiveness:** The assessment of the ability of a weapon system to accomplish its intended mission under realistic conditions, where regular military personnel operates the systems and operation is conducted as defined in the doctrine.
- **Operational Suitability:** The assessment for the degree to which a weapon system can be satisfactorily used in the field considering reliability, maintainability, interoperability, compatibility, safety, supportability, availability, and transportability

Consequently, the main questions addressed in OTE can be visualized as in Figure 1.

DTE and OTE can be conducted as follow-on processes or outputs of DTE can be used as inputs in OTE. However, respectively or jointly performing these two processes may not be possible for weapon systems that are procured as COTS or NDI. Moreover, an OTE might be required for a weapon system that have been procured in the past in order to develop a new concept or doctrine. Consequently, considering the fact that most countries procure weapon systems as COTS and NDI, and new OTE requirements emerge for concept development processes, typically DTE and OTE are performed as separate processes. Having these in mind, a comparison of DTE and OTE is presented in Table 1.

Test and Evaluation Process

The success of a test and evaluation process depends heavily on a good planning. Test and evaluation should provide the required information, hence, should be structured to answer questions posed by the decision makers. A well-designed test and evaluation helps developing agencies to diagnose early defi-

Figure 1. The main questions addressed in OTE
Taken from Claxton et al. (2005).

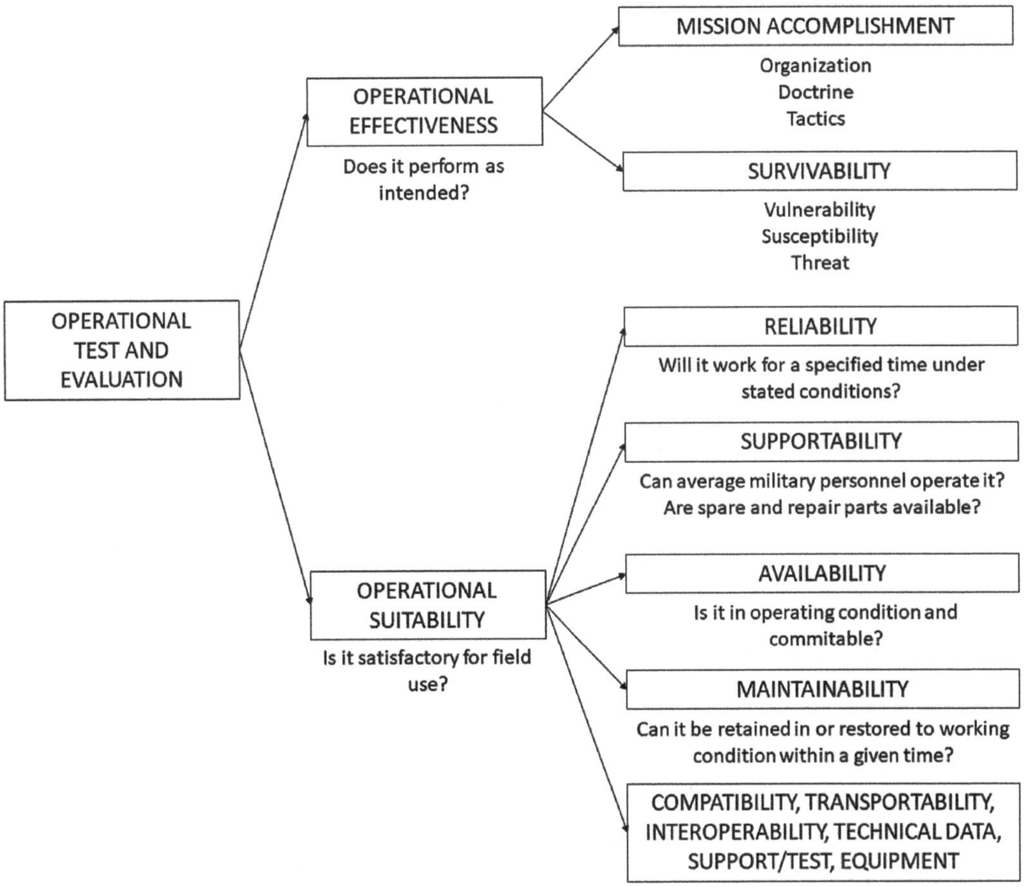

Table 1. Comparison of DTE and OTE

SCOPE	DTE	OTE
Coordinator	Contractor/Developing agency	User/procuring agency
Main focus	Meeting capability requirements	Mission accomplishment
Test type	One-on-one Instrumented	Many-on-many Operational scenario
Performing personnel	Technicians/engineers	Operational/military personnel
Environment	Controlled Fully test conditions	Uncontrolled Realistic conditions
Data	Mostly quantitative	Qualitative and quantitative
Analysis	Mostly statistical	Level of satisfaction Statistical

ciencies and improve product quality, while helping military agencies to make better decisions in terms of procuring the right weapon system or developing better concepts. Additionally, a test and evaluation process designed using statistical experimental design principles enables us to perform experimentation with fewer tests. In this respect, test and evaluation should be analytical in nature and should rely on quantitative data as much as possible.

Test and evaluation plan serves as a road map that shows capability requirements or mission needs, evaluation criteria, tests required for measuring how much the system meets these requirements, and the analysis plan. Hence, this plan translates a test concept and statistical/analytical test design into concrete resources, procedures and responsibilities (Claxton, Cavoli, & Johnson, 2005). Having these properties, a generic test and evaluation can be considered as a 6-step process as shown in Figure 2 (Claxton, Cavoli, & Johnson, 2005).

In Step 1 of the process, the requirement for the test and evaluation is identified. In this step, mission need and capability requirements to be met by the weapon system are clarified. In line with the stated mission need and capability requirements, critical issues are determined. Critical issues are the questions about the weapon system that we desire to answer. These issues can be classified as technical or operational issues. Technical issues are associated with technical or engineering characteristics of the system. These are generally considered in DTE. Conversely, Critical Operational Issues (COIs) address to effectiveness and suitability properties of the system. For instance, "Will the Close In Weapon System (CIWS) successfully defend the ship against guided missiles in the combat?" is an effectiveness COI, while "Will the CIWS operate compatibly with the track radar in the combat?" is an suitability COI.

Figure 2. Generic Test and Evaluation Process

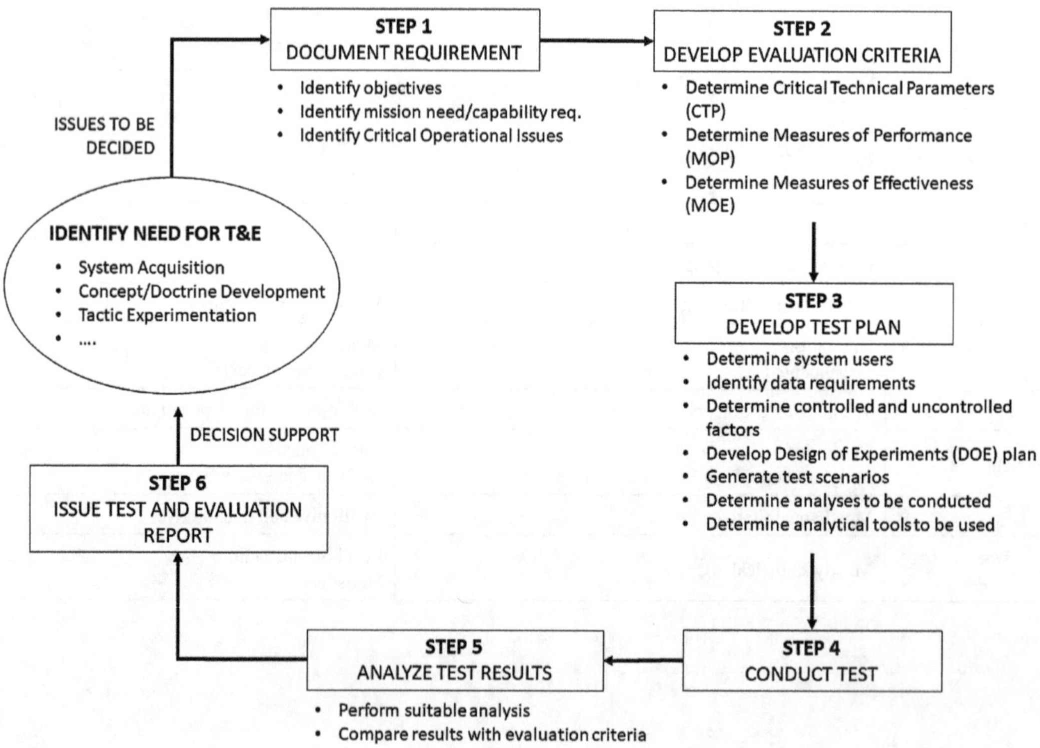

In Step 2, qualitative or quantitative evaluation criteria that will be used for answering the questions posed by the critical issue statements are determined. These criteria are the standards that we expect the weapon system to accomplish in terms of technical specifications, operational effectiveness and suitability. In this step the following are formulated:

- **Critical Technical Parameters (CTP):** These are engineering level parameters that the system under evaluation is required to meet or exceed. CTPs are derived from Capability Development Documents (CDD), critical system characteristics and system engineering documents (McShea, 2010).
- **Measure of Effectiveness (MOE):** MOE is a measure which expresses the extent to which a weapon system accomplishes the designated task. MOEs are designed to address both effectiveness and suitability issues. Suitability MOEs are sometimes called Measure of Suitability (MOS). MOS focuses on the ability to accomplish mission in terms of reliability, maintainability, interoperability, compatibility, safety, supportability, availability, and transportability. These measures drive the DOE and resource requirements, and type of analysis. Each MOE should be associated with a COI, however, a COI can be answered by more than one MOEs. A good MOE should be relevant, expressed clearly, quantifiably measurable (if possible), and mutually exclusive. Probability of survival or probability of kill can be given as examples for MOEs.
- **Measure of Performance (MOP):** MOP is a qualitative or quantitative measure that is used to assess a weapon system's performance for a particular function. MOPs are usually associated with physical performance of the system. In this respect, they are mostly defined quantitatively, such as, Mean Time Between Failure (MTBF), False Alarm Rate, Mean Detection Range, Miss Distance etc. Data requirements are determined particularly for each MOP. Moreover, specific tests are designed for collecting relevant data. Each MOP should be traced and provide input to a MOE.

Consequently, for developing evaluation criteria, COIs that address to mission statement or capability requirements are determined firstly. Secondly, MOEs that are required to answer COIs are formulated. At last, MOPs that are needed to provide input for MOEs are determined and associated to relevant MOEs. This hierarchical process of structuring evaluation criteria is shown in Figure 3 (McShea, 2010).

In Step 3, test plan is developed based on critical issues to be answered and evaluation criteria formulated. In this step data requirements are determined, firstly. In this framework, data sources, data type and format are documented. Factors that are expected to effect on the performance of the system are determined. Some of these factors are those that we would like to measure their effect on the system. Hence, in order to measure their effect, we replicate the test for different levels of these factors. Others can be included in the test as blocking factors. Based on data requirements, designated factors and required replications, experimental design is built, and test scenarios are generated. Test scenarios are detailed in terms of user personnel, run profiles and other scenario elements such as equipment, doctrine and environment. In line with the issues to be decided, required analyses and tools needed for the analysis are determined. Additionally, all responsibilities regarding data collection and format for data sheets should be decided.

Evaluation criteria, relationship between them and the tests to be used for evaluation can be organized in a matrix form. This matrix makes all these components visible and tractable. An example matrix is shown in Table 2.

Figure 3. Hierarchical Process of Structuring Evaluation Criteria

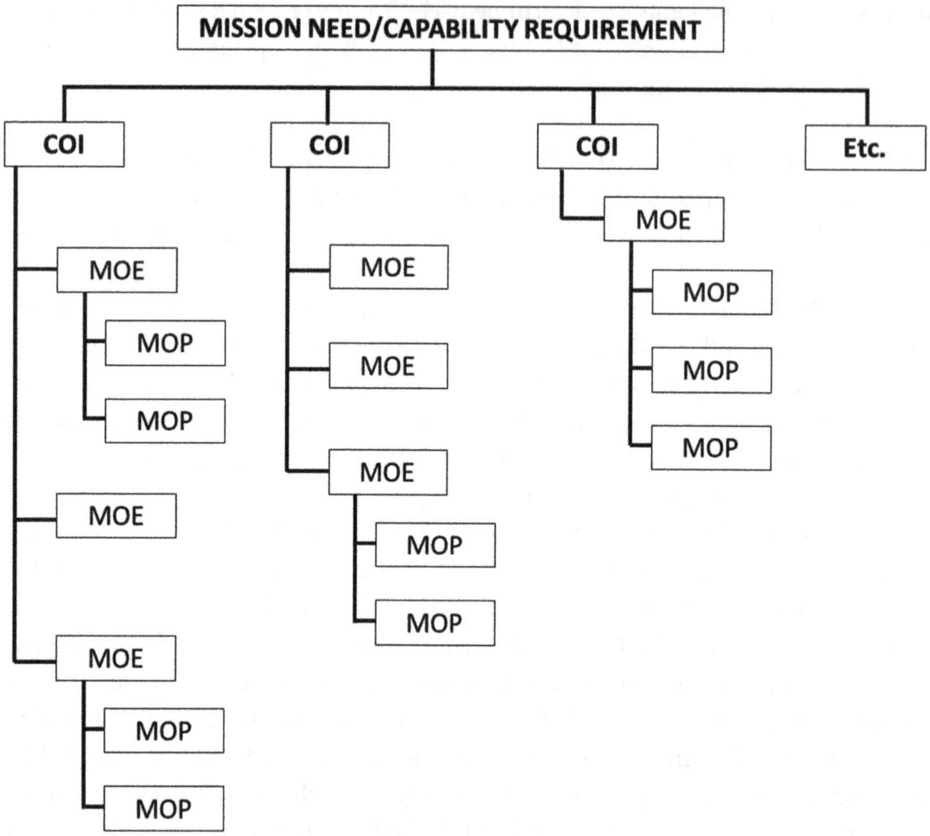

Table 2. An Example of Evaluation Criteria and Test Matrix

COI	MOE	MOP	Criteria	Test
Anti Air Warfare Performance	Defense capability	Probability of survival	More than 0.9	E-1
Reliability	Operating time	MTBF	More than 24 hours	E-2
Detection	Detection capability	Detection range	More than 200 miles	E-3
		Bearing accuracy	Less than 2 degrees	

Step 4 is the test implementation phase. All measures are taken for sensitive data collection. Tests are performed as directed in Step 3. Test runs that are conducted delinquently are replicated as needed in order to provide healthy data collection. At the end of the test execution, all data are screened for completeness, accuracy and validity before proceeding to the next step (Claxton, Cavoli, & Johnson, 2005).

In Step 5, all data gathered in Step 4 are analyzed for evaluation. Statistical methods are used extensively in this phase. Measured outcomes are compared with expected outcomes which are formulated as MOEs and MOPs. The most important issue in comparison is to determine if deviation of measured outcomes from their target values are due to randomness or an indication that target values are not attained. If there exist non-normal measurements, test conditions and procedures used for these measurements should be examined in detail.

Test and Evaluation Report is issued in Step 6. All test procedures implemented, and test results obtained are explained in detail so that the procedures and analyses could be repeatable and falsifiable. Even though the report should contain all fine-details regarding test conduct, results should be expressed in a way that decision makers could take away the message saliently that the analyses reveal.

MODELING AND SIMULATION IN TEST AND EVALUATION

Modelling and simulation has a special role in both DTE and OTE particularly when real-life tests are not possible due to high costs and difficulty in providing security. Running an operational or developmental test on computer is safer and cheaper than doing a test in real. Imagine a case where a CIWS is to be tested in an operational area. A significant amount of safety precautions must be taken in advance to avoid unintended consequences which in case of their happening will incur high costs. Simulation, as a computer-based method, can be used to execute tests safely and with relatively low cost. Although a simulation can never be as real as a real trial, to some extend simulation models can represent the reality and therefore can be used to supplement real tests. Level of representation is a matter of model accuracy and fidelity (Balci, 1994).

Review of M&S in Test and Evaluation

In the U.S., Department of Defense (1997) developed "Simulation, Test, and Evaluation Process (STEP)" concept which integrates simulation into the test and evaluation process. Although STEP was first introduced in late 1990s, it was a milestone for modelling and simulation since M&S was in its early stages for being used in acquisition process. M&S has changed the way test and evaluation is conducted, from "test-fix-test" to "model-test- model". This switch eventually has contributed to reduce time, resources, and risks associated with the test and evaluation process. STEP concept proposes the use of simulation at any level; engineering, engagement, mission, and campaign.

After the introduction of STEP, it evolved and is being accepted by the test and evaluation community. Meilunas et al (2007) reports STEPNET, a project which incorporates high performance computing and collaborative simulation to STEP. The idea behind STEPNET is also to make the historical, design, simulation, and test data available to users so that they can be utilized for test and evaluation with less effort.

Froggett and Liu (2001) describe how engineering and discrete event simulation models are used for the development of Command, Control, Communications, Computers, Intelligence, Surveillance, and Reconnaissance (C4ISR) and weapon systems. As for the test and evaluation, they utilized STEP approach to minimize cost and risk. Likewise, Lazić and Velašević (2004) used STEP for an automated target tracking radar system. They report a significant reduction in live experiments for testing the software for the device.

Other than STEP, simulation is positioned in test and evaluation efforts in other ways. Modelling and simulation is adopted by Cho et al (2007) for test and evaluation of weapon systems in Korea. They developed Real-Time Distributed Simulation Environment (RDSE) for testing and performance analysis of a surface-to-air missile system and reported its success.

Test and evaluation is not only needed for military but also for other domains such as transportation. Tiernan et al (2017) is a recent example of why we need test and evaluation procedures. For the adoption of autonomous vehicles, plans for track and road tests must be rigorously done for the safety, before actual tests. Tierman et al (2017) suggests a test and evaluation framework which also includes use of simulation.

Use of Simulation Models

Simulation models can be developed at different levels of detail (LoD) which is determined by the purpose or the intended use. In general, we can classify LoD in four levels;

- **Engineering Level:** On the lowest level of use, simulation models can be used in designing weapon systems. Generally, these models are CAD/CAM models which can not only evaluate physical attributes but can also be used to analyze dynamics of weapon systems. Some other simulation models can support such models by extending them to chemical and impact analysis.

This level of use in test and evaluation is perceived by weapon system manufacturers and therefore is associated with DTE. In the design phase of a system, simulation models support testing against design requirements.

- **Engagement Level:** On top of the engineering level use, there exist simulation models with higher level of detail. These models, however, benefits from engineering level models and therefore is an extension to engineering level. An engagement is a weapon's firing to a single target and therefore target's physical attributes are included in tests. This level of model also support DTE since a weapon system's design requirement can be tested under thread scenarios.
- **Mission Level:** On top of the engagement level, collection of engagements forms a mission and mission level models can simulate a system's performance under more complex operational environments. In this level, systems other than weapon systems are included in simulation models. Simulation models in this level can support OTE since weapon manufacturers are not interested in operational/platform level issues. Whole platform performance can be tested with models in this level.
- **Campaign Level:** This type of models are highest level models which includes elements of a force and environment. OTE is possible using campaign level models. A weapon system's performance can be evaluated under more realistic scenarios including weather and sea conditions, coordinated firing etc.

CONCLUSION

Decisions for selecting the right weapon system are inevitable difficult which requires elaborate evaluations. Developing or procuring a weapon system for a naval platform has different T&E procedures, mainly categorized as developmental and operational. The developmental T&E looks for a system which meets technical specifications determined by the procuring authority. This type of T&E is for demonstrating that the design is fit for the purpose and is ready for operational T&E. Operational T&E, on the other hand, is used to evaluate the level which the system accomplishes a mission specified by the potential users. The system's performance is explained as operational effectiveness and suitability. Main questions to be answered are "does the system perform as intended?" and "Is the system satisfactory for field use?".

T&E process should be planned and reply on quantitative techniques for creating experiments. The process starts with identifying the needs for T&E and documenting the system requirements. Evaluation criteria which includes critical technical parameters, measures of performance and effectiveness should be determined. Experimental design concepts help determine which tests should be conducted. By analyzing test results, issues and gaps are identified and, if needed, more tests are conducted. In this regard, the T&E process is a cyclic process which aims at testing all system requirements.

In this chapter, we discussed T&E principal concepts for weapon systems, and process of its implementation. Our discussion could be extended with statistical methods for evaluating test results.

REFERENCES

Balci, O. (1994). Principles of Simulation Model Validation, Verification, and Testing. Technical Report. Virginia Polytechnic Institute & State University.

Buede, D. M., & Bresnick, T. A. (1992). Applications of Decision Analysis to the Military Systems Acquisition Process. *Interfaces*, *22*(6), 110–125. doi:10.1287/inte.22.6.110

Cho, B., Kim, D. Y., Kim, S. H., & Youn, C. (2007). Real-Time Distributed Simulation Environment for Air Defense System Using A Software Framework. *Journal of Defense Modeling and Simulation.*, *4*(2), 64–79. doi:10.1177/154851290700400201

Claxton, J. D., Cavoli, C., & Johnson, C. (2005). *Test and evaluation management guide*. Defense Acquisition Univ Ft Belvoir VA.

Council, N. R. (1998). *Statistics, Testing, and Defense Acquisition: New Approaches and Methodological Improvements* (M. L. Cohen, J. E. Rolph, & D. L. Steffey, Eds.). Washington, DC: The National Academies Press.

Dahmann, J. S., & Baldwin, K. J. (2008). Understanding the Current State of US Defense Systems of Systems and the Implications for Systems Engineering. *2008 2nd Annual IEEE Systems Conference*, 1-7.

Fox, B., Boito, M., Graser, J. C., & Younossi, O. (2004). Test and evaluation trends and costs for aircraft and guided weapons (No. RAND/MG-109-AF). RAND Corp.

Froggett, S., & Liu, S. (2001). System Performance Modeling in C4ISR/Weapon System Design and Development. *Naval Engineers Journal, 113*(2), 55-64. DOI:10.1111/j.1559-3584.2001.tb00034.x

Greiner, M. A., Fowler, J. W., Shunk, D. L., Carlyle, W. M., & McNutt, R. T. (2003, May). A hybrid approach using the analytic hierarchy process and integer programming to screen weapon systems projects. *IEEE Transactions on Engineering Management*, *50*(2), 192–203. doi:10.1109/TEM.2003.810827

Kass, R. A. (2006). The logic of warfighting experiments. Assistant Secretary of Defense (C3I/Command Control Research Program). doi:10.21236/ADA457164

Lazić, Lj., & Velašević, D. (2004, December). Applying simulation and design of experiments to the embedded software testing process. Journal of Software: Testing. *Verification and Reliability.*, *4*(14), 257–282. doi:10.1002tvr.299

Mathew, T., & Webb, D. W. (2005). Generalized p Values and Confidence Intervals for Variance Components: Applications to Army Test and Evaluation. *Technometrics*, *47*(3), 312–322. doi:10.1198/004017005000000265

McShea, B. (2010). Test and Evaluation of Aircraft Avionics and Weapons Systems. *Sciences et Techniques (Paris)*.

Meilunas, Ng, Rajagopal, & Chaudhury. (2007). *Overview of the Simulation, Test and Evaluation Process Network (STEPNET) Program.* American Institute of Aeronautics and Astronautics (AIAA) Atmospheric Flight Mechanics Conference and Exhibit, Hilton Head, SC. doi:10.2514/6.2007-6572

Tiernan, T. A., Richardson, N., Azeredo, P., Najm, W. G., & Lochrane, T. (2017). *Test and Evaluation of Vehicle Platooning Proof-of-Concept Based on Cooperative Adaptive Cruise Control. Research Report. U.S. Department of Transportation John A.* Volpe National Transportation Systems Center Advanced Vehicle Technology Division.

US Department of Defense. (1997). *Simulation, Test, and Evaluation Process Step Guidelines*. Retrieved from http://www.dtic.mil/get-tr-doc/pdf?AD=ADA341111

Chapter 6
Optimization in Military Planning:
Resource Allocation Problems

Mehmet Gokhan Metin
Marmara University, Turkey

Serol Bulkan
Marmara University, Turkey

ABSTRACT

In recent years, advancing technology has introduced a wide range of resources for military forces. This rapid improvement in the number and types of military resources also led to difficulties in the management of resource allocation issues in combat missions. The general resource allocation problem can be defined as determining the optimal sequence of resource usage allocations while maximizing the kills on the enemy's forces or minimizing casualties. In this chapter, the authors investigate the resource allocation problems and solution approaches observed in planning military missions. Theoretical background information and some examples based on real-time data are provided to illustrate the chosen problem types.

INTRODUCTION

Today, the conflicts and the precariousness which the countries have been faced are more complex than compared to the past. Nowadays, asymmetrical threats and terrorist activities have become more an issue than the conventional wars. Therefore, the counter-measures against these situations are substantially different.

The technology advances quickly, and as a result, both the offensive and defensive systems have been evolved. With these rapid advances in military systems technologies, more capabilities are available to the armed forces and planners. All these capabilities and the various types of threats bring out many different types of problems as well. One of the most important problem types among those is the optimization and allocation of the force resources.

DOI: 10.4018/978-1-5225-5513-1.ch006

In this chapter we will focus on the general resource allocation problem and its applications in the militray context. For this reason we first provide a background on the studies related to resource allocation problems. Then, we discuss on specific types of problems observed in military organizations. Next, we discuss both analytic and heuristic methodologies developed to solve those problems, and finally conclude.

BACKGROUND

Force deployment and optimal resource allocation have been an area of considerable research interest in conventional warfare for a long time (Danshin, 1967; Shubik, 1983; Colegrave, & Hyde, 1993; Bracken, Kress, & Rosenthal, 1995; Cruz, Simaan, Gacic, Jiang, Leitellier, Li, & Liu, 2001; Krichman, Ghose, Speyer, & Shamma, 2001; Ghose, Speyer, & Shamma, 2002; Ghose, Krichman, Speyer, & Shamma, 2002).

Due to the complexities of the situations which the operation planners encounter, especially the human lives are in question, planning and optimizing the resources rapidly, in a cost-effective manner and with minimum mistake is vital. Because the slightest mistake that the planner might do may cause many lives in the operational theatre. Many real-world decision problems in the operational theatre exhibit the presence of different, contrary objectives in process of evaluating alternatives, as well as the necessity for making adjustments or trade-offs concerning the outcomes of changing the course of action (Karatas, 2017; Karatas, 2018; Karatas & Yakici, 2018; Razi & Karatas, 2016; Razi, Karatas, & Gunal, 2016). Further, most of these same resource allocation decision problems are encountered under inherent, concealed uncertainties.

In our era, with rapid and significant advances in communication and computation technologies, sophisticated decision-making in these situations has become feasible. And also these improvements have generated renewed interest in formulating decision-making problems in these areas and seeking optimal solutions. (Sheeba, & Ghose, 2005)

Resources are limited and time is short. The resources must be utilized in the most effective way in order to encounter the threats. To address these problems, Operations Research methods, a sub-level research topic of Industrial Engineering, has been used since World War II. Operations Research's foundations invented to address the strategical and tactical problems in air / naval defense and many different algorithms have been invented up today. Resource allocation and optimization problems are one of these research areas.

One of the main problems of combat planning is resource allocation problem, which defines the operations against enemy targets and aims to optimize the resource allocation efficiency and effectivity. Research on this problem types emerged with the adoption of scientific approach in the organization management field. While dealing strategic and tactical problems of military operations in World War II, need for academic study and scientific support raised and Operations Research (OR) activity officially began with created military OR teams. (Hillier, & Lieberman, 2004)

RESOURCE ALLOCATION PROBLEM

In general; resource allocation problems involve the process of assigning and managing valuable assets in a way that supports an organization's strategic missions. They manages the assignment of resources to the most promising solution among the competitive ones for minimizing total costs or maximizing total return. Resource allocation problems have the following elemental factors: a set of resources available in given amounts; a set of missions to be achieved, each mission requires a specified amount of resources; and a set of costs or returns for each mission and resource. The problem is to determine how much of each resource should be allocated to each mission for achieving desired goals (Ackoff, 2017).

In the military terms, resource allocation problems can be categorized as follows:

1. Weapon Mix Problem
2. Weapon Deployment Problem
3. Weapon Target Allocation Problem
4. Mission Allocation Problem
5. Sealift Problem (Jaiswal, 1997)

Weapon Mix Problem

This type of resource allocation problem involves with the identification of the convenient combination of weapons and forces in our possession to perform a successful military offensive mission. There are two main questions related to this topic: for example in naval warfare, the first question to be answered is what kind of naval platform are requested to achieve the mission, given the set of circumstances considered most probable and the second one would be, what would happened if the availability of these platform changed? (Brooks, Bankes, & Bennett, 1997)

Weapon Deployment Problem

With an assumption of a given weapon systems combination and candidate zones where these systems can be installed, seeking a weapon deployment plan to maximize the destruction on a given set of targets (Jaiswal, 1997).

Weapon Target Allocation Problem

Given weapon systems combination installed at specified zones, determine the allocation of weapons to each target to expose desired damage (Jaiswal, 1997).

Mission Allocation Problem

Given the number of submarine attack missions starting from specified naval bases (or some pre-specified initial coordinates) and targets of different types of enemy ships (or submarines, land targets etc.), how many attack missions from each naval base should be allocated to each target in an naval attack mission to maximize the total value of targets killed or minimize the total value of surviving assets? This problem is similar to the weapon target allocation problem discussed above (Jaiswal, 1997).

Sealift Problem

The military sealift problem deals with effectively routing a fleet of naval logistics ships (or class of LHD, LPD etc.) to deliver group of companies, materials and goods (troops, armored vehicles, aircrafts, food and other forms of freight) from different starting points to different landing zones as quickly as possible under a variety of constraints. (Wu, Powell, & Whisman, 2008)

ANALYTIC OPTIMIZATION MODELS

Description

Many different optimization models are available. A model to solve resource allocation problems can be considered one of them. Resource allocation models deals with providing an optimization in usage the resources to achieve a defined target subject to different constraints. Examples of these models are the ones for solving problems such as weapon deployment problem are mentioned in above section. In addition to these examples, transportation and assignment problems models are also considered as resource allocation models.

The above models can be worked out by optimization techniques. Some of these techniques are clarified below.

Linear Programming

A model, which is used for optimum allocation of scarce or limited resources to competing products or activities under such assumptions as certainty, linearity, fixed technology, and constant profit per unit, is linear programming (Murthy, 2005).

Linear programming is one of the optimization models which is applied in solving optimization problems where the objective function and the constraints are both linear functions.

Linear programming methodology is widely used in operations research. Elements of linear programming are given below:

1. **Linear Objective Function:** Value to be optimized is called the objective function. This function should be represented as a linear equation, such as:

$$Maximize: c_1x_1 + c_2x_2$$

2. **Constraints:** Linear inequalities of the problem domain that bounds the solution space are called constraints. An optimum solution that meets these constraints' requirements is searched by the objective function. Each constraint should be represented as a linear equation, such as:

$$a_{1,1}x_1 + a_{1,2}x_2 \leq b_1$$

$$a_{2,1}x_1 + a_{2,2}x_2 \leq b_2$$

$$a_{3,1}x_1 + a_{3,2}x_2 \leq b_3$$

3. **Decision Variables:** Both objective function and constraints are based on decision variables x_i whose optimal values are searched with simplex methods in linear programming.

The application of linear programming has greatly increased following the development of the simplex method. The simplex algorithm uses only function value information. One of the advantages among many others, the simplex algorithm requires a small number of calculations for initializing the algorithm.

Let x_i ($j=1, 2, ..., n$) be the number of units (frigates, attack boats, submarines, air defense guns, missiles, torpedoes etc.) of jth type deployed to achieve a mission and c_j ($j=1, 2, ..., n$) be the performance measure coefficient of the one deployed unit of jth type (expressed in terms of damage or value of surviving assets). Let $c_j x_j$ be the outcome from the deployment of x_j units. It must be noted that in military situations, the outcome does not generally increase linearly with x_j. But in non-military situations, this counts because the outcome is generally expressed in terms of profit or cost which is assumed to increase linearly.

Let's assume that the outcome linearly additive. So that total outcome z from the allocation of x_j units can be expressed as:

$$z = \sum_{j=1}^{n} c_j x_j$$

We may like to maximize or minimize z subject to certain constraints such that limited amount of manpower, budget and number of (or capacity of) weapons and due to the operational restrictions such as weather conditions. If a_{ij} represents the ith type of resource required by the jth type of unit in its deployment, then the total resource of ith type required for the allocation is:

$$\sum_{j=1}^{n} a_{ij} x_j$$

This should not exceed b_i, the total resource of the ith type available, so:

$$\sum_{j=1}^{n} a_{ij} x_j \leq b_i, i = 1,2,..,m$$

Finally, the problem can be stated as:

$$Maximize\, z = \sum_{j=1}^{n} c_j x_j \tag{1}$$

$$subject\, to \sum_{j=1}^{n} a_{ij} x_j \leq b_i \left(i = 1,2,..,m\right) \tag{2}$$

$$x_j \geq 0 \left(j = 1,2,..,n \right) \tag{3}$$

To express the problem in matrix form, 2 vectors should be defined as follows:

$$\boldsymbol{C} = \left(c_1, c_2, .., c_n \right)$$

$$\boldsymbol{X} = \left(x_1, x_2, .., x_n \right)^T$$

$$\boldsymbol{b} = \left(b_1, b_2, .., b_m \right)^T$$

and

$$\boldsymbol{A} = \begin{bmatrix} a_{11} & a_{12} & \dots a_{1n} \\ a_{21} & a_{22} & \dots a_{2n} \\ \vdots & \vdots & \vdots \\ a_{m1} & a_{m2} & \dots a_{mn} \end{bmatrix} \tag{4}$$

Now, the above formulation can be expressed as:

$$Maximize\, z = \boldsymbol{CX} \tag{5}$$

$$subject\ to\ \boldsymbol{AX} \leq \boldsymbol{b} \tag{6}$$

$$\boldsymbol{X} \geq 0 \tag{7}$$

It is also possible that instead of maximization in Equation (5) we can have a minimization problem (Jaiswal, 1997).

The Weapon-Mix Problem

Consider an important naval base (NB) whose protection by F type of air defense ship is being planned. It is assumed that not more than one air defense ship will be used against an aircraft. The problem is to determine a mix of ships to maximize the average number of kills of the enemy aircraft approaching the NB in a single formation. Let us define the following:

k_f = Probability of killing an enemy aircraft by an air defense ship of type f ($f = 1,2,..,F$)
c_f = Cost of deploying an air defense ship of type f ($f = 1,2,..,F$)
C = Total budget for deploying the weapons at NB
m_f = Number of crew required to operate an air defense ship of type f ($f = 1,2,..,F$)
M = Total number of crew available at the NB assuming that each crew member can operate each type of ship
N_f = Total number of air defense ship of f type available ($f = 1,2,..,F$)

Let x_f be the number of air defense ships deployed at the naval base NB. Then the average number of aircrafts those are shot down by the fth type of air defense ship is $k_f x_f$. If the average number of aircrafts killed by each f type air defense ship can be added then the total number of eliminated aircrafts is calculated by:

$$z = \sum_{f=1}^{F} k_f x_f$$

and this value should be maximized. Hence the problem can be expressed as:

$$Maximize\ z = \sum_{f=1}^{F} k_f x_f \tag{8}$$

$$subject\ to \sum_{f=1}^{F} c_f x_f \leq C\left(Budget\ constraint\right) \tag{9}$$

$$\sum_{f=1}^{F} m_f x_f \leq M\left(Manpower\ constraint\right) \tag{10}$$

$$x_f \leq N_f\left(Air\ defense\ ship\ availability\ constraint\right) \tag{11}$$

$$x_f \geq 0\left(Non-negativity\ constraint\right) \tag{12}$$

The value of x_f can be determined which maximizes the value of z subject to constraints (9) to (12) (Jaiswal, 1997).

Example 1: A navy air defense ship commander has been assigned the task of defending a valuable naval logistics ship from enemy air attack. It has three types of air defense systems; 9 long range ground-to-air missiles of type I, 6 short range ground-to-air missiles of type II and 2 close range Close-In Weapon System (CIWS) of type III are available for deployment. Each type I long range missile costs 8.5 units for deployment, each type II short range missile costs 7 units for deployment and similarly each type III CIWS costs 5 units for deployment. The total budget available is 60 units. Each type I long range missile requires 6 persons for handling and each type II short range missile requires 4 persons for handling whereas each type III CIWS can be handled by 2 persons. There are only 35 trained persons to handle the weapon systems on board. If the naval logistics ship is not defended, the enemy aircraft are estimated to destroy 95 percent of the asset value. If one type I long range missile is deployed from the air defense ship, it is expected to save 13 percent of the asset value. Similarly, deployment of type II short range missile is estimated to save 10 percent of the asset value whereas deployment of type III CIWS is estimated to save 5 percent of the asset value. In other words, the enemy aircraft which were earlier capable of destroying 95 percent of the asset value are able to destroy 82 percent of the asset value in presence of one long range missile of type I, 85 percent of the asset value in presence of one short range missile of type II and 90 percent of asset value in presence of CIWS of type III. Determine the mix of weapon systems (missiles and CIWS) that provides maximum protection to the asset against an attack of enemy aircraft aiming simultaneously.

Let x_1, x_2 and x_3 denote the number of weapon systems of type I, type II and type III respectively chosen by the air defense ship's commander to defend the asset. Then the objective function becomes:

$$Maximize\, z = 13x_1 + 10x_2 + 5x_3$$

subject to

$$8.5x_1 + 7x_2 + 5x_3 \le 60\left(Budget\ constraint\right)$$

$$6x_1 + 4x_2 + 2x_3 \le 35\left(Manpower\ constraint\right)$$

$$x_1 \le 9\left(Weapon\ availability\ constraint\right)$$

$$x_2 \le 6\left(Weapon\ availability\ constraint\right)$$

$$x_3 \le 2\left(Weapon\ availability\ constraint\right)$$

$$x_1 \ge 0, x_2 \ge 0, x_3 \ge 0\left(Non-negativity\ constraint\right)$$

The solution is $x_1 = 1.46, x_2 = 6, x_3 = 1.12$ and $z = 84.58$. Thus the navy air defense ship's commander should use a mix of 1.46 missiles of type I, 6 missiles of type II and 1.12 of CIWS of type III to provide maximum protection to the naval logistics ship against the enemy attack.

It may be noted that the solution obtained here by linear programming may lead to fractional values of the decision variables, for example, $x_1 = 1.46$ or $x_3 = 1.12$ in the above example. In case the decision variable can only take an integer value, the results may be truncated or round off the value. However, this may lead to a suboptimal solution requiring the use of integer programming technique that will be discussed briefly later in this chapter (Jaiswal, 1997).

The Weapon Deployment Problem

Let D types of all weapon systems be available for deployment at S navy ships in a tactical mission group. Let the total budget for installing the weapon systems be C and let M_d be the total manpower available to handle d^{th} ($d = 1,2,...,D$) type of weapon systems. Further, let N_d be the maximum number of the weapon systems of type d that can be deployed. The problem is to determine the deployment of appropriate weapon systems combination at each of the navy ships ($s = 1,2, ...,S$) to maximize the average number of kills of the enemy's ships. Let us define the following:

x_{ds} = Number of weapon systems of type d ($d = 1,2,..,D$) deployed at the navy ship s ($s = 1,2,..,S$)
k_{ds} = Probability of killing an enemy navy ship by a weapon system of type d ($d = 1,2,..,D$) which is deployed from one of our navy ship of type s ($s = 1,2,..,S$)
m_{ds} = Number of crew required to operate a weapon system of type d ($d = 1,2,..,D$) at the navy ship of type s ($s = 1,2,..,S$)
c_d = Cost of deploying a weapon system of type d ($d = 1,2,..,D$)
C = Total budget for deploying all the weapon systems available in the tactical group
M_d = Total number of crew available to operate the weapon system of type d ($d = 1,2,..,D$)
N_d = Total number of the weapon systems of type d available

Then the problem can be expressed as:

$$Maximize\ z = \sum_{d=1}^{D}\sum_{s=1}^{S} k_{ds} x_{ds} \tag{13}$$

$$subject\ to \sum_{d=1}^{D}\sum_{s=1}^{S} c_d x_{ds} \leq C \left(Budget\ constraint\right) \tag{14}$$

$$\sum_{s=1}^{S} m_{ds} x_{ds} \leq M_d \left(Manpower\ constraint\right) \tag{15}$$

$$\sum_{s=1}^{S} x_{ds} \leq N_d \left(Weapon\ system\ availability\ constraint\right) \tag{16}$$

$$x_{ds} \geq 0 \left(Non - negativity\ constraint\right) \tag{17}$$

The value of x_{ds} can be determined which maximizes the value of z subject to constraints (14) to (17) (Jaiswal, 1997).

Example 2: A navy commander has been assigned the task of defending two valuable naval logistics support centers from enemy air attack. It has two types of air defense missiles and one type of automatic air defense gun; 5 missiles of type I, 7 missiles of type II and 4 automatic air defense guns are available for deployment. Each type I missile costs 7 units for installation and each type II missile costs 8.5 units for installation. Similarly, one automatic air defense gun costs 3 units for installation. The total budget available is 90 units. Each type I missile requires 6 persons for handling and each type II missile requires 4 persons for handling whereas each automatic air defense gun can be handled by 2 persons. There are only 35 trained persons to handle the weapon systems at both sites. If these logistics centers are not defended, the enemy aircraft are estimated to destroy 95 percent of the assets' value. If one type I missile is deployed on one of the sites, it is expected to save 16 percent of the asset value and deployment of one type II missile is estimated to save 10 percent of the asset value. Similarly, deployment of one automatic air defense gun is estimated to save 4 percent of the asset value. In other words, the enemy aircraft which were earlier capable of destroying 95 percent of the asset value are able to destroy 79 percent of the asset value in presence of one missile of type I, 85 percent of the asset value in presence of one missile of type II and 91 percent of asset value in presence of one automatic air defense gun.

Let us assume that the area for deployment of each of the air defense weapon systems is same. Also, the total area available for deployment at the two naval logistics centers I and 2 is in the ratio of 70:50. The combat management and communication systems can either support a maximum of 3 missiles of type I, 5 missiles of type II and 7 automatic air defense guns at site I. The corresponding numbers of weapon systems at site 2 are 4 of type I, 6 of type II or 8 of automatic air defense guns.

Determine the mix of air defense weapon systems that provide maximum protection to both naval logistics centers against an attack of enemy aircraft aiming simultaneously.

Let x_{ij} denote the number of weapon systems of type i ($i = 1,2,3$) at the naval logistics center j ($j = 1,2$). The problem can be stated as follows:

$$Maximize\ z = 16x_{11} + 16x_{12} + 10x_{21} + 10x_{22} + 4x_{31} + 4x_{32}$$

subject to

$$x_{11} + x_{12} \leq 5 \left(Weapon\ availability\ constraint\right)$$

$$x_{21} + x_{22} \leq 7 \left(Weapon\, availability\, constraint \right)$$

$$x_{31} + x_{32} \leq 11 \left(Weapon\, availability\, constraint \right)$$

$$7\left(x_{11} + x_{12}\right) + 8.5\left(x_{21} + x_{22}\right) + 3\left(x_{31} + x_{32}\right) \leq 90 \left(Budget\, constraint \right)$$

$$6\left(x_{11} + x_{12}\right) + 4\left(x_{21} + x_{22}\right) + 2\left(x_{31} + x_{32}\right) \leq 35 \left(Manpowerconstraint \right)$$

$$\frac{x_{11} + x_{21} + x_{31}}{x_{12} + x_{22} + x_{32}} \leq \frac{60}{40} \left(Area\, constraint \right)$$

$$x_{11} \leq 3 \left(System\, support\, constraint \right)$$

$$x_{21} \leq 5 \left(System\, support\, constraint \right)$$

$$x_{31} \leq 7 \left(System\, support\, constraint \right)$$

$$x_{12} \leq 4 \left(System\, support\, constraint \right)$$

$$x_{22} \leq 6 \left(System\, support\, constraint \right)$$

$$x_{32} \leq 8 \left(System\, support\, constraint \right)$$

$$x_{11} \geq 0, x_{21} \geq 0, x_{31} \geq 0, x_{12} \geq 0, x_{22} \geq 0, x_{32} \geq 0 \left(Non - negativity\, constraint \right)$$

The solution of the problem becomes:

$x_{11} = 3, x_{21} = 2, x_{31} = 0.75, x_{12} = 0.50, x_{22} = 0, x_{32} = 0, z = 92.50$

Thus by deploying 3 missiles of type I, 2 missiles of type II, 0.75 unit of automatic air defense guns on naval logistics support center 1 and 0.50 missiles of type I, any missiles of type II and any unit of automatic air defense guns on site naval logistics support center 2, 92.50% of the value of the assets can be saved. This is an optimal deployment in the sense that no other deployment plan satisfying the constraints will provide better overall protection of the two sites (Jaiswal, 1997).

The Weapon Target Allocation Problem

If the enemy's assault tactics are known in prior, then the problem becomes to specify the allocation of the air defense systems in our arsenal to confront the enemy attack for maximizing the overall protection of our sites. If x_{dsa} represents the dth ($d = 1,2,..,D$) type of the air defense system installed at the site s ($s = 1,2,..,S$) against the ath ($a = 1,2,..,A$) type of the enemy's assault weapon, the problem is to determine x_{dsa}'s value which maximizes the total protection of our sites or the kill potential against the enemy assets subject to the constraints. The formulation in example 1 and 2 can be extended to this case and is therefore not given here. The objective function in this case, if formulated on a more realistic basis, will be nonlinear and the algorithm will be discussed briefly later on this chapter (Jaiswal, 1997).

The Mission Allocation Problem

Let us next consider the problem of mission allocation of group of submarines of type i ($i= 1,2,..,A$) required to attack an enemy tactical mission group of naval ships (including aircraft carriers, missile frigates etc.), assuming t_j targets of type j ($j = 1,2,..,T$). We assume either different group of submarines leave from the same naval base or if the same group of submarines leaves from the different naval bases which are labeled distinctly. Let us define the following:

v_j = The assessed value of the target of type j
p_{ij}= Probability of killing one target of type j by a submarine of type i
A_i = Group of submarines of type i available ($i =1,2, ...,A$).

Let us assume x_{ij} as the number of attacks of type i allocated to attack targets of type j. The average number of targets of type j killed by all attacks is given by:

$$y_j = \sum_{i=1}^{A} e_{ij} x_{ij} \tag{18}$$

where $e_{ij} = p_{ij}t_j$, the average number of targets of type j killed by a attack of submarine of type i. Then if the objective is to maximize the average value of the targets killed, we can state the problem mathematically as:

$$Maximize \sum_{j=1}^{T} v_j y_j = \sum_{j=1}^{T} v_j \sum_{i=1}^{A} e_{ij} x_{ij} \tag{19}$$

$$subject\ to \sum_{j=1}^{T} x_{ij} \leq A_i \ for\ i = 1,2,..,A \tag{20}$$

$$x_{ij} \geq 0 \tag{21}$$

This model was not found suitable since it assigns the submarine attack of type i to target of type j if j maximizes $v_i e_{ij}$. Hence attacks of all types of submarines will tend to be allocated to the same target. This may be true in a target-rich environment but need not be necessarily true in general. Hence the above formulation was modified by adding the constraint $y_j \leq t_j$, the number of targets of type j. This constraint ensures that the number of jth type of targets killed is less than or equal to the number of jth type of targets available. This formulation can be generalized by additional constraints such as weather conditions, weapons availability etc (Jaiswal, 1997).

Sealift Problem

The Navy possessing a limited number of logistics transport ships has been assigned the task of sealifting troops, some material and delivering it in the forward landing area. The forward landing area can be divided into small zones such that supply at a single point will provide all the requirements in that zone. Each logistics transport ship has a limited carrying capacity. The various transport missions are affected by the prevailing weather condition. An estimate of the likely good, partially good and bad weather days in a month from the meteorological department is available. The Navy has also identified the number of cruises of each type that can take-off from each naval base in good, partially good and bad weather days. The objective of the problem is to determine an optimal plan for the sealift operation corresponding to the demand at various zones at minimum cost.

This problem can be formulated as an LP problem. Let i ($i = 1,2,..,b$) be the index on naval bases, j ($j = 1,2,..,m$) on landing zones and k ($k = 1,2,..,n$) on logistics transport ships. Let D_j denote the tonnage required by landing zone j in a month and T_k be the carrying capacity of kth logistics transport ship. Let α_i and β_i be the good and partially good weather days in a month at naval base i and let m_{ik} and n_{ik} be the number of cruises possible per day on good and partially good weather days by logistics transport ship of type k from base i. For bad weather days, it is assumed that no mission is possible.

Let x_{ijk} be the decision variable denoting the number of cruises by k^{th} logistics transport ship from naval base i to landing zone j during a month and c_{ijk} be the corresponding operating cost. Then the objective function can be stated as:

$$Minimize \sum_i \sum_j \sum_k c_{ijk} x_{ijk}$$

The allocation should meet the demand at the zones and as such:

$$\sum_i \sum_k x_{ijk} T_k \geq D_j$$

for each $j = 1,2,..,m$. We can express the maximum possible load S_{ik} that can be transported from the naval base i using logistics transport ship k on a monthly basis as:

$$S_{ik} = \left(\alpha_i m_{ik} + \beta_i n_{ik}\right) T_k \tag{22}$$

The problem can then be stated as:

$$Minimize \sum_i \sum_j \sum_k c_{ijk} x_{ijk} \tag{23}$$

where Q_i represents the quantity available for transportation on the naval base *i*. Inequality (25) ensures that quantity transported is bounded above by the net carrying capacity of the cruises in the prevailing weather condition. Inequality (26) ensures that quantity transported cannot exceed supply level at a particular base (Jaiswal, 1997).

Example 3: Consider a situation in which two types of logistics transport ships have to be used to transport material from 2 naval logistics bases to 3 overseas destinations. The demand at each overseas destination D_j ($j = 1,2,3$) is $D_1 = 132$, $D_2 = 138$ and $D_3 = 144$ units. The number of good weather days and the partially good weather days (in the month under consideration) at the ith naval base ($i = 1,2$) are given by:

$$\text{Base1}: \alpha_1 = 23, \beta_1 = 6$$

$$\text{Base2}: \alpha_2 = 17, \beta_2 = 8$$

Corresponding to this, the number of cruises possible per day by each logistics transport ship from naval base *i* ($i = 1,2$) to various landing zones is given in Table 1.

The first logistics transport ship has a carrying capacity of $T_1 = 7$ units per cruise and the second logistics transport ship $T_2 = 5$ units per cruise. The costs of executing each cruise from different naval bases to different landing zones by the two logistics transport ships are given as follows:

$c_{111} = 3.12$ $c_{211} = 0.67$

$c_{112} = 2.55$ $c_{212} = 0.77$

Table 1. Number of cruises per day by each logistics transport ship from a naval base

In Partially Good Weather Days (n_{ik})	In Good Weather Days (m_{ik})	
	Naval Base 1 ($i = 1$)	
1	3	Transport Ship type 1 ($k = 1$)
1	2	Transport Ship type 2 ($k = 2$)
	Naval Base 2 ($i = 2$)	
1	2	Transport Ship type 1 ($k = 1$)
1	4	Transport Ship type 2 ($k = 2$)

$c_{121} = 0.72$ $c_{221} = 1.82$

$c_{122} = 1.02$ $c_{222} = 2.37$

$c_{131} = 0.70$ $c_{231} = 1.85$

$c_{132} = 1.04$ $c_{232} = 1.92$

Determine an optimal plan for sealifting the material to meet the demand at minimum cost when the supply levels at the naval bases 1 and 2 are given by 250 and 300 units respectively.

From Equation (22), we get

$S_{11} = (23\ x\ 3 + 6\ x\ 1)\ x\ 7 = 525$ units

$S_{12} = (23\ x\ 2 + 6\ x\ 1)\ x\ 5 = 260$ units

$S_{21} = (17\ x\ 2 + 8\ x\ 1)\ x\ 7 = 294$ units

$S_{22} = (17\ x\ 4 + 8\ x\ 1)\ x\ 5 = 380$ units.

The problem can then be stated as:

Also,

$$\sum_i Q_i = 550, \sum_j D_j = 414$$

so that

$$\sum_i Q_i \geq \sum_j D_j$$

i.e., the total amount that can be carried by logistics transport ships from all the naval bases is more than the total demand at overseas destinations as it should be. If this condition is not satisfied, the optimal solution does not exist since the constraint defined by the inequality (24) is not satisfied. The solution is given by $x_{111} = 0$, $x_{112} = 0$, $x_{121} = 15.143$, $x_{122} = 0$, $x_{131} = 20.5714$, $x_{132} = 0$, $x_{211} = 18.8571$, $x_{212} = 0$, $x_{221} = 4.5714$, $x_{222} = 0$, $x_{231} = 0$, $x_{232} = 0$ with a cost of 46.2571 units.

Integer Programming

Linear programming may lead the solutions to have some fractional parts. In the real combat cases, generally, this situations cannot provide appropriate solutions to the problems. For example, the mission planner may have a solution to deploy 4.11 torpedoes for protecting a valuable asset given by using linear programming. Because the torpedoes cannot be divided, the mission planner must plan to deploy either 4 or 5 torpedoes. This allocation plan, however, may not be feasible or optimal. To overcome such difficulties, a scientific approach called integer programming has been invented. There are several algorithms in integer programming to solve such problems. Cutting plane or branch and bound algorithms are two of these methods.

Multi-Objective Programming

Linear or integer programming methods have a single objective function. However, there may be more than one goals for mission planners to optimize. For example, in weapon target allocation problem the planner must take into account both to maximize the total destruction of the enemy's assets and to minimize our total losses while achieving the maximization goal. In multi-objective programming, there are more than one overlapping objectives by which improving one objective will decrease the other goal. Thus the planner must make trade-offs between the solutions. There is an assumption that none of the single solutions will optimize all the objectives at the same time. If the objective functions are linear, the problem reduced to become a multi-objective linear programming problem or a weighted linear problem.

Dynamic Programming

Dynamic programming is a method for efficiently solving a broad range of search and optimization problems based on overlapping subproblems and optimal substructure.

A very important feature of the dynamic programming is the principle of optimality.The definition of this principle is as follows: an optimal policy is defined at the initial state and decision, after that all the remaining decisions must be aligned with this optimal policy. The local solutions which cannot accommodate this optimal policy should finally be abandoned. The local solutions are selected in accordance with the optimal policy to ensure that the global solution is optimal in every step. This explanation indicates that the whole decision process refers to a partitioning into distinct stages of decisions.

Nonlinear Programming

If the objective function or any one of the constraints is nonlinear then the mathematical model is called nonlinear programming model. There are three possibilities that may occur in nonlinear programming model:

1. The objective function is linear while some or all of the constraints are nonlinear.
2. The objective function is nonlinear while all the constraints are linear.
3. The objective function is nonlinear and some or all of the constraints are nonlinear.

Nonlinear programming models deal with continuous variables as in linear programming models.

HEURISTIC OPTIMIZATION MODELS

Description

There is a series of mixed problems that are of high complexity and require huge amounts of computational power to solve. The exact solution may be found by combining the solution techniques which are mentioned above. If the problem is not suitable for these optimization models because of the high computational time demand, the solution can be obtained by sacrificing the optimality, in other words, the solution may be a sub-optimal one. These methods are called heuristics models. Some of the important techniques are mentioned below.

Greedy Algorithms

A greedy algorithm is an iterative process that chooses the current most promising solution at each step for detailed, multi-staged problems.

The algorithm is called greedy because the process provides an optimal solution to each smaller phase but does not appraise the complete problem as a whole. Once a choice is made, the algorithm never reconsider it. The benefit of using a greedy algorithm is that solutions to smaller pieces of the problem can be apparent and easy to apprehend. The disadvantage is that it is likely to be that the most optimal short-term solutions may lead to the worst possible long-term outcome.

In comparison with modeling techniques like dynamic programming which are discussed in section "Optimization Models", a greedy algorithm may fail to find the global optima, but it usually finds an appropriate optimal solution. It also performs much faster than an algorithm which uses dynamic programming technique.

Simulated Annealing

Simulated annealing is an algorithm which is used to reach a global optimum of a multi-dimensional function despite the presence of several local maxima and minima. It is promising and advantageous because unlike conventional optimization techniques like random walk or hill climbing, this algorithm does not get stuck at a local optimum. It has the capability to roam through an array of local optima in the investigation of a globally optimal solution and to comprehend the reach of the global extreme point.

Genetic Algorithms

A genetic algorithm is used for search and optimization problems as an evolutionary computational method. It is widely used and very powerful technology for a variety of applications. Its basic foundations are based on the principle of genetics and natural selection. Genetic algorithm does not produce a solution based on the problem but proposes a generic formulation for all kind of problems by investigating the solution space with defined pertinence calculation. In every calculation, different solutions may emerge because of randomness and statistics. Therefore, more than one execution can be performed for converging the optimal solution.

Artificial Neural Networks

Artificial Neural Networks (ANN) are computing systems which are designed to imitate the human brain activities in a mathematical modeling way for analyzing and processing information. ANN form a foundation for Artificial Intelligence (AI) and self-learning. This modeling technique allows us to solve problems which are hardly possible for human or basic statistical standards. ANN provides a scientific model to systematize the process and it has self-learning capabilities which provide it to create better results as additional data becomes available.

Tabu Search

Tabu Search is a meta-heuristic that is invented to govern a confined heuristic search algorithm for inspecting the solution space away from local optimality. One of the distinctive features of the tabu search is the utilization of adaptive memory that a more amenable search behavior is created.

In this method, the search space that is created by all feasible solutions is analyzed moderately by a series of moves. At each repetition of search, the most favorable move is executed among all available ones. There is also a function which helps the algorithm to avoid from local minima (Brandsma & Griffioen-Young, 2006).

CONCLUSION

Resource Allocation is an inseparable part of the operational planning. Its intention is the verify, maintain and promote the introduced operational actions in the real-world war theatres. Even if the planning and allocating the resources in advance is important, these efforts are seldomly realized as they are planned. Because the nature of military operations is highly complicated and planning such complexities is not sufficient to guarantee the victory. It is, however, a necessary condition to develop a carefully analyzed and robust plan to achieve accomplished operation.

The variation of the current conditions in a war zone is very high, so the fast development and verification of an alternative resource optimizing plan are crucial. Rapid advances in the Artificial Intelligence area become more an issue herein. With their increasing computational powers, the automated intelligent planning systems will reduce the human interference at the planning phase of the military operations in the future.

Despite the fast rise of the computerized systems, still, the human is the most important factor in planning the operational theatre. Developing fast and accurate algorithms can only be achieved by the human mind. At this point, advances in the fuzzy logic and evolutionary solution techniques may cover the uncertainties of the different inputs to the operation planners. Employing heuristics, metaheuristics and exact solutions based on fuzzy logic and memetic algorithms will produce better results in the military resource allocation optimization area.

REFERENCES

Ackoff, R. L. (n.d.). *Resource Allocation.* Retrieved August 2017, from https://www.britannica.com/topic/operations-research/Resource-allocation

Bracken, J., Kress, M., & Rosenthal, R. E. (1995). *Special Issue on Air-Land-Naval Warfare Models* (Vol. 42). Naval Research Logistics.

Brandsma, M.G., Vliet, A.J., & Griffioen-Young, H.J. (2006). *Analysis Methods and Models for Small Unit Operations.* TNO-DV 2006 A228, TNO Report.

Brooks, A., Bankes, S., & Bennett, B. (1997). *Weapon Mix and Exploratory Analysis, A Case Study. Project Air Force.* RAND.

Colegrave, R. K., & Hyde, J. M. (1993). The Lanchester square-law model extended to a (2,2) conflict. *IMA Journal of Applied Mathematics, 51*(2), 95–109. doi:10.1093/imamat/51.2.95

Cruz, J. B., Simaan, M. A., Gacic, A., Jiang, H., Leitellier, B., Li, M., & Liu, Y. (2001, October). Game-theoretic modeling and control of a military air operation. *IEEE Transactions on Aerospace and Electronic Systems, 37*(4), 1393–1405. doi:10.1109/7.976974

Danskin, J. M. (1967). *The Theory of Max-Min and its Application to Weapons Allocation Problems.* Berlin: Springer-Verlag. doi:10.1007/978-3-642-46092-0

Ghose, D., Krichman, M., Speyer, J. L., & Shamma, J. S. (2002, May). Modeling and analysis of air campaign resource allocation: A spatio-temporal decomposition approach. *IEEE Transactions on Systems, Man and Cybernetics. Part A, 32*(3), 403–418.

Ghose, D., Speyer, J. L., & Shamma, J. S. (2002). A game-theoretical multiple resource interaction approach to temporal resource allocation in an air campaign. *Annals of Operations Research, 109*(1/4), 15–40. doi:10.1023/A:1016335731847

Hillier, F. S., & Lieberman, G. J. (2004). *Introduction to Operations Research.* McGraw-Hill Science/Engineering/Math.

Jaiswal, N. K. (1997). Optimization Problems. In *Military Operations Research: Quantitative Decision Making* (pp. 129-167). Springer Science+Business Media, LLC.

Karatas, M. (2017). A multi-objective facility location problem in the presence of variable gradual coverage performance and cooperative cover. *European Journal of Operational Research, 262*(3), 1040–1051. doi:10.1016/j.ejor.2017.04.001

Karatas, M. (2018). Optimal Deployment of Heterogeneous Sensor Networks for a Hybrid Point and Barrier Coverage Application. *Computer Networks, 132*(26), 129–144. doi:10.1016/j.comnet.2018.01.001

Karatas, M., & Yakıcı, E. (2018). An iterative solution approach to a multi-objective facility location problem. *Applied Soft Computing*, *62*, 272–287. doi:10.1016/j.asoc.2017.10.035

Krichman, M., Ghose, D., Speyer, J. L., & Shamma, J. S. (2001). Theater level campaign resource allocation. *Proceedings of the American Control Conference*, 4716-4721. 10.1109/ACC.2001.945727

Murthy, P. R. (2005). Operations Research (2nd ed.). New Delhi, India: New Age International (P) Limited, Publishers.

Razi, N., & Karatas, M. (2016). A multi-objective model for locating search and rescue boats. *European Journal of Operational Research*, *254*(1), 279–293. doi:10.1016/j.ejor.2016.03.026

Razi, N., Karatas, M., & Gunal, M. M. (2016). A combined optimization and simulation based methodology for locating search and rescue helicopters. In *Proceedings of the 49th Annual Simulation Symposium* (p. 5). Society for Computer Simulation International.

Sheeba, P. S., & Ghose, D. Optimal Resource Partitioning in a Military Conflict based on Lanchester Attrition Models. In *Proceedings of the 44th IEEE Conference on Decision and Control, and the European Control Conference 2005*. Seville, Spain: IEEE. 10.1109/CDC.2005.1583098

Shubik, M. (Ed.). (1983). *Mathematics of Conflict*. New York: North-Holand.

Wu, T.T., Powell, W.B., & Whisman, A. (2008, September). The Optimizing-Simulator: An Illustration using the Military Airlift Problem. *ACM Transactions on Computational Logic*.

Chapter 7
Attrition Models and Applications

Ilknur Karacan
Marmara University, Turkey

Ismet Karacan
Marmara University, Turkey

Serol Bulkan
Marmara University, Turkey

ABSTRACT

Frederick W. Lanchester proposed simple ordinary differential equations that plainly model the attrition of fighting forces in a battlefield. With this insight, researchers studied extensions of these equations to model various battles for years. Novel studies include the application of these equations to miscellaneous field apart from battles that comprise reciprocal contention of opponents. If well-defined, these models can assist decision makers in revealing the shortcomings of a war strategy and discovering the bottlenecks that should be optimized. The recent studies prove that the insights gained from these models can also be utilized in other fields such as economy, biology, engineering, etc. This chapter includes the classic Lanchester equations, significant extensions of classical models, and a number of important application examples.

INTRODUCTION

During the planning and execution phases of a war, decision makers are mostly concentrated on achieving the victory. Obviously, winning a battle may not be simply based on the amount of friendly or enemy forces. Many parameters, such as morale, maneuver, C4ISR (Command, Control, Communications, Computers, Intelligence, Surveillance and Reconnaissance) capability, weapon technology, crew training level etc. have significant effect on trajectory of the battle states. Even in today's world, defense industries intensified their work on game-changing technologies. Stealth technologies avoid detection of fighting units. A laser gun becomes a source of a continuous destruction. Unmanned vehicles are

DOI: 10.4018/978-1-5225-5513-1.ch007

expendable units that may accomplish critical tasks like detecting a key enemy unit or destruction of a critical facility. Therefore, the fusion of these parameters complicates the simple modeling process of a war. To overcome this challenge, a number of high-level mathematical models have been developed and those models are stochastically solved by simulation softwares.

The state of a victory is mostly correlated with the number of remaining friendly and enemy forces and their potential capabilities. Since WWI, researchers and military authorities tried to build simple models to observe the fluctuation of fighting forces. This descriptive information involved the crucial interpretation of the course of the action. The most commonly acknowledged models are proposed by Frederick W. Lanchester during WWI. He proposed ordinary differential equations that simulate the flow the war deterministically and/or stochastically until a defined end state. These models are concentrated on the loss of force and have a large set of assumptions to keep the models easy. Hence, a complete war may be easily simulated with any spreadsheet.

This chapter will include preliminary information on the attrition models that explain a war holistically with differential equations and their extensions that are proposed by researchers in course of time. In addition to land operations, these models may also be applied to naval and air operations. Some important examples of attrition model applications include the Battle of Britain and the Battle of Atlantic. Even in recent studies, the equations developed for attrition models are applied to any competitive field such as market share, public goods game, manufacturing strategies, etc. The aim of the chapter is to provide insight for attrition models, describe the origin and the idea behind models, exemplify the types of the models and refer to substantial studies and applications.

BACKGROUND

A war (or a combat) may be modeled for various purposes. Regardless of its purpose, a combat model is expected to produce the required output to decision-makers to assist them in analyzing the performance of both sides of the war. Tactical warfare models can be grouped under three category as follows (Bonder, 1981):

- War Games
- Simulations
- Analytical Models

Each category possesses its own pros and cons in dimensions of realism, simplicity, accessibility etc. War games are common frameworks for a broad class of applications that encapsulate the games that progress upon the decisions of the opposing decision-makers. The decision-makers conclude their decisions according to the information that is derived from the domain of the game. War games are played for centuries even only for relaxation (Shephard,1963). Chess is a valuable worldwide example. However, war gaming services especially to military tacticians for boosting their capability to achieve crucial decisions, analyze their tactics, practice the war environment etc. Kriegsspiel is a tangible domain for war games that is developed and heavily practiced in 18th and 19th centuries. With the aid of computers, complex war games may be played through smart systems. Simulations are realistic runs of human reasoning and behavior by actions of automated and/or semi-automated forces (Ilachinski, 2004). They provide a collective output of reactions provided by consistently partitioned event or agents.

Analytical models ensure numerical description of warfare within defined assumptions. One of the earliest attempts to model the dynamics of military combat was the classical work of Lanchester (1916). Lanchester described the attrition of opposing forces by ordinary differential equations. Attrition models by Lanchester derive from analytical models that usually require primitive calculations. The idea behind the models is simply to demonstrate the attrition of own and opposite forces as a function of time. These models may be categorized by various aspects. In this chapter, we branch the attrition models with respect to the certainty criteria they involve, i.e. deterministic and stochastic Lanchester models.

DETERMINISTIC LANCHESTER MODELS

Lanchester (1956) considered "concentration" as a fundamental criterion to analyze a war. Concentration masks the capability of fighting sides of a combat holistically. It comprises both the stability of the objective and the capability of a force as moral concentration and material concentration, respectively. Both concentrations are covered with a common frame. However, they express completely different ideas and enclose peculiar conceptions. The underlying principles of material concentration are strictly related to major components of combat, namely attack and defense. Hence, focusing on principles of concentration rather than concentration itself assists to reveal factors of warfare.

In literature, symbolization of such a process exists in various forms. The forces are named as red and blue forces, friendly and opposing forces, and they are dddenoteddd as *X* and *Y* forces, respectively. The constants may emerge as *a* and *b*, α and β etc. To provide integrity, the following notation is used throughout the chapter:

$x(t)$ and $y(t)$: the expected number of surviving *X* and *Y* forces, respectively at time *t* after the engagement begins.

x_0 and y_0: Initial size of X and Y forces, respectively.

α: the constant rate at which a single *Y* unit kills *X* units (*X* units destroyed per unit of time by each surviving *Y* units)

β: the constant rate at which a single *X* unit kills *Y* units (*Y* units destroyed per unit of time by each surviving *X* units)

Figure 1 simply indicates the flow of a war by time (Taylor, 1980). Two opposing forces, namely *X* and *Y*, have attrition in their number of forces as a function of time in the scene of action. These numbers

Figure 1. Schematic of homogeneous combat

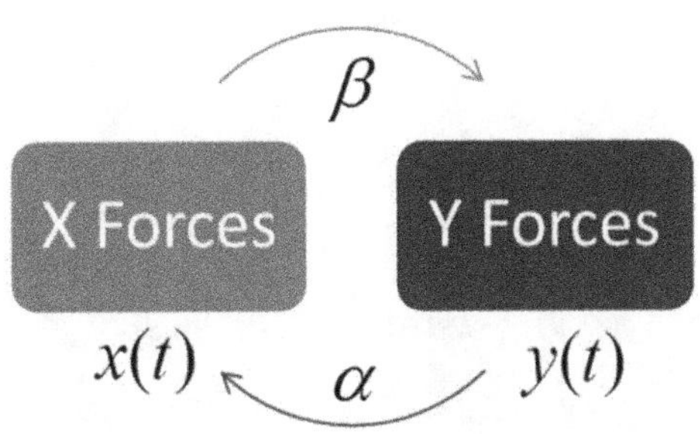

are represented by $x(t)$ and $y(t)$. The constant rates α and β describe the influence of a unit on opposing force. More precisely, α stands for the rate at which a single Y unit can neutralize X units within a unit time. β represents the vice versa effect, the rate at which a single X unit can neutralize Y units. The sizes of the forces shrink by attrition rates. The attrition rate of a force may be determined by the perspective of modeler on mutual influence. Many extensions to these assumptions are introduced in other studies of Lanchester. In this section, we give a brief discussion of the most commonly-used models in the literature.

Lanchester's Linear Law

According to Lanchester (1916), the fashion of ancient war separated from the modern war. Ancient war represents the efforts of hand to hand combat. In this regard, defense with a weapon returned a direct effect for protection. To exemplify, a swordsman would use his sword and shield to eliminate an attack of the opponent. By virtue of this property, this type of fire is called unaimed or area fire. Following assumptions are consistent for the Linear Law.

Linear Law assumptions:

1. Each shot aimed at the general operating area of the opposing units.
2. Each shot damages a specified percentage of the opposing units.
3. All units are within firing range of each other.
4. Rates of fire are fixed.

Fighting of units consisted of one-to-one attack and defense actions. Hence, the most significant parameter that leads to victory in historical wars was numerical superiority of forces. This fact clarifies Lanchester's Linear Law as shown in following equations.

Attrition of opposing forces by two sets of simultaneous differential equations:

$$\frac{dx}{dt} = -\alpha y(t)x(t) \tag{1}$$

$$\frac{dy}{dt} = -\beta x(t)y(t) \tag{2}$$

and,

$$\frac{dx}{dt} = -\alpha y(t) \tag{3}$$

$$\frac{dy}{dt} = -\beta x(t) \tag{4}$$

If the fire of blue force is directed into reds operating area rather than at a specific red unit, red's attrition rate is proportional to both red and blue forces. With this fact, Linear Law is derived with following equations. When we eliminate time by taking the ratio of derivatives of Equations (1) and (2), we get,

$$\frac{dx}{dy} = \frac{\alpha}{\beta} \tag{5}$$

Separate variables and integrate,

$$\int_{x_0}^{x} \beta dx = \int_{y_0}^{y} \alpha dy \tag{6}$$

$$\beta(x - x_0) = \alpha(y - y_0) \tag{7}$$

x_0 and y_0 are known. When x wins, then $y = 0$, and

$$y = 0 \Rightarrow x = -\frac{\alpha}{\beta} y_0 + x_0 \tag{8}$$

this occurs when $\beta x_0 \geq \alpha y_0$. When y wins, then $x = 0$, and

$$x = 0 \Rightarrow y = -\frac{\beta}{\alpha} x_0 + y_0 \tag{9}$$

this occurs when $\beta x_0 \leq \alpha y_0$.

As clearly indicated in the above equations, the attrition of a force is negatively related to the fighting capability of units in opposing force. The size of the force X dissolves with the rate of $-\alpha$, while the attrition rate for force Y is $-\beta$. With ancient conditions, the decisive factor for the result of war is the number of units, if each force has equal fighting strength. The main reason for this fact is that the "duels" result in attrition of a force. As an outcome, it has little or no importance whether the forces fight simultaneously with their whole units on the same scene or the combat drives intermittently with an unbalanced number of units. To explain more clearly, suppose a war in which forces have 500 unit with same fighting strength. For X force, it is equivalent to fight with whole 500 units of Y force or initially eliminate 250 of units, then re-engage to remaining 250 units. This inference is the main reason for calling these equations as the "Linear Law".

Example 1

Suppose, $\alpha = 0,03$ and $\beta = 0,015$. The inequality, $\beta x_0 < \alpha y_0$ declares the victory of Y forces. The time plot in Figure 2 shows the attrition of forces by time. Although having a larger initial force, X forces, due to their incapability of fighting, have a faster attrition and eventually lose the fight. Another notable point is that the lines of attritions have a linear trend. When the combat ends, the number of remaining units of Y force is 100.

Lanchester's Square Law

Classical Square Law symbolizes modern war. According to Lanchester (1916), modern war includes firearms that are aimed at any of the opposing equivalent units. That is, artilleries fight artilleries, while riflemen try to neutralize opposing riflemen.

Square Law assumptions:

1. Each shot is aimed at a particular opposing unit.
2. Each shot damages its target with a specified probability.
3. All units are within firing range of each other.
4. Rates of fire are fixed.

Rather than having duels of an ancient war, modern war has fire shots that aim opponents. Hence, Square Law is called alternatively as aimed fire. The defensive action against a shot is indirect avoidance. This avoidance can rise up only if the defending side can neutralize the enemy unit before it shoots. Thus, every unit is counted as capable of firing shots and every shot of each individual unit in a force influences the attrition of enemy. In this way, modern war differs from an ancient war, since in addition to fighting capability, the number of fighting units is related to the attrition of enemy. The Square Law may be derived with following equations keeping Equations 3 and 4 in mind.

When we eliminate time by taking the ratio of derivatives, we get,

Figure 2. Attrition of Forces by Time

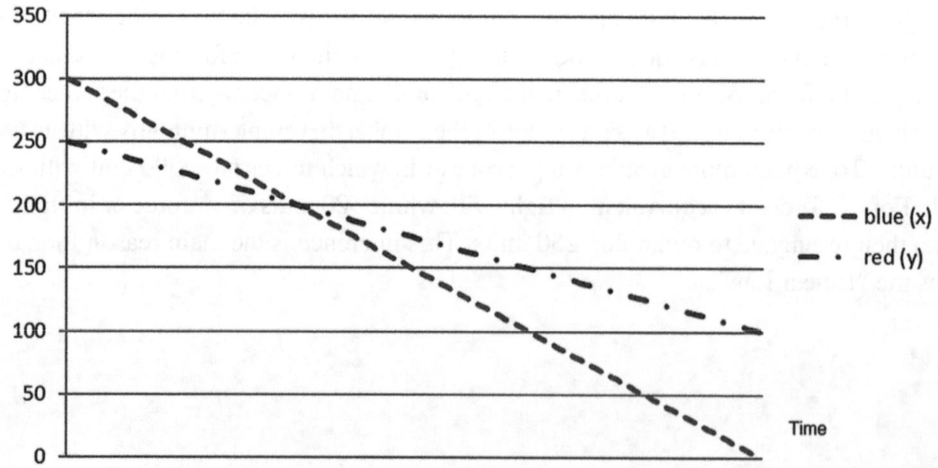

$$\frac{dx}{dy} = \frac{\alpha y}{\beta x} \quad (10)$$

Separate variables and integrate,

$$\int_{x_0}^{x} \beta x dx = \int_{y_0}^{y} \alpha y dy \quad (11)$$

$$\beta(x^2 - x_0^2) = \alpha(y^2 - y_0^2) \quad (12)$$

x_0 and y_0 are known. When either x or y decrease to 0 ;

$$y = 0 \Rightarrow x = \sqrt{x_0^2 - \frac{\alpha}{\beta} y_0^2} \quad (13)$$

Equation (13) occurs when $\alpha y_0^2 \leq \beta x_0^2$, so blue forces wins.

$$x = 0 \Rightarrow y = \sqrt{y_0^2 - \frac{\beta}{\alpha} x_0^2} \quad (14)$$

Equation (14) occurs when $\alpha y_0^2 \geq \beta x_0^2$, so red forces wins.

The quantities αy_0^2 and βx_0^2 are called the fighting strengths in aimed fire and determine the final outcome of the battle. Fighting strengths are proportional to the square of the initial strengths.

Besides fighting capability, the number of friendly units is negatively related to attrition of foe forces. This leads to a square effect on attrition. Hence, the concentration of higher numbers grants an immediate superiority and enemy force gets under a heavier fire.

Example 2

Consider example 1. Remodeling the example with Square Law yields the results as shown in Figure 3. Again, the Y force has superiority upon completion of the war. However, the attrition rate for Y force is decreasing by time. The reason for this decrease is the effect of the number of remaining force. Remember that each shot influences the attrition rate. Hence, the attrition rates follow a non-linear path. The number of remaining units of Y force upon completion of war is 132.

A shortcoming of these models is the way of solutions. Models are analytically solvable. However, ordinary differential equations are solved numerically in practice. To provide a simple way for numerical calculations, Euler method is applied with following equations (Washburn, 2000).

If the time increment is Δt, the difference equation counterpart to (3) and (4) is

Figure 3. Attrition of Forces by Time

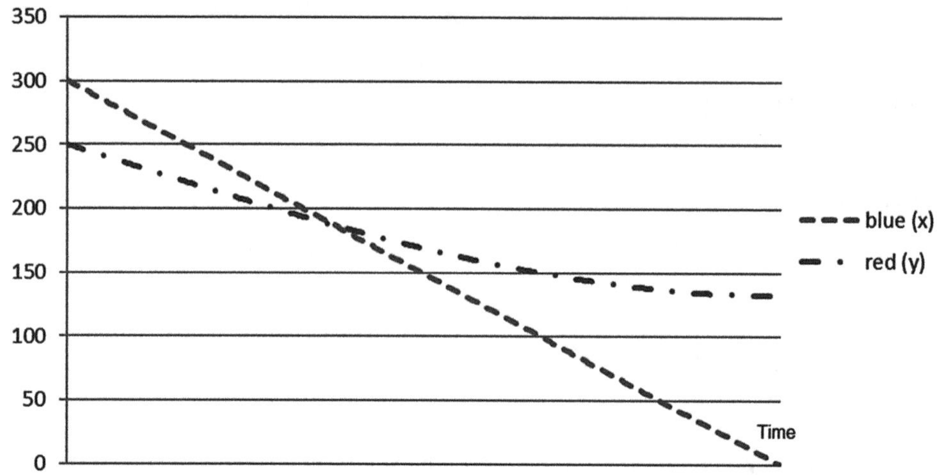

$$x(t+\Delta t) = x(t) - \alpha \Delta t y(t) \tag{15}$$

$$y(t+\Delta t) = y(t) - \beta \Delta t x(t) \tag{16}$$

When $x(0)$ is given,

$$\begin{aligned} &x(\Delta t) = x(0) - \alpha \Delta t y(0) \\ &x(2\Delta t) = x(\Delta t) - \alpha \Delta t y(\Delta t) \\ &\ldots \\ &\text{until termination} \end{aligned} \tag{17}$$

When $y(0)$ is given,

$$\begin{aligned} &y(\Delta t) = y(0) - \beta \Delta t x(0) \\ &y(2\Delta t) = y(\Delta t) - \beta \Delta t x(\Delta t) \\ &\ldots \\ &\text{until termination} \end{aligned} \tag{18}$$

Euler method assures demonstration of attrition by defined time increments.

EXTENSIONS OF CLASSICAL LANCHESTER MODELS

Lanchester defined a way to model concentration of a war with simple ordinary differential equations. By time, researchers tried to extent his view and provide novel attrition rates in various forms. In this heading, some of the prominent studies are included.

Mixed Combat Models

Basic assumptions used in Lanchester Models limit the applicability of the approach to only simple problem types. However, within some alterations of parameters, various types of combats may be modeled from different perspectives. Guerrilla-Counterguerrilla fight is an example for a custom combat (Deitchman, 1962). Analysis by Deitchman proved that the only chance for attacking guerillas to win a fight is division of both sides into smaller groups and having numerical superiority to attack within local groups. However, if guerillas decide to conduct ambushes, this benefit reduces the requirement for local superiority greatly and defender requires to have very high local force ratios or significant weapons to react. Ambushes modify the characteristic of the combat. While searching an area, all defenders are visible by guerillas. The guerrillas try to repel the fire of defenders by ambushes. When the fight begins, all defenders are visible to guerillas. Hence, the attrition rate of defenders is proportional to the number of firing guerrillas. However, the guerrillas are hidden and defenders fire blindly to the area where the guerilla fire comes from. Thus, the attrition rate for guerillas is proportional to the number of firing defenders and the number of guerrillas in the area. Hence a mixture of linear and square laws is proposed by Deitchman. Equation (19) holds for a combat with ambushes.

$$\frac{\beta}{2}(x_0^2 - x^2) = \alpha(y_0 - y) \tag{19}$$

Logarithmic Law

Another extension of classical laws is the Logarithmic Law by Peterson (1967). The idea behind the law is the cumulative effect of non-combat factors. More precisely, before the close combat begins, there are reasons for attrition of own forces that are not caused by enemy (Taylor, 1980). Examples for such losses are diseases, desertions, failure of weapons etc. Without being capable of clarifying a whole combat, logarithmic law can make sense for the initial losses of forces during the startup phase of a war. In most cases with forces of large sizes, the logarithmic law does not hold when the battle actually begins. The formulations are illustrated in the following equations.

$$\frac{dx}{dt} = -\alpha x \; and \; \frac{dy}{dt} = -\beta y \tag{20}$$

The state equation for the logarithmic law is:

$$\beta \ln\left(\frac{x_0}{x}\right) = \alpha \ln\left(\frac{y_0}{y}\right) \tag{21}$$

Helmbold Equations

Helmbold (1965) proposed that the reciprocal sizes of forces may affect the fighting capability of a force. If the force ratio x/y is excessively large, then X cannot effectively manage its units for allocation of enemy forces (Caldwell, Hartman, Parry, Washburn, & Yungren, 2000). Hence, X cannot actively include its whole fighting power to the scene. Hembold demonstrates this fact with Equation 22:

$$\frac{dx}{dt} = -\alpha\left(\frac{x}{y}\right)^{1-\omega} y \; and \; \frac{dy}{dt} = -\beta\left(\frac{y}{x}\right)^{1-\omega} x \tag{22}$$

In Equation 22, ω represent the measure of efficiency. Three numerical value of ω can clarify its effect on the model.

- For $\omega = 0$, the model turns into Logarithmic Law
- For $\omega = 1$, the model turns into Square Law
- For $\omega = 0.5$, the model turns into Linear Law

The coefficient ω modifies the model by adding the capability of target acquisition by the ratio force sizes.

Amendment in Force Structure

In addition to tracking the attrition of two forces on the field, the overall trend of a war is also of interest. While the units in the forces dissolve by time, reinforcement units arrive to relieve them. Other actions for a force is replacement and withdrawal of units. With a holistic point of view, forces have additional units by a production process, however, all three actions diminish the operational capability of units. The reason of this loss is the adaptation requirement of new units to war, adaptation of replacement units to the new field, and missing units by withdrawals. The loss in operational capability causes a decrease in fighting capability. Morse and Kimball (1950) called such a model as Generalized Lanchester equations and formulated as in Equation 23.

$$\frac{dx}{dt} = P - \alpha y - ax \; and \; \frac{dy}{dt} = Q - \beta x - by \tag{23}$$

P and Q are reproduction rates for units in X and Y forces, respectively. While $-\alpha y$ and $-\beta x$ illustrate the loose rates caused by the enemy, $-ax$ and $-by$ stand for the rates of operational loss.

Effect of Range

The duels in an ancient war were carried out with contact of bodies. However, modern firearms have ranges that enable shooters to avoid physical contact to the opponent. Classical Lanchester models neglect range effect of weapons. Bonder proposed an updated attrition rate that is specified as a function of range (Howland & Bonder, 1963).

$$\frac{dx}{dt} = -\alpha(r)x \; and \; \frac{dy}{dt} = -\beta(r)y \tag{24}$$

The range r denotes the distance from the firer to the target. The functions of range may be defined in various forms. Power, exponential, cosine, piecewise-constant are examples of custom attrition rate coefficients (Taylor, 1980). The power attrition rate is included in Equation 25.

$$\alpha(r) = \begin{cases} \alpha_0 \left(1 - \frac{r}{r_e}\right)^{\mu} & for \; 0 \le r \le r_e \\ 0 & for \; r_e \le r \end{cases} \tag{25}$$

α_0 stands for maximum attrition rate. r is the variable that defines the distance between shooter and target. r_e denotes the maximum effective range of the weapon. μ is the variable coefficient power that is determined by the modeler. The maximum range of a weapon describes the maximum distance at which a bullet can reach. Nevertheless, the effectiveness and accuracy of a weapon differ from the maximum effective range of a weapon. The maximum effective range is the maximum distance within that the bullet can make its intended move. Thus, the maximum effective ranges of weapons are counted in the formulations.

Figure 4 represents the distance-based attrition rate curves by different μ values. It is clearly notable that higher μ values cause more reduction in the attrition rate values. The maximum effective range of the weapons is calculated as 500 meters. Maximum attrition rate is defined as 0.4.

Heterogeneous Models

Classical attrition models consider forces as consisting of homogenous units. This assumption is valid only if forces have identical types of units. Nevertheless, modern warfare includes various types of units in distinct environments. While fighters try to neutralize enemy fighters in the air, warships try to sink opposing units at sea etc. In this case, single attrition rates for forces cannot make sense for the entire battle.

Figure 5 illustrates the variety of a battle between heterogeneous forces. Each group of weapon systems has its own attrition rate caused by opposing groups of weapon systems. Hence the attrition rates have two indices. The attrition rate formula for weapon systems of X force is stated in Equation 26.

$$\frac{dx_i}{dt} = -\sum_{j=1}^{n} \alpha_{ij}, i = 1 \, to \, m \tag{26}$$

Figure 4. Change in Attrition Rates by different μ values

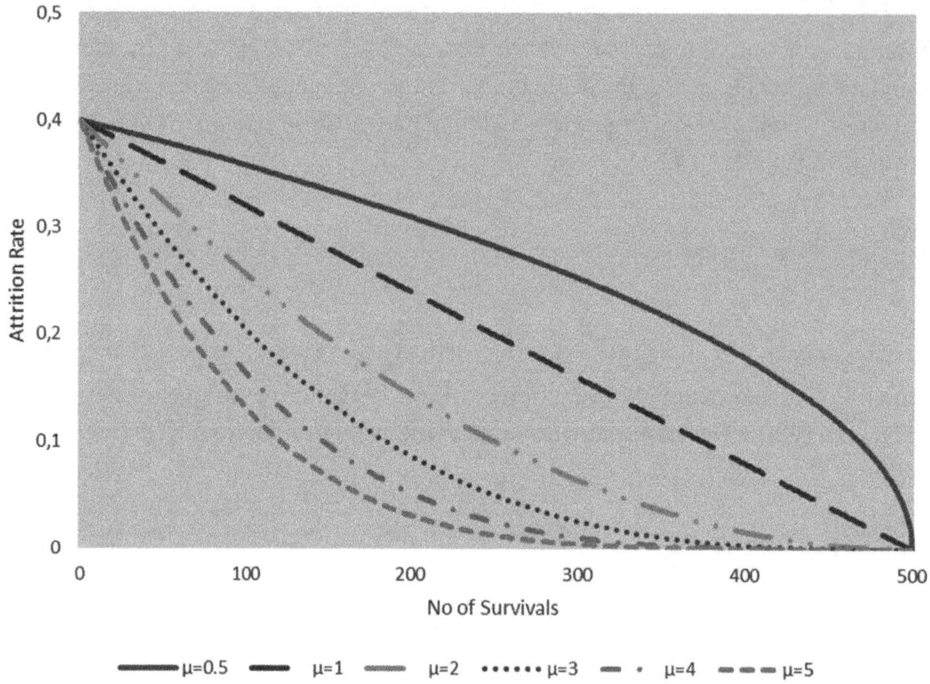

Figure 5. Attrition in Heterogenous Models

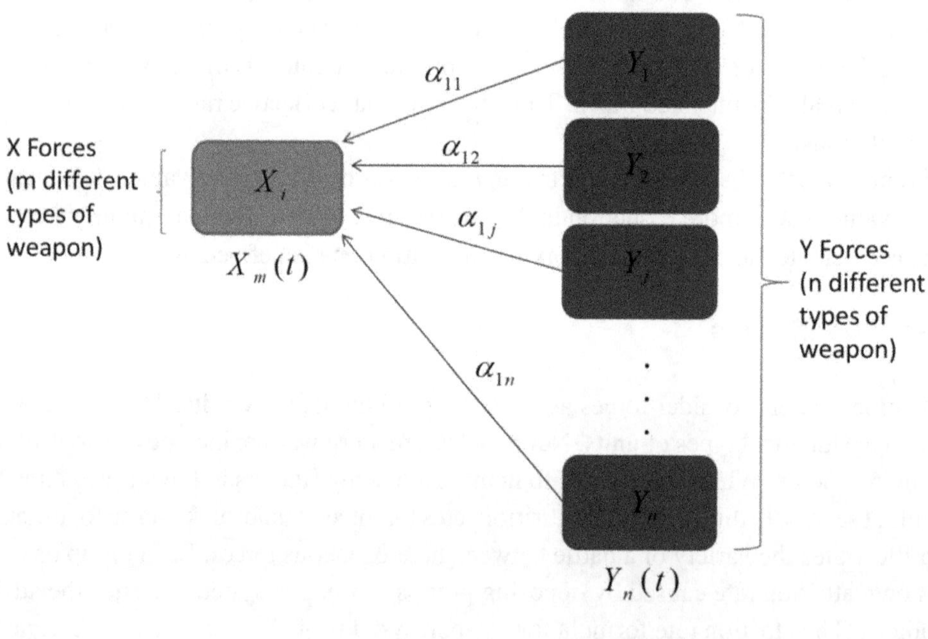

x_i denotes the size of ith weapon system of X force. α_{ij} is the attrition rate of ith weapon system of X force that is caused by j^{th} weapon system of Y force. However, this form does not adequately represent the effect of heterogeneous forces. Equation 27 illustrates this effect explicitly (Bonder & Farrell, 1970).

$$\frac{dx_i}{dt} = -\sum_{j=1}^{n} \alpha_{ij} p_{ij} y_j, i = 1\, to\, m$$
$$\frac{dy_j}{dt} = -\sum_{i=1}^{m} \beta_{ji} r_{ji} x_i, j = 1\, to\, n \tag{27}$$

In this formulation, p_{ij} stands for the proportion of time in which j^{th} weapon system of Y engages the ith weapon system of X. Not all units of y_j engage x_i all time. The proportion p_{ij} assures the inclusion of accurate ratio of fighting units. Many applications of heterogeneous models are proposed in the literature. However, enumerating and solving these models get complex and crucial as the number *m* and *n* grows. The modeler should provide 2*nm* attrition rates to complete the model. For simplicity, the model has the assumption of additivity. Additivity dictates that the weapon systems do not synergize for the units in the battlefield. More clearly, the power of tanks does not cause an enhancement for the attrition rate of riflemen. Revealing the proportion time of engagements of weapon systems is another hard process. With a comprehensive view, it is nearly impractical to completely deploy and solve heterogeneous models. However, they provide insight for the course of action.

STOCHASTIC LANCHESTER EQUATIONS

Being ordinary differential equations, deterministic Lanchester equations are solvable by Euler's method with time increments of Δt. For each time increment of Δt, the casualties of X and Y forces may be calculated with simple formulas. However, these values make no sense for some cases. Let Δt as a small number. Hence, the number of casualties may be fractional, for instance, 0.2 for a force. For large forces, this number may be neglectable. Nevertheless, this fractional number does not clearly identify the state for forces with small sizes (Washburn, 2000). Another shortcoming of a deterministic model is its insufficiency on revealing different outcomes of a battle. The model only returns the expected result of a battle. While the model always returns the same force as a conqueror, the opposing force may have a probability, though small, to win the battle. A stochastic model may discover different outcomes and win probabilities of forces at the end of the war.

Stochastic process of Lanchester models follows continuous time Markov Chains (CTMC) (Ross, 2014). CTMC dictate transition from state to state from with known transition probabilities within a limited time in which an event occurs. The transitions of Stochastic Lanchester Models are illustrated in Figure 6.

Figure 6. Transitions of Continuous Time Markov Chains

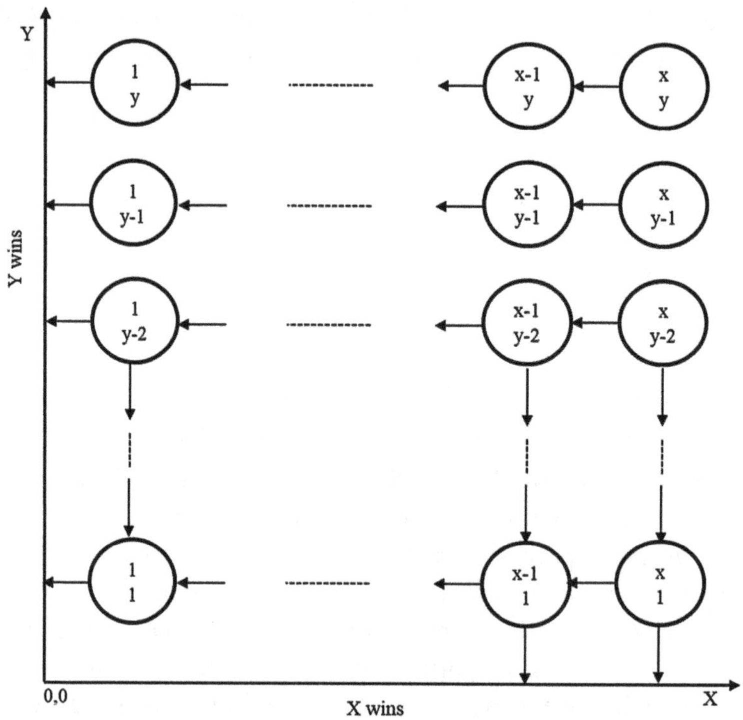

The best method to run the model is Monte Carlo Simulation. Eventually, Monte Carlo Simulation supplies a planar graph that expresses occurrence probabilities of each end state. To enrich the outcome of a dice roll of Monte Carlo simulation, Jisong et al.(2017) suggested utilizing Combat Result Table (CRT) in which an outcome is defined for each result of a dice or simulated draw. CRT consists of a matrix where vertical numbers represent the outcome of the dice and horizontal numbers represent the ratios of fighting strengths. The ratios of fighting strengths generate a required bias for the success of attacker with higher fighting strength.

Stochastic Lanchester models are not studied widely compared to deterministic models. However, various studies include different solution approaches to models. Pettit et.al (2003) deploy Bayesian Statistical Inference for some simple cases of both linear and square law. The authors claim that there are very limited Bayesian Inference studies on stochastic attrition models. Hence, the study empirically employs various distributions to estimate parameters of attrition models and illustrates on a generated tank battle.

APPLICATION EXAMPLES

Numerous attrition models based on Lanchester models are applied not only to combat models but also to real-life problems for a century. Since the models of Lanchester defined a combat with a simple point of view, it is of interest for researchers to contribute new properties to models and apply to novel fields. In this section, famous models for great wars are illustrated. Besides, novel applications for real-life problems are introduced to provide insight into the various fields to which models may be applied.

Battle of Iwo Jima

The model introduced by Engel (1954) simply exemplifies a battle that involves reinforcement units as stated in Equation 28.

$$\frac{dx}{dt} = m(t) - \alpha y; x>0$$
$$\frac{dy}{dt} = -\beta x; y>0 \quad (28)$$

In this model, X force stands for US troops and Y force stands for Japanese troops. The variable $m(t)$ declares the rate of reinforcement of US troops by time.

The battle was a WWII event and happened in 1945. The war was ferocious since the Japanese troops were trained to fight until no survivor remains. The island was strictly fortified and the terrain complicated the war for US Marines. However, by paying a high price, US Marines achieved to capture the island.

If x defines the number of active US troops that are not killed, wounded or missing, then the best fitting value for α is 0.0544. Investigating the journals of the war, the best value for β is 0.0106. The ratio α/β for this model proves that Japanese troops, also by holding the terrain advantage, were nearly 5 times mortal then US troops. However, US troops gradually kept the numerical superiority by regular reinforcements. The exact number of Japanese troops is not known, but logs of the battle show

Figure 7. Lanchester model of the Iwo Jima battle. M(t) denotes the number of United States troops in action
Source: Engel, 1954.

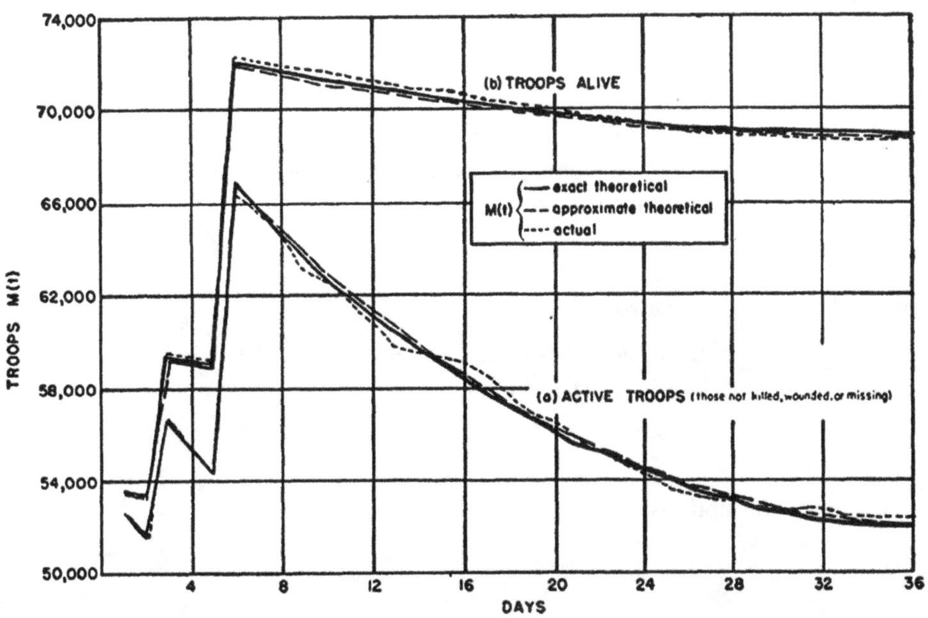

that the number was nearly 19000 and except a neglectable number of captives, all were killed by US troops. The corresponding number of US Troops was nearly 29000 (Burrell, 2011). The US troops achieved victory by numerical superiority in spite of being less lethal.

Battle of Atlantic

Battle of Atlantic was based on the domination of sea lanes of communication in Atlantic during WWII. Morse and Kimball (1950) modeled the escort operations carried out by allied forces. While the allied forces tried to transfer military power by escorted convoys, U-Boats (submarines) of Germany aimed to impede the passage of the convoys by any means. The submarines kept their position within defined "wolf packs" for an ordered period of time. This event of interest is quietly easy to understand. However, modeling requires many parameters and approximations. Morse and Kimball made the assumptions that $5w/f$ and $wf/100$ merchant ships and submarines, respectively, will be sunk in a conflict consisting f escort units that try to protect the convoy and w attack submarines. As an inference from the exchange ratio of submarines to escort units, $5/1$, the rate of attrition for escort units may be approximated as $wf/500$. Remaining parameters and definitions for the model are as stated below.

MS : cumulative merchant ships sunk
RS : remaining submarines
RE : remaining escorts
p_{cw} : probability that a given convoy is attacked by a given patrolling wolf pack on the way over = 0,01
t_{ce} : time a convoy must be escorted counting trips in both directions = 30 days
T_{ce} : cycle time for an escort = 50 days
r_{ce} : ratio of escorts escorting convoys = t_{ce}/T_{ce} =0,6
t_{sp} : patrol time for a submarine = 20 days
T_{sp} : cycle time for a submarine = 50 days
r_{sp} : ratio of submarines on patrol = t_{sp}/T_{sp} =0,4
r_c : rate at which convoys leave = 1/day
c : convoy size = 40
w : wolf pack size = 4
p_{sc} : probability of loss per cycle for a submarine due to causes other than escorts = 0,04
R_{SR} : rate of replacement of submarines = 0,7/day
R_{ER} : rate of replacement of escorts = 0,5/day
$\frac{RS\,r_{sp}}{w}$: number of wolfpacks on patrol
e_r : engagement rate counting engagements both ways = $(2r_c)(p_{cw})(RS\,r_{sp}/w)$
f : escorts/convoy = $(RE\,r_{ce})/(r_c t_{ce}) = RE/(r_c T_{ce})$

Given all the above assumptions,

$$\frac{dMS}{dt} = e_r(5w/f) = 10r_c^2 T_{ce} p_{cw} r_{sp} (RS/RE); MS>0$$
$$\frac{dRS}{dt} = -e_r(wf/100) - p_{sc} RS/T_{sp} + R_{SR} = -RS([2 p_{cw} r_{sp}/(100 T_{ce})]RE + p_{sc}/T_{sp}) + R_{SR}; RS>0$$
$$\frac{dRE}{dt} = -e_r(wf/500) + R_{ER} = -RS[2 p_{cw} r_{sp}/(500 T_{ce})]RE + R_{ER}; RE>0$$

The assumptions of the model eased the decision process of tactician by revealing the factors that were the origin of the actual values of the campaign. The relation between the rate of sails and convoy sizes was the most significant inference. The rate of loss of merchant ships is proportional to the square of the sailing rate whereas the size of convoys does not have an impact on the losses. This inference was the origin of the decision to increase the convoy sizes and reduce the frequency of sails. As a result, the number of sunk merchant ships drastically increased since a lower number of conflicts actually occurred.

Other Examples

Battle of Trafalgar occurred between British and Spanish/French navies in 1805. At the cost of the life of Admiral Nelson, British navy achieved to defeat the opposing navy and achieved to become the dominant naval power in the region.

Instead of having a single line of formation, Admiral Nelson divided the British Navy into two divisions and divided opponent navy into three partitions. The purpose of such classification was his divide and conquer strategy. Fox (2017) modeled the battle by fitting to Square Law of Lanchester. Simply fitting one state equation does not hold and evaluates Spanish/French navy as a conqueror. With a comprehensive view, modeling the war with three equations reveals the effect of divide and conquer. The corresponding model is illustrated in Figure 9.

Upper series of points in Figure 9 demonstrate the divisions of British Navy. As clearly notable, model consists of three battles. When the number of survivors in one opponent gets close to zero, the next battle is triggered to begin with new initial values.

Lanchester models are mostly studied in battles carried out by humans. Leastwise, there exist application fields that are imitated to the battles between human opponents. Plowes (2005) modeled the mortality during the battle of fire ants. Both linear and square laws are tried to observe numerical results. The study included Bayesian statistical inference to define attrition rates of groups. An interesting finding of the study is the synergy of pair or trio of groups. The experiment of the study reveals high mortality and aggressiveness of these groups. It is common to apply Lanchester's Laws on social animals that have fights within/inter groups. Franks and Partridge (1993) deployed the equations over ants to clarify their predictions and intuitive movements in combats. The study by Adams and Mesterton-Gibbons reveals the individual and group attack strategies of social animals. Batchelor and Briffa (2011) represent another study in which emotional effect of the size of the fighting wood ant groups considered. Jonhson and MacKay (2015) take a step further and try to fit Attrition Models on Human Evolution. The study highlights Lanchester's Square Law and explains its effect on physiological behavior along the evolution,

Figure 8. The Formation of Navies during Battle of Trafalgar
Source: Gardiner, 1891.

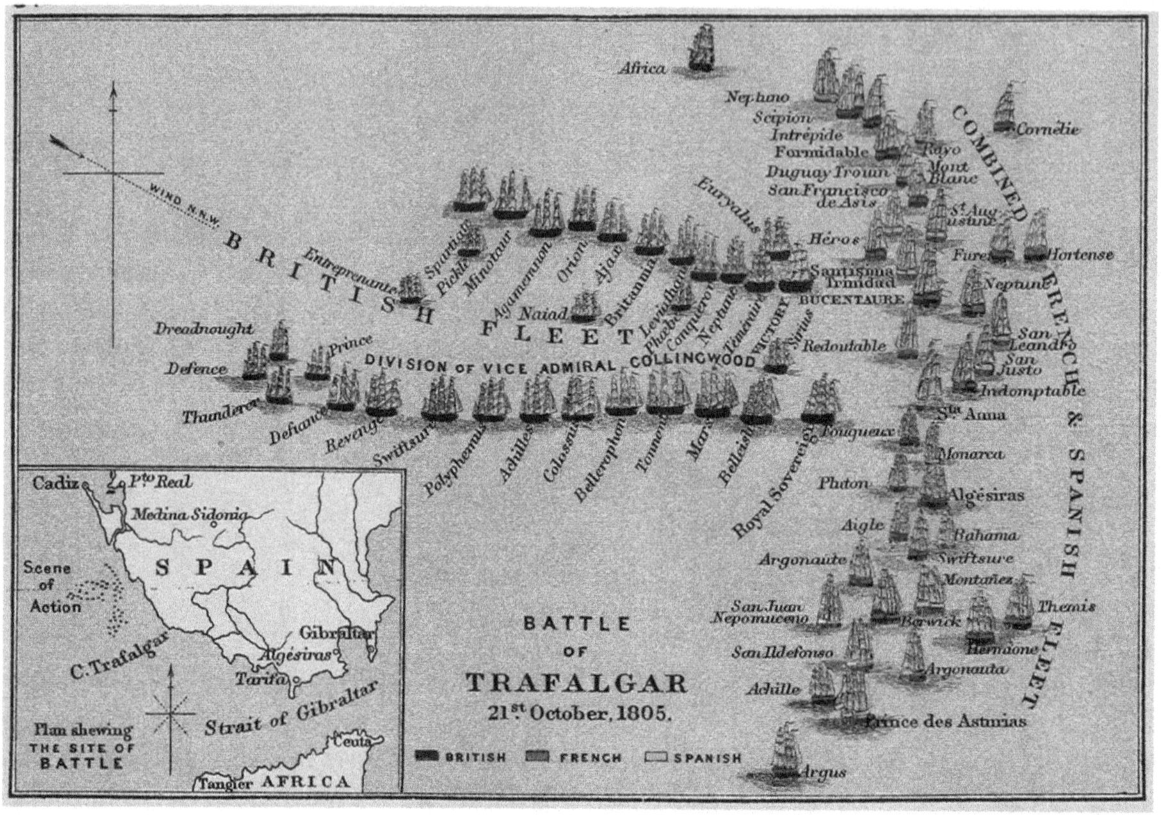

Figure 9. Lanchester Models According to Divide and Conquer Strategy of British Navy
Source: Fox, 2017.

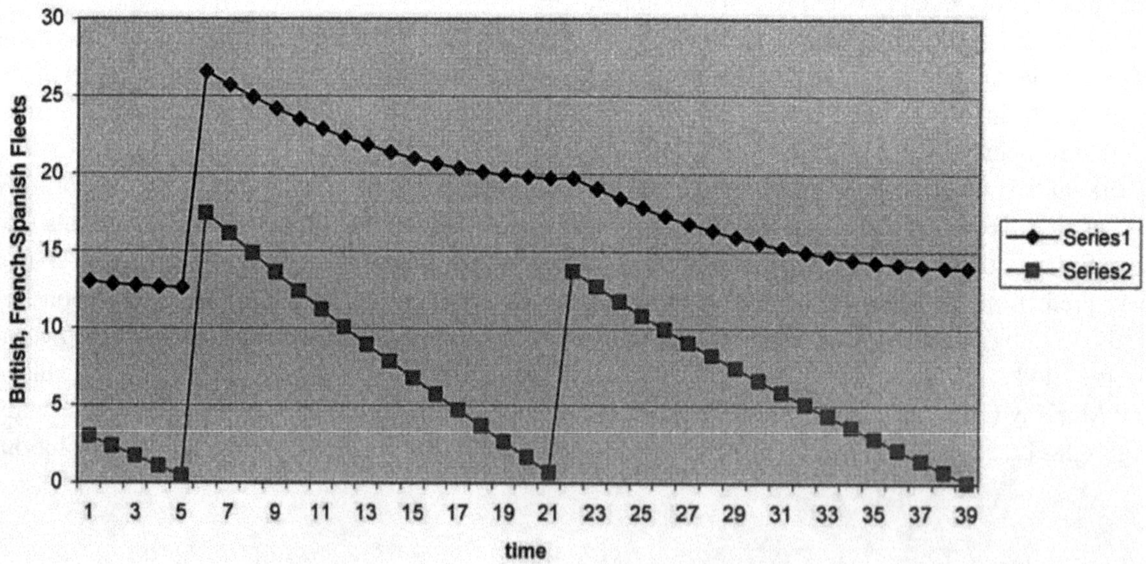

i.e. individual attacks of primitives, tendency for grouping, raids and ambushes, behaving overconfident or overestimating the enemy, organized conflicts etc. The study points out similarities between human vs. chimpanzees and bonobos which are viewed as closest animals to human.

Kimball (1957) summarized four attrition models that are purely designed for military purposes. The study points out that is designed primarily for military purposes, nearly all operations research models have real-life application areas. Hence, the author relates these attrition models with the non-military field, namely two opposing first-order chemical reactions, biological situations, and the effect of advertisement. Another attractive set of models are the equations by Erickson (1992) to model the advertisement strategies of Coca Cola vs Pepsi and Anheuser-Busch vs Miller. Final decision variable that is implicitly formulated within this study is market share. The study examines the effects of open-loop and closed-loop strategies. The resulting model dictate that closed-loop equilibrium fits better to actual data. As a recent study, Kaszkurewicz and Bhaya (2018) modeled Market Share Dynamics that include both advertisement effect and Word-of-Mouth (WOM) interactions. Instead of having a specific case, the study has a generalized focus that brings out additional complexities on proposing universal equations. The authors overcome these complexities by two actions. The first action is defining the roles well as satisfied customers, dissatisfied defectors, and undecided customers. The second action is suggesting a hybrid model that deploys Vidale–Wolfe and Lanchester's Equations to describe the attrition of roles and Lotka–Volterra-type interaction terms to include the effect of WOM.

Recent studies might be found on real-time strategy (RTS) games to rapidly determine the results of conflicts that include complex types of units. An appealing study is the simulations conducted by Stanescu et al. (2015) on the famous RTS game StarCraft. The study includes simulations on both simulated data and tournament environments. The model covers assumptions for various types of units with different fighting capabilities. A contribution of the study is learning combat effectiveness that is achieved by updating the attrition rates on runtime.

In some studies, Lanchester models are to model some functions of the war rather than modeling whole attrition of the forces. In a sample study, Coulson (2018) modeled the effect of operational intelligence rather than modeling the model itself. Operational intelligence requires collection and analysis of enemy-related data and helps to infer predictions on intentions of enemies. Hence, operational intelligence can be viewed as a game-changing effect for the side that has reliable source data and can draw rational predictions on enemy movements. The model in the study is built on Squared Law and can be deployed if any side or both sides use intelligence. Chen et al. (2011) added control variables to classical attrition models and tried to optimize reinforcement strategies for both sides by controlling these variables. Liu et al. (2013) discussed the importance of network-centric warfare for modern war and introduced network attack-defense model. The model is built upon both linear and square laws. The attack on a network may occur in an uncertain time. However, the defender would have a limited response time to eliminate the threat whenever an attack occurred. Hence, time action factors are included in models to accurately simulate the behavior both attacker and defender. Since the emergent equations are hard to solve, simulation-based experiments are carried out to verify the models.

Classical attrition models dictate involvement of only two sides, ignoring whether they are individual identities or coalitions. Kress et al. (2017) extended classical models and introduced the "multi-duel" attrition models that enable to express dynamics of multilateral wars. The models have two expected end states. The first outcome is the victory of the strongest side that has the capability to overcome all the enemies. The second outcome is a deadlock which induces reciprocal extinction. The conflict in Syria since 2011 which has multiple players rather than two sides, inspires the study. In a similar study, Lin and MacKay (2014) tried to discover optimum fire distribution of a homogenous blue force over a heterogenous red force. In this study, it is assumed that blue force with a single type of unit tries to defeat a red force that has diverse types of units. The authors prove that the optimum firing policy should aim to neutralize the remainder red units with highest fighting strength at any given time. Kress et al. (2018) introduced another compelling study, in which Lanchester's Square Law is modified to model three-way combat instead of duels by working on eigenspaces and Jacobian matrices. The study indicates the complexity of deriving models that simultaneously emit attrition of three sides.

As previously mentioned Admiral Nelson divided his force to have more effective fighting capability against opposing warships and won Battle of Trafalgar. Fox (2017) modeled the flow of the attrition by extending classical view and dividing the war into phases. Rather than modeling an existing battle, Jing et al. (2017) sought for the optimal decision of dividing a large force to have a better fighting capability. The motivation of the study is the complexity of Command, Control, Communication (C3) functions that get more difficult while the size of the force grows. Rather than dividing the war into phases, Gonzales and Villena (2011) related the well-known Reaction-Diffusion Equations to Classical Lanchester's Laws in order to acquire spatial dimensions for units. Unlike previous studies, this study provides clear linkage between equations and satisfying formulations that are supported by numerical examples and three-dimensional charts.

FUTURE RESEARCH DIRECTIONS

Lanchester introduced simple attrition equations that are easily adaptable to fit many cases with defined assumptions. In the course of time, researchers modified these equations to fulfill gaps while modeling different types of wars. Examples are Logarithmic Law that links cumulative effect of non-combat factors and deploying multiple equations that are practiced to model Battle of Trafalgar. New equations or solution approaches are proposed with modeling distinct types of wars. These distinct war models forced researchers to remove some assumptions and apply altered equations to achieve accuracy. In this point of view, it would be useful to try modeling an unrevealed war to encounter new states and fit new equations. For instance, attrition models ignore mutual synergy of different type of units. More clearly, it is assumed that the actions of a friendly artillery do not boost the fighting capability of the infantry units. The equations are capable of displaying attrition of artillery and infantry units mutually independent. Defining relation-based equations for multi-unit war would be an interesting area of research.

Attrition models are proposed with intention of modeling real wars to ease decisions and analyze the course of action. However, novel studies apply the equations to many fields by ensuring that an 'attrition' available. As mentioned previously, cola advertisement strategies, market shares, RTS games, fight among social enemies are significant examples. These studies prove the potential of application to novel fields, e.g., modeling sales of smartphones by leading brands may help to drive new strategies on the management level.

CONCLUSION

In this chapter, Lanchester based attrition models are discussed. Since a century, the models are diversified by researchers to fit into many types of battles and other fields of application. Literature review reveals that these models are mostly easy to build, however, it's difficult to solve and verify the model in complex cases. The most complicated task is to determine the attrition rates for forces. This challenge occurs as a consequence of missing data or conflicting fluctuation of attrition due to factors like reinforcement, diseases, desertions etc. However, if well defined, these models ease to reveal the simple precautions that alter the tempo of a battle. An outstanding example is the Battle of Atlantic. Enlarging the convoy sizes and extending the frequencies of sails by allied units helped them to save many of their merchant ships against submarine attack. This decision reduced the number of encounters by merchant ships and German submarines that patrol in wolf packs and trimmed the probability for a submarine to hit merchant ships. Eventually, Lanchester equations are still attractive for researchers to propose an extension or apply to a novel field. Novel studies include modeling of great battles like Trafalgar, Kursk, Ardennes, Atlantic etc.

Different approaches are introduced that extend the classical equations. These studies prove that Lanchester's Laws may be used to model different aspects of a war or be applied to fields other than war, with some alterations. At this point, it is clearly visible that the equations have potential to be applied to novel applications and may be coincided in most recent studies.

REFERENCES

Adams, E. S., & Mesterton-Gibbons, M. (2003). Lanchester's attrition models and fights among social animals. *Behavioral Ecology*, *14*(5), 719–723. doi:10.1093/beheco/arg061

Batchelor, T. P., & Briffa, M. (2011). Fight tactics in wood ants: individuals in smaller groups fight harder but die faster. *Proceedings of the Royal Society of London B: Biological Sciences, 278*(1722), 3243-3250. 10.1098/rspb.2011.0062

Bonder, S. (1981). Mathematical modeling of military conflict situations. In Proceedings of Symposia in Applied Mathematics (Vol. 25, pp. 1-51). American Mathematical Society.

Bonder, S., & Farrell, R. (1970). *Development of models for defense systems planning (No. SRL-2147-TR-70-2)*. Michigan Univ Ann Arbor Systems Research Lab.

Burrell, R. S. (2011). *The Ghosts of Iwo Jima* (Vol. 102). Texas A&M University Press.

Caldwell, B., Hartman, J., Parry, S., Washburn, A., & Yungren, M. (2000). *Aggregated Combat Models*. Academic Press.

Chen, X., Jiang, N., Jing, Y., Stojanovski, G., & Dimirovski, G. M. (2011). Differential game model and its solutions for force resource complementary via Lanchester square law equation. *IFAC Proceedings Volumes, 44*(1), 14229-14233. 10.3182/20110828-6-IT-1002.01042

Coulson, S. G. (2018). Lanchester modelling of intelligence in combat. *IMA Journal of Management Mathematics*. doi:10.1093/imaman/dpx014

Deitchman, S. J. (1962). A Lanchester model of guerrilla warfare. *Operations Research, 10*(6), 818–827. doi:10.1287/opre.10.6.818

Engel, J. H. (1954). A verification of Lanchester's law. *Journal of the Operations Research Society of America, 2*(2), 163–171. doi:10.1287/opre.2.2.163

Erickson, G. M. (1992). Empirical analysis of closed-loop duopoly advertising strategies. *Management Science, 38*(12), 1732–1749. doi:10.1287/mnsc.38.12.1732

Fox, W. P. (2017). *The Battle of Trafalgar: An Example of Modeling with Technology*. Academic Press.

Franks, N. R., & Partridge, L. W. (1993). Lanchester battles and the evolution of combat in ants. *Animal Behaviour, 45*(1), 197–199. doi:10.1006/anbe.1993.1021

Gardiner, S. R. (Ed.). (1891). *A School Atlas of English History*. Longmans, Green.

González, E., & Villena, M. (2011). Spatial Lanchester models. *European Journal of Operational Research, 210*(3), 706–715. doi:10.1016/j.ejor.2010.11.009

Helmbold, R. L. (1965). Letter to the Editor—A Modification of Lanchester's Equations. *Operations Research, 13*(5), 857–859. doi:10.1287/opre.13.5.857

Howland, D., & Bonder, S. (1963). *The Tank Weapon System. Report No. RF 573 AR 63-1 (U). Systems Research Group*. Columbus, OH: Ohio State Univ.

Ilachinski, A. (2004). *Artificial war: Multiagent-based simulation of combat*. World Scientific. doi:10.1142/5531

Jing, Y., Zhang, S., & Dimirovski, G. M. (2017). Complexity of warfare command, communication and control systems simplified: Optimal resource partitioning via Lanchester equations. In *Smart Technologies, IEEE EUROCON 2017-17th International Conference on* (pp. 611-617). IEEE.

Jisong, F., Hui, R., & Chengxiang, T. (2017). An Analysis of Wargame Rules Simulation Based on Stochastic Lanchester Models. In *Proceedings of the 2017 VI International Conference on Network, Communication and Computing* (pp. 135-139). ACM. 10.1145/3171592.3171612

Johnson, D. D., & MacKay, N. J. (2015). Fight the power: Lanchester's laws of combat in human evolution. *Evolution and Human Behavior, 36*(2), 152–163. doi:10.1016/j.evolhumbehav.2014.11.001

Kaszkurewicz, E., & Bhaya, A. (2018). *Modeling Market Share Dynamics Under Advertising Effort and Word-of-Mouth Interactions Between Customers. IEEE Transactions on Computational Social Systems.*

Kimball, G. E. (1957). Some industrial applications of military operations research methods. *Operations Research*, *5*(2), 201–204. doi:10.1287/opre.5.2.201

Kress, M., Caulkins, J. P., Feichtinger, G., Grass, D., & Seidl, A. (2018). Lanchester model for three-way combat. *European Journal of Operational Research*, *264*(1), 46–54. doi:10.1016/j.ejor.2017.07.026

Kress, M., Lin, K., & MacKay, N. J. (2017). The Attrition Dynamics of Multilateral War. *Operations Research.*

Lanchester, F. W. (1916). *Aircraft in warfare: The dawn of the fourth arm.* Constable limited.

Lanchester, F. W. (1956). Mathematics in warfare. *The World of Mathematics, 4*, 2138-2157.

Lin, K. Y., & MacKay, N. J. (2014). The optimal policy for the one-against-many heterogeneous Lanchester model. *Operations Research Letters*, *42*(6-7), 473–477. doi:10.1016/j.orl.2014.08.008

Liu, J. X., Zhao, D. M., & Wang, F. (2013). Networks Attack-Defense model based on the improved Lanchester equation. In *Machine Learning and Cybernetics (ICMLC), 2013 International Conference on* (Vol. 3, pp. 1083-1086). IEEE.

Morse, P., & Kimball, G. (1950). *Method of Operations Research.* New York: John Wiley & Sons.

Peterson, R. H. (1967). Letter to the Editor—On the "Logarithmic Law" of Attrition and its Application to Tank Combat. *Operations Research*, *15*(3), 557–558. doi:10.1287/opre.15.3.557

Pettit, L. I., Wiper, M. P., & Young, K. D. S. (2003). Bayesian inference for some Lanchester combat laws. *European Journal of Operational Research*, *148*(1), 152–165. doi:10.1016/S0377-2217(02)00462-9

Ross, S. M. (2014). *Introduction to probability models.* Academic Press.

Shephard, R. W. (1963). War gaming as a technique in the study of operational research problems. *The Journal of the Operational Research Society*, *14*(2), 119–130. doi:10.1057/jors.1963.20

Stanescu, M., Barriga, N., & Buro, M. (2015). Using Lanchester attrition laws for combat prediction in StarCraft. *Eleventh Annual AAAI Conference on Artificial Intelligence and Interactive Digital Entertainment (AIIDE)*, 86-92.

Taylor, J. G. (1980). *Lanchester-Type Models of Warfare* (Vol. 1). Naval Postgraduate School Monterey CA.

Taylor, J. G. (1980). *Lanchester-Type Models of Warfare* (Vol. 2). Naval Postgraduate School Monterey CA.

Washburn, A. (2000). *Lanchester systems.* Naval Postgraduate School Monterey CA. doi:10.21236/ADA383409

KEY TERMS AND DEFINITIONS

Ancient War: Historical wars in which classical weapons such as sword, axe, and arrows are used.

Attrition: Gradually making something weaker and destroying it, especially the strength or confidence of an enemy by repeatedly attacking it.

Attrition Rate: The speed at which attrition happens or changes, or the amount or number of times it happens or changes in a particular period.

Combat: A fight, especially during a war.

Concentration: Cumulative fighting capability of a military force.

Forces: The military organizations for land, air, and sea.

Modern War: A war in which directed firearms such as artilleries, cannons, etc. are deployed.

Chapter 8
Multi-Criteria Decision Making:
A Cast Light Upon the Usage in Military Decision Process

Tolga Temucin
Turkish Naval Forces, Turkey

ABSTRACT

Multi-criteria decision making (MCDM) is a discipline that explicitly considers assessing alternatives in a decision problem with respect to multiple criteria. Those methods are frequently used to solve real-life decision problems that incorporate multiple, conflicting, and incommensurate criteria. Considering the chaotic, complex, and ambiguous nature and the dynamics of the military operations, most decision problems observed in military organizations also follow a similar structure involving multiple criteria. This chapter gives an overview of the basic decision-making problem types and decision processes observed in military organizations and provides information on the MCDM methodologies adopted to solve those problems.

INTRODUCTION

Like all other decision processes, the decision process in military operations and; their integral part military problems, mainly concerns with selection of the best alternative under present situation. However, the complex structure of military problems makes the determination of the best alternative nearly impossible in most cases due to the limited capability and emotional characteristic of human. In this context; Multi-criteria Decision Making (MCDM) methods fits to the chaotic, complex and ambiguous nature of military problems.

If we go deeper, we will easily notice that this complicated but inalienable decision-making process starts with the life in our world, and at the beginning, humans used it in very simple decisions for basic requirements like where to sleep and what to eat. In initial periods of humanity, namely The Stone and Iron Ages there was limited input that had to be considered during decision-making activities. However, the complexity of decision processes has changed in the course of time. Technological improvements in science, improved theories in mathematics and other changes speed up the progress of military systems

DOI: 10.4018/978-1-5225-5513-1.ch008

as well as human and exposed new problem fields. Beginning from this moment decision processes wouldn't be easy as they were before due to the increased number of inputs, conflicting criteria, importance weights, etc. Thus, the evolution of MCDM methods has started.

MCDM methods propose different methodologies to help decision maker(s). The main concept of these methods is maximizing the benefit criteria while minimizing the cost criteria to determine the best option (alternative). Some of these methods determine the best option while some others order the options from best to worst. Although solution process of most MCDM methods does not seem complicated, they become complexes when the number of alternatives and conflicting criterions increase; which, unfortunately, is a common situation that decision-makers face in most military decision problems. Consequently, this deficiency led to the existence of computer-based decision support systems to simplify the procedures.

The aim of this chapter is to provide a basic knowledge of the decision process, MCDM methods and light a way for the usage of MCDM in military decision problems. Information about some common MCDM techniques, e.g. ELECTRE, TOPSIS, PROMETHEE will also be illustrated to support the idea of global applicability of MCDM for possible users in the field of military.

DECISION MAKING AND MULTI-CRITERIA DECISION MAKING

Human is a living creature that makes a decision to determine the best choice by compiling lots of input presented in a raw status to meet different kinds of requirements.

It is appropriate to define "decision process" which humans execute in every second of life as below:

- Decision making is an act of choice of an individual or a group of individuals. It is the exercising of one's free will to choose a single alternative or a hierarchy of alternatives among available options (Gunasekera, 2010),
- Decision making is about asking clear questions and obtaining clear and definite answers, for example, "where should we go?", "what options have we got?", "what should we do?" and "what should our strategy be?" and so on (Drummond, 2001),
- Decision-making process which is related to the selection or the preference of a choice when there are more than one is the sum of physical and mental efforts (Tosun, 1992),
- A decision implies the end of deliberation and the beginning of an action (Buchanan & O'Connell, 2006), and
- Decision making is a procedure to find the best alternative among a set of feasible alternatives (Cebi & Kahraman, 2010; Karatas, 2017b).

To make a decision means jumping from one state of mind to another. In this exercise, the mind follows a certain process. The objective of this process is to find the best alternative. Sometimes the number of alternatives increases a lot. Besides, the decision-maker has to take into consideration the conflicting criteria having different importance weights and assess alternatives with respect to each of these criteria. There is also a growing tendency that, in many real-world decision problems, the planners and decision makers are likely to pursue multiple and possibly conflicting objectives (Karatas, 2017a; Karatas & Yakici, 2018).

The decision-maker has to consider lots of criteria to fulfill the constraints such as economic concern, environmental conditions, technical constraints, organizational rules, etc. A decision-maker refers to multi-criteria decision-making methods whenever human perception is not enough to reach a logical solution for the problems encountered with. Multi-criteria decision making is one of the most widely used decision methodologies in science, business, government, and engineering worlds (Karacan, Tozan, & Karatas, 2016). MCDM methods can help to improve the quality of decisions by making the decision process more explicit, rational and efficient (Montazer, Saremi, & Ramezani, 2009). MCDM is a powerful tool which is widely used for evaluating problems that contain multiple, usually conflicting criteria (Paksoy, Pehlivan, & Kahraman, 2012). It is an operational evaluation and decision support approach suitable for addressing complex problems featuring high uncertainty, conflicting objectives, multi-interests and perspectives (Kaya & Kahraman, 2011).

Common aims in MCDM methods are as below:

- To evaluate lots of messy information,
- To execute decision process systematically,
- To combine expert opinions with subjective evaluations of decision-makers,
- To analyze complicated subjects which are difficult to understand absolutely, and
- To create a common platform where there are lots of decision maker, to ease the communication and make debates possible.

A BRIEF HISTORY

Today there are numerous tools which can be used during a decision-making process unlike a few hundred years ago. It is because of the progress concerning numbers which couldn't complete their evolution until the seventeenth century. In early ages numbers were expressed by scratches on clay tablets or notches on tree branches. In Ancient Egypt, numbers were specified by some figures like "I" for 1, "^" for 10, "?" for 100 and etc. In Ancient Mesopotamia "D" and "O" were used instead of 1 and 10, respectively. Both in Ancient Egypt and Ancient Mesopotamia, there was no figure symbolizing 0 (zero). Moreover, Romans used letters to symbolize numbers such as "I" for 1, "V" for 5, "X" for 10, "L" for 50, "C" for 100, "D" for 500, and "M" for 1000. Just like the numbers used in Ancient Egypt and Ancient Mesopotamia, Roman numbers were not appropriate for arithmetic operations because of the deficiency of zero. Hindu numeral system was the first system containing zero. This progress simplified arithmetic operations and enticed philosophers to investigate the nature of numbers.

Arabs, through their invasion of India, became familiar with Hindu numeral system which enabled them to incorporate eastern intellectual advances into their own scholarship, and scientific research. The results were momentous, first for the Arabs and then for the West. Thus, new methods of calculation replaced the abacus, which for centuries had been the only tool for arithmetic operations everywhere from the Mayans to India and the Orient (Bernstein, 1998).

The first known work was written by al-Khowarizmi where we get the word "algorithm", a mathematician lived around 825. He was the first to establish rules for adding, subtracting, multiplying, and dividing with the new Hindu numerals. The story of numbers in west begins around 1200s with the book titled as "Liber Abaci" which was written by Leonardo Pisano (1170 – 1250) who was known for most of his life as Fibonacci, the name by which he is known today. Bernstein (1998) claims that "Liber Abaci" made people aware of a new world in which numbers could be substituted for the Greek and Roman systems that used letters for counting and calculating. In that period one of the interesting inventions of Leonardo Pisano was the "Fibonacci Series" that permits people to make predictions.

Although it was obvious that Hindu-Arabic numeral system was more useful than Greek, and Roman systems in computations, it was the beginning of the 1500s that the new numeral system started to spread with the invention of printing. The improvement continued with the studies of Luca Paccioli (1446/7 – 1517), a Franciscan monk lived around the 1490s. His masterwork "Summa de arithmetic" set out the basic principles of algebra and contained multiplication tables all the way up to 60 x 60. Later one physician named Girolamo Cardano (September 24, 1501 – September 21, 1576) published "Ars Magna (The Great Art)" in 1545. It was the first major work of Renaissance concentrating on algebra. Cardano's another study named as "Liber de Ludo Aleae (Books on Games of Chance)" was the first serious effort to develop statistical principles of probability. In his autobiography, Cordano says that he wrote the book in 1525 and rewrote it in 1565. However, the book was published long after his death when the impressive progress in the theory of probability had been made by others who were unaware of Cardano's efforts. Galileo (February 15, 1564 – January 8, 1642) wrote a short essay on gambling titled as "Sopra le Scoperte dei Dadi (On Playing Dice)" in the beginning of 1620s. Bernstein (1998) claims that by the end of the seventeenth century, about a hundred years after the death of Cardano and less than fifty years after the death of Galileo, the major problems in probability analysis had been resolved. The next step was to tackle the question of how humans recognize and respond to the probabilities they confront. This, ultimately, is what risk management and decision making are all about.

Neither Cardano nor Galileo realized that they had discovered the most powerful tool of risk management ever to be invented: Probability. After the death of Galileo, two Frenchmen, Blaise Pascal (June 19, 1623 – August 19, 1662) and Pierre de Fermat (August 17, 1601, or 1607/8 – January 12, 1665) had continued to improve in probability analysis. The correspondence between them in 1654 was the messenger of the future improvements in probability branch. According to Buchanan and O'Connell (2006), Blaise Pascal and Pierre de Fermat developed a way to determine the likelihood of each possible result of a simple game.

From the start of humanity, humans made decisions with limited data. It is not logical to refer to millions of people if the new product meets the expectations or not. In such cases sampling is essential. John Graunt (April 24, 1620 – April 18, 1674) had written "Natural and Political Observations Made upon the Bills of Mortality" which is interested in sampling methods and the calculation of probabilities. Following him, William Petty (May 27, 1623 – December 16, 1687) and Edmund Halley (November 8, 1656 – January 14, 1742) applied the concept of probability to the analysis of raw data. Afterward, Jacob Bernoulli (December 27, 1654 – August 16, 1705) proposed central limit theorem.

One article named as "Specimen Theoriae Novae de Mensura Sortis (Exposition of a New Theory on the Measurement of Risk)" is emerged in 1738. The writer was the Swiss mathematician Daniel Bernoulli (February 8, 1700 – March 8, 1782). The introduced concept was that "the value of an item must not be based on its price, but rather on the utility that it yields." It was the times when people were making decisions by utilizing the expected value concept. According to Bernoulli, the concept was limited to predict the future in order to make decisions. Bernoulli claims that to make a decision requires two distinct elements. The first one is the objective facts and the second one is the subjective view that is influenced by the thoughts of decision maker about the consequences. For example, though it is well known that flying is far safer than any kind of land traveling, some people still prefer to use buses instead of airplanes for safety purposes. It is the traveler's weight determined for the consequences of different types of traveling which is helpful to explain the question why some people prefer the less-secure option. The brilliance of Bernoulli's formulation lies in his recognition that, while the role of facts is to provide a single answer to the expected value (the facts are the same for everyone), the subjective process will produce as many answers as there are humans involved. This was the first time when the measurement is applied to something that can't be counted. Other studies in probability and sampling are made by Abraham de Moivre (May 26, 1667 – November 27, 1754) and Thomas Bayes (1701 – April 7, 1761).

Carl F. Gauss (April 30, 1777 – February 23, 1855), the German mathematician published the book named as "Disquisitiones Arithmeticae" in 1801 which was about the theory of numbers. Afterward, he published the book named as "Theoria Motus (Theory of Motion)" in 1809 which is about probability. The bell curve structure (normal distribution) that reflects modern world events excellently is one of his findings in probability. Without the bell curve structure, today it will be impossible to determine the accuracy of the information in hand or to estimate the probability that an event will occur.

The Belgian scientist Lambert Adolphe Jacques Quetelet (February 22, 1796 – February 17, 1874) was influential in introducing statistical methods to social sciences. One of his studies named as "A Treatise on Man and the Development of His Faculties" which is published in 1835 established his reputation. In his study, he tried to define the characteristics of the average man (or woman) in some instances. He saw the bell curve structure everywhere he looked. In almost every instance, the errors, or deviations from the average, obediently distributed themselves according to the predictions of Gauss (Bernstein, 1998). Francis Galton (February 16, 1822 – January 17, 1911), English explorer, the inventor is well-known in statistics and risk management. His analysis led to the concept of correlation.

Numerous improvements influenced by lots of scientists, mathematicians, astronomers, inventors, and statisticians in history made it possible to encounter with the rough way of science letting to establish the foundations of decision making.

According to Tsoukias (2008), the origin of decision making can be fixed to sometime just before the Second World War, in the studies conducted by the British army on their new radar system installation and their efforts to break the German secret communication code.

Two inseparable parts of decision making developed during the time between the 1940s and 1970s: Computer-based Information Systems and Operations Research/Management Science. These branches let the evolution of Decision Support Systems.

At the time of the Second World War, the first fully electronic digital computers were developed (Silver, 1991). The inspiration for such devices was mostly supported by military needs in those years. The first commercial computer was delivered to the Bureau of the Census in 1951 (Silver, 1991). Later in those years, some other firms adopted computer-based systems for such transaction processing tasks (TPS) as payroll and billing. These systems allowed firms to profit from costs and encounter with business activities in a minimum time that wouldn't be possible if they were not created.

In the 1960s, the time when computation speed and the volume of processed information are increased, the notion to use the computerized systems in the managerial area emerged. The improvements led to the birth of what was commonly referred to as "Management Information Systems (MIS)". However, these computerized systems had presented managers some standardized periodical reports that looked like a mass of information. Most of the managers couldn't find whatever they looked for in those stack information. As a result, most managers didn't like these systems. According to Silver (1991), some managers quipped that "these systems were not at all Management Information Systems, systems providing useful information to management, but Information Management Systems, systems that simply managed large quantities of information."

The main problems demonstrating these systems as useless machines are:

- Reports didn't include much of the information that management required, and;
- Required information constituted only a small fraction of the whole report which made management to search for hours to reach the right information.

More specifically, predefined standardized reports were not responsive to the needs of managers to retrieve relevant information as they analyze problems and make decisions. Therefore, the solution may be found in designing systems that provide manager more timely access to information base and ensure computational capabilities specifically relevant to the manager's particular decision-making tasks (Silver, 1991). In the late 1960s, some of the systems were created having this philosophy (or understanding) which was commonly referred to as "Decision Support Systems (DSS)."

On the other hand; during the time between the 1940s and 1970s, several fundamental contributions in Operations Research and Management Science appeared. Linear programming, simplex algorithm, game theory, expected utility theory, decision theory, fuzzy logic, artificial intelligence, optimization, simulation, and etc. can be counted in the group of improvements.

Some other improvements came after the 1970s can be summarized as (Buchanan and O'Connell, 2006):

- John D.C. Little developed the underlying theory and advanced the capabilities of Decision Support Systems in 1970;
- Irving Janis (May 26, 1918 – November 15, 1990) coined the term "groupthink" in 1972;
- Michael D. Cohen, James G. March, and Johan P. Olsen published "A Garbage Can Model of Organizational Choice" to advice organizations to search their informational trash bins for solutions thrown out earlier for lack of a problem in 1972;
- Henry Mintzberg described several kinds of decision makers and decision-making positions within the context of managerial work in 1973;
- Victor Vroom and Philip Yetton developed the Vroom–Yetton model which explains how different leadership styles can be used to solve different types of problems in 1973;
- Amos Tversky (March 16, 1937 – June 2, 1996) and Daniel Kahneman published "Prospect Theory" which demonstrates that the rational model of economics fails to describe how people arrive at decisions when they face with the uncertainties of real life in 1979;
- John Rockart explored the specific data needs of chief executives, leading to the development of executive information systems in 1979;
- W. Carl Kester raised corporate awareness of real options by suggesting that managers must think of investment opportunities as options on the company's future growth in 1984;
- Howard Dresner introduced the term "business intelligence" to describe a set of methods that support sophisticated analytical decision making for improving business performance in 1989, and;
- Malcolm Gladwell explored the notion that our instantaneous decisions are sometimes better than those based on the lengthy rational analysis in 2005.

DECISION MAKING PROCESS

Decision making should start with the identification of the decision maker(s) and stakeholder(s) which will eventually be helpful to reduce the possible disagreement about problem definition, requirements, goals, and criterions (Baker, Bridges, Hunter, Johnson, Krupa, & Murphy, 2001).

The general decision-making process shown in Figure 1 can be summarized as follows:

Define the Problem

Problem definition is the crucial first step in any decision-making process. The point is to express the problem with a clear one sentence statement that describes both the initial and the desired conditions. Problem definition forms the basis of the process which will definitely affect the latter steps. A perfect definition for the wrong problem will not be helpful to encounter the core problem.

Problems demonstrate themselves with some symptoms. For example; fever or headache are symptoms of illness. Unfortunately; people mostly deal with symptoms which will provide temporary solutions instead of the core problem. Baker et al. (2001) mention that the key to developing an adequate problem statement is to ask enough questions about the problem to ensure that the final report will clearly answer the questions of reviewers and stakeholders. Problem definition can be summarized as in Figure 2.

Determine Requirements

Requirements are the minimum limits of conditions that any solution has to meet in order to be evaluated as acceptable. Any requirement that is far away to answer the purpose can be eliminated directly.

In mathematical form, requirements are the constraints that describe the set of the feasible solutions of the decision problem. Hence, it is very important that even if subjective evaluations may occur in the latter steps, requirements must be stated in exact quantitative forms, i.e. for any possible solution it has to be decided unambiguously if it meets the requirements or not.

Figure 1. General Decision Making Process
Baker, Bridges, Hunter, Johnson, Krupa, & Murphy, 2001.

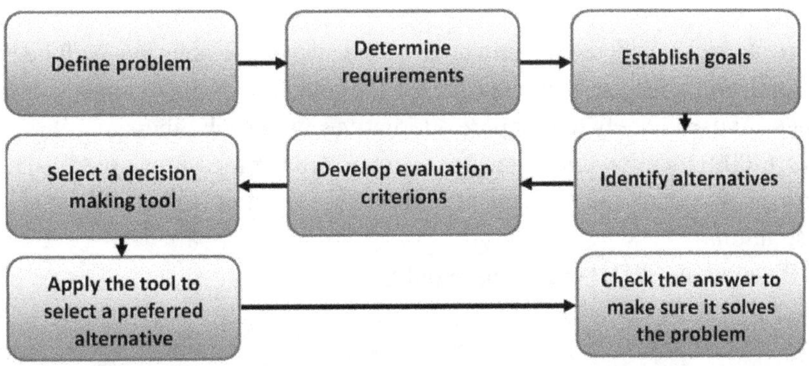

Figure 2. Problem Definition
Baker, Bridges, Hunter, Johnson, Krupa, & Murphy, 2001.

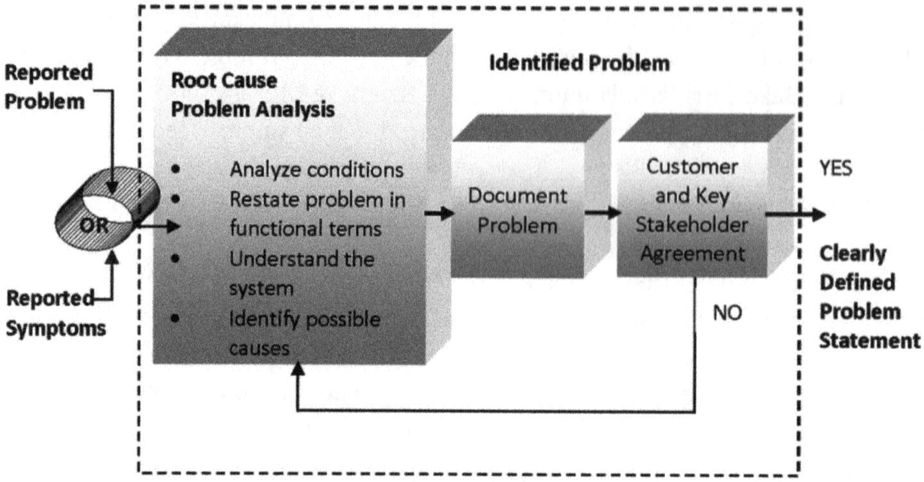

Establish Goals

Goals are the broad statements that explain the desired conclusions. Sometimes they may conflict but this is neither unusual nor cause for concern (Baker et al., 2001).

Identify Alternatives

Alternatives are elements that help decision maker to reach the desired condition. In this context; alternatives that don't meet the requirements should not be considered (Tozan, 2011; Yanar & Tozan, 2012). According to Baker et al. (2001) if an alternative does not meet the requirements, three actions are available:

- The alternative is discarded;
- The requirement is altered or eliminated; and
- The requirement is restated as a goal.

Define Criterion(s)

When all requirements are considered, mostly none of the alternative solutions will be best with respect to each. Otherwise; it won't be wise to waste time for the selection of the obvious solution. This is why MCDM techniques are used widely to compare alternatives with each other.

Verderber et al. (2009) mention that criterions are standards or measures used for judging the merits of proposed solutions. They provide a blueprint for how the decision maker will evaluate the virtues of each alternative solution.

According to Baker et al. (2001) criterions should be:

- Able to discriminate alternatives;
- Complete to include all goals;
- Non-redundant;
- Few in number, and;
- Meaningful.

Some methods used to determine criterions are as follows:

- **Brainstorming:** The method, found by Alex F. Osborn (May 24, 1888 – May 13, 1966), takes nearly ninety minutes in which the number of ideas is important as well as quality. Verderber et al. (2009) mention that it is an uncritical, non-evaluative process of generating possible solutions by being creative, suspending judgment, and combining or adapting ideas.
- **Gordon's Method:** William J. J. Gordon (September 09, 1919 – June 30, 2003) is the father of this method. Gordon's method is a kind of teamwork just like brainstorming; however, in this method, the fact about the problem is known by only the leader of the teamwork. In this method, the evaluation phase is processed by the leader of the teamwork. Hence; the success of this technique mostly depends to the leader.
- **Buzz Meetings:** It is a kind of teamwork in which brainstorming is conducted by four to six people.
- **Nominal Grouping:** The ideas of the team members are demanded in black and white in this method. Later these ideas are put to the vote without any discussion.
- **Delphi Method:** In this method members of teamwork do not have a face to face discussion, unlike the Nominal Grouping method.
- **Reverse Direction:** Team members consider available alternatives, identify differences among them, and develop criterion that reflects these differences (Baker, Bridges, Hunter, Johnson, Krupa, & Murphy, 2001).
- **Synectics:** The method, developed by George M. Prince (April 05, 1918 – June 09, 2009) and William J. J. Gordon, approaches to the problem with a comparison to an example in biology scope.
- **Philips 66 Method:** Firstly, a major team is divided into minor teams in which there exist five to six people. Each team determines a leader, searches for possible solutions, and determines the team decision. In the session in which the participants are team leaders, each team leader expresses his / her team's decision.

Select a Decision-Making Tool

Method selection phase must be based on the complexity of the problem and the experience of the team. Nevertheless; if it is appropriate decision maker should prefer the methods that have simple procedures (Toker, Tozan, & Vayvay, 2013).

Evaluate Alternatives With Respect to Each Criterion

Alternatives can be evaluated in one of the three ways as below:

- Only quantitatively,
- Only qualitatively, and
- Both quantitatively and qualitatively.

Validate Solution(s) Against the Problem Statement

After the evaluation process has selected a preferred alternative, the solution should be checked to ensure that it truly solves the identified problem. A final solution should fulfill the desired state, meet requirements, and achieve the aims (Baker, Bridges, Hunter, Johnson, Krupa, & Murphy, 2001).

Although the characteristics of the military decision-making process are the same as the all other decision-making processes, from the point of risks, uncertainty and probability losses due to the made decisions, differs from the others. The following figure illustrates the military decision-making process (http://www.au.af.mil/au/awc/awcgate/army/fm101-5_mdmp.pdf).

Figure 3. 'Military Decision Making Process'
Source: http://www.au.af.mil/au/awc/awcgate/army/fm101-5_mdmp.pdf

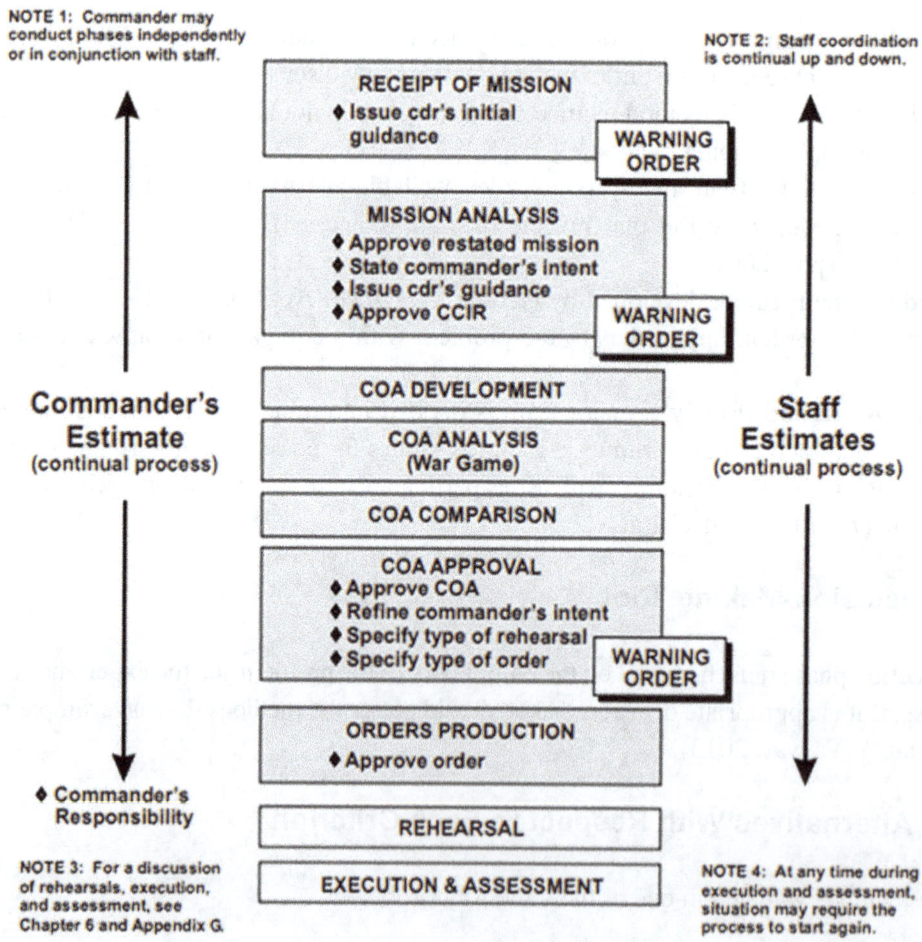

DECISION SUPPORT SYSTEMS

Decision Support Systems term emerged in 1971 when G. Anthony Gorry and Michael S. S. Morton published the paper named as "A Framework for Management Information Systems". DSS are computerized systems mostly using the methods in Operations Research and Management Science to help decision makers in their decisions with a minimum time usage. There are several MCDM methodologies in the literature which are basically developed for assisting decision-makers and planers in their decision making or ranking problems. These methods permit them to evaluate different alternatives for achieving their goals (Razi & Karatas, 2016).

There are many definitions of decision support systems. Little (1970) defines them as a "model-based set of procedures for processing data and judgments to assist a manager in his decision making." Turban and Aronson (2001) define them as they were aimed at decisions where the judgment was required or at decisions that could not be completely supported by algorithms. Karataş and Gecili (2012) define decision support systems as "knowledge-based systems" since they attempt to formalize related knowledge to utilize within mechanized reasoning. Kou et al. (2011) define them as interactive computer-based systems that support decision-making activities. Although there is a common sense that these systems are one of the subsets of "Computer Based Information Systems (CBIS)", there is no consensus for explanations defining them as shown in Table 1.

When different definitions are aggregated, some common characteristics and capabilities can be summarized as (see Figure 4) (Turban & Aronson, 2001; Ozturk & Tozan, 2015; Temucin & Tozan 2016):

- These systems support executives working in each level;
- Not only individuals but also groups of individuals benefit from these systems;
- Not only independent decisions but also decisions following each other (sequential) can be supported by these systems;
- Decision support systems contain intelligence, design, choice, and implementation, different phases of decision-making process;
- Decision support systems can be adapted to any kind of situations;
- They are user-friendly systems;
- In these systems user controls the machine; and
- Decision support systems try to improve the effectiveness of decision making.

Table 1. Concepts Underlying DSS Definitions

Source	DSS Defined in Terms of
Gorry and Scott-Morton (1971)	Problem type, system function (support)
Little (1970)	System function, interface characteristics
Alter (1980)	Usage pattern, system objectives
Moore and Chang (1980)	Usage pattern, system capabilities
Keen (1980)	Development process
Bonczek et al. (1989)	System components

Turban & Aronson, 2001.

Figure 4. Characteristics and Capabilities of DSS
Turban & Aronson, 2001.

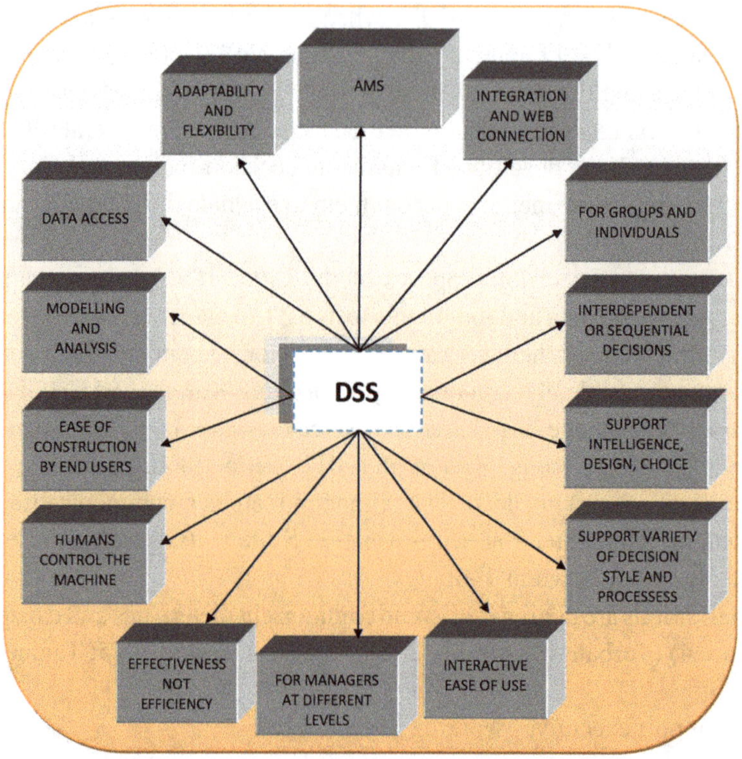

Nowadays there are lots of decision support systems utilizing different methods of MCDM. A comprehensive comparison of some decision support systems is presented in Table 2 and Table3.

Eshlaghy and Homayonfar (2011) had studied 628 scientific papers in order to classify recent researches on MCDM. According to study they explained that only 61.5 percent of the papers include an application area and divided them into 12 specific areas, namely Environment Management, Water Management, Business and Financial Management, Transportation and Logistics, Manufacturing and Assembly, Energy Management, Agricultural and Forestry Management, Managerial and Strategic Planning, Project Management and Evaluation, Social Service, Military Service and Other Topics Moreover there seems to be an upward trend on the number of papers if we come closer to the present day. The frequency of papers by application area and by year is presented in Figure 5 and Figure 6, respectively.

Another study is conducted by Behzadian et al. (2010). The study was concerned with the literature review of the PROMETHEE method. For this study, 217 papers were analyzed and classified into two groups as papers having an application (89.9%) and others (10.1%). They classified application papers into nine specific areas, namely Environment Management, Hydrology and Water Management, Business and Financial Management, Chemistry, Logistics and Transportation, Manufacturing and Assembly, Energy Management, Social, and Other Topics. Distribution of papers by application areas is presented in Table 4. Moreover, papers are classified by publication year as shown in Table 5.

Table 2. Features of Some of Decision Support Systems – I

No.	Software Name	Supported MCDM Method(s)	Fuzzy Methods Supported	Sensitivity Analysis	Group Decision Making
1	1000Minds	PAPRIKA	NO	YES	YES
2	AliahThink	AHP (Analytical Hierarchy Process)	NO	YES	YES
3	Choice Reaults	AHP	NO	YES	YES
4	Criterium DecisionPlus	AHP, SMART	NO	YES	NO
5	Decision Lab 2000	PROMETHEE	NO	YES	NO
6	Decision Lens	AHP	NO	YES	YES
7	Decision Manager	Utility Theory	NO	YES	NO
8	Decision Oven	AHP	NO	NO	NO
9	SETED 1.0	ELECTRE I, TOPSIS, PROMETHEE II, Fuzzy ELECTRE I, Fuzzy TOPSIS	YES	NO	NO
10	DESTEC 1.0	AHP, ANP, FAHP, FANP	YES	NO	NO
11	DecisionPad	AHP, MAUT	NO	YES	NO
12	D-Sight	PROMETHEE, MAUT	–	YES	YES
13	ELECCALC	ELECTRE II	NO	NO	NO
14	ELECTRE III-IV	ELECTRE III AND IV	NO	NO	NO
15	ELECTRE IS	ELECTRE I	NO	NO	NO
16	ELECTRE TRI	ELECTRE TRI	NO	NO	NO
17	ERGO	MAUT	NO	YES	NO
18	Expert Choice	AHP	NO	YES	YES
19	HiPriority	SMART	NO	NO	NO
20	Hiview3	MACBETH	NO	YES	YES
21	IRIS	ELECTRE TRI	NO	NO	NO
22	Logical Decisions	AHP	NO	YES	NO
23	MACBETH	MACBETH	NO	YES	NO
24	MakeItRational	AHP	NO	YES	YES
25	MindDecider	AHP	NO	YES	YES
26	NAIADE	NAIADE	YES	NO	NO
27	Open Decision Maker	AHP	NO	YES	NO
28	Prime Option	MFEP	NO	YES	NO
29	RightChoiceDSS	AHP	NO	YES	YES
30	SelectPro	AHP	NO	YES	YES
31	Tribium Decisions	Utility Theory	NO	YES	YES
32	Web-HIPRE	AHP	NO	YES	YES

Yanar, 2011.

Table 3. Features of Some of Decision Support Systems – II

No.	Software Name	Limitations	Operating System
1	AIM	50 Alternatives and 10 Criterions	DOS
2	CRITERIUM DECISION PLUS 3.0	50 Alternatives and 7 Levels of Hierarchy for Criterions	Windows
3	Decision Lab 2000	150 Alternatives and 150 Criterions	Windows
4	DEFINITE	–	Windows
5	ELECCALC	–	DOS
6	ELECTRE IS	–	DOS
7	ELECTRE III\IV (Version 3.1)	–	Windows
8	EXPERT CHOICE 11.5	Less Than Nine Alternatives and 9^5 Criterions	Windows
9	HIPRE3+ (Version 3.14)	20 Hierarchical Levels and 50 Elements per Level	DOS \ Windows
10	HIVIEW & EQUITY	–	Windows
11	LDW	–	Windows
12	MACBETH	–	Windows
13	QUALIFLEX	500 Alternatives and 20 Criterions	DOS
14	UTA PLUS	–	Windows
15	VIMDA	500 Alternatives and 10 Criterions	Windows
16	VISA	–	Windows

Figure 5. Frequency of Papers by Application Area
Eshlaghy & Homayonfar, 2011.

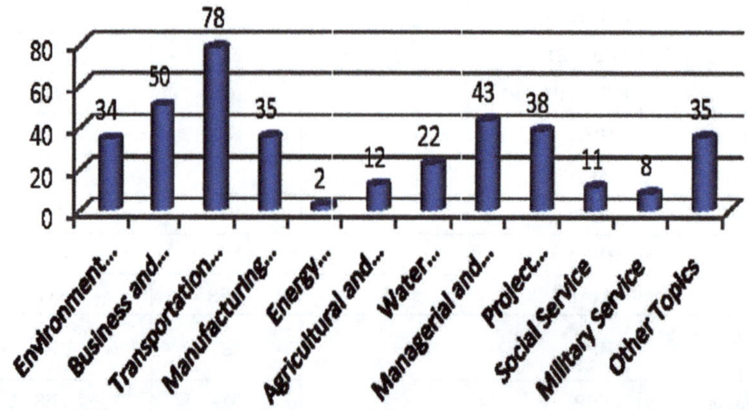

Figure 6. Frequency of Papers by Year
Eshlaghy & Homayonfar, 2011.

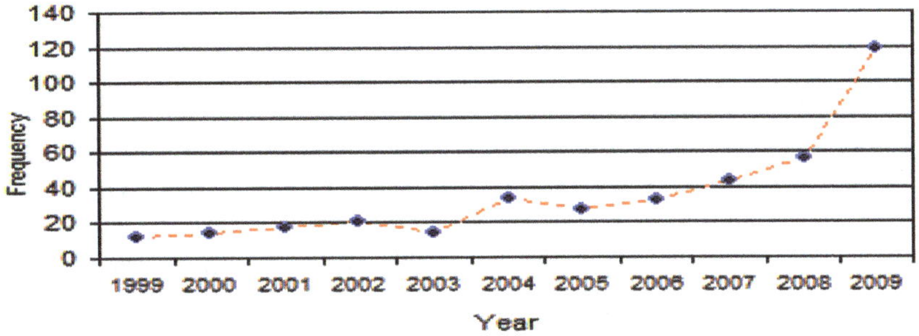

Table 4. Distribution of Papers by Application Areas

Application Areas	Number	%
Environment Management	47	24.1
Business and Financial Management	25	12.8
Hydrology and Water Management	28	14.4
Chemistry	24	12.3
Logistics and Transportation	19	9.7
Energy Management	17	8.7
Manufacturing and Assembly	19	9.7
Social	7	3.6
Other Topics	9	4.6
Total	195	100

Behzadian, Kazemzadeh, Albadvi, & Aghdasi, 2010.

Table 5. Frequency of Papers by Year

Year	Number	%
Prior to 1990	11	5.1
1991-1993	20	9.2
1994-1996	17	7.8
1997-1999	27	12.4
2000-2002	28	12.9
2003-2005	48	22.1
Since 2006	66	30.4
Total	217	100

Behzadian, Kazemzadeh, Albadvi, & Aghdasi, 2010.

ELECTRE

Georgopoulou et al. (2003) utilized the ELECTRE in order to define national priorities for a National Action Plan for greenhouse gases mitigation in the energy sector and to formulate a relevant time-schedule for actions' implementation. Qin et al. (2008) developed an MCDM-based expert system to facilitate assessment of climate-change impacts on socio-economic and environmental sectors, as well as the formulation of relevant adaptation policies in terms of water resources management.

Papadopoulos and Karagiannidis (2008) used the ELECTRE III method to rank some energy choices such as wind energy, solar energy, geothermal energy, and biomass in terms of the values of the economic criteria such as CO2, and the cost for not covering the demand for the purpose of electricity generation. Wang and Triantaphyllou (2008) tested the performances of the ELECTRE II and III methods and reported some ranking irregularities for both of the methods. Xidonas et al (2009) employed the ELECTRE Tri method for selecting the attractive equities, through the evaluation of the overall corporate performance of the corresponding firms. Montazer et al. (2009) discussed the architecture of a fuzzy system including both modules, evaluation and ranking ones, utilizing fuzzy concept for dealing with the uncertainty of the problem.

ELECTRE I

Roy (1991) reports that ELECTRE I was the first decision-aid method using the concept of outranking relation.

The ELECTRE method for choosing the best action(s) from a given set of actions was thus devised in 1965, and was later referred to as ELECTRE I. In that same year (July 1965) the new multiple criteria outranking method was presented for the first time at a conference (les journ´ees d'´etudes sur les m´ethodes de calcul dans les sciences de l'homme), in Rome, Italy. Nevertheless, the original ideas of ELECTRE methods were first merely published as a research report in 1966, the notorious Note de Travail 49 de la SEMA. Shortly after its appearance, ELECTRE I was found to be successful when applied to a vast range of fields, but the method did not become widely known until 1968 when it was published in RIRO, la Revue d'Informatique et de Recherche Op´erationnelle.

Steps of ELECTRE I can be summarized as follows:

Step 1 - Construct Decision (*A*) and Weight (*W*) Matrices: The decision matrix, *A*, which consists of three components is described as follows:

These components are viable alternatives which are defined by a_1, $a_2,\ldots,$ a_i, $a_k,\ldots,$ a_m (rows); criterions which are defined by c_1, $c_2,\ldots,$ $c_j,\ldots,$ c_n (columns) and the performances of alternatives for given criterions which are defined by $a_{ij}\left(i=1,2,\ldots,m\right)\left(j=1,2,\ldots,n\right)$. The decision matrix is the starting matrix developed by the decision maker.

$$A = \begin{pmatrix} a_{11} & \cdots & a_{1n} \\ \vdots & \ddots & \vdots \\ a_{m1} & \cdots & a_{mn} \end{pmatrix}$$

In this step it is also required from decision maker to decide the weights of criterions which are defined by w_1, w_2,..., w_j,..., w_n for each criterion, $\sum_{j=1}^{n} w_j = 1$,

Step 2 - Convert Criterions to the Same Type: Conversion is done by dividing "1" with the evaluation values in the column of criterion which is desired to be converted.

Step 3 - Construct Normalized Decision Matrix (*X*): The data of the decision matrix, *A*, comes from different sources, so it is necessary to normalize it in order to transform it into a dimensionless matrix which allows the comparison of the various criteria. The normalized decision matrix which is described as below is calculated with the formula given below by using the elements of decision matrix, A.

$$X = \begin{pmatrix} x_{11} & \cdots & x_{1n} \\ \vdots & \ddots & \vdots \\ x_{m1} & \cdots & x_{mn} \end{pmatrix}, x_{ij} = \frac{a_{ij}}{\sqrt{\sum_{k=1}^{m} a_{kj}^2}}$$

Step 4 - Construct Weighted Normalized Decision Matrix (Y): The importance degrees of criterions can be different for the decision maker. So this difference has to be reflected to the normalized decision matrix. The weighted normalized decision matrix which is described as below is calculated by the formula (2).

$$Y = \begin{pmatrix} y_{11} & \cdots & y_{1n} \\ \vdots & \ddots & \vdots \\ y_{m1} & \cdots & y_{mn} \end{pmatrix}, y_{ij} = x_{ij} * w_j$$

Step 5 - Determine Concordance (C_{ik}) and Discordance (D_{ik}) Sets: In general, the criteria can be classified into two types: benefit and cost. The benefit criterion means that a higher value is better while for the cost criterion is valid the opposite (Krohling & Campanharo, 2011). The type of criterions makes it mandatory for decision maker to determine which formula to use while designating the elements of concordance and discordance sets.

If we suppose that the criterion we are dealing with is a benefit criterion, we can form the concordance set with criterions in which one of the alternatives of pairwise comparisons that we compare with another alternative has the higher performance value. In addition; the discordance set includes the criterions of pairwise comparisons which do not belong to concordance set. The formulas that we can use during the formation of concordance and discordance sets are as below:

$$C_{ik} = \{\forall j \rightarrow y_{ij} \geq y_{kj}\}, D_{ik} = \{\forall j \rightarrow y_{ij} < y_{kj}\}$$

In case the criterion we are dealing with is a cost criterion, the formulas that we can use during the formation of concordance and discordance sets are as below:

$$C_{ik} = \{\forall j \rightarrow y_{ij} < y_{kj}\}, D_{ik} = \{\forall j \rightarrow y_{ij} \geq y_{kj}\}$$

The number of the concordance set as well as the discordance set in a multi-criteria decision-making problem is equal to $(m * m - m)$ where m defines the number of alternatives. The number of elements in a concordance and discordance set varies from 0 to n where n defines the number of criteria.

Step 6 - Construct Concordance (C) and Discordance (D) Matrices: Concordance sets are used to form the concordance matrices. The concordance matrix is mxm dimensional and does not take any value when $i = k$. Both the concordance matrix and the formula which helps to calculate the elements of it are given as below:

$$C = \begin{pmatrix} - & c_{12} & c_{13} & \cdots & c_{1m} \\ c_{21} & - & c_{23} & \cdots & c_{2m} \\ . & & & & . \\ . & & & & . \\ . & & & & . \\ c_{m1} & c_{m2} & c_{m3} & \cdots & - \end{pmatrix}, c_{ik} = \sum_{j \in C_{ik}} w_j$$

Similarly, discordance sets are used to form the discordance matrices. The discordance matrix is mxm dimensional and does not take any value when $i = k$. Both the discordance matrix and the formula which helps to calculate the elements of it are given as below:

$$D = \begin{pmatrix} - & d_{12} & d_{13} & \cdots & d_{1m} \\ d_{21} & - & d_{23} & \cdots & d_{2m} \\ . & & & & . \\ . & & & & . \\ . & & & & . \\ d_{m1} & d_{m2} & d_{m3} & \cdots & - \end{pmatrix}, d_{ik} = \frac{\sum_{j \in D_{ik}} \left| y_{ij} - y_{kj} \right|}{\sum_{\forall j} \left| y_{ij} - y_{kj} \right|}$$

Step 7 - Construct Dominance Concordance (F) and Dominance Discordance (G) Matrices: The mxm dimensional dominance concordance matrix is formed by the comparisons between the elements of concordance matrix, c_{ik}, and concordance threshold value, $\bar{c}$. Concordance threshold value, $\bar{c}$, is calculated by the formula as below:

$$\bar{c} = \frac{1}{m\left(m-1\right)} \sum_{i=1}^{m} \sum_{k=1}^{m} c_{ik}$$

where m symbolizes the number of alternatives. In other words; $\bar{c}$ is equal to the arithmetic mean of the elements which form the concordance matrix. The dominance concordance matrix and the elements, f_{ik}, forming the matrix which can be either 0 or 1 are expressed as below:

$$F = \begin{pmatrix} - & f_{12} & f_{13} & \cdots & f_{1m} \\ f_{21} & - & f_{23} & \cdots & f_{2m} \\ . & & & & . \\ . & & & & . \\ . & & & & . \\ f_{m1} & f_{m2} & f_{m3} & \cdots & - \end{pmatrix}, f_{ik} = \begin{Bmatrix} 0 & \text{if} & c_{ik} < \bar{c} \\ 1 & \text{if} & c_{ik} \geq \bar{c} \end{Bmatrix}$$

The mxm dimensional dominance discordance matrix is formed by the comparisons between the elements of discordance matrix, d_{ik}, and discordance threshold value, $\bar{d}$. Discordance threshold value, $\bar{d}$, is calculated by the formula as below:

$$\bar{d} = \frac{1}{m\left(m-1\right)} \sum_{i=1}^{m} \sum_{k=1}^{m} d_{ik}$$

where m symbolizes the number of alternatives. In other words; $\bar{d}$ is equal to the arithmetic mean of the elements which form the discordance matrix. The dominance discordance matrix and the elements, g_{ik}, forming the matrix which can be either 0 or 1 are expressed as below:

$$G=\begin{pmatrix} - & g_{12} & g_{13} & \cdots & g_{1m} \\ g_{21} & - & g_{23} & \cdots & g_{2m} \\ . & & & & . \\ . & & & & . \\ . & & & & . \\ g_{m1} & g_{m2} & g_{m3} & \cdots & - \end{pmatrix}, g_{ik}=\begin{Bmatrix} 0 & \text{if} & d_{ik} \geq \bar{d} \\ 1 & \text{if} & d_{ik} < \bar{d} \end{Bmatrix}$$

Step 8 - Construct Total Dominant Matrix (E): Elements of the total dominant matrix, e_{ik}, can be calculated by multiplying f_{ik} and g_{ik} mutually. The *mxm* dimensional total dominant matrix whose elements can be either 0 or 1 is described as below:

$$E=\begin{pmatrix} - & e_{12} & e_{13} & \cdots & e_{1m} \\ e_{21} & - & e_{23} & \cdots & e_{2m} \\ . & & & & . \\ . & & & & . \\ . & & & & . \\ e_{m1} & e_{m2} & e_{m3} & \cdots & - \end{pmatrix}$$

Step 9 - Determine Importance Sequence of Alternatives: Consider that elements in total dominant matrix, e_{ik}, is calculated as below:

$$E=\begin{pmatrix} - & 0 & 0 \\ 1 & - & 0 \\ 1 & 1 & - \end{pmatrix}$$

where $e_{21}=e_{31}=e_{32}=1$. This situation presents the absolute advantage of;

- Second and third alternatives to first one; and
- Third alternative to the second one.

In this case; if we symbolize the alternatives as $A_i\left(i=1,2,3\right)$, the importance sequence of alternatives will be as A_3, A_2, A_1. At this point, it will be appropriate to mention about one weak characteristic of ELECTRE I. If there is only one alternative in the kernel at the end of solution procedure, it means that the applicable alternative is determined. However, if there is more than one alternative in the kernel, the question "which alternative is the applicable one?" recurs to the mind. "Net Concordance Indexes" and "Net Discordance Indexes" have to be calculated to fix this problem. The decision maker can find the dominant alternative with the help of these indexes. The alternative having the maximum net concordance index and minimum net discordance index is the applicable one among others. The formulas to calculate these indexes are given below:

$$C_p = \sum_{\substack{k=1 \\ k \neq p}}^{m} C_{pk} - \sum_{\substack{k=1 \\ k \neq p}}^{m} C_{kp}$$

$$D_p = \sum_{\substack{k=1 \\ k \neq p}}^{m} D_{pk} - \sum_{\substack{k=1 \\ k \neq p}}^{m} D_{kp}$$

TOPSIS

Chen and Tzeng (2004) combined the grey relation model with the concepts of the TOPSIS method to evaluate and select the best expatriate host country. Tzeng et al. (2005) applied the TOPSIS and VIKOR methods to determine the best compromise alternative fuel mode in public transportation buses. Chen et al (2006) presented a fuzzy decision-making approach based on the TOPSIS method concepts to deal with the supplier selection problem in supply chain management. A hybrid model in which the vendor evaluation problem is formulated by the combined use of the MCDM approach and a proposed five-step hybrid process and then the modified TOPSIS method is adopted to rank competing products in terms of their overall performances is proposed by Shyur and Shih (2006) for supporting the vendor selection process. Gamberini et al. (2006) presented a new heuristic that is based on the TOPSIS method and the well-known Kottas and Lau heuristic approach for solving the assembly line re-balancing problem.

A fuzzy TOPSIS method based on alpha level sets is proposed by Wang and Elhag (2006). They also show that the proposed fuzzy TOPSIS method performs better than the other versions of the TOPSIS method. The optimization of the Berth Allocation Problem is transformed into a multiple stage decision making procedure and a new multiple stage search method, namely stochastic beam search algorithm as well as improved beam search scheme, two-phase node goodness estimation, and stochastic node selection criteria techniques are proposed by Wang and Lim (2007) to solve the transformed problem. Wang and Chang (2007) developed an evaluation approach based on the TOPSIS method to choose optimal initial training aircraft in a fuzzy environment. Farahani and Asgari (2007) investigated the locating problem of some warehouses as distribution centers in a real-world military logistics system to fulfill the two objectives: To find the least number of distribution centers and locate them in the best possible locations.

Büyüközkan et al. (2008) proposed an MCDM approach in which fuzzy logic is adopted as the subjective considerations to effectively evaluate e-logistics-based strategic alliance partners. Sheu (2008) presented a hybrid neuro-fuzzy methodology to identify appropriate global logistics operational modes used for global supply chain management.

Anagnostopoulos et al. (2008) proposed a multi-criteria approach that tries to model human thinking. A fuzzy MCDM method to evaluate the financial performance of airlines is applied by Wang (2008). Lin and Gen (2008) proposed a multiobjective hybrid genetic algorithm approach for solving combinatorial optimization problems. A decision support system to choose the appropriate coating, excellent solution when resistance to corrosion, oxidation or low friction is demanded is presented by Athanasopoulos et al.

(2009). Wu et al. (2009) proposed a fuzzy MCDM approach for banking performance evaluation. Zammori et al. (2009) presented a framework to extend the capabilities of the traditional project-scheduling approach and define the resulting critical path. Huang et al. (2009) dealt with personnel assignment problem.

Shanian and Savadogo (2009) utilized some of the most widely potential MCDM models for solving a material selection problem of highly sensitive components. A hybrid approach to ensure the competitiveness requirements for major Turkish container ports to manage strategic decision making is proposed by Çelik et al (2009). Dağdeviren et al. (2009) dealt with weapon selection problem. İç and Yurdakul (2009) developed a decision support system, namely MACSEL, to help the decision makers in their machining center selection decisions. An expert selection system of how ventures can be evaluated and a favorable city can be selected to invest is developed by Lin and Tsai (2009). Yu and Hu (2010) developed an integrated MCDM approach that combines the voting method and the fuzzy TOPSIS method to evaluate the performance of multiple manufacturing plants. Sadi-Nezhad and Damghani (2010) presented a TOPSIS approach based on preference ratio and a fuzzy distance measurement. Yang et al. (2011) studied the robustness of different supply chain strategies under various uncertain environments.

PROMETHEE

Vaillancourt and Waaub (2004) tried to allocate equitable international greenhouse gases emission entitlements. Khalil et al. (2005) applied the PROMETHEE method and GAIA to facilitate the selection of the most suitable hydrothermal conditions for sludge destruction. Mavrotas et al. (2006) dealt with project prioritization problem under policy restrictions.

Wang and Yang (2007) proposed the use of AHP to analyze the structure of the outsourcing problem as well as determine weights of the criteria and PROMETHEE for final ranking in making information systems outsourcing decisions. Araz et al. (2007) dealt with outsourcing management problem. Some multiple criteria analysis techniques such as Weighted Summation, Range of Value, PROMETHEE II, Evamix, and Compromise Programming are compared by Hajkowicz and Higgins (2008). Kowalski et al. (2009) and Tsoutsos et al. (2009) dealt with sustainable energy.

Some of the sample studies on ELECTRE, TOPSIS, and PROMETHEE are presented in Table 6.

CONCLUSION

Every second mankind deals with distinct decision problems to perpetuate his / her life. As humans, it is our talent to solve these problems in a sensible way within insignificant time periods. However, when the problem size (the alternatives and criterions that have to be considered) increases, it becomes harder to arrive a logical conclusion. Additionally, the problems get even harder for the military decision problems which incorporate multiple and possibly conflicting assessment criteria (benefit and cost criterions in the same decision problem). In such cases, the conclusion reached for the problem may look suspicious due to the capacity of human judgment; which is a risk that should not be taken singlehanded in military operations.

Table 6. Some of the Sample Studies on ELECTRE, TOPSIS, and PROMETHEE

Author(s)	Published Year	Application Area	Used Method(s)
Georgopoulou et al.	2003	National Priorities for GHG Emission	ELECTRE Tri
Chen and Tzeng	2004	Selection of an Expatriate Host Country	TOPSIS
Vaillancourt and Waaub	2004	Equity in International Greenhouse Gases Abatement Scenarios	PROMETHEE II
Albadvi	2004	Formulation of National Information Technology Strategies	PROMETHEE II
Tzeng et al.	2005	Analysis of Alternative-Fuel Buses for Public Transportation	TOPSIS
Khalil et al.	2005	Selection of Hydrothermal Pre-treatment Conditions of Waste Sludge Destruction	PROMETHEE GAIA
Chen et al.	2006	Supplier Evaluation and Selection in Supply Chain Management	TOPSIS
Shyur and Shih	2006	Strategic Vendor Selection Problem	TOPSIS
Gamberini et al.	2006	Assembly Line Re-balancing Problem	TOPSIS
Wang and Elhag	2006	Bridge Risk Assessment	TOPSIS
Mavrotas et al.	2006	Project Prioritization Under Policy Restrictions	PROMETHEE V
Farahani and Asgari	2007	Facility Location Problem	TOPSIS
Wang and Chang	2007	Evaluation of Initial Training Aircraft	TOPSIS
Wang and Lim	2007	Berth Allocation Problem	TOPSIS
Albadvi et al.	2007	Decision Making in Stock Trading	PROMETHEE I
Araz et al.	2007	Outsourcing Management	PROMETHEE I-II
Wang and Yang	2007	Information Systems Outsourcing	PROMETHEE II
Wang and Triantaphyllou	2008	Ranking Irregularities	ELECTRE II - III
Papadopoulos & Karagiannidis	2008	Optimization of Decentralized Energy Systems	ELECTRE III
Qin et al.	2008	Climate Change Impact Assessment and Adaptation Planning	ELECTRE
Büyüközkan et al.	2008	Strategic Alliance Partner in Logistics Value Chain	TOPSIS
Lin and Gen	2008	Human Resource Allocation Problem	TOPSIS
Wang	2008	Evaluation of Financial Performances of Domestic Airlines	TOPSIS
Anagnostopoulos et al.	2008	Location Site Selection Problem	TOPSIS
Sheu	2008	Mode Choice of Global Logistics Management	TOPSIS
Hajkowicz and Higgins	2008	Water Resource Management	PROMETHEE II
Xidonas et al.	2009	Equity Selection Using Financial Analysis	ELECTRE Tri
Montazer et al.	2009	Vendor Selection Problem	ELECTRE III
Athanasopoulos et al.	2009	Coating Selection Problem	TOPSIS
Wu et al.	2009	Evaluation of Banking Performances	TOPSIS
Zammori et al.	2009	Critical Path Definition	TOPSIS
Huang et al.	2009	Bi-objective Personnel Assignment Problem	TOPSIS

continued on following page

Table 6. Continued

Author(s)	Published Year	Application Area	Used Method(s)
Shanian and Savadogo	2009	Material Selection of Highly Sensitive Components	TOPSIS
Shanian and Savadogo	2009	Material Selection of Highly Sensitive Components	TOPSIS
Çelik et al.	2009	Turkish Container Ports in Maritime Transportation Network	TOPSIS
Dağdeviren et al.	2009	Weapon Selection Problem	TOPSIS
İç and Yurdakul	2009	Development a DSS for Machining Center Selection	TOPSIS
Lin and Tsai	2009	Choosing Ideal Cities for Medical Centers Ventures	TOPSIS
Seçme et al.	2009	Performance Evaluation in Turkish Banking Sector	TOPSIS
Önüt et al.	2009	Long Term Supplier Selection Problem	TOPSIS
Ertuğrul and Karakaşoğlu	2009	Performance Evaluation of Turkish Cement Firms	TOPSIS
Lee et al.	2009	Selection of Technology Acquisition Mode	TOPSIS
Kowalski et al.	2009	Sustainable Energy	PROMETHEE
Tsoutsos et al.	2009	Sustainable Energy Planning	PROMETHEE I-II
Sun and Lin	2009	Evaluation of the Competitive Advantages of Shopping Websites	TOPSIS
Nikolic et al.	2009	Environmental Management in the Vicinity of Copper	PROMETHEE V
Yu and Hu	2010	Performance Evaluation of Multiple Manufacturing Plants	TOPSIS
Sadi-Nezhad and Damghani	2010	Assessment of Traffic Police Centers Performance	TOPSIS
Yang et al.	2011	Evaluation of Robustness of Supply Chain Information Sharing Strategies	TOPSIS
Temucin et al.	2013	Non Traditional Machining Process	TOPSIS
Temucin et al.	2014	Non Traditional Machining Process	ELECTRE,PROMETHEE I, TOPSIS

MCDM methods emerged because of the necessity mentioned above. During the 1940s and 1970s, a rapid progress demonstrated itself in the computer science field which led the development of Decision Support Systems. Decision Support Systems are user-friendly computerized systems dealing with lots of information piles to help the managerial staff in military organizations during the decision-making process under a limited time with high risk and uncertainty in decision environment by using distinct scientific subjects such as Multi-Criteria Decision Making methods. This chapter clears up the basis of decision making and process and provides basic information about MCDM for their possible in military decision problems.

REFERENCES

Albadvi, A., Chaharsooghi, S. K., & Esfahanipour, A. (2007). Decision Making in Stock Trading: An Application of PROMETHEE. *European Journal of Operational Research*, *177*(2), 673–683. doi:10.1016/j.ejor.2005.11.022

Anagnostopoulos, K., Doukas, H., & Psarras, J. (2008). A Linguistic Multicriteria Analysis System Combining Fuzzy Sets Theory, Ideal and Anti-Ideal Points for Location Site Selection. *Expert Systems with Applications*, *35*(4), 2041–2048. doi:10.1016/j.eswa.2007.08.074

Araz, C., Ozfırat, P. M., & Ozkarahan, I. (2007). An Integrated Multicriteria Decision-Making Methodology for Outsourcing Management. *Computers & Operations Research*, *34*(12), 3738–3756. doi:10.1016/j.cor.2006.01.014

Athanasopoulos, G., Riba, C. R., & Athanasopoulou, C. (2009). A Decision Support System for Coating Selection Based on Fuzzy Logic and Multi-Criteria Decision Making. *Expert Systems with Applications*, *36*(8), 10848–10853. doi:10.1016/j.eswa.2009.01.016

Baker, D., Bridges, D., Hunter, R., Johnson, G., Krupa, J., & Murphy, J. (2001, December). Retrieved January 02, 2012, from Guidebook to Decision-Making Methods, WSRC-IM-2002-00002, Department of Energy, USA: http://www.dss.dpem.tuc.gr/pdf/Decision%20Making%20Guidebook_2002.pdf

Behzadian, M., Kazemzadeh, R. B., Albadvi, A., & Aghdasi, M. (2010). PROMETHEE: A Comprehensive Literature Review on Methodologies and Applications. *European Journal of Operational Research*, *200*(1), 198–215. doi:10.1016/j.ejor.2009.01.021

Bernstein, P. L. (1998). *Against the Gods: The Remarkable Story of Risk*. New York: John Wiley & Sons, Inc.

Buchanan, L., & O'Connell, A. (2006). A Brief History of Decision Making. Harvard Business School Publishing.

Büyüközkan, G., Feyzioğlu, O., & Nebol, E. (2008). Selection of the Strategic Alliance Partner in Logistics Value Chain. *International Journal of Production Economics*, *113*(1), 148–158. doi:10.1016/j.ijpe.2007.01.016

Cebi, S., & Kahraman, C. (2010). Developing a Group Decision Support System Based on Fuzzy Information Axiom. *Knowledge-Based Systems*, *23*(1), 3–16. doi:10.1016/j.knosys.2009.07.005

Çelik, M., Cebi, S., Kahraman, C., & Er, I. D. (2009). Application of Axiomatic Design and TOPSIS Methodologies Under Fuzzy Environment for Proposing Competitive Strategies on Turkish Container Ports in Maritime Transportation Network. *Expert Systems with Applications*, *36*(3), 4541–4557. doi:10.1016/j.eswa.2008.05.033

Chen, C.-T., Lin, C.-T., & Huang, S.-F. (2006). A Fuzzy Approach for Supplier Evaluation and Selection in Supply Chain Management. *International Journal of Production Economics*, *102*(2), 289–301. doi:10.1016/j.ijpe.2005.03.009

Chen, M.-F., & Tzeng, G.-H. (2004). Combining Grey Relation and TOPSIS Concepts for Selecting an Expatriate Host Country. *Mathematical and Computer Modelling*, *40*(13), 1473–1490. doi:10.1016/j.mcm.2005.01.006

Dagdeviren, M., Yavuz, S., & Kılınç, N. (2009). Weapon Selection Using the AHP and TOPSIS Methods Under Fuzzy Environment. *Expert Systems with Applications*, *36*(4), 8143–8151. doi:10.1016/j.eswa.2008.10.016

Drummond, H. (2001). *The Art of Decision Making Mirrors of Imagination, Masks of Fate*. West Sussex, UK: John Wiley & Sons ltd.

Ertuğrul, İ., & Karakaşoğlu, N. (2009). Performance Evaluation of Turkish Cement Firms with Fuzzy Analytic Hierarchy Process and TOPSIS Methods. *Expert Systems with Applications, 36*(1), 702–715. doi:10.1016/j.eswa.2007.10.014

Eshlaghy, A. T., & Homayonfar, M. (2011). MCDM Methodologies and Applications: A Literature Review from 1999 to 2009. *Research Journal of International Studies, 21*, 86–137.

Farahani, R. Z., & Asgari, N. (2007). Combination of MCDM and Covering Techniques in a Hierarchical Model for Facility Location: A Case Study. *European Journal of Operational Research, 176*(3), 1839–1858. doi:10.1016/j.ejor.2005.10.039

Gamberini, R., Grassi, A., & Rimini, B. (2006). A New Multi-Objective Heuristic Algorithm for Solving the Stochastic Assembly Line Re-Balancing Problem. *International Journal of Production Economics, 102*(2), 226–243. doi:10.1016/j.ijpe.2005.02.013

Georgopoulou, E., Sarafidis, Y., Mirasgedis, S., Zaimi, S., & Lalas, D. P. (2003). A Multiple Criteria Decision-Aid Approach in Defining National Priorities for Greenhouse Gases Emissions Reduction in the Energy Sector. *European Journal of Operational Research, 146*(1), 199–215. doi:10.1016/S0377-2217(02)00250-3

Gorry, G. A., & Morton, M. S. (1971). *A Framework for Management Information Systems. Alfred P.* Cambridge, MA: Sloan School of Management.

Gunasekera, N. V. (2010). Collaboration of the Left and the Right Brain in Decision Making. *Chartered Institute of Management Accountants, 2*(2).

Hajkowicz, S., & Higgins, A. (2008). A Comparison of Multiple Criteria Analysis Techniques for Water Resource Management. *European Journal of Operational Research, 184*(1), 255–265. doi:10.1016/j.ejor.2006.10.045

http://www.au.af.mil/au/awc/awcgate/army/fm101-5_mdmp.pdf; (21/11/2017)

Huang, D. K., Chiu, H. N., Yeh, R. H., & Chang, J. H. (2009). A Fuzzy Multi-Criteria Decision-Making Approach for Solving a Bi-Objective Personnel Assignment Problem. *Computers & Industrial Engineering, 56*(1), 1–10. doi:10.1016/j.cie.2008.03.007

İç, Y. T., & Yurdakul, M. (2009). Development of a Decision Support System for Machining Center Selection. *Expert Systems with Applications, 36*(2), 3505–3513. doi:10.1016/j.eswa.2008.02.022

Karacan, İ., Tozan, H., & Karatas, M. (2016). Multi Criteria Decision Methods in Health Technology Assessment: A Brief Literature Review. *Eurasian Journal of Health Technology Assessment, 1*(1), 12–19.

Karatas, M. (2017a). A multi-objective facility location problem in the presence of variable gradual coverage performance and cooperative cover. *European Journal of Operational Research, 262*(3), 1040–1051. doi:10.1016/j.ejor.2017.04.001

Karatas, M. (2017b). Multiattribute Decision Making Using Multiperiod Probabilistic Weighted Fuzzy Axiomatic Design. *Systems Engineering*, *20*(4), 318–334. doi:10.1002ys.21397

Karataş, M., & Gecili, H. (2012). The role of decision support systems in steel industry. *Engineering Science & Technology, an International Journal*, *15*(1).

Karatas, M., & Yakici, E. (2018). An Iterative Solution Approach to a Multi-Objective Facility Location Problem. *Applied Soft Computing*, *62*, 272–287. doi:10.1016/j.asoc.2017.10.035

Kaya, T., & Kahraman, C. (2011). Multicriteria Decision Making in Energy Planning Using a Modified Fuzzy TOPSIS Methodology. *Expert Systems with Applications*, *38*(6), 6577–6585. doi:10.1016/j.eswa.2010.11.081

Khalil, W. A.-S., Shanableh, A., Rigby, P., & Kokot, S. (2005). Selection of Hydrothermal Pre-Treatment Conditions of Waste Sludge Destruction Using Multicriteria Decision Making. *Journal of Environmental Management*, *75*(1), 53–64. doi:10.1016/j.jenvman.2004.11.007 PMID:15748803

Kou, G., Shi, Y., & Wang, S. (2011). Multiple Criteria Decision Making and Decision Support Systems. *Decision Support Systems*, *51*(2), 247–249. doi:10.1016/j.dss.2010.11.027

Kowalski, K., Stagl, S., Madlener, R., & Omann, I. (2009). Sustainable Energy Futures: Methodological Challenges in Combining Scenarios and Participatory Multi-Criteria Analysis. *European Journal of Operational Research*, *197*(3), 1063–1074. doi:10.1016/j.ejor.2007.12.049

Krohling, R. A., & Campanharo, V. C. (2011). Fuzzy TOPSIS for Group Decision Making: A Case Study for Accidents with Oil Spill in the Sea. *Expert Systems with Applications*, *38*(4), 4190–4197. doi:10.1016/j.eswa.2010.09.081

Lee, H., Lee, S., & Park, T. (2009). Selection of Technology Acquisition Mode Using the Analytic Network Process. *Mathematical and Computer Modelling*, *49*(5-6), 1274–1282. doi:10.1016/j.mcm.2008.08.010

Lin, C.-M., & Gen, M. (2008). Multi-Criteria Human Resource Allocation for Solving Multistage Combinatorial Optimization Problems Using Multiobjective Hybrid Genetic Algorithm. *Expert Systems with Applications*, *34*(4), 2480–2490. doi:10.1016/j.eswa.2007.04.016

Lin, C.-T., & Tsai, M.-C. (2009). Development of an Expert Selection System to Choose Ideal Cities for Medical Service Ventures. *Expert Systems with Applications*, *36*(2), 2266–2274. doi:10.1016/j.eswa.2007.12.056

Mavrotas, G., Diakoulaki, D., & Caloghirou, Y. (2006). Project Prioritization Under Policy Restrictions. A Combination of MCDA with 0-1 Programming. *European Journal of Operational Research*, *171*(1), 296–308. doi:10.1016/j.ejor.2004.07.069

Montazer, G. A., Saremi, Q. H., & Ramezani, M. (2009). Design a New Mixed Expert Decision Aiding System Using Fuzzy ELECTRE III Method for Vendor Selection. *Expert Systems with Applications*, *36*(8), 10837–10847. doi:10.1016/j.eswa.2009.01.019

Nikolic, D., Jovanovic, I., Mihajlovic, I., & Zivkovic, Z. (2009). Multi-Criteria Ranking of Copper Concentrates According to Their Quality - An Element of Environmental Management in the Vicinity of Copper - Smelting Complex in Bor, Serbia. *Journal of Environmental Management*, *91*(2), 509–515. doi:10.1016/j.jenvman.2009.09.019 PMID:19833429

Ozturk, N., & Tozan, H. (2015). A fuzzy-based decision support model for selecting the best dialyser flux in haemodialysis. *Journal of Healthcare Engineering, Vol*, *6*(:3), 303–324. doi:10.1260/2040-2295.6.3.303 PMID:26753437

Paksoy, T., Pehlivan, N. Y., & Kahraman, C. (2012). Organizational Strategy Development in Distribution Channel Management Using Fuzzy AHP and Hirarchical Fuzzy TOPSIS. *Expert Systems with Applications*, *39*(3), 2822–2841. doi:10.1016/j.eswa.2011.08.142

Papadopoulos, A., & Karagiannidis, A. (2008). Application of the Multi-Criteria Analysis Method ELECTRE III for the Optimisation of Decentralised Energy Systems. *Omega*, *36*(5), 766–776. doi:10.1016/j.omega.2006.01.004

Qin, X. S., Huang, G. H., Chakma, A., Nie, X. H., & Lin, Q. G. (2008). A MCDM-Based Expert System for Climate-Change Impact Assessment and Adaptation Planning - A Case Study for the Georgia Basin, Canada. *Expert Systems with Applications*, *34*(3), 2164–2179. doi:10.1016/j.eswa.2007.02.024

Razi, N., & Karatas, M. (2016). A multi-objective model for locating search and rescue boats. *European Journal of Operational Research*, *254*(1), 279–293. doi:10.1016/j.ejor.2016.03.026

Roy, B. (1991). The Outranking Approach and the Foundations of ELECTRE Methods. *Theory and Decision*, *31*(1), 49–73. doi:10.1007/BF00134132

Sadi-Nezhad, S., & Damghani, K. K. (2010). Application of a Fuzzy TOPSIS Method Based on Modified Preference Ratio and Fuzzy Distance Measurement in Assessment of Traffic Police Centers Performance. *Applied Soft Computing*, *10*(4), 1028–1039. doi:10.1016/j.asoc.2009.08.036

Seçme, N. Y., Bayrakdaroğlu, A., & Kahraman, C. (2009). Fuzzy Performance Evaluation in Turkish Banking Sector Using Analytic Hierarchy Process and TOPSIS. *Expert Systems with Applications*, *36*(9), 11699–11709. doi:10.1016/j.eswa.2009.03.013

Shanian, A., & Savadogo, O. (2009). A Methodological Concept for Material Selection of Highly Sensitive Components Based on Multiple Criteria Decision Analysis. *Expert Systems with Applications*, *36*(2), 1362–1370. doi:10.1016/j.eswa.2007.11.052

Sheu, J.-B. (2008). A Hybrid Neuro-Fuzzy Analytic Approach to Mode Choice of Global Logistics Management. *European Journal of Operational Research*, *189*(3), 971–986. doi:10.1016/j.ejor.2006.06.082

Shyur, H.-J., & Shih, H.-S. (2006). A Hybrid MCDM Model for Strategic Vendor Selection. *Mathematical and Computer Modelling*, *44*(7-8), 749–761. doi:10.1016/j.mcm.2005.04.018

Silver, M. S. (1991). *Systems that Support Decision Makers: Description and Analysis*. West Sussex, UK: John Wiley & Sons.

Sun, C.-C., & Lin, G. T. (2009). Using Fuzzy TOPSIS Method for Evaluating the Competitive Advantages of Shopping Websites. *Expert Systems with Applications*, *36*(9), 11764–11771. doi:10.1016/j.eswa.2009.04.017

Temuçin, T., & Tozan, H. (2016). A fuzzy based decision support system for air conditioner selection and an application to Turkish construction sector. *Technical Gazette-TEHNIČKI VJESNIK, 23*(:4), 1117–1122.

Temuçin, T., Tozan, H., Valíček, J., & Harničárova, M. (2013). A fuzzy based decision support model for nontraditional machining process selection. *Technical Gazette-TEHNIČKI VJESNIK, 20*(Iss:3), 683–687.

Temuçin, T., Tozan, H., Vayvay, O., Valíček, J., & Harničárova, M. (2014). A fuzzy based decision model for nontraditional machining process selection. *International Journal of Advanced Manufacturing Technology, Vol, 70*(9-12), 2275–2282. doi:10.100700170-013-5474-z

Toker, D., Tozan, H., & Vayvay, O. (2013). A decision model for pharmaceutical marketing and a case study in Turkey. *Ekonomska istraživanja-Economic Research, 26*(1), 101-114.

Tosun, K. (1992). *İşletme Yönetimi (Genel Esaslar) (6th ed.)*. Ankara: Savaş Yayınları.

Tozan, H. (2011). Fuzzy AHP based decision support system for technology selection in abrasive water jet cutting processes. *Technical Gazette-TEHNIČKI VJESNIK, Vol, 18*(:2), 187–191.

Tsoukias, A. (2008). From Decision Theory to Decision Aiding Methodology. *European Journal of Operational Research*, *187*(1), 138–161. doi:10.1016/j.ejor.2007.02.039

Tsoutsos, T., Drandaki, M., Frantzeskaki, N., Losifidis, E., & Kiosses, L. (2009). Sustainable Energy Planning by Using Multi-Criteria Analysis Application in the Island of Crete. *Energy Policy*, *37*(5), 1587–1600. doi:10.1016/j.enpol.2008.12.011

Turban, E., & Aronson, J. E. (2001). Decision Support Systems and Intelligent Systems (6th ed.). Prentice-Hall, Inc.

Vaillancourt, K., & Waaub, J.-P. (2004). Equity in International Greenhouse Gases Abatement Scenarios: A Multicriteria Approach. *European Journal of Operational Research*, *153*(2), 489–505. doi:10.1016/S0377-2217(03)00170-X

Verderber, R. F., Verderber, K. S., & Sellnow, D. D. (2009). Communicate! (13th ed.). Boston: Wadsworth Publishing.

Wang, F., & Lim, A. (2007). A Stochastic Beam Search for the Berth Allocation Problem. *Decision Support Systems*, *42*(4), 2186–2196. doi:10.1016/j.dss.2006.06.008

Wang, J.-J., & Yang, D.-L. (2007). Using a Hybrid Multi-Criteria Decision Aid Method for Information Systems Outsourcing. *Computers & Operations Research*, *34*(12), 3691–3700. doi:10.1016/j.cor.2006.01.017

Wang, T.-C., & Chang, T.-H. (2007). Application of TOPSIS in Evaluating Initial Training Aircraft Under a Fuzzy Environment. *Expert Systems with Applications*, *33*(4), 870–880. doi:10.1016/j.eswa.2006.07.003

Wang, X., & Triantaphyllou, E. (2008). Ranking Irregularities When Evaluating Alternatives by Using Some ELECTRE Methods. *Omega*, *36*(1), 45–63. doi:10.1016/j.omega.2005.12.003

Wang, Y.-J. (2008). Appliying FMCDM to Evaluate Financial Performance of Domestic Airlines in Taiwan. *Expert Systems with Applications*, *34*(3), 1837–1845. doi:10.1016/j.eswa.2007.02.029

Wang, Y.-M., & Elhag, T. M. (2006). Fuzzy TOPSIS Method Based on Alpha Level Sets with An Application to Bridge Risk Assessment. *Expert Systems with Applications*, *31*(2), 309–319. doi:10.1016/j.eswa.2005.09.040

Wu, H.-Y., Tzeng, G.-H., & Chen, Y.-H. (2009). A Fuzzy MCDM Approach for Evaluating Banking Performance Based on Balanced Scorecard. *Expert Systems with Applications*, *36*(6), 10135–10147. doi:10.1016/j.eswa.2009.01.005

Xidonas, P., Mavrotas, G., & Psarras, J. (2009). A Multicriteria Methodology for Equity Selection Using Financial Analysis. *Computers & Operations Research*, 36.

Yanar, L. (2011). *A Fuzzy Hybrid Decision Support Software Development for Multi-Criteria Decision Making Problems and an Application to Abrasive Water Jet*. İstanbul: Turkish Naval Academy.

Yanar, L., & Tozan, H. (2012). A fuzzy hybrid decision support system for interceptor baywatch boat propulsion system selection. *Technical Gazette-TEHNIČKI VJESNIK, 19*(2), 407–413.

Yang, T., Wen, Y.-F., & Wang, F.-F. (2011). EValuation of Robustness of Supply Chain Information - Sharing Strategies Using a Hybrid Taguchi and Multiple Criteria Decision-Making Method. *International Journal of Production Economics*, *134*(2), 458–466. doi:10.1016/j.ijpe.2009.11.018

Yu, V. F., & Hu, K.-J. (2010). An Integrated Fuzzy Multi-Criteria Approach for the Performance Evaluation of Multiple Manufacturing Plants. *Computers & Industrial Engineering*, *58*(2), 269–277. doi:10.1016/j.cie.2009.10.005

Zammori, F. A., Braglia, M., & Frosolini, M. (2009). A Fuzzy Multi-Criteria Approach for Critical Path Definition. *International Journal of Project Management*, *27*(3), 278–291. doi:10.1016/j.ijproman.2008.03.006

Chapter 9
A Fuzzy MCDM Framework for Weapon Systems Selection

Melis Almula Karadayi
Istanbul Medipol University, Turkey

Yeliz Ekinci
Istanbul Bilgi University, Turkey

Hakan Tozan
Istanbul Medipol University, Turkey

ABSTRACT

The weapon system selection problem is a crucial issue for military logistics managers and decision makers. In most real-world cases, such critical selection problems include many alternatives, and those alternatives have to be assessed with respect to multiple criteria. In this chapter, the use of multi-criteria decision making (MCDM) approaches to tackle the weapon selection problem is discussed. Next, a fuzzy MCDM framework that is based on the hierarchical fuzzy technique for order preference by similarity to ideal solution (HFTOPSIS) method is proposed to solve the problem. The proposed approach is capable of incorporating both crisp and fuzzy data. The authors demonstrate the performance of the proposed methodology on a missile system selection problem which incorporates fuzzy environment elements.

INTRODUCTION

One of the critical tools that is important in determining the outcome of a war is the selection of the weapon systems that will be used during the war. Selection of a weapon system requires an assessment of those systems with respect to multiple and possibly conflicting criteria (Lee, Kang, Rosenberger, & Kim, 2010). This makes the selection and evaluation process a difficult task for decision-makers. The two most common criteria used in weapon system assessment are the weapon performance and cost. The performance and cost depend on the improvement of science/technology and economic resources. Technology is improved by ideas and resources; politics support ideas for their acceptance and so on. These factors directly rely on the expectations of the decision-makers, and on some other factors (Cheng, Yang, & Hwang, 1999).

DOI: 10.4018/978-1-5225-5513-1.ch009

Since technology develops rapidly in recent years, the number of weapon systems, i.e. number of alternatives, increases and each of the newly developed system has superiorities in one or more factor and one system cannot be named as the best one in terms of all factors. Therefore, the problem includes several alternatives and these alternatives must be assessed in terms of several conflicting criteria/factors in a weapon selection problem (Dağdeviren, Yavuz, & Kılınç, 2009). Some of these criteria are defined by ambiguous phrases and some quantitative requirements are expressed by quality measures (Cheng, Yang, & Hwang, 1999). Hence in this study, the fuzzy theory is utilized to deal with some vague or not well-defined language variables and qualitative requirements. Real-world problems mostly involve conflicting, multiple, implicit or explicit criteria (Karatas, 2007b; Karatas & Akman, 2014; Karacan, Karatas, Tozan, & Sulukan, 2015, Karatas & Yakıcı, 2018). In those cases, decision-makers generally tend to consider all criteria simultaneously to increase the efficiency of their operations (Karatas, 2017a). Problems which incorporate multiple criteria are handled with Multi-Criteria Decision Making (MCDM) methods. There exist several military applications which involve multiple criteria or objectives, such as search and rescue boat and helicopter location and allocation planning (Razi & Karatas, 2016; Razi, Karatas, & Gunal, 2016; Karatas, Razi, & Gunal, 2017).

In classic MCDM methods, the ratings and weights of the criteria are assumed to be known precisely. However, some critical criteria are usually evaluated subjectively for each of the evaluated criteria. The fuzzy MCDM methodology is more appropriate to help decision-makers make their evaluation of available alternatives and find the best one (Wang & Lee, 2009). Furthermore, the selection of the best weapon system involves the consideration of numerous criteria, hence a multiple level hierarchy structure is required in order to make an effective analysis. A methodology, which is developed for complex, uncertain and vague characteristics of the weapon systems selection problem, is needed. By that manner, the Hierarchical Fuzzy Technique for Order Preference by Similarity to Ideal Solution (HFTOPSIS) model proposed by Kahraman et al. (2007) has been used for evaluating alternative missile systems for the air force in this study as a military application.

This chapter includes four more sections. The following section summarizes the existing studies on the selection of weapon systems. The other two sections present the HFTOPSIS method and include the application of the method to the missile systems selection problem. The last section finalizes the paper with the conclusion and future research directions.

LITERATURE REVIEW ON SELECTION OF WEAPON SYSTEMS

The existing studies on the selection of weapon systems are grouped into three. The first group includes MCDM methods, while the second group includes fuzzy applications of these methods. The third group consists of hybrid methods, usually combining an MCDM method and a linear programming model.

In the first group of studies, Mon et al. (1994) state that performance evaluation and optimal design of weapon systems are MCDM problems. Ashari and Parsaei (2014) solve the infantry rifle selection problem by using MCDM methods. They determine the available alternatives and criteria through Simos method and use ELECTRE (Elimination Et Choix Traduisant la Realité) III to rank various alternatives. Wang et al. (2014) propose a mathematical model based on the Response Surface Method (RSM) and the Grey Relational Analysis (GRA), to solve the weapon selection problem. RSM is used to obtain the experimental points and analyze the factors that have a significant impact on the selection results. GRA is used to analyze the trend relationship between alternatives and reference series. And then an RSM model is obtained, which can be used to calculate all alternatives and obtain ranking results. Genc (2015) uses ELECTRE III and PROMETHEE (Preference Ranking Organization Method for Enrichment Evaluations) II to evaluate military tanks. Göleç et al. (2016) select the best military cargo aircraft using 4 different methods, namely Analytic Hierarchy Process (AHP), Simple Additive Weighting (SAW), ELECTRE and TOPSIS methods.

In the second group of studies, Cheng and Lin (2002) state that real-world problems are usually fuzzy by their nature and, the probability applications have not been very effective in a vast of the cases. Crisp data are inadequate to model real-world problems but the fuzzy set theory can be used in real-world decision-making problems to quantify the qualitative data. Moreover, the application of fuzzy set theory in real-world MCDM problems has given satisfactory results. These facts motivate researchers to utilize fuzzy theory in their studies. Hence, there are many studies that employ fuzzy MCDM methods in selecting weapon systems. Mon et al. (1994) point out that the descriptions and judgments used in this type of problem are usually linguistic and fuzzy. Therefore, they use fuzzy AHP (FAHP) method based on entropy weight calculation. In order to overcome the complexity of the entropy weight calculations, Chen (1996) proposes using simplified fuzzy number arithmetic operations in weapon selection problem. Cheng (1997) evaluates naval tactical missile systems by using FAHP based on the grade value of membership function.

Cheng et al. (1999) evaluate attack helicopters by FAHP based on linguistic variable weight. In their study, the membership function that is used to calculate the performance scores is built by the experts' opinions. They measure the importance of relative weight and use it to determine the centralization or dilation power of linguistic hedge. When the power of linguistic hedge is determined, it is used to find the results of the problem. Cheng and Lin (2002) evaluate the best main battle tank by using fuzzy Delphi method to adjust the fuzzy rating of every expert to achieve the consensus condition. After fuzzy decision matrices are generated, the aggregate fuzzy numbers are calculated by multiplying the fuzzy decision matrix with the corresponding fuzzy attribute weight. The final numbers are the ranking fuzzy numbers problem, afterward a procedure is proposed for ranking fuzzy numbers to rank aggregate fuzzy numbers.

In the third group, there are hybrid methods for selecting weapon systems. Lee et al. (2010) propose a hybrid approach for the problem, which includes AHP and principal component analysis (PCA) to determine the weights to assign to the criteria under consideration. These weights are used in a goal programming (GP) model to select the best alternative among the weapon systems alternatives. Greiner and Fowler (2003) suggest a hybrid approach for evaluating new weapon systems development projects. They integrate an AHP with a 0–1 integer portfolio optimization model. They indicate that the AHP component allows the decision-maker to incorporate qualitative and intangible criteria into the decision-making process. The priority rankings, which are the outcomes of the AHP are used to represent a measure of value in the 0–1 integer model objective function. Similarly, Chen and Shyu (2006) solve the weapon system selection problem by using the 0-1 goal programming approach and Analytic Network Process (ANP). Dağdeviren et al. (2009) use AHP to calculate the weights of the criteria, and fuzzy TOPSIS (FTOPSIS) method is utilized to obtain the final ranking of the alternatives.

PROPOSED METHODOLOGY: HIERARCHICAL FUZZY TOPSIS

In the literature, there exist many FTOPSIS methods for multi-attribute decision- making problems. Kahraman et al. (2007) apply HFTOPSIS for evaluating and selecting logistics information technologies. Wang and Chang (2007) propose FTOPSIS approach to select the best initial aircraft for Taiwan Air Force Academy. Wang and Lee (2007) generalize classical TOPSIS methodology for fuzzy multi-criteria group decision-making. Mahdavi et al. (2008) propose an FTOPSIS approach to assign the relative importance of the criteria utilizing fuzzy numbers. Önüt and Soner (2008) employ AHP and FTOPSIS algorithms for solid waste transshipment site selection problem.

Sun and Lin (2009) evaluate the competitive advantages of shopping websites in Taiwan utilizing FTOPSIS method. Yurdakul and Iç (2009) suggest FTOPSIS methodology for selecting the best machine tools for a manufacturing company. Kannan et al. (2009) integrate Interpretive Structural Modeling and FTOPSIS in order to evaluate fifteen third-party reverse logistics providers. Saremi et al. (2009) extend TOPSIS to evaluate three Total Quality Management consultant candidates under a fuzzy environment.

Amiri (2010) employs the AHP and FTOPSIS methods in order to select the best project for National Iranian Oil Company. Cavallaro (2010) applies FTOPSIS algorithm to evaluate different heat transfer fluids via comparing their costs and benefits. Iç and Yurdakul (2010) develop a credibility scoring system using FTOPSIS for the banks. The proposed system is tested with real cases. Sun (2010) combines FAHP and FTOPSIS methodologies for performance evaluation according to six dimensions, namely; manufacturing capability, supply chain capability, innovation capability, financial capability, human resource capability and service quality capability. Awasthi et al. (2011) evaluate three urban transportation systems alternatives in France using FTOPSIS. Kaya and Kahraman (2011) propose a modified FTOPSIS method for energy planning decision-making problems. Singh and Benyoucef suggest an FTOPSIS based methodology in order to evaluate fifteen suppliers for B2B e-sourcing sector. Kelemenis et al. (2011) develop an extended FTOPSIS approach. A case study was carried out to illustrate the application of the proposed method. Torlak et al. (2011) perform a comparative evaluation employing FTOPSIS for Turkish domestic airline industry. Four different airline companies in Turkey are analyzed in the study.

Iç (2012) develops an integrated approach based on the FTOPSIS and linear programming to assess credit risk limits of banks. A real case banking application was also presented in the study. Rouhani et al. (2012) propose an FTOPSIS approach for managers to evaluate enterprise systems in business intelligence aspects. Zouggari and Benyoucef (2012) apply simulation based FTOPSIS method to improve supplier selection process. Kim et al. (2013) select the best sites for treated wastewater instream using FTOPSIS among ten candidate sites in a South Korean urban watershed. Tian et al. (2013) utilize chi-squared test in order to measure relative closeness instead of using Euclidean distance in FTOPSIS for an information source selection problem.

Kannan et al. (2014) suggest a framework using FTOPSIS to select green suppliers for a Brazilian electronics. Lima-Junior et al. (2014) compare FAHP and FTOPSIS methodologies for supplier selection problem. Mahdevari et al. (2014) apply FTOPSIS to determine risks associated with human health for three hazardous coal mines in Iran. Taylan et al. (2014) employ FAHP and FTOPSIS methodologies to evaluate risks of construction projects at King Abdulaziz University in Saudi Arabia. Wang et al. (2014) analyze financial performance of Taiwan container shipping firms via integrating grey relation analysis and FTOPSIS.

Azizi et al. (2015) evaluate four automotive suppliers using FTOPSIS. The results reveal that automotive suppliers need to implement more effective strategies for finding the bottlenecks. Guo and Zhao (2015) use FTOPSIS method for electric vehicle charging station site selection problem in Beijing, China. Sang et al. (2015) develop an analytical solution procedure to FTOPSIS. The proposed methodology was illustrated on a personnel selection problem. Şengül et al. (2015) suggest a decision support system using FTOPSIS for evaluating different alternative energy supply systems in Turkey. Mavi et al. (2016) integrate Shannon entropy method and FTOPSIS for a spare parts supplier selection problem. Morteza et al. (2016) combine the ANP and FTOPSIS for tourism site selection problem. The proposed methodology is adapted in Queshm Island.

Essay-Ossandon et al. (2017) evaluate different solid waste treatment alternatives using FTOPSIS in the Canary Islands. Hatami-Marbini and Kangi (2017) develop three versions of TOPSIS method, namely; conventional TOPSIS, adjusted TOPSIS and modified TOPSIS where a new fuzzy distance measure is used to incorporate confidence levels of experts and fuzzy performance ratings. The proposed framework is illustrated through an application in the Tehran stock exchange. Walczak and Rutkowska (2017) propose a modified FTOPSIS method for ranking projects in a participatory budget. In the suggested method, the criteria are taken into account as fuzzy or categorical values and the distances are calculated based on the proposed classifications.

In the literature, the researchers have used FHTOPSIS method in order to solve complex MCDM problems efficiently, especially in order to address the criteria hierarchies. Among the existing studies, some studies solely use FHTOPSIS to solve MCDM problems while some combine this method with other MCDM methods.

Wang et al. (2009) state that FHTOPSIS method is well suited for evaluating fuzziness and uncertainty problems and also can provide more objective and accurate criterion weights, compared to other methods. They apply the method to the popular problem, supplier selection. Bao et al. (2010) propose an FHTOPSIS method to combine the multilayer safety performance indicators into one overall index by incorporating experts' opinions. Lessan et al. (2010) suggest FHTOPSIS method as an optimization model for railway signaling system selection issue. In their proposed method, fuzzy normalized judgment matrix is constructed by implementing a new normalization method. Afterward, the alternatives' fuzzy scores are determined by using a non-linear programming model. Final rankings of the alternatives are found by a fuzzy ranking method.

Lee et al. (2012) evaluate e-service quality by applying FHTOPSIS method, the results can be used by travel website managers to understand their rankings among their competitors. Paksoy et al. (2012) develop an organizational strategy to manage distribution channels of a vegetable oil producer, using FAHP and FHTOPSIS. Wang and Chan (2013) evaluate different green initiatives for companies using the method. Baykasoğlu et al. (2013) integrate fuzzy DEMATEL and FHTOPSIS to evaluate trucks of a land transportation company. In a more recent study, Karacan et al. (2016) discuss the implementations of MCDM methods for health technology selection purposes in the presence of multiple and conflicting assessment criteria.

Baykasoğlu and Gölcük (2015) solve the SWOT (Strengths, Weaknesses, Opportunities, and Threats) based strategy selection problem by integrating the FHTOPSIS and Fuzzy Cognitive Maps. Qi et al. (2015) use FHTOPSIS model to assess the performance of hotel websites. In the study of Zhao et al (2016), the evaluations of test equipment used for preventive maintenance are conducted by fuzzy DEMATEL and FHTOPSIS methods. Fuzzy DEMATEL is used to calculate the criteria weights and FHTOPSIS is used to rate the importance degrees of alternative components. Masudin and Saputro (2016) evaluate B2C websites by FHTOPSIS.

In the scope of this study, in order to take hierarchy into account, the HFTOPSIS model of Kahraman et al. (2007) is employed. Since the selection of the best missile system alternative involves the consideration of numerous criteria, a multi-level hierarchy structure should be conducted to handle the complexity of the problem. FAHP has the capability of taking pairwise comparisons of attributes into consideration with a hierarchical structure. Besides, the reason for choosing HFTOPSIS method is because it has the aforementioned ability without making pairwise comparisons.

Assume that we have n main attributes, m sub-attributes, k alternatives. Each main attribute has r_i sub-attributes where the total number of sub-attributes m is equal to $\sum_{i=1}^{n} r_i$.

The method can be described using the following steps:

Step 1: The experts use the linguistic variables "very low (VL)", "low (L)", "moderate (M)", "high (H)" and "very high (VH)" to assess the importance of the attributes and sub-attributes, and also the ratings of alternatives with respect to various subjective attributes and sub-attributes.

Step 2: Construct the first matrix $\tilde{I}_{MA}$ using the weights of main attributes with respect to the goal.

$$\tilde{I}_{MA} = \begin{matrix} & Goal \\ \begin{matrix} MA_1 \\ MA_2 \\ \vdots \\ MA_p \\ \vdots \\ MA_n \end{matrix} & \begin{bmatrix} \tilde{w}_1 \\ \tilde{w}_2 \\ \\ \tilde{w}_p \\ \\ \tilde{w}_n \end{bmatrix} \end{matrix} \tag{1}$$

Step 3: Construct the second matrix $\tilde{I}_{SA}$ using the weights of sub-attributes with respect to the main attributes. Write the weights vector obtained from $\tilde{I}_{MA}$ above this $\tilde{I}_{SA}$ as follows:

$$
\tilde{I}_{SA} = \begin{array}{c|cccccc|}
 & \tilde{w}_1 & \tilde{w}_2 & \cdots & \tilde{w}_p & \cdots & \tilde{w}_n \\
 & MA_1 & MA_2 & \cdots & MA_p & \cdots & MA_n \\
\hline
SA_{11} & \tilde{w}_{11} & 0 & \cdots & 0 & \cdots & 0 \\
SA_{12} & \tilde{w}_{12} & 0 & \cdots & 0 & \cdots & 0 \\
\vdots & \vdots & \vdots & & \vdots & & \vdots \\
SA_{1r_1} & \tilde{w}_{1r_1} & 0 & \cdots & 0 & \cdots & 0 \\
SA_{21} & 0 & \tilde{w}_{21} & \cdots & 0 & \cdots & 0 \\
SA_{22} & 0 & \tilde{w}_{22} & \cdots & 0 & \cdots & 0 \\
\vdots & \vdots & \vdots & & \vdots & & \vdots \\
SA_{2r_1} & 0 & \tilde{w}_{2r_1} & \cdots & 0 & \cdots & 0 \\
\vdots & \vdots & 0 & & & & \vdots \\
SA_{pl} & 0 & 0 & \cdots & \tilde{w}_{pl} & \cdots & 0 \\
\vdots & \vdots & & & & & 0 \\
SA_{n1} & 0 & 0 & \cdots & 0 & \cdots & \tilde{w}_{n1} \\
SA_{n2} & 0 & 0 & \cdots & 0 & \cdots & \tilde{w}_{n2} \\
\vdots & \vdots & \vdots & & \vdots & & \vdots \\
SA_{nr_n} & 0 & 0 & \cdots & 0 & \cdots & \tilde{w}_{nr_n}
\end{array} \tag{2}
$$

Step 4: Construct the decision matrix $\tilde{I}_A$ using the scores of the alternatives with respect to the sub-attributes. Write the weights vector obtained from $\tilde{I}_{SA}$ above this $\tilde{I}_A$ as follows:

$$
\tilde{I}_A = \begin{array}{c|cccccccc|}
 & \tilde{W}_{11} & \tilde{W}_{12} & \cdots & \tilde{W}_{1r_1} & \cdots & \tilde{W}_{pl} & \cdots & \tilde{W}_{nr_n} \\
 & SA_{11} & SA_{12} & \cdots & SA_{1r_1} & \cdots & SA_{pl} & \cdots & SA_{nr_n} \\
\hline
A_1 & \tilde{c}_{111} & \tilde{c}_{112} & \cdots & \tilde{c}_{11r_1} & \cdots & \tilde{c}_{1pl} & \cdots & \tilde{c}_{1nr_n} \\
A_2 & \tilde{c}_{211} & \tilde{c}_{212} & \cdots & \tilde{c}_{21r_1} & \cdots & \tilde{c}_{2pl} & \cdots & \tilde{c}_{2nr_n} \\
\vdots & \vdots & \vdots & & \vdots & & \vdots & & \vdots \\
A_q & \tilde{c}_{q11} & \tilde{c}_{q12} & \cdots & \tilde{c}_{q1r_1} & \cdots & \tilde{c}_{qpl} & \cdots & \tilde{c}_{qnr_n} \\
\vdots & \vdots & \vdots & & \vdots & & \vdots & & \vdots \\
A_k & \tilde{c}_{k11} & \tilde{c}_{k12} & \cdots & \tilde{c}_{k1r_1} & \cdots & \tilde{c}_{kpl} & \cdots & \tilde{c}_{knr_n}
\end{array} \tag{3}
$$

where,

$$\tilde{W}_{pl} = \sum_{j=1}^{n} \tilde{w}_{p} \cdot \tilde{w}_{pj} \tag{4}$$

Since $\tilde{w}_{pl} = 0$ for $j \neq 1$, use Eq. (5) instead of Eq. (4)

$$\tilde{W}_{pl} = \tilde{w}_{p} . \tilde{w}_{pl} \tag{5}$$

Step 5: Normalize the Decision Matrix using Eq.(6) and Eq. (7) for crisp and fuzzy data respectively. Obtain r_{ij}. The original data with respect to each criterion will be normalized to obtain unit-free and comparable values. The normalized values for crisp data regarding benefit-related as well as cost-related quantitative sub-criteria are calculated via a linear scale transformation as:

$$y'_{ijk} = \begin{cases} \dfrac{y_{ijk} - y^{-}_{jk}}{y^{*}_{jk} - y^{-}_{jk}}, & k \in CB_{j} \\ \dfrac{y^{*}_{jk} - y_{ijk}}{y^{*}_{jk} - y^{-}_{jk}}, & k \in CC_{j}, \end{cases} \tag{6}$$

where y'_{ijk} denotes the normalized value of y_{ijk}, which is the crisp value assigned to alternative *i* with respect to the sub-criterion *k* of criterion *j*, *m* is the number of alternatives, *n* is the number of criteria, CB_j is the set of benefit-related crisp sub-criteria of criterion *j* and CC_j is the set of cost-related crisp sub-criteria of criterion *j*, $y^{*}_{jk} = \max_{i} y_{ijk}$ and $y^{-}_{jk} = \min_{i} y_{ijk}$.

The normalized values for crisp data can be represented as $\tilde{y}_{ijk} = (y'_{aijk}, y'_{bijk}, y'_{cijk})$ in triangular fuzzy number format, where $y'_{aijk} = y'_{bijk} = y'_{cijk} = y'_{ijk}$. For fuzzy data denoted by triangular fuzzy numbers as $(a_{ijk}, b_{ijk}, c_{ijk})$, the normalized values for benefit-related criteria $\left(k \in CB_{j}\right)$ and cost- related criteria $\left(k \in CC_{j}\right)$ are:

$$\tilde{y}_{ijk} = (y'_{aijk}, y'_{bijk}, y'_{cijk}) = \begin{cases} \left(\dfrac{a_{ijk} - a^{-}_{jk}}{c^{*}_{jk} - a^{-}_{jk}}, \dfrac{b_{ijk} - a^{-}_{jk}}{c^{*}_{jk} - a^{-}_{jk}}, \dfrac{c_{ijk} - a^{-}_{jk}}{c^{*}_{jk} - a^{-}_{jk}}\right), k \in CB_{j} \\ \left(\dfrac{c^{*}_{jk} - c_{ijk}}{c^{*}_{jk} - a^{-}_{jk}}, \dfrac{c^{*}_{jk} - b_{ijk}}{c^{*}_{jk} - a^{-}_{jk}}, \dfrac{c^{*}_{jk} - a_{ijk}}{c^{*}_{jk} - a^{-}_{jk}}\right), k \in CC_{j}, \end{cases} \tag{7}$$

where $c^{*}_{jk} = \max_{i} c_{ijk}$ and $a^{-}_{jk} = \min_{i} a_{ijk}$.

Step 6: Compute the Weighted Normalized Decision Matrix as follows:

$$\tilde{V}_{ij} = \tilde{r}_{ij} \cdot \tilde{W}_{ij}, \forall i, j \tag{8}$$

Step 7: Obtain the Positive Ideal Solution (PIS), A* and the Negative Ideal Solution (NIS). PIS and NIS are defined as:

$$A* = \left[v_1^*, \ldots, v_n^*\right] \tag{9}$$

$$A^- = \left[v_1^-, \ldots, v_n^-\right] \tag{10}$$

where $v_j^* = \max_i v_{ij}$ and $v_j^- = \min_i v_{ij}$.

In case of fuzzy data, v_j^* and v_j^- are the fuzzy numbers with the largest generalized mean and the smallest mean respectively. Since a triangular fuzzy number (*a,b,c*) which are used in this application can be represented in trapezoidal form as (*a,b,b,c*).

The generalized mean for fuzzy number $v_j^-, \forall j$ is computed as:

$$M(v_{ij}) = \frac{-a_{ij}^2 + d_{ij}^2 - a_{ij}b_{ij} + b_{ij}d_{ij}}{3(-a_{ij} + d_{ij})} \tag{11}$$

Step 8: Calculate the Seperation Measures S_i^* and S_i^- as follows:

$$S_i^* = \sum_{j=1}^{n} D_{ij}^*, i = 1, \ldots, m \tag{12}$$

And:

$$S_i^- = \sum_{j=1}^{n} D_{ij}^-, i = 1, \ldots, m \tag{13}$$

For crisp data, the difference measures D_{ij}^* and D_{ij}^- are obtained as:

$$D_{ij}^* = \left|v_{ij} - v_j^*\right| \tag{14}$$

$$D_{ij}^{-} = \left| v_{ij} - v_j^{-} \right| \quad (15)$$

In the case of fuzzy data difference measures are obtained as the following:

$$D_{ij}^{*} = \begin{cases} 1 - \dfrac{c_{ij} - a^{*}}{b^{*} + c_{ij} - a^{*} - b_{ij}} \; for \, b_{ij} \prec b^{*} \\ 1 - \dfrac{c^{*} - a_{ij}}{b_{ij} + c^{*} - a_{ij} - b^{*}} \; for \, b^{*} \prec b_{ij} \end{cases} \quad \forall i, j \quad (16)$$

$$D_{ij}^{-} = \begin{cases} 1 - \dfrac{c^{-} - a_{ij}}{b_{ij} + c^{-} - a_{ij} - b^{-}} \; for \, b^{-} \prec b_{ij} \\ 1 - \dfrac{c_{ij} - a^{-}}{b^{-} + c_{ij} - a^{-} - b_{ij}} \; for \, b_{ij} \prec b^{-} \end{cases} \quad \forall i, j \quad (17)$$

where $v_j^{*} = \left(a^{*}, b^{*}, c^{*}\right)$ and $v_j^{-} = \left(a^{-}, b^{-}, c^{-}\right)$ are the fuzzy numbers with the largest generalized mean and the smallest generalized mean, respectively.

Step 9: Compute the Relative Closeness to Ideals as:

$$C_i = S_i^{-} / (S_i^{*} + S_i^{-}) \quad (18)$$

Step 10: Rank the alternatives in descending order of C_i values. Identify the alternative with the highest C_i as the best alternative.

CASE STUDY: APPLICATION OF THE HIERARCHICAL FUZZY TOPSIS MODEL FOR MISSILE SYSTEMS SELECTION

In this paper, a case study on surface-to-air missile system selection is presented to illustrate the application of the proposed approach. The data in the study of Lee et al (2010) is used to select the optimal alternative. Lee et al. (2010) consider six candidate missile systems as decision variables and evaluate them based on three criteria and 19 relevant subcriteria to conduct a comprehensive evaluation of the alternatives. Each subcriterion, identified and structured in the previous stage, has its own characteristic data about the candidate missile systems (Table 2). The decision process of selecting an appropriate alternative usually has to take numerous factors into considerations. The hierarchy in Figure 1 will be considered in the scope of this study.

Figure 1. The hierarchical structure for the missile systems selection
Source: (Lee, Kang, Rosenberger, & Kim, 2010).

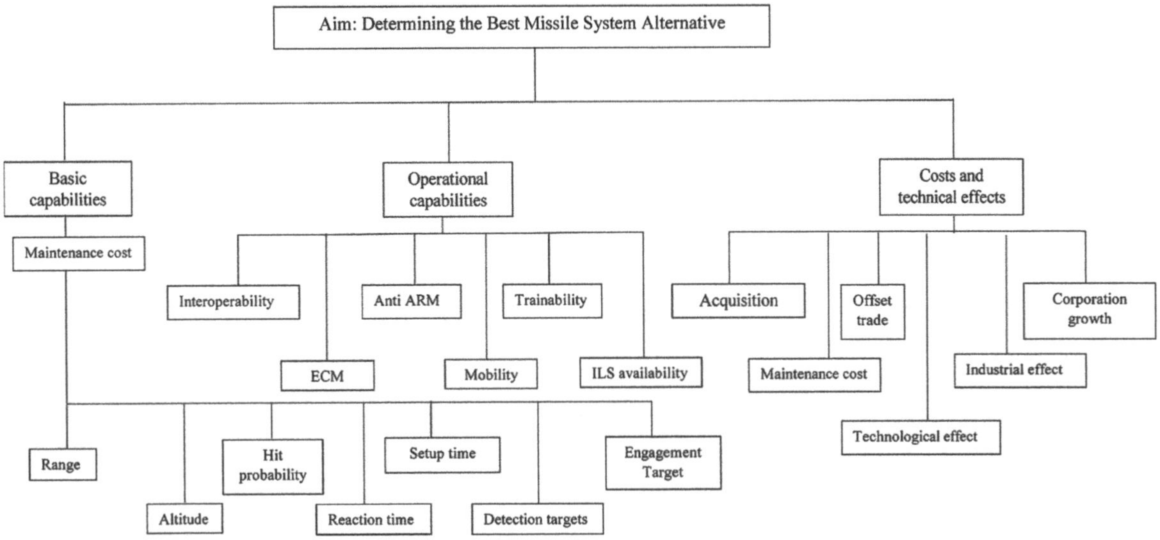

In this study, the expert used the linguistic variables "very low (VL)", "low (L)", "moderate (M) ", "high (H)" and "very high (VH)" in order to assess the importance of criteria and ratings of qualitative sub-criteria. These linguistic variables can be explained via triangular fuzzy numbers as presented in Figure 2. The evaluations are represented in Table 1.

APPLICATION OF THE HIERARCHICAL FUZZY TOPSIS METHOD

The evaluations of the experts are made based on Table 1. By using equations (1) and (2), we write $\tilde{I}_{MA}$ and $\tilde{I}_{SA}$ for the case study application. $\tilde{I}_{MA}$ and $\tilde{I}_{SA}$ are given in Table 3 and Table 4, respectively.

Figure 2. A linguistic term set where VL = (0, 0, 0.25), L = (0, 0.25, 0.5), M = (0.25, 0.5, 0.75), H = (0.5, 0.75, 1), VH = (0.75, 1, 1)

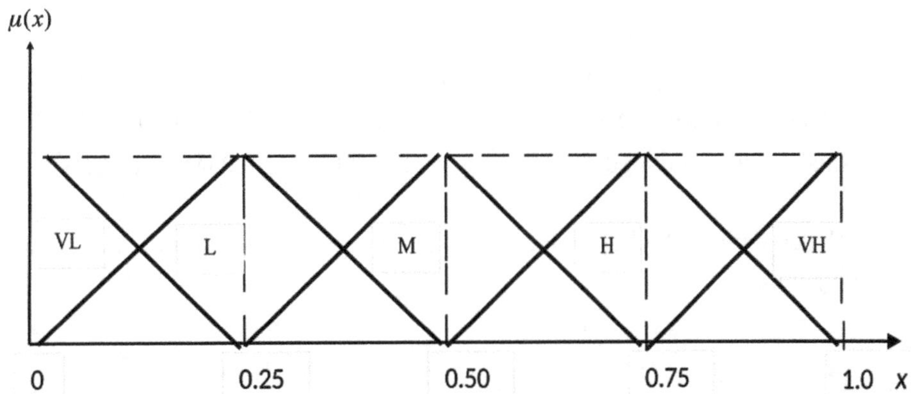

Table 1. Importance of the evaluation criteria and the related sub-criteria

Criteria/Sub-Criteria	Importance Weight
Basic capabilities	VH
Range	VH
Altitude	M
Hit probability	VH
Reaction time	M
Setup time	H
Detection targets	H
Engagement targets	H
Operational capabilities	VH
Interoperability	H
ECM	VH
Anti-ARM	VH
Mobility	H
Trainability	M
ILS Availability	H
Cost and technical effects	M
Acquisition cost	VH
Maintenance cost	H
Offset trade	L
Technological effect	M
Industrial effect	M
Corporation growth	M

Table 5 gives the generalized mean fuzzy numbers $M(v_{ij})$, for our case study.

Table 6 summarizes the D_{ij}^{*} values. They are the distance values from the positive ideal solution. It is good to have small D_{ij}^{*} values, hence, the "0" values are the ideal ones.

Table 7 summarizes the D_{ij}^{-}. values. They are the distance values from the negative ideal solution. Thus, it is good to have large D_{ij}^{-} values.

Separation measures S_i^+ and S_i^- values are given in Table 8.

Table 9 gives the C_i values for each missile system alternative. C_i is computed by dividing S_i^- to $(S_i^* + S_i^-)$. The final ranking obtained due to the hierarchical fuzzy MCDM algorithm is $A_5 > A_2 > A_1 > A_4 > A_6 > A_3$.

A_5 is the best choice with the highest value (0.722). A_3 has the lowest value (0.338) and it is the least favorable missile system.

Table 2. Data related to the alternative six missile systems

	A1	A2	A3	A4	A5	A6
Basic Capabilities						
Range	150	160	135	140	155	170
Altitude	24	28	22	24	28	30
Hit probability	0.75	0.80	0.75	0.75	0.80	0.80
Reaction time	12	9	13	12	10	9
Setup time	5.5	5	6	5.5	5	5
Detection targets	95	110	85	95	100	100
Engagement targets	6	9	6	6	8	8
Operational Capabilities						
Interoperability	0.75	0.8	0.65	0.65	0.7	0.7
ECM	0.65	0.75	0.65	0.65	0.75	0.75
Anti-ARM	0.75	0.8	0.65	0.7	0.7	0.8
Mobility	0.65	0.65	0.75	0.75	0.75	0.7
Trainability	0.75	0.75	0.7	0.65	0.65	0.65
ILS Availability	0.8	0.8	0.75	0.75	0.75	0.75
Cost and Technical Effects						
Acquisition cost	1100	1250	950	1050	1050	1100
Maintenance cost	12	14	8	8	8	12
Offset trade	0.5	0.45	0.75	0.6	0.6	0.45
Technological effect	0.9	0.9	1.1	1	0.9	0.9
Industrial effect	0.8	0.8	1.2	1.1	1	0.8
Corporation growth	0.9	0.9	1.1	1	1	0.8

Source: (Lee et al., 2010).

Table 3. $\tilde{I}_{MA}$ *for the case study*

Basic Capabilities	(0.75, 1, 1)
Operational Capabilities	(0.75, 1, 1)
Cost and Technical Effects	(0.25, 0.5, 0.75)

CONCLUSION

Weapon selection problem is one of the critical problems to be solved in military operations. The main reason is that the selected weapon system will affect the outcome of a war. In this paper, the problem of evaluating the surface-to-air missile systems is solved by HFTOPSIS. The existing literature on these types of problems shows that they are MCDM problems and should be solved by MCDM methods. HFTOPSIS method has the advantage of defining the complexity of the problem via its hierarchical

Table 4. $\tilde{I}_{SA}$ *for the case study*

	Basic	Operational	Cost
Range: *C1*	(0.75,1, 1)	0	0
Altitude: *C2*	(0.25,0.5, 0.75)	0	0
Hit probability: *C3*	(0.75,1, 1)	0	0
Reaction time: *C4*	(0.25, 0.5, 0.75)	0	0
Setup time: *C5*	(0.5, 0.75, 1)	0	0
Detection targets: *C6*	(0.5, 0.75, 1)	0	0
Engagement targets: *C7*	(0.5, 0.75, 1)	0	0
Interoperability: *C8*	0	(0.5, 0.75, 1)	0
ECM: *C9*	0	(0.75,1, 1)	0
Anti-ARM: *C10*	0	(0.75,1, 1)	0
Mobility: *C11*	0	(0.5, 0.75, 1)	0
Trainability: *C12*	0	(0.25, 0.5, 0.75)	0
ILS Availability: *C13*	0	(0.5, 0.75, 1)	0
Acquisition cost: *C14*	0	0	(0.75,1, 1)
Maintenance cost: *C15*	0	0	(0.5, 0.75, 1)
Offset trade:*C16*	0	0	(0, 0.25, 0.5)
Technological effect: *C17*	0	0	(0.25, 0.5, 0.75)
Industrial effect: *C18*	0	0	(0.25, 0.5, 0.75)
Corporation growth: *C19*	0	0	(0.25, 0.5, 0.75)

Table 5. $M(v_{ij})$ *values for the case study application*

	C1	C2	C3	C4	C5	C6	C7	C8	C9	C10
A_1	0.366	0.120	0	0.120	0.254	0.283	0	0.472	0	0.569
A_2	0.610	0.359	0.854	0.479	0.508	0.708	0.708	0.708	0.854	0.854
A_3	0	0	0	0	0	0	0	0	0	0
A_4	0.122	0.120	0	0.120	0.254	0.283	0	0	0	0.285
A_5	0.488	0.359	0.854	0.359	0.508	0.425	0.472	0.236	0.854	0.285
A_6	0.854	0.479	0.854	0.479	0.508	0.425	0.472	0.236	0.854	0.854
	C11	**C12**	**C13**	**C14**	**C15**	**C16**	**C17**	**C18**	**C19**	
A_1	0	0.479	0.708	0.240	0.139	0.028	0	0	0.097	
A_2	0	0.479	0.708	0	0	0	0	0	0.097	
A_3	0.708	0.240	0	0.479	0.417	0.167	0.292	0.292	0.292	
A_4	0.708	0	0	0.319	0.417	0.083	0.146	0.219	0.194	
A_5	0.708	0	0	0.319	0.417	0.083	0	0.146	0.194	
A_6	0.354	0	0	0.240	0.139	0	0	0	0	

Table 6. D^*_{ij} values for the case study application

	C1	C2	C3	C4	C5	C6	C7	C8	C9	C10
A_1	1.306	1.000	2.286	1.000	0.750	0.947	2.000	0.462	2.286	0.762
A_2	0.653	0.250	0	0	0	0	0	0	0	0
A_3	2.286	1.600	2.286	1.600	2.000	2.000	2.000	2.000	2.286	2.286
A_4	1.959	1.000	2.286	1.000	0.750	0.947	2.000	2.000	2.286	1.524
A_5	0.980	0.250	0	0.250	0	0.571	0.462	1.091	0	1.524
A_6	0	0.000	0	0	0	0.571	0.462	1.091	0	0
	C11	**C12**	**C13**	**C14**	**C15**	**C16**	**C17**	**C18**	**C19**	
A_1	2.000	0	0	0.571	0.667	0.625	1.333	1.333	0.571	
A_2	2.000	0	0	1.600	1.500	1.000	1.333	1.333	0.571	
A_3	0	0.571	2.000	0	0	0	0	0	0	
A_4	0	1.600	2.000	0.348	0	0.250	0.364	0.148	0.211	
A_5	0	1.600	2.000	0.348	0	0.250	1.333	0.364	0.211	
A_6	0.750	1.600	2.000	0.571	0.667	1.000	1.333	1.333	1.333	

Table 7. D^-_{ij} values for the case study application

	C1	C2	C3	C4	C5	C6	C7	C8	C9	C10
A_1	2.286	1.600	0	1.600	2.000	2.000	0	2.000	0	2.286
A_2	2.286	1.600	2.286	1.600	2.000	2.000	2.000	2.000	2.286	2.286
A_3	0	0	0	0	0	0	0	0	0	0
A_4	2.286	1.600	0	1.600	2.000	2.000	0	0	0	2.286
A_5	2.286	1.600	2.286	1.600	2.000	2.000	2.000	2.000	2.286	2.286
A_6	2.286	1.600	2.286	1.600	2.000	2.000	2.000	2.000	2.286	2.286
	C11	**C12**	**C13**	**C14**	**C15**	**C16**	**C17**	**C18**	**C19**	
A_1	0	1.600	2.000	1.600	1.500	1.000	0	0	1.333	
A_2	0	1.600	2.000	0	0	0	0	0	1.333	
A_3	2.000	1.600	0	1.600	1.500	1.000	1.333	1.333	1.333	
A_4	2.000	0	0	1.600	1.500	1.000	1.333	1.333	1.333	
A_5	2.000	0	0	1.600	1.500	1.000	0	1.333	1.333	
A_6	2.000	0	0	1.600	1.500	0	0	0	0	

structure. Moreover, it does not require pairwise comparisons like AHP, which makes it simple to apply. The reason for using fuzzy set theory is that crisp data are inadequate to model real-world problems but fuzzy numbers are effective in real-world decision-making problems for quantifying the qualitative data. Besides having the capability of considering numerous attributes that are structured in a multi-level hierarchy, the applied methodology enable the decision-maker to use linguistic terms.

Table 8. Separation Measures

A_i	S_i^+	S_i^-
A_1	19.900	22.805
A_2	10.241	25.276
A_3	22.914	11.700
A_4	20.672	21.871
A_5	11.233	29.110
A_6	12.712	25.443

Table 9. Ranking of Missile System Alternatives

A_i	C_i	Ranking
A_1	0.534	3
A_2	0.712	2
A_3	0.338	6
A_4	0.514	4
A_5	0.722	1
A_6	0.378	5

The data in the study of Lee et al (2010) is used for the case study in this chapter. Hence, six candidate missile systems are considered as decision variables and evaluated based on three criteria and 19 relevant subcriteria. The method gives a ranking of the six alternative missile systems. The fifth alternative outranks all the alternatives while alternative 2 is a close follower of it. The other alternatives are not close followers of them, which means that the fifth alternative should be selected. However, in a case that if the winner alternative (fifth alternative) was not selected somehow, the second alternative can be selected.

This study shows that HFTOPSIS is an effective method in solving the problem of selection of missile systems. However, the study can be enlarged by using some other MCDM methods and by comparing the results of them all.

REFERENCES

Amiri, M. P. (2010). Project selection for oil-fields development by using the AHP and fuzzy TOPSIS methods. *Expert Systems with Applications*, *37*(9), 6218–6224. doi:10.1016/j.eswa.2010.02.103

Ashari, H., & Parsaei, M. (2014). Application of the multi-criteria decision method ELECTRE III for the weapon selection. *Decision Science Letters*, *3*(4), 511–522. doi:10.5267/j.dsl.2014.6.001

Awasthi, A., Chauhan, S. S., & Omrani, H. (2011). Application of fuzzy TOPSIS in evaluating sustainable transportation systems. *Expert Systems with Applications*, *38*(10), 12270–12280. doi:10.1016/j.eswa.2011.04.005

Azizi, A., Aikhuele, D. O., & Souleman, F. S. (2015). A Fuzzy TOPSIS Model to Rank Automotive Suppliers. *Procedia Manufacturing*, *2*, 159–164. doi:10.1016/j.promfg.2015.07.028

Cavallaro, F. (2010). Fuzzy TOPSIS approach for assessing thermal-energy storage in concentrated solar power (CSP) systems. *Applied Energy*, *87*(2), 496–503. doi:10.1016/j.apenergy.2009.07.009

Chen, R. S., & Shyu, J. Z. (2006). Selecting a weapon system using zero-one goal programming and analytic network process. *Journal of Information and Optimization Sciences*, *27*(2), 379–399. doi:10.1080/02522667.2006.10699697

Chen, R.-S., & Shyu, J. Z. (2006). Selecting a weapon system using zero-one goal programming and analytic network process. *Journal of Information & Optimization Sciences*, *27*(2), 379–399. doi:10.1080/02522667.2006.10699697

Chen, S. M. (1996). Evaluating weapon systems using fuzzy arithmetic operations. *Fuzzy Sets and Systems*, *77*(3), 265–276. doi:10.1016/0165-0114(95)00096-8

Cheng, C. H. (1997). Evaluating naval tactical missile systems by fuzzy AHP based on the grade value of membership function. *European Journal of Operational Research*, *96*(2), 343–350. doi:10.1016/S0377-2217(96)00026-4

Cheng, C. H., & Lin, Y. (2002). Evaluating the best main battle tank using fuzzy decision theory with linguistic criteria evaluation. *European Journal of Operational Research*, *142*(1), 174–186. doi:10.1016/S0377-2217(01)00280-6

Cheng, C. H., Yang, K. L., & Hwang, C. L. (1999). Evaluating attack helicopters by AHP based on linguistic variable weight. *European Journal of Operational Research*, *116*(2), 423–435. doi:10.1016/S0377-2217(98)00156-8

Dağdeviren, M., Yavuz, S., & Kılınç, N. (2009). Weapon selection using the AHP and TOPSIS methods under fuzzy environment. *Expert Systems with Applications*, *36*(4), 8143–8151. doi:10.1016/j.eswa.2008.10.016

Estay-Ossandon, C., Mena-Nieto, A., & Harsch, N. (2017). Using a fuzzy TOPSIS-based scenario analysis to improve municipal solid waste planning and forecasting: A case study of Canary archipelago (1999–2030). *Journal of Cleaner Production.*

Genc, T. (2015). Application of ELECTRE III and PROMETHEE II in evaluating the military tanks. *International Journal of Procurement Management*, *8*(4), 457–475. doi:10.1504/IJPM.2015.070743

Göleç, A., Gürbüz, F., & Şenyiğit, E. (2016). Determination of Best Military Cargo Aircraft with Multi-Criteria Decision-Making Techniques. *MANAS Journal of Social Studies*, *5*(5), 87–101.

Greiner, M. A., Fowler, J. W., Shunk, D. L., Carlyle, W. M., & McNutt, R. T. (2003). A hybrid approach using the analytic hierarchy process and integer programming to screen weapon systems projects. *IEEE Transactions on Engineering Management*, *50*(2), 192–203. doi:10.1109/TEM.2003.810827

Guo, S., & Zhao, H. (2015). Optimal site selection of electric vehicle charging station by using fuzzy TOPSIS based on sustainability perspective. *Applied Energy*, *158*, 390–402. doi:10.1016/j.apenergy.2015.08.082

Hatami-Marbini, A., & Kangi, F. (2017). An extension of fuzzy TOPSIS for a group decision making with an application to tehran stock exchange. *Applied Soft Computing*, *52*, 1084–1097. doi:10.1016/j.asoc.2016.09.021

İç, Y. T. (2012). Development of a credit limit allocation model for banks using an integrated Fuzzy TOPSIS and linear programming. *Expert Systems with Applications*, *39*(5), 5309–5316. doi:10.1016/j.eswa.2011.11.005

İç, Y. T., & Yurdakul, M. (2010). Development of a quick credibility scoring decision support system using fuzzy TOPSIS. *Expert Systems with Applications*, *37*(1), 567–574. doi:10.1016/j.eswa.2009.05.038

Kahraman, C., Ateş, N., Çevik, S., Gülbay, M., & Erdoğan, S. (2007). Hierarchical Fuzzy TOPSIS Model for Selection Among Logistics Information Technologies. *Journal of Enterprise Information Management*, *20*(2), 143–168. doi:10.1108/17410390710725742

Kannan, D., Jabbour, A. B. S. L., & Jabbour, C. J. C. (2014). Selecting green suppliers based on GSCM practices: Using fuzzy TOPSIS applied to a Brazilian electronics company. *European Journal of Operational Research, 233*(2), 432–447. doi:10.1016/j.ejor.2013.07.023

Kannan, G., Pokharel, S., & Kumar, P. S. (2009). A hybrid approach using ISM and fuzzy TOPSIS for the selection of reverse logistics provider. *Resources, Conservation and Recycling, 54*(1), 28–36. doi:10.1016/j.resconrec.2009.06.004

Karacan, İ., Karatas, M., Tozan, H., & Sulukan, E. (2015). Evaluating the Energy Management Performance in Country Scale by Using a Multi Criteria Decision Making Methodology. *5th International 100% Renewable Energy Conference and Exhibition*.

Karacan, İ., Tozan, H., & Karatas, M. (2016). Multi Criteria Decision Methods in Health Technology Assessment: A Brief Literature Review. *Eurasian Journal of Health Technology Assessment, 1*(1), 12–19.

Karatas, M. (2017a). A multi-objective facility location problem in the presence of variable gradual coverage performance and cooperative cover. *European Journal of Operational Research, 262*(3), 1040–1051. doi:10.1016/j.ejor.2017.04.001

Karatas, M. (2017b). Multiattribute Decision Making Using Multiperiod Probabilistic Weighted Fuzzy Axiomatic Design. *Systems Engineering, 20*(4), 318–334. doi:10.1002ys.21397

Karatas, M., & Akman, G. (2014). An Extension to Multi-Attribute Decision Making Method: Dynamic Fuzzy Axiomatic Design Approach. *Joint International Symposium on CIE44 and IMSS'14 Proceedings*.

Karatas, M., Razi, N., & Gunal, M. M. (2017). An ILP and simulation model to optimize search and rescue helicopter operations. *The Journal of the Operational Research Society, 68*(11), 1335–1351. doi:10.105741274-016-0154-7

Karatas, M., & Yakıcı, E. (2018). An iterative solution approach to a multi-objective facility location problem. *Applied Soft Computing, 62*, 272–287. doi:10.1016/j.asoc.2017.10.035

Kaya, T., & Kahraman, C. (2011). Multicriteria decision making in energy planning using a modified fuzzy TOPSIS methodology. *Expert Systems with Applications, 38*(6), 6577–6585. doi:10.1016/j.eswa.2010.11.081

Kelemenis, A., Ergazakis, K., & Askounis, D. (2011). Support managers' selection using an extension of fuzzy TOPSIS. *Expert Systems with Applications, 38*(3), 2774–2782. doi:10.1016/j.eswa.2010.08.068

Kim, Y., Chung, E.-S., Jun, S.-M., & Kim, S. U. (2013). Prioritizing the best sites for treated wastewater instream use in an urban watershed using fuzzy TOPSIS. *Resources, Conservation and Recycling, 73*, 23–32. doi:10.1016/j.resconrec.2012.12.009

Lee, J., Kang, S. H., Rosenberger, J., & Kim, S. B. (2010). A hybrid approach of goal programming for weapon systems selection. *Computers & Industrial Engineering, 58*(3), 521–527. doi:10.1016/j.cie.2009.11.013

Lima, F. R. Junior, Osiro, L., & Ribeiro Carpinetti, L. C. (2014). A comparison between Fuzzy AHP and Fuzzy TOPSIS methods to supplier selection. *Applied Soft Computing, 21*, 194–209. doi:10.1016/j.asoc.2014.03.014

Mahdavi, I., Mahdavi-Amiri, N., Heidarzade, A., & Nourifer, R. (2008). Designing a model of fuzzy TOPSIS in multiple criteria decision making. *Applied Mathematics and Computation*, *206*(2), 607–617. doi:10.1016/j.amc.2008.05.047

Mahdevari, S., Shahriar, K., & Esfahanipour, K. (2014). Human health and safety risks management in underground coal mines using fuzzy TOPSIS. *The Science of the Total Environment*, *488–489*, 85–99. doi:10.1016/j.scitotenv.2014.04.076 PMID:24815558

Mavi, R. K., Goh, M., & Mavi, N. K. (2016). Supplier selection with Shannon entropy and fuzzy TOPSIS in the context of supply chain risk management. *Procedia: Social and Behavioral Sciences*, *235*, 216–225. doi:10.1016/j.sbspro.2016.11.017

Mon, D. L., Cheng, C. H., & Lin, J. C. (1994). Evaluating weapon system using fuzzy analytic hierarchy process based on entropy weight. *Fuzzy Sets and Systems*, *62*(2), 127–134. doi:10.1016/0165-0114(94)90052-3

Morteza, Z., Reza, F. M., Seddiq, M. M., Sharareh, P., & Jamal, G. (2016). Selection of the optimal tourism site using the ANP and fuzzy TOPSIS in the framework of Integrated Coastal Zone Management: A case of Qeshm Island. *Ocean and Coastal Management*, *130*, 179–187. doi:10.1016/j.ocecoaman.2016.06.012

Önüt, S., & Soner, S. (2008). Transshipment site selection using the AHP and TOPSIS approaches under fuzzy environment. *Waste Management (New York, N.Y.)*, *28*(9), 1552–1559. doi:10.1016/j.wasman.2007.05.019 PMID:17768038

Razi, N., & Karatas, M. (2016). A multi-objective model for locating search and rescue boats. *European Journal of Operational Research*, *254*(1), 279–293. doi:10.1016/j.ejor.2016.03.026

Razi, N., Karatas, M., & Gunal, M. M. (2016). A combined optimization and simulation based methodology for locating search and rescue helicopters. In *Proceedings of the 49th Annual Simulation Symposium* (pp. 1-8). Society for Computer Simulation International.

Rouhani, S., Ghazanfari, M., & Jafari, M. (2012). Evaluation model of business intelligence for enterprise systems using fuzzy TOPSIS. *Expert Systems with Applications*, *39*(3), 3764–3771. doi:10.1016/j.eswa.2011.09.074

Sang, X., Liu, X., & Qin, J. (2015). An analytical solution to fuzzy TOPSIS and its application in personnelselection for knowledge-intensive enterprise. *Applied Soft Computing*, *30*, 190–204. doi:10.1016/j.asoc.2015.01.002

Saremi, M., Mousavi, S. F., & Sanayei, A. (2009). TQM consultant selection in SMEs with TOPSIS under fuzzy environment. *Expert Systems with Applications*, *36*(2), 2742–2749. doi:10.1016/j.eswa.2008.01.034

Şengül, Ü., Eren, M., Shiraz, S. E., Gezder, V., & Şengül, A. B. (2015). Fuzzy TOPSIS method for ranking renewable energy supply systems in Turkey. *Renewable Energy*, *75*, 617–625. doi:10.1016/j.renene.2014.10.045

Singh, R. K., & Benyoucef, L. (2011). A fuzzy TOPSIS based approach for e-sourcing. *Engineering Applications of Artificial Intelligence*, *24*(3), 437–448. doi:10.1016/j.engappai.2010.09.006

Sun, C.-C. (2010). A performance evaluation model by integrating fuzzy AHP and fuzzy TOPSIS methods. *Expert Systems with Applications*, *37*(12), 7745–7754. doi:10.1016/j.eswa.2010.04.066

Sun, C.-C., & Lin, G. T. R. (2009). Using fuzzy TOPSIS method for evaluating the competitive advantages of shopping websites. *Expert Systems with Applications*, *36*(9), 11764–11771. doi:10.1016/j.eswa.2009.04.017

Taylan, O., Bafail, A. O., Abdulaal, R. M. S., & Kabli, M. R. (2014). Construction projects selection and risk assessment by fuzzy AHP and fuzzy TOPSIS methodologies. *Applied Soft Computing*, *17*, 105–116. doi:10.1016/j.asoc.2014.01.003

Tian, J., Yu, D., & Ma, S. (2013). A fuzzy TOPSIS model via chi-square test for information source selection. *Knowledge-Based Systems*, *37*, 515–527. doi:10.1016/j.knosys.2012.09.010

Torlak, G., Sevkli, M., Sanal, M., & Zaim, S. (2011). Analyzing business competition by using fuzzy TOPSIS method: An example of Turkish domestic airline industry. *Expert Systems with Applications*, *38*(4), 3396–3406. doi:10.1016/j.eswa.2010.08.125

Walczak, D., & Rutkowska, A. (2017). Project rankings for participatory budget based on the fuzzy TOPSIS method. *European Journal of Operational Research*, *260*(2), 706–714. doi:10.1016/j.ejor.2016.12.044

Wang, P., Meng, P., & Song, B. (2014). Response surface method using grey relational analysis for decision making in weapon system selection. *Journal of Systems Engineering and Electronics*, *25*(2), 265–272. doi:10.1109/JSEE.2014.00030

Wang, T.-C., & Chang, H.-S. (2007). Application of TOPSIS in evaluating initial training aircraft under a fuzzy environment. *Expert Systems with Applications*, *33*(4), 870–880. doi:10.1016/j.eswa.2006.07.003

Wang, T. C., & Lee, H. D. (2009). Developing A Fuzzy TOPSIS Approach Based on Subjective Weights and Objective Weights. *Expert Systems with Applications*, *36*(5), 8980–8985. doi:10.1016/j.eswa.2008.11.035

Wang, Y.-J. (2014). The evaluation of financial performance for Taiwan containershipping companies by fuzzy TOPSIS. *Applied Soft Computing*, *22*, 28–35. doi:10.1016/j.asoc.2014.03.021

Wang, Y.-J., & Lee, H.-S. (2007). Generalizing TOPSIS for fuzzy multiple-criteria group decision-making. *Computers & Mathematics with Applications (Oxford, England)*, *53*(11), 1762–1772. doi:10.1016/j.camwa.2006.08.037

Yurdakul, M., & İç, Y. T. (2009). Analysis of the benefit generated by using fuzzy numbers in a TOPSIS model developed for machine tool selection problems. *Journal of Materials Processing Technology*, *209*(1), 310–317. doi:10.1016/j.jmatprotec.2008.02.006

Zouggari, A., & Benyoucef, L. (2012). Simulation based fuzzy TOPSIS approach for group multi-criteria supplier selection problem. *Engineering Applications of Artificial Intelligence*, *25*(3), 507–519. doi:10.1016/j.engappai.2011.10.012

Chapter 10
Obsolescence Management for Sustainment-Dominated Military Systems:
Multiple Criteria Decision-Making Approach Using Evolutionary Algorithms

Barış Egemen Özkan
Marmara University, Turkey

Serol Bulkan
Marmara University, Turkey

ABSTRACT

As more and more Commercial Off The Shelf (COTS) parts are being used in sustainment-dominated systems where average product lifecycle is above 25 years, obsolescence management becomes a greater concern for program managers. The lack of management and poor planning for obsolescence cause companies, governments, and military organizations to spend progressively more to deal with aging systems. For a successful obsolescence management, program managers should consider both the cost-effectiveness and system availability issues simultaneously and should implement appropriate solution approaches. In this chapter, the authors first define the obsolescence management in sustainment-dominated systems and then give a brief summary of the related literature. They finally discuss and propose multiple criteria decision-making methodologies and evolutionary algorithms to tackle the management problem.

INTRODUCTION

Ever since the end of cold war, consumer electronics, and integrated circuit market became dominant and higher market share moved from military products to small-scale consumer products due to the advances in the consumer markets and fast-changing nature of the dynamic consumer market.

DOI: 10.4018/978-1-5225-5513-1.ch010

Fast technological advances created new consumer markets for which the needs are constantly being changed by either producers or consumers; thus, the life cycle of consumer products have become very short. This fact created a new type of product, called Commercial Off The Shelf (COTS). COTS stands for software and hardware products or services available in the market for public use that are free of charge or that have a cost associated with them. These products are attractive with their low cost and availability within their life cycle time. Additionally, they are available for use in any system design.

Following Perry Notice (Defence, 1994), as many nations, regardless of the proportion of the military budget, have placed as much emphasis as possible on the use of COTS products, we had experienced more COTS products in the military systems. Many state acquisition agencies and departments now strongly advise a policy that focuses on a gradual withdrawal from military specifications (Mil-Specs) and the use of COTS systems if applicable.

While there are numerous advantages to using COTS products, there are also a number of disadvantages as well. Some of those disadvantages include product volatility, obsolescence, and security. Even though reduced maintenance costs and vast availability are considered one of the advantages of using COTS products, they have turned into a barrier for maintenance agencies of systems with longer life cycle periods when we consider the disadvantages mentioned above. Normally the life cycle of a COTS part is typically 2-5 years. However, using such parts in systems with 25+ years of life cycle creates a big problem for maintenance agencies. Those systems with longer life cycle periods are not fully ready to adapt itself to the new phenomenon. In addition to that, legacy acquisition systems are also not ready for such a change.

Industries including military, aerospace, nuclear energy plants and railways still have life cycles of 25+ years. Due to high investment, testing and certification costs these systems seem to remain within their long life cycle periods. These systems are considered to be “sustainment dominated systems”.

Obsolescence is a major problem for such sustainment-dominated systems. In certain cases, obsolescence of a COTS part in a sustainment-dominated system requires a change in the original design. Due to the high complexity and environment in which they are used, they have higher standards for requalification and recertification for a design change. Sometimes, certain COTS products go obsolete even before production starts. Therefore, obsolescence is a major problem for systems with a long life cycle and this problem has to be addressed by an Obsolescence Management Plan (OMP) as early as possible. The lack of management and poor planning for obsolescence cause companies and governments to spend progressively more to deal with aging systems.

In order to underpin the severity of the obsolescence problem, here are a few of the examples with figures: A US Navy shipboard sonar system development project experienced the obsolescence of 70% of the parts listed in Bill Of Material (BOM) even before the system had completely developed. Also, the US Air Force spent $81 million USD to change and redesign the electronic parts for the F-22 program. The redesign of a system used in F-16 fighters radar systems costs $500 million USD for the US Air Force. The Electronic Industries Alliance (EIA) Manufacturing Operations and Technology Committee reported that the cost for the redesign of an obsolete electronic part is between $26,000 and $2 million USD. In 1997, the US DoD (Department of Defense) spent $264.000 USD for Life-of-Type Buy for an obsolete part. The US DoD is reported to have spent $10 billion USD to manage and to mitigate obsolescence problems.

As each study draws attention to obsolescence, the proposed methods, methodologies and models for obsolescence are ultimately formed around strategies forming a three-tiered solution:

- Reactive management strategy methods;
- Pro-active management strategy methods;
- Strategic management methods.

In a reactive strategy, no action is taken until obsolescence occurs. Once obsolescence occurs, different types of reactive mitigation techniques may be applied. Some of those techniques are Life-Time Buy (LTB), Bridge Buy, Reverse Engineering, Redesign, and Emulation. Each one of these techniques comes with an associated cost. An optimal solution may include one or a combination of these techniques.

Reactive strategies cost least among others if we consider the individual part obsolescence mitigation costs. However, if we consider the aggregate cost for all parts mitigation cost within a system, the cost might increase dramatically. In some rare cases, reactive strategies may be the optimal solution to follow until a proactive or strategic solution milestone such as a Design Refresh date is devised. Several considerations are taken into account while selecting the techniques such as the number of obsolete parts in the system, expected lifetime of the system, future demand, testing and re-certification costs.

Proactive strategies favor the monitoring of parts before they actually become obsolete and aim to prevent obsolescence driven risks. As monitoring all the materials in the system may not be feasible, only peculiarly selected parts are closely monitored and managed by a group of experts on obsolescence. Clustering parts based on the obsolescence risk and adjusting the level of management based on clusters is another to apply the strategy. Proactive management is performed during design, production and after-sale phases. Each selected part is analyzed against obsolescence risk and graded. The Material Risk Index and Progress Indicator are two of the metrics used for such purposes. This is not a one-time process but it is performed continuously until a decision is made not to do so.

Strategic obsolescence management is a holistic approach, which includes both reactive and proactive strategies as well as strategic level managerial actions. It starts even before the product design phase. Systems must be designed for obsolescence management. One of the most common ways of achieving this is to design loosely coupled components with standard interfaces so obsolete parts can easily be replaced with others. Strategic level management requires planning for obsolescence. This plan must be continuously reviewed and is subject to change if necessary. All systems and system parts need to be planned, designed, produced and maintained with the mindset that product lifecycle costs are always considered. It requires forecasts for both the demand and obsolescence of parts while estimating the best point in the product lifecycle to refresh the design; if necessary it combines the most feasible reactive methods till the design refresh date. It requires institutional level obsolescence awareness and the development of an organizational culture for obsolescence. Supply change management is another major part of strategic level management in order to prevent obsolescence. Long-term agreements with suppliers are considered to be one of the most effective ways to prevent obsolescence.

US Defense Standardization Program Office defined the obsolescence management framework with the following processes: Prepare, Identify, Assess, Analyze, Implement. Prepare process defines developing an obsolescence management plan, establishing an obsolescence management team and tailoring obsolescence management operational processes in accordance with the organization's needs and priorities. Next step is the identification process. Within this process, all necessary data is collected as well as the tools that is needed for a successful obsolescence management. Systems are prioritized, necessary data is collected and prepared to feed into monitoring tools, each item is analyzed based on availability and eventually, health assessment is developed. The next step is the assessment of obsolescence. This process includes identifying the data needs and assessing the possible impact of an obsolescence issue.

Once the possible obsolescence is predicted in the identification process, assessment process seeks for answers to the questions of whether the problem shall be pursued, should there be more than one obsolescence problem, which ones must be handled first and what level of resolution shall be applied. Assessment is a continues work throughout the lifecycle of the system and may be initiated upon as a result of predictive tools, vendor surveys, receiving a vendor warning and a change on health assessment of a system. The next step is to implement analysis to find the optimum (best feasible solution found if not optimum) resolution. This is an iterative process and aims to find out the best solution based on the objective functions, in our case minimum cost and maximum availability of the system. The last step is to implement the resolution that has been defined in the previous step.

Strategic obsolescence management covers all the processes defined above. Proactive obsolescence management covers partially identification and whole assessment, analysis processes. Since reactive management processes are implemented when the obsolescence occurs, this management strategy covers the analysis process.

In this process suit, Operations Research (OR) related areas are mainly centered on forecasting and optimization techniques. Forecasting obsolescence is one of the major areas in which OR is applied because it is essential for both strategic and proactive management strategy. At the same time when an obsolescence is notified or forecasted, various objectives are needed to be optimized some of which are minimum cost for replacement or design refresh, maximum availability of the system. Cost includes various factors some of which are inventory cost, replacement cost, training cost, administrative cost, recertification cost and etc. Design refresh time selection covers both areas; first, by using forecasting the future obsolescence, and then calculating the optimum design refresh date by using various optimization techniques.

In this chapter of the book, we will study the COTS obsolescence problem for sustainment-dominated systems. While there are many OR employed in this problem area, we will mainly focus on multi-objective decision making by using various meta-heuristics.

BACKGROUND

Preliminary studies on obsolescence management are mostly focused on reactive obsolescence strategies, which heavily relied on the recognition of obsolescence notification.

Due to its simplicity, the most commonly used methods in military and commercial industries are Bridge Buy or Life-Time-Buy (LTB) (Feng, 2007). Bridge Buy and LTB problems are cost minimization problems. There are several proposed models that include individual buy models and life cycle cost minimization models. Feng proposed two models and the first one is based on the data mining of historical buys (Peter Sandborn, Prabhakar, & Feng 2007). The latter one shows resemblance to classical newsvendor problem and is an extension of the Fortuin model (Feng, 2007). It is based on stochastic analysis and the Life of Type Evaluation (LOTE) tool that was devised to optimize lifetime buy quantities by minimizing the lifecycle cost and it takes procurement cost, inventory cost, disposal cost, and penalty cost into account (Feng, Singh, & Sandborn 2007). Fortuin's model also takes procurement, inventory, disposal, and penalty costs into account however it was constructed on user-defined time intervals as a discrete time optimization model (Teunter & Fortuin 1999).

Another method for obsolescence is uprating that allows a part to be used outside the specification limits determined by the original manufacturer, especially in thermal limits (Pecht & Humphrey 2008).

Rai and Terpenny proposed Piggybacking as another method and stated that it *enables renewed functionality of a technologically obsolete product through the integration or add-on of a secondary device or component* (Rai & Terpenny, 2008). Kooten proposed a model to calculate final orders for repairable spare parts by using the transient Markov chain (Van Kooten & Tan 2007). Since there are many options for obsolescence mitigation techniques, Shen proposed a framework for selecting the most appropriate reactive technique among a subset of reactive methods (LOT Buy, Part Substitution and Line Redesign) till a design refresh date and forecasting the optimum time for design refresh (Shen & Willems 2014). A similar comparison was made by Sandborn for Life Time Buy and Bridge Buy to find optimum quantities to procure (Peter Sandborn & Goudrazi, 2015).

Proactive approaches are better than reactive approaches however proactive approach requires forecasting obsolescence. Sandborn et.al. proposed one of the first studies to propose a forecasting model for the obsolescence zone ([mean+2stdDev - mean+3 stdDev]) and they used peak sale data for electronic parts. They tried to forecast the life cycle based on that data (Solomon, Sandborn, and Pecht 2000). Later on Singh et.al. modified the same model to include additional adjustments by adding the electronic part as a secondary attribute (Singh, Sandborn, Lorenson, & Geiser, 2002; Singh & Sandborn 2006).

Sandborn et.al. augmented the previous obsolescence zone forecasting method by using an analytical approach (Sandborn, Mauro, & Knox, 2007). While the previous method used peak sales data to estimate a part's life cycle, the analytical approach took the market trends and technology into account and used data mining based algorithms to forecast obsolescence dates with an increased accuracy. The combined approach gives a window of obsolescence based on a desired confidence level. Sandborn stated that better obsolescence forecast results only come with continuous forecasting. However, this method is good for electronic parts with identifiable parametric drivers. Sandborn later iterated the forecasting method using data mining for electronic parts with no evolutionary parametric drivers (Sandborn, Prabhakar, & Ahmad, 2011).

Another approach to forecast obsolescence is regression models. Josias proposed an obsolescence forecast model that uses regression model (Josias, Terpenny, & McLean, 2004). While this model resembles the model of Sandborn in terms of the independent variables, as they are similar to the attributes defined in the Sandborn model, there is a nuance between two studies. Sandborn used those attributes one after another, while Josias used all independent variables altogether. The dependent variable is the introduction date of the given part. This value is of no use directly but the study proposes to use that variable to derive obsolescence date indirectly.

Konozo proposed a model to uncover the End of Repair/Maintenance period by using discrete event simulation technique and the model used the existing inventory and future demands of a given systems parts (Konoza 2012; Konoza & Sandborn 2012; Konoza & Sandborn, 2013).

Design refresh time is a common term in product lifecycle management and it is "*the ultimate obsolescence mitigation approach where obsolete parts are designed out of the system in favor of newer, non-obsolete parts*" (Sandborn, Singh, Geiser, & Lorenson, 2002). Design refresh is a challenge with its large non-recurring costs for sustainment dominated systems due to recertification and other additional costs such as documentation changes, new training needs etc. Therefore the decision to initiate a design refresh necessitates careful planning but conditions in this era make design refresh essential (Plunkett, & Verma, 2003) and that makes proactive design and open system architecture is a MUST with standard commercial interfaces.

Sandborn proposed a design refresh date selection model and made this model available for use with his tool, Mitigation of Obsolescence Cost Analysis (MOCA). The difference between the Porter's (Porter,

1998) approach and Sandborn's model is that Porter compared only Last Time Buy reactive strategy to design refresh while Sandborn used more mitigation strategies. The MOCA tool supports the obsolescence forecast functionality for each part in the system by using the primary and secondary attributes via part's life cycle. A part's life cycle is determined by using the statistical analysis on the peak sales data.

According to Sandborn, as shown in Figure 1, every obsolescence problem, once notified, always starts with a reactive technique for short-term solutions and followed by a design refresh activity for the longer term as a strategic level technique (Sandborn, 2013).

Sandborn's design refresh model was extended by Singhe and Zheng. Singh extended the MOCA by adding Bayesian Belief Network to the model (Pameet Singh, Sandborn, & Singh, 2004) while Zheng added the Integer Programming method (Zheng, Terpenny, & Sandborn, 2015) to the model.

Since redesign technique is considered to be an expensive method, Barton supports the idea of decreasing the number of redesigns by paying more attention to the specification phases of the program both in the analysis and design processes (Barton & Chawla, 2003).

If the remaining life of the system is long or the technology change rate is high, selection of design refresh time or the amount of LTB do not usually provide efficient solutions. Kumar and Saranga proposed a model for program managers to select the best obsolescence mitigation strategy, namely design refresh model and Life-of-Type Buy selection (Kumar & Saranga, 2008) by using Bandit models. The Bandit models allocate scarce resources to competing selections when there is not much information during allocation period, however such information becomes more available as time passes. Bandit method in obsolescence management models provides a framework for program managers to make better decisions as time passes and more information becomes available.

Meng is another researcher to utilize the Last Time Buy and redesign techniques to minimize the obsolescence cost (Meng, Thornberg, & Olsson, 2014). He devised a mathematically strategic and proactive model to find the minimum cost for system obsolescence management cost. The model is composed of two parts: graph model and optimization model. The graph model first creates a data table to be used by the optimization model. The optimization model uses MILP by making use of transshipment model.

From a holistic point of view, obsolescence is also affiliated with integrated logistics systems (Marshall & Lambert, 2008). Since obsolescence problem is more heavily felt in sustainment dominated systems

Figure 1. Design refresh planning analysis timeline
Source: (Singh, Pameet, Sandborn, & Singh, 2002).

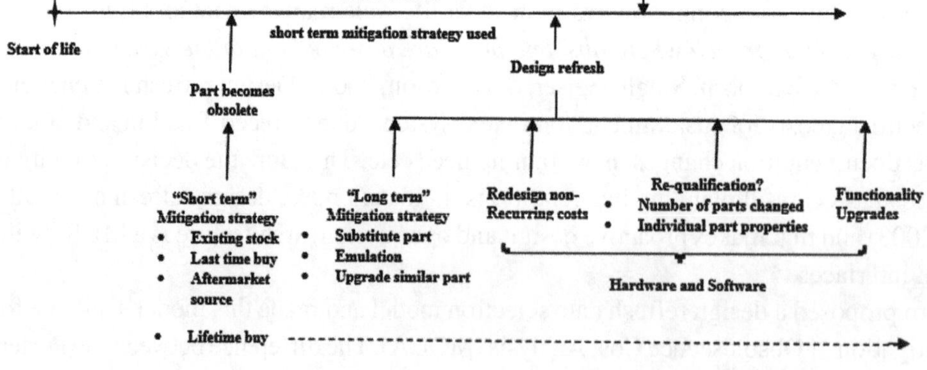

than others Marshall asserts that obsolescence and its impacts need to be taken into account in Integrated Logistics Systems and relevant solutions and advice DMSMS to be included in the contract terms.

Since Performance Based Logistics (PBL) transfers the burden of obsolescence mitigation responsibility to the contractor, the estimation of the cost of obsolescence of parts inside a given sustainment dominated system during the bidding period is subject to a wide range of uncertainty and a major concern for contractors. Certain models and frameworks are proposed to estimate obsolescence cost during bidding phase (Rojo & Romero, 2012; Feldman & Sandborn, 2007). The effective solution is designing the system to obsolescence so that less reactive obsolescence management techniques need to be used (Feldman & Sandborn, 2007; Baker, 2013). This necessitates the use of BOM and running this list against obsolescence forecasting methods that are followed by strategic obsolescence management techniques such as design refresh.

Sandborn proposed a model to incorporate a Public Private Partnership approach to the military PBL contracts that had produced successful results in civil infrastructure areas (Sandborn, 2014). The model uses Petri-Nets as well as affine controller function points and can be used by both contractors and states to increase their mutual profits.

Prabhakar approached the obsolescence problem from the supply chain management aspect and proposed second sourcing and hoarding among other reactive strategies to be considered as another way of mitigating the probability of obsolescence (Prabhakar, Varun, & Sandborn, 2013; Prabhakar & Sandborn, 2013).

All those strategies have strong sides as well as shortfalls. For that reason, Sandborn proposed the idea of using reactive, proactive and strategic management techniques together within a top-level umbrella approach to minimize the life cycle costs of a given sustainment dominated system (Sandborn, Herald, Houston, & Singh, 2003; Sandborn, 2008).

Various studies on obsolescence management which include forecasting uses different approaches, and necessitate a common language. Zheng proposed an ontology-based knowledge representation for parts, components, resolution techniques and forecasting to enable the integration of different techniques (Zheng & Liyu, 2011; Zheng, Liyu, Nelson, Sandborn, & Terpenny, 2013). The motivation for such a study is fully necessary however this study needs much more elaboration in the future.

Cuculoski proposed a model based on the reliability and maintainability of a given system (Cuculoski, 2013) in which he stressed the need for data and a proper analysis method performed by capable teams. The study claims that if appropriate obsolescence management methodologies are followed, keeping the existing system with refreshes would be more cost-effective than acquiring a new system. However, he did not take the high recertification cost of a redesigned system into account when comparing the cost of a new system to a redesigned one.

Until this point, all studies are conducted with certain assumptions and loose constraints. Studies on taking constraints into account (Sandborn, Terpenny, & Zheng, 2011) in obsolescence management models are later on expanded upon in another study to turn implicit constraints into explicit ones (Nelson III, Raymond, & Sandborn, 2013).

Rojo et.al. proposed a model to estimate the cost of component obsolescence (Rojo & Romero, 2012) by using different aspects such as “Obsolescence Management Levels”, “Obsolescence Complexity Levels”, “Resolution Profiles” and “Cost Metrics”. He took those aspects into account and estimated the obsolescence cost of a given system by utilizing a tool that provides assistance during the bidding process but prior to the contract.

Another model to calculate life cycle cost is proposed by Raghavan. It is a three-step model that starts with the estimation of the failure rates of a given component followed by the estimation of the impact of obsolescence and preventative measures and ultimately end with cost calculation (Raghavan & Member, 2012). He, later on, improved the model by adding state diagrams and utilizing the Monte Carlo simulation as well as using a modified Markov Chain approach (Raghavan & Chowdhury, 2015).

As it can be easily derived from all those studies on obsolescence management, almost all of the studies conducted in this area are seeking for an optimal solution for a single objective function: minimizing cost. However, obsolescence management is a multi-dimensional problem and all studies so far have considered the various aspects of subsets of the problem. The first study to draw attention of obsolescence management problem from a Multi-Criteria Decision Making (MCDM) view is prepared by Pingle (Pingle, 2015).

In the following section of the chapter, the authors will review the MCDM methodologies and evolutionary algorithms that are implemented to solve Obsolescence Management in the context of MCDM.

MULTI-CRITERIA DECISION MAKING IN OBSOLESCENCE MANAGEMENT

While optimization of a single objective can be quite cumbersome and a challenging task to solve, it is not rare to face with a case in which there are multiple objectives (Razi & Karatas, 2017; Karatas, 2017). Most of the time, those objectives are conflicting each other and that makes our life even harder, if not impossible. In this case, the decision maker has to satisfy all objectives with some degree of trade-offs based on subjective criteria and preferences. Criteria are reflections of the desires of the decision maker, which indicates the direction to the better

The aim of the MCDM problems is to minimize (maximize)

$$\min\left(\max\right)\left\{f_1\left(\overline{x}\right), f_2\left(\overline{x}\right), \ldots., f_k\left(\overline{x}\right)\right\}$$

a set of objective functions, subject to certain equality and inequality constraints

$$g_i\left(\overline{x}\right) \geq 0; i = 1, 2, \ldots., m$$

$$h_i\left(\overline{x}\right) = 0; i = 1, 2, \ldots., n$$

In MCDM problems, there are k amount of objective functions, most probably conflicting each other. $\vec{x}$ is the decision variables vector. x_1^*, x_2^*,... x_n^* are decision variables that yield to optimum solutions for given each objective.

Rather than finding a single optimum point in the solution space, multi-objective problems try to define a set of candidate solutions in solution space that is called “Pareto Front”. Pareto Front holds the Pareto Optimal solutions. Pareto optimality refers to a condition in which a decision variable $\overline{x}^*$ is pareto optimal if there does not exist another $\overline{x}$ such that $f_i\left(\overline{x}\right) \leq f_i\left(\overline{x}^*\right)$ for all i=1,2,….,k and $f_j\left(\overline{x}\right) < f_j\left(\overline{x}^*\right)$ for at least one *j*.

This means that a Pareto optimal solution cannot increase the value of a given objective unless there is a decrease on at least one of the other objectives (Karatas & Yakıcı, 2018). The solutions that reside on the Pareto Front are called Non-Dominated Solution and they are equally good to each other. A solution is a non-dominated solution if none of the objective functions can be improved in value at the expense of degrading at least one another objective value.

The objective of MCDM problems is to find most diversified and converged Pareto optimal front that holds Pareto optimal solutions. The decision maker can either trade off among multiple objectives on the Pareto optimal solutions or selects one of the solutions as a single solution based on his/her preferences.

When there is no preference information regarding to the optimal solution set, a method called Global Criteria Method is used to minimize the difference between objective function and optimal solution for individual objective functions. However, in most of the cases, the decision makers get involved in the solution process before, after or during the solution.

When decision makers involve in the solution before it starts, preference information is provided by the decision maker and then a solution that is best satisfying these preferences is found. This approach is called Prior Approach and may use one of the following methods:

- Utility function method
- Lexicographic method
- Goal programming

There are cases in which decision makers get involved in the solution after it finishes. Decision makers are presented with a set of alternative solutions to select one of them with his/her subjective preferences. This approach is called Posterior Approach. Some of methods that is being used in this approach are listed below:

- Mathematical programming
    - Normal Boundary Intersection (NBI),
    - Modified Normal Boundary Intersection (NBIm)
    - Normal Constraint (NC)
    - Successive Pareto Optimization (SPO)
    - Directed Search Domain (DSD)
- Evolutionary Algorithms
    - Non-dominated Sorting Genetic Algorithm-II
    - Strength Pareto Evolutionary Algorithm 2 (SPEA-2)

In the third approach, the decision maker iteratively searches for the most preferred solution. At each iteration of the interactive method, the decision maker is shown with Pareto optimal solutions. New Pareto optimal solutions are generated by taking the information provided by the decision maker into account. In this way, at each iteration, the decision maker learns more insight about the solution space and can focus on alternatives that interest him/her. The decision maker may stop the search whenever he/she wants to.

When there is no exact method available for a given problem, approximation methods are preferred. Approximation methods use certain heuristics. Some of those heuristics are problem specific. Problem independent heuristic methods are meta-heuristics. Meta-heuristics can be used for both single and multi-objective optimization problems to find both single and population solutions.

Obsolescence Management is one of the areas where decision makers have to find an optimum solution for various objectives, including but not limited to, minimization the life cycle cost, maximization of the system efficiency. Life cycle cost factor includes various cost parameters including initial procurement cost, inventory cost, design refresh cost, sustainment cost and etc. However, it is not a trivial task to represent the efficiency of the system compared to other currency based objective. It is problem and system specific to make such a conversion to compare the objectives.

Obsolescence management within the context of multiple objective optimizations will require a set of solutions from which the decision maker can select the best option based on his/her preference. For that reason, we have to apply population-based algorithms rather than a single solution algorithm to be able to populate Pareto front. In the next section, you can find an introduction to the evolutionary algorithms that is one of the methods to populate Pareto front.

EVOLUTIONARY ALGORITHMS

Evolutionary algorithms are based on natural selection that was originally defined by Charles Darwin (Darwin, 1929). Those algorithms use biological inspirations as their bases. Evolution is an adaptation of any given living species to its environment to avoid extinction. In this process, strong features remain from generation to generation while weak features fade away. Those features are encoded in the genes that eventually compose chromosomes. The algorithms successively apply evolution operations on the initial population to generate new population and the best individual is selected as the solution. Any evolution algorithm uses the following operations (Michalewicz, 1996):

- Initialization
- Selection
- Genetic Operations
- Termination

Initialization is the selection of initial candidate solutions. It is usually populated randomly. Selection process determines which candidate solutions are to be used to generate next generation. There are various methods to select the candidate solutions among which include Fitness Proportionate Selection (Roulette Wheel Selection), Tournament Selection, and Truncation Selection etc. Following the selection process comes genetic operators: Crossover and Mutation. The main idea of the genetic operators is to create next generation out of the best candidates of the previous generation. Some of the non-best candidates are involved in the creation of the next generation to maintain genetic diversity. There have been various debates over which genetic operator has the most influence on the generation of the next population. Decision is highly coupled with the nature of the problem; hence the crossover type and probability, mutation type and probability shall be selected carefully. Those parameters affect the convergence rate and accuracy of the solution. One point crossover, two-point crossover, uniform crossover are some of the methods used in crossover operation. The algorithm terminates when it reaches to termination criteria. A termination criterion is usually selected as a maximum number of iteration or a satisfactory margin of fitness value is achieved.

EVOLUTIONARY ALGORITHMS IN MCDM

In the context of multi-criteria decision-making, the operators of evolutionary algorithms need to be adapted to address multiple objectives. In the following section, we will review the evolutionary algorithms operators for multiple objective problems (Talbi, 2009; Karatas, 2018).

- Initialization
- Fitness Evolution
- Diversity Check
- Replacement
- Elitism
- Termination Control
- Selection
- Variation

Initialization and variation operators are problem dependent. Fitness calculation, diversity check and elitism operators are specific to multiple objective and finally, replacement, selection and termination control operators are specific to meta-heuristic that is being used.

Since initialization and representation have a great impact on the following operators, the following section will explicitly focus on initialization, fitness evolution, diversity check, elitism operators. Replacement and selection operators are going to be mentioned implicitly within each section since those operations are embedded inside each algorithm defined below.

Initialization

Almost all evolutionary algorithms use random initialization method. However, in some cases, some deterministic methods can be used to initialize the population pending to the nature of the problems. The number of population is one of the parameters that needs to be selected carefully. Each individual member of the solution is called phenotype. However, it is not always possible to fully correspond the phenotypes in evolutionary algorithms. Hence, we use genotypes in evolutionary algorithms to map phenotypes in the solution space. Representation of the phenotype as a genotype is a major part of initialization and it is problem specific (Back, Fogel, & Michalewicz, 2000). This step hugely affects the efficiency of the search algorithms; for that reason, a quite consideration shall be given in this phase to select the best representation.

There are various methods that can be used during the representation of the initial population. The most trivial and most commonly used one is binary string representation. It is very straightforward to apply genetic algorithms for such representations. Since one bit change can affect the whole value of string by doubling of halving the overall value, grey coding is another method used in binary representation. With grey coding, there is always one-bit change between two consecutive strings. When real values are to be used in the string, special consideration must be given to other evolutionary algorithm operators such as crossover and mutation.

In Evolutionary Strategy and Evolutionary Programming methods, real values of variables can be directly used. In those methods, mutation and crossover operators must be applied to each variable separately. In some cases, the variables may be in different format. While some variables are binary, others may be integer, decimal, a member of a nominal set and etc. This kind of representation is called combined representation and evolutionary operators must be applied separately for each given variable type.

Initialization for obsolescence management depends on the decision variables. Cost minimization will certainly be one of the objectives. Some of the cost factors affecting the total cost value are given below. Researchers can select various variables for each of the cost factors depending on the fidelity of their models. Randomly selected decision variables values within their ranges in the decision space create the initial objective values in the objective space. You can find a non-exhaustive list of variables for each cost factors below.

- Initial Procurement Cost
    - Number of parts to procure
    - Part
    - Vendor
- Holding Cost
    - Inventory on hand
        - Pilfering rate
        - Degradation rate in inventory
        - Demand rate (derived from MTBF)
- Disposal Cost
    - Inventory on hand
- Non-availability Cost
    - Number of shortage parts

Fitness Evaluation

Fitness evaluation is the representation of the objective function(s) of the original problem in the evolutionary algorithms. This step guides the search towards to the optimal solution(s) direction for a better convergence. It is considered to be the degree of the quality of a solution that is the reflection of an option in the decision space.

In rare cases, developers may opt to pick another form of fitness function than the objective of the original problem. In that case, reversing the genotype with the highest fitness back to phenotype (decoding) would require a special consideration.

In a single objective problem, it is straightforward to design the fitness evolution function. However in multi-objective problem, for each genotype selected, there are multiple numbers of fitness values for each objective. In order to achieve a cumulative value of a given genotype, various methods are being applied.

Different authors have clustered the fitness evolution methods with various approaches. One-way of clustering the fitness evolution methods is based on non-Pareto and Pareto-based techniques (Ehrgott & Gandibleux, 2002). Pareto based techniques are aggregation methods, Vector Evaluated Genetic Algorithm (VEGA), Lexicographic Ordering, the ε constraint, Target-Vector approaches and game theory. Pareto-based techniques are Pure Pareto technique, Multi-Objective Genetic Algorithm (MOGA), Non-Dominated Sorting Genetic Algorithm (NSGA), Niched Pareto Genetic Algorithm (NPGA) and non-generated approaches.

Another way of classification of the fitness evolution methods are scalar approaches, criterion, dominance and indicator-based strategies (Talbi 2009). Scalar methods may be listed as weighted aggregation, goal programming, the ε constraint, achievement function and goal attainment. Criterion-based strategies are parallel and lexicographic approach. Dominance-based strategies are Pareto, weak, strict, ε, g, guided, fuzzy and stochastic dominance.

This section will give a review of most commonly used methods in fitness evolution that can be used for obsolescence management. Readers are encouraged to refer to (Ehrgott & Gandibleux, 2002; Sanghamitra, Bandyopadhyay, & Saha, 2013; Talbi, 2009) to review the exhaustive list and detail of each fitness evolution method.

Aggregation is a type of scalar fitness calculation in which the weights are distributed over the objectives based on the preferences of the decision maker. This is the most straightforward method to turn multi-objective problem into a single objective that is a combination of all objectives. This method is derived from Kuhn-Tucker conditions (Kuhn & Tucker, 1951). Pending the decision maker's preference, appropriate weight values may be selected in close coordination with the decision maker for each objective of obsolescence management such as minimum cost and maximum operational efficiency.

$$\min\left(\max\right)\sum_{i=1}^{k} w_i f_i\left(\vec{x}\right).$$

where $\sum_{i=1}^{k} w_i = 1$

Another method to use fitness calculation is goal programming. Goal is defined as target values in the direction of the objective. For each objective, a set of goals are defined and the deviation from each goal is minimized. Solution found with goal programming is not necessarily the global optimum or Pareto optimal solution but instead, it is the best solution that meets the goals defined for each objective with minimum, preferably zero deviation.

The decision maker may determine target values for each objective. Those target values may be a ceiling budget for a given system and the operational requirement for the system to be operationally available with a given minimum efficiency.

ε constraint method is another scalar approach and it is similar to preemptive goal programming. It first picks the most important objective as a single objective function and regards all other objectives as a constraint that should be less than a carefully selected ε value. At each iteration, different ε values are run in the algorithm to create the whole Pareto front. Similarly, the decision maker selects the higher priority among all the objectives and ε values for other objectives for obsolescence management.

$$\min\left(\max\right) f_n\left(\vec{x}\right).$$

$$\text{s.t. } f_i\left(\vec{x}\right) \leq \varepsilon_i \, for \, i = 1,2,\dots n-1, n+1,\dots k, i \neq n.$$

VEGA method is considered to be non-Pareto population-based algorithm (Ehrgott & Gandibleux, 2002) in which population is composed of subpopulations that are created by taking each objective into account separately. At each iteration of the genetic algorithm, selection operation is made based on each objective to create subpopulations. Then all sub-populations are shuffled to make up the whole population.

The lexicographic approach is a non-Pareto population-based criterion-based method (Ehrgott & Gandibleux, 2002). Decision maker prioritizes the objectives based on his/her preferences and the algorithm starts with minimizing (maximizing) the highest priority objective. Once an objective value is found for higher priority objective, it is added as an equality constraint to the problem until the last objective is calculated. Depending on the phase of the product lifecycle, decision-makers may have a different priority for a given military system. During the initial procurement and implementation phase, cost minimization may be the highest priority but during operational use phase, under the certain geopolitical situation, operational efficiency may become a higher priority for decision maker than cost minimization.

MOGA, NPGA, and NSGA are three Pareto based, non-elitist, population-based algorithms that are used for fitness calculation (Bandyopadhyay & Saha, 2013). MOGA assigns a rank to each solution based on the number of other solutions that dominate the selected solution. All non-dominated solutions assume a rank value of one. Once all solutions are assigned a rank value, the population is sorted and fitness value is calculated by interpolation between the highest and lowest rank value. NPGA is similar to MOGA but uses a tournament selection in which any randomly selected two instances in the population are compared to a niche of the whole population, preferably 10% of the population. NSGA method clusters all non-dominated solution into one set and this set is then removed from the population with a given set value. This operation is iteratively applied until all solutions are clustered accordingly.

Strength Pareto Evolutionary Algorithm (SPEA), SPEA2, Pareto Archived Evolutionary Strategy (PAES), Pareto Envelope-based Selection Algorithm (PESA), PESA-II, NSGA-II are considered to be Pareto-based, elitist, population-based algorithms (Bandyopadhyay & Saha, 2013).

SPEA is the simplest method among others mentioned above. Non-dominated solutions are moved into another explicit set to archive at each iteration, the ranking value of any given solution is calculated based on the number of solutions within the archived solutions that it dominates. This method can be considered as MOGA method with the elitist selection.

SPEA2 has been developed to overcome the deficiency of SPEA method in case all solutions are dominated by the same archived solution, which would eventually assign the same ranking value to all solutions in the population.

PAES algorithm uses a simple strategy in which a single parent generates a single child. If the parent dominates child, then the child is discarded. If the child dominates the parent, then it is taken as the new offspring. In case child and parent cannot dominate each other, then the child is compared to the archived set to see if it can dominate any solution in the achieved set. Depending on the controls, the child is either taken into achieved set or discarded.

Diversity Check

The purpose of any given search problem is to ensure that search space is adequately explored and the search is not stuck within a local minimum. This is specifically important for multi-objective decision-making problems since Pareto front must be uniformly created overall objectives. If diversity is not addressed properly evolutionary algorithms may opt to favor a solution due to phenomena called genetic drift.

There are generally three main methods used for diversity checks as given below (Talbi, 2009). Diversity check is embedded in the fitness calculation method in most of the algorithms.

- Kernel Methods
- Nearest Neighbor method
- Histogram method

Kernel methods use a kernel function to find out the neighborhood value of a given solution by calculating the distance to another point. MOGA, NSGA and NPGA use kernel method to check the diversity of the solutions in the population (Zitzler, 2002).

The Nearest Neighbor method uses *k*th nearest neighbors to calculate the density of a given solution in the population. SPEA2 uses this method to check and ensure diversity within the population.

Histogram method uses a histogram in which the space is divided into neighbors on a hyper-grid. The density of a given solution is calculated based on the number of other solutions within the same grid. PAES uses this kind of check to ensure diversity within the population.

Elitism

Elitism is defined as preserving the best-found solutions set while the algorithm continues to form the Pareto front. At each generation, a set of best solutions is copied onto a separate set called archive. Populating the archive and managing the archive members is algorithm specific.

SPEA, SPEA2, PAES, PESA, PESA-II are elitist methods in which a set of solutions are stored separately to preserve the best-found solutions (Zitzler, 2002). NSGA-II is a special case for elitist algorithms since it has not separate archive to store non-dominated solutions but instead, the algorithm preserves the non-dominated solutions with the selection operation. While there are differences between the way the archive is established and how a replacement of a member in the archive with a new one is made, the main idea is the same: preserve the non-dominated solutions in the population to maintain the movement towards establishing Pareto front.

FUTURE RESEARCH DIRECTIONS

Obsolescence management within the context of MCDM is still an open area for future research. Even though there are a few models developed to seek for solutions for Obsolescence Management problem by using MCDM, we see a vast opportunity to increase the fidelity level of those models to be more viable and usable in real life applications.

CONCLUSION

This chapter has reviewed the obsolescence management in sustainment-dominated systems, namely military systems. Most of the studies conducted on obsolescence management take the minimization of a cost factor as a single objective. However, many problems that we encounter in our daily life have multiple aspects that need to be taken into account. Obsolescence management in sustainment-dominated systems is one of them and decision makers have to find a balance between cost minimization with operational efficiency objectives of a given system. Following the exhaustive review of studies on obsolescence management, we have outlined a review of multiple criterion decision making and how evolutionary algorithms are used to form a Pareto front that holds the non-dominated solutions. MCDM frameworks.

REFERENCES

Back, Fogel, & Michalewicz. (2000). *Evolutionary Computation 1 Basic Algorithms and Operators*. Institute of Physics Publishing.

Baker, A. (2013). Configurable Obsolescence Mitigation Methodologies. *Procedia CIRP*, *11*, 352–356. doi:10.1016/j.procir.2013.07.013

Bandyopadhyay, S., & Saha, S. (2013). *Unsupervised Classification* (1st ed.). Springer-Verlag Berlin Heidelberg. doi:10.1007/978-3-642-32451-2

Bandyopadhyay, S., & Saha, S. (2013). Some Single- and Multiobjective Optimization. In *Unsupervised Classification* (pp. 17–58). Springer. doi:10.1007/978-3-642-32451-2_2

Barton, D. L., & Chawla, P. (2003). Designing Systems for Future Obsolescence. *Digital Avionics Systems Conference, 2003. DASC '03. The 22nd, 2*, 1–11.

Cuculoski, V. (2013). Obsolescence Management of Electronic and Control Systems. *Australian Control Conference,* (7), 342–47. 10.1109/AUCC.2013.6697296

Darwin, C. (1929). *The Origin of Species by Means of Natural Selection or the Preservation of Favored Races in the Struggle for Life*. New York: Random House.

Defence, The Secretary of. (1994). *Specifications & Standards - A New Way of Doing Business*. Retrieved from http://sw-eng.falls-church.va.us/perry94.html

Ehrgott, M., & Gandibleux, X. (2002). Multiple Criteria Optimization, State of the Art Annotated Bibliographic Surveys. Springer US.

Feldman, K., & Sandborn, P. (2007). Integrating Technology Obsolescence Considerations into Product Design Planning. *Information in Engineering Conference*, 1–8. 10.1115/DETC2007-35881

Feng, D. (2007). *Optimizing Lifetime Buy Quantities to Minimize Lifecycle Cost*. Academic Press.

Feng, D., Singh, P., & Sandborn, P. (2007). Lifetime Buy Optimization to Minimize Lifecycle Cost. *Proceedings of the 2007 Aging*. Retrieved from http://www.enme.umd.edu/ESCML/Papers/AgingAircraft07-LTB.pdf

Josias, Terpenny, & McLean. (2004). Component Obsolescence Risk Assessment. *Proceedings of the 2004 industrial engineering research conference (IERC), 15*, 2–7. Retrieved from http://www.enme.umd.edu/ESCML/ObsSGER/IERC2004-Josias-Terpenny-McLean-376.pdf

Karatas, M. (2017). A multi-objective facility location problem in the presence of variable gradual coverage performance and cooperative cover. *European Journal of Operational Research, 262*(3), 1040–1051. doi:10.1016/j.ejor.2017.04.001

Karatas, M. (2018). Optimal Deployment of Heterogeneous Sensor Networks for a Hybrid Point and Barrier Coverage Application. *Computer Networks, 132*(26), 129–144. doi:10.1016/j.comnet.2018.01.001

Karatas, M., & Yakıcı, E. (2018). An iterative solution approach to a multi-objective facility location problem. *Applied Soft Computing, 62*, 272–287. doi:10.1016/j.asoc.2017.10.035

Konoza, A., & Sandborn, P. (2012). An Evaluation of End of Maintenance Dates for Electronic Assemblies. In *Aircraft Airworthiness*. Sustainment Conference.

Konoza, A. J. (2012, April). An Evaluation of End of Maintenance Dates and Lifetime Buy Estimations for Electronic Systems Facing Obsolescence. *ProQuest Dissertations and Theses*, 165. Retrieved from http://search.proquest.com/docview/1346194418?accountid=14511%5Cnhttp://sfx.ucl.ac.uk/sfx_local?url_ver=Z39.88-2004&rft_val_fmt=info:ofi/fmt:kev:mtx:dissertation&genre=dissertations+%26+theses&sid=ProQ:ProQuest+Dissertations+%26+Theses+Global&atitle=&titl

Konoza, S. (2013). Evaluating the End of Maintenance Dates for Electronic Assemblies Composed of Obsolete Parts. *Journal of Mechanical Design*, *c*, 1–35.

Kuhn, H. W., & Tucker, A. W. (1951). Nonlinear Programming. In *Proceeding of Second Berkeley Symposium on Mathematical Statistics and Probability*. Berkeley, CA: University of California Press.

Kumar & Saranga. (2008). *Optimal Selection of Obsolescence Mitigation Strategi ~ Using a Class of Bandit Models*. Academic Press.

Marshall, M. M., & Lambert, K. R. (2008). Insights into Supporting Complex Systems under Conditions of Obsolescence. PICMET: Portland International Center for Management of Engineering and Technology, Proceedings, 1918–23. doi:10.1109/PICMET.2008.4599811

Meng, X., Thornberg, B., & Olsson, L. (2014). Strategic Proactive Obsolescence Management Model. *IEEE Transactions on Components, Packaging, and Manufacturing Technology, 4*(6), 1099–1108. doi:10.1109/TCPMT.2014.2316212

Michalewicz, Z. (1996). *Genetic Algorithms + Data Structures = Evolution Programs* (3rd ed.). Springer-Verlag. doi:10.1007/978-3-662-03315-9

Nelson III & Sandborn. (2013). Strategic Management of Component Obsolescence Using Constraint-Driven Design Refresh Planning. *International Journal of Product Lifecycle Management, 6*(2), 99–120. Retrieved from http://search.proquest.com/docview/1328233150?accountid=26636%5Cnhttp://link.periodicos.capes.gov.br/sfxlcl41?url_ver=Z39.88-2004&rft_val_fmt=info:ofi/fmt:kev:mtx:journal&genre=article&sid=ProQ:ProQ:pqdieng&atitle=Strategic+management+of+component+obsoles

Pecht & Humphrey. (2008). Addressing Obsolescence - The Uprating Option. *IEEE Transactions on Components and Packaging Technologies, 31*(3), 741–45.

Pingle, P. (2015). *Selection of Obsolescence Resolution Strategy Based on a Multi Criteria Decision Model*. Iowa State University.

Plunkett & Verma. (2003). *Technology Refreshment and Obsolescence Avoidance Strategy and Plan*. Academic Press.

Porter, Z. (1998). An Economic Method for Evaluating Electronic Component Obsolescence Solutions. *Boeing Information Space & Defence Systems*, 1–9. Retrieved from http://scholar.google.com/scholar?hl=en&btnG=Search&q=intitle:An+Economic+Method+for+Evaluating+Electronic+Component+Obsolescence+Solutions#0

Prabhakar, V. J., Allison, H., Sandborn, P., & Eriksson, B. (2013). Optimizing Part Sourcing Strategies for Low-Volume, Long Life Cycle Products Using Second Sourcing Ad Part Hoarding. International Design Engineering Technical Conferences.

Prabhakar, V. J., & Sandborn, P. (2013). A Model for Comparing Sourcing Strategies for Parts in Long Life Cycle Products Subject to Long-Term Supply Chain Disruptions. *International Journal of Product Lifecycle Management*, *6*(3), 228–249. doi:10.1504/IJPLM.2013.055880

Raghavan, S., & Chowdhury, B. (2015). State Diagram-Based Life Cycle Management Plans for Power Plant Components. *IEEE Transactions on Smart Grid*, *6*(2), 965–972. doi:10.1109/TSG.2015.2390232

Raghavan, S., & Member, S. (2012). Developing Life Cycle Management Plans For Power Plant Components. *North American Power Symposium (NAPS)*. 10.1109/NAPS.2012.6336327

Rai, R., & Terpenny, J. (2008). Product Obsolescence. *IEEE Transactions on Components and Packaging Technologies*, *31*(4), 880–889. doi:10.1109/TCAPT.2008.2005115

Razi, N., & Karatas, M. (2016). A multi-objective model for locating search and rescue boats. *European Journal of Operational Research*, *254*(1), 279–293. doi:10.1016/j.ejor.2016.03.026

Rojo, F. J. R. (2012). A Cost Estimating Framework for Electronic, Electrical and Electromechanical (EEE) Components Obsolescence within the Use-Oriented Product-Service Systems Contracts. *Proceedings of the Institution of Mechanical Engineers. Part B, Journal of Engineering Manufacture*, *226*(1), 154–166. doi:10.1177/0954405411406774

Sandborn, Terpenny, & Zheng. (2011). Modeling Constraints in Design Refresh Planning. *ASME 2011 International Design Engineering Technical Conferences*, 1–11.

Sandborn, P. (2008, June). Strategic Management of DMSMS in Systems. *DSP Journal*, 24–30.

Sandborn, P. (2013). Design for Obsolescence Risk Management. *Procedia CIRP*, *11*, 15–22. doi:10.1016/j.procir.2013.07.073

Sandborn, P., & Goudrazi, N. (2015). A Real Options Approach to Quantity and Cost Optimization for Lifetime and Bridge Buys of Parts. *Acquisition Research Symposium*.

Sandborn, P., Prabhakar, V., & Ahmad, O. (2011). Forecasting Electronic Part Procurement Lifetimes to Enable the Management of DMSMS Obsolescence. *Microelectronics and Reliability*, *51*(2), 392–399. doi:10.1016/j.microrel.2010.08.005

Sandborn, P., Prabhakar, V., & Feng, D. (2007). DMSMS Lifetime Buy Characterization Via Data Mining of Historical Buys. *DMSS Conference*. Retrieved from http://www.enme.umd.edu/ESCML/obsolescence.htm

Sandborn, P., Singh, P., Geiser, T., & Lorenson, D. (2002). Electronic Part Obsolescence Driven Product Redesign Optimization. *International Journal of Advanced Manufacturing Systems*, *7*(2), 1–13. http://www.enme.umd.edu/ESCML/Papers/AgingAircraft.pdf

Sandborn, P. A., Herald, T. E., Houston, J., & Singh, P. (2003). Optimum Technology Insertion into Systems Based on the Assessment of Viability. *IEEE Transactions on Components and Packaging Technologies*, *26*(4), 734–738. doi:10.1109/TCAPT.2003.820984

Sandborn, P. A., Mauro, F., & Knox, R. (2007). A Data Mining Based Approach to Electronic Part Obsolescence Forecasting. *IEEE Transactions on Components and Packaging Technologies*, *30*(3), 397–401. doi:10.1109/TCAPT.2007.900058

Sandborn. (2014). *A New "Availability-Payment" Model for Pricing Performance-Based Logistics Contracts*. Academic Press.

Shen, Y., & Willems, S. P. (2014). Modeling Sourcing Strategies to Mitigate Part Obsolescence. *European Journal of Operational Research*, *236*(2), 522–533. doi:10.1016/j.ejor.2014.01.025

Singh, Sandborn, & Singh. (2004). Forecasting Technology Insertion Concurrent with Design Refresh Planning for COTS-Based Electronics Systems. *Logistics Spectrum, 38*(2), 349–54. Retrieved from http://www.enme.umd.edu/ESCML/Papers/solemocapaper.pdf%5Cnhttp://ieeexplore.ieee.org/lpdocs/epic03/wrapper.htm?arnumber=1408387

Singh, P., & Sandborn, P. (2006). Obsolescence Driven Design Refresh Planning for Sustainment-Dominated Systems. *The Engineering Economist*, *51*(2), 115–139. doi:10.1080/00137910600695643

Singh, P., Sandborn, P., Lorenson, D., & Geiser, T. (2002). Determining Optimum Redesign Plans for Avionics Based on Electronic Part Obsolescence Forecasts. *World Avioation Congress and Exposition*. 10.4271/2002-01-3012

Solomon, R., Sandborn, P. A., & Pecht, M. G. (2000). Electronic Part Life Cycle Concepts and Obsolescence Forecasting. *IEEE Transactions on Components and Packaging Technologies*, *23*(4), 707–717. doi:10.1109/6144.888857

Talbi. (2009). *Metaheuristics: From Design to Implementation*. Academic Press.

Teunter, R. H., & Fortuin, L. (1999). End-of-Life Service. *International Journal of Production Economics*, *59*(1-3), 487–497. doi:10.1016/S0925-5273(98)00112-1

Van Kooten, J. P. J., & Tan, T. (2007). The Final Order Problem for Repairable Spare Parts under Condemnation. *Operations Research Society*, *53*(9), 1689–1699.

Zheng, Nelson, & Sandborn, & Terpenny. (2013). Ontology-Based Knowledge Representation for Obsolescence Forecasting. *Journal of Computing and Information Science in Engineering*, 13.

Zheng, L. (2011). Knowledge Representation and Decision Support for Managing Product Obsolescence. *ProQuest Dissertations and Theses*, 174. Retrieved from http://search.proquest.com/docview/1020897067?accountid=14511%5Cnhttp://sfx.ucl.ac.uk/sfx_local?url_ver=Z39.88-2004&rft_val_fmt=info:ofi/fmt:kev:mtx:dissertation&genre=dissertations+&+theses&sid=ProQ:ProQuest+Dissertations+&+Theses+Global&atitle=&title=Kn

Zheng, L., Terpenny, J., & Sandborn, P. (2015). Design Refresh Planning Models for Managing Obsolescence. *IIE Transactions*, *47*(12), 1407–1423. doi:10.1080/0740817X.2014.999898

Zitzler, E. (2002). *Evolutionary Algorithms for Multiobjective*. Academic Press.

Chapter 11
Preserving Logistical Support for Deployed Battle Groups in Hostile Environments:
A Decentralized Approach

Brian Colburn
US Navy, USA

Emily Craparo
Naval Postgraduate School, USA

ABSTRACT

The U.S. Navy's at-sea replenishment system is a mobile supply line designed to support the deployed carrier task force (CTF)/cruiser/destroyer (CRUDES) surface action group (SAG) and forward deployed units while at sea. In the Pacific, the main component of the mobile supply line, the combat logistics force (CLF) ship, has become a possible target with the development of the anti-ship ballistic missile. With the ability to target and disable a CLF, an enemy can now disable a deployed CTF/CRUDES fleet by eliminating its required resources. With the goal of preserving the CLF's capabilities to perform its mission while avoiding ASBM threat, the authors consider the possibility of utilizing a "mini-CLF" to shuttle fuel between CLFs operating in a safe environment and warships operating in a threat zone. The authors perform two analyses: they (1) analyze the feasibility of using the Littoral combat ship/joint high-speed vessel, reconfigured as a shuttle to transport resources, and (2) analyze requirements for development of a new class of ships to support the CTF/CRUDES SAG while deployed in the Pacific.

DOI: 10.4018/978-1-5225-5513-1.ch011

INTRODUCTION

As the U.S. military and its leaders continue to evaluate how to ensure that the United States has a superior force, and to preserve or regain our military advantages, the Navy has identified a possible roadmap to its future.

In the words of Rowden, Gumataotao, and Fanta (2015, p. 18), "of Chief of Naval Operations Admiral Jonathan Greenert's three tenets, 'Warfighting First' is at the top, and that is no accident." With senior military officers such as Vice Admiral Rowden and Rear Admirals Gumataotao and Fanta publishing concepts like "distributed lethality" as the way to employ our naval assets in the future, we must give attention to how to support this distributed mode of operations. "For more power in more places, the Navy should increase the offensive might of the surface force and employ ships in dispersed formations known as 'hunter-killer surface action groups'" (Rowden, Gumataotao, & Fanta, 2015, p. 18).

The concept of "distributed logistics" to support such "distributed lethality" is already being addressed by students at the Naval Postgraduate School in Monterey, CA. As identified by LCDR Ellis in a paper for the Joint Campaign Analysis class, Legacy logistics solutions are ill-suited for supporting a large fleet of small units operating in an Anti-Access, Area Denial (A2AD) region (Ellis, 2013). A Distributed Logistics concept is proposed, serving as a hybrid of traditional established logistics bases (ELBs) and Combat Logistics Force (CLF) routes, augmented with a network of small and temporary Expeditionary Support Bases (ESBs). These ESBs are supplied with the Combat Support Shuttles (CSSs), which could be new ships, or a potential use for existing platforms such as the Littoral Combat Ship (LCS) or Joint High Speed Vessel (JHSV) (Ellis, 2013).

BACKGROUND

The U.S. Navy's at-sea replenishment system is a mobile supply line designed to support the deployed Carrier Task Force (CTF)/Cruiser/Destroyer (CRUDES) Surface Action Group (SAG) and forward deployed units while at sea. In the 7th Fleet area of responsibility (AOR) the main component of the mobile supply line, the CLF ship, has become a possible target with the development of the Dong-Feng-21D (DF-21D), a Chinese-developed anti-ship ballistic missile (ASBM). With the claimed ability to target and disable our current CLF fleet with a DF-21D, if within range, an enemy can now disable a deployed CTF/CRUDES SAG by eliminating its required replenished resources, rendering it combat ineffective and more vulnerable to attack. With the goal of preserving the CLF's ability to perform its mission while not subjecting it to an ASBM threat, this chapter considers the possibility of utilizing a "mini-CLF" to shuttle fuel between CLFs operating in a safe environment and warships operating in a threat zone. This study: (1) analyzes the feasibility of using existing assets such as LCS/JHSV, reconfigured as shuttles to transport underway replenishment requirements from the CLFs to the CTF/CRUDES SAG while deployed in the Western Pacific, and; (2) analyzes requirements for development of a new class of ships to support the CTF/CRUDES SAG while deployed in the Western Pacific. Examination of a shuttle's performance and ability to support the deployed CTF/CRUDES SAG is based solely on the potential for a shuttle to connect with and transfer commodities from a port or CLF outside of the threat area to the deployed CTF/CRUDES SAG in the assigned AOR.

It has been suggested that LCS/JHSV could act as a shuttle supplying the deployed CTF/CRUDES SAG adequately. This work seeks to determine whether that is possible for our scenario, and if not, what would be the configuration requirements for a shuttle to perform this mission. Such a shuttle could act as an intermediate CLF less vulnerable to attack in an A2AD environment due to decreased size and possibly increased speed relative to traditional CLF vessels. To accomplish this, this work develops an optimization model that prescribes a refueling schedule for a set of warships, shuttles, and CLF ships, with the goal of maximizing on-station time for the warships while keeping the CLF ships out of harm's way.

This work contributes a valuable network performance assessment and alternative to the 7th Fleet and Operational Navy (OPNAV) leadership. The tools developed for this study can be easily adapted to evaluate shuttle performance under a host of other employment parameters. The resultant understanding of how shuttle deployment affects a supply network's ability to support mission critical replenishment will be invaluable in future systems analysis and CLF ship development.

SCOPE, LIMITATIONS, AND MODEL ASSUMPTIONS

This work supports unpublished, ongoing research by U.S. Pacific Fleet (USPACFLT) in efforts to provide feasible options to resolve the identified issues of continued logistic support to deployed forces in the Western Pacific and the South China Sea. This research analyzes the potential for and design requirements of a shuttle as a delivery ship between a CLF ship or port and deployed CTF/CRUDES SAG assets in the 7th Fleet AOR while preserving the safety of the CLF ship now potentially subject to the threat presented by the A2AD environment.

Our model only accounts for the fuel commodity required by a deployed CTF/CRUDES SAG during a predetermined, fixed time horizon. It does not account for the consumption of stores commodities or ordnance that may require replenishment due to expenditure. Our scenarios consider a predetermined fleet composition in an identified AOR, excluding the addition or subtraction of assets and the possibility of external demands on the CLF/Shuttle force in the model.

This study assumes the following in order to produce a mathematical model and a realistic scenario for evaluating the effectiveness of numerous design criteria of a shuttle ship to be the "delivery boy" between current CLF and deployed CTF/CRUDES assets:

- The DF-21D threat keeps large ships from safely operating in the South China Sea/Western Pacific. Large ships include the Aircraft Carrier (Nuclear) (CVN), Amphibious Assault Ship (LHD), and Auxiliary Dry Cargo and Ammunition Ship (T-AKE)/Fleet Oiler (T-AO). This drives the design requirement for the production of a smaller shuttle ship to more safely operate within the DF-21D threat range based solely on the size restriction.
- A deployed CTF/CRUDES SAG requires resupply to continue operations at the forward operating stations (FOSs). This drives the requirement for fuel commodities to be transported through the network of arcs and nodes for consumption by the deployed CTF/CRUDES assets.
- For simplicity, the model only accounts for liquid fuel in this initial analysis.
- The quantity of JP5 (aviation fuel) required by the airwing to conduct aircraft carrier flight operations is calculated into the liquid fuel requirements by combining Diesel Fuel, Marine (DFM) and JP5 quantity requirements into a single liquid fuel quantity measured in barrels (bbls). This allows

for the complete accounting of throughput of liquid fuel to a CTF as an aircraft carrier is a nuclear powered vessel and does not consume large quantities of DFM.

- All ships in an assigned battle group stay within close proximity of each other unless an individual ship's burnable fuel quantity falls below the safety fuel quantity level, if this occurs, the individual ship can detach from the CTF/CRUDES SAG and operate independently to navigate to a forward replenishment at sea (RAS) lane (FRL) or port to receive fuel.
- Our initial analysis considers peacetime operations. Peacetime operations typically have a lengthier time horizon, therefore testing the model in a sustained optimization environment as opposed to a shorter time horizon wartime scenario.

This study contributes to the ongoing study of CLF ships and their utilization in the fleet. It develops the Dual Lane RAS Model (DL-RASM), an optimization model designed to evaluate a possible method to preserve logistic support to a deployed CTF/CRUDES SAG operating within an anti-access, area denial (A2AD) environment. Through optimization and sensitivity analysis, this study identifies critical design characteristics for a next generation of CLF ship to provide improved safety to the current CLF fleet configuration, provide sufficient support to deployed assets operating in an A2AD environment, improve resiliency in the RAS logistics network, and provide improved future support of a changing fleet configuration as adversaries move to master the seas in the littorals.

LITERATURE REVIEW

Several prior studies have investigated design and utilization of CLF ships to support U.S. assets afloat. These include both theses and operational logistics models. While this prior work has been extremely useful and has brought about improvements in the at-sea replenishment process, none of these works have considered the threat to the highly valued assets in the CLF ship fleet. Work has been completed focusing on the next composition of the U.S. Navy taking into account the developed A2AD threat and the move of the at-sea battle into littoral waters, but it has not yet provided answers of how to support this new fleet while at sea. This section gives an overview of prior work that considers aspects of these subjects; no prior study combines the topics of at-sea replenishment of the current or future naval fleet and the threat of the A2AD environment.

CLF Fleet Composition Comparison

Givens (2002) used an optimization-based assessment model to compare two top CLF alternatives published by the Commander in Chief, U.S. Atlantic Fleet (CINCLANTFLT) report in 2001. It was based on Borden (2001) and was the first to compare and optimize the configuration and loading of CLF ships to determine the CLF fleet composition based on a deterministic model. The results reveal the best way to utilize CLF assets to achieve full force potential comparing details that include stock levels maintained, prepositioning of CLF ships, and the minimum number of ships required. While these models include a global view of navigable world sea routes, forward positioned commodities, and the network between theaters, they do not consider the developing threat to the CLF ship and the impact it could have on the ability of deployed CTF/CRUDES assets to be replenished at sea if a CLF was lost.

CLF Planner

With the success of the early optimization models, additional models were developed that assess specific aspects of logistics support to deployed naval forces. Models such as Global Fleet Station Mission Planner (GFSMP) (Spitz, 2007) and "Navy Combat Logistics Force" (CLF) Planner (Brown & Carlyle, 2008) are just two (Brown, Carlyle, Kelton, Kline & Salmeron, 2009). However, the CLF Planner became the more adopted tool and has been applied to exercises developed by U.S. Commander, Second Fleet, exhibiting significant improvements over manual planning (Brown, Carlyle, Kelton, Kline, & Salmerón, 2009). Morse (2008) uses this model in its most recent refinement developed by Doyle (2006), in parallel to a concurrent OPNAV N-42 Combat Logistics Force Zero-Baseline Review, to provide additional insights into asset allocation and utilization for specific scenarios. It demonstrates that the model can be used as a decision analysis tool and that it could be adapted to any scenario, fleet-specific or global, to provide CLF fleet analysis as well as operational CFL requirements planning (Morse, 2008). In a time when the CLF was comprised of 31 ships of five basic ship types and was being transformed to 30 ships of three basic types by 2014, Morse's work was pivotal in providing essential decision making information to OPNAV N-42. While Morse's thesis does a thorough job of examining past analysis for CLF fleet sizing and expands the model and information base pertinent to the subject, it does not take into account the developing threat of the current A2AD environment, then in its infancy. Our optimization model could be adapted and applied as Morse's model was and address a solution to the A2AD threat to our active CLF. It also provides a next generation analysis tool to be expanded upon and utilized in the future of CLF planning.

Replenishment At Sea Planner

Brown, Carlyle and Burson's Replenishment At Sea Planner (RASP) (2010) was the next generation of optimization model designed more for the everyday CLF scheduler, and it is based on CLF Planner. However, where CLF Planner addresses the strategic question of how many and what type of supply ships to build in the future on a macro level, RASP evaluates CLF employment at an operational level (Stewart, 2013). It evaluates the ability of a single ship in a region to replenish multiple ships over a four-hour window. In 2012, Naval Postgraduate School (NPS) Research Associate, Anton Rowe, author of much of the detailed code that makes the Replenishment at Sea Planner (RASP) work, stated that "the Replenishment at Sea Planner creates a schedule that minimizes the distance that supply ships have to travel and identifies routes that allow vessels to travel at speeds optimal to fuel conservation" (Stewart, 2013). While RASP optimizes a schedule for CLF support of naval assets afloat in a region, saving the military millions in fuel costs, it only identifies the most efficient means based on its inputs to meet the deployed fleet's needs. Our model accomplishes this while accounting for the safety of the CLF ship in its daily operations.

Development of the A2AD Environment

The present era of warfare at sea is defined by missile technology. DF-21D has become the primary threat to the larger ships in our fleet, and with its range and targeting capability, it is a substantial threat the U.S.'s ability to continue to operate its navy in the manner in which it is accustomed. While DF-21D is not the only missile threat to our fleet afloat (Ross & Harmon, 2012), it is the threat that defines

Figure 1. Range rings for the Chinese DF-21D (listed as CSS-5) ASBM and other conventional anti-access capabilities
From Collins & Erickson, 2010.

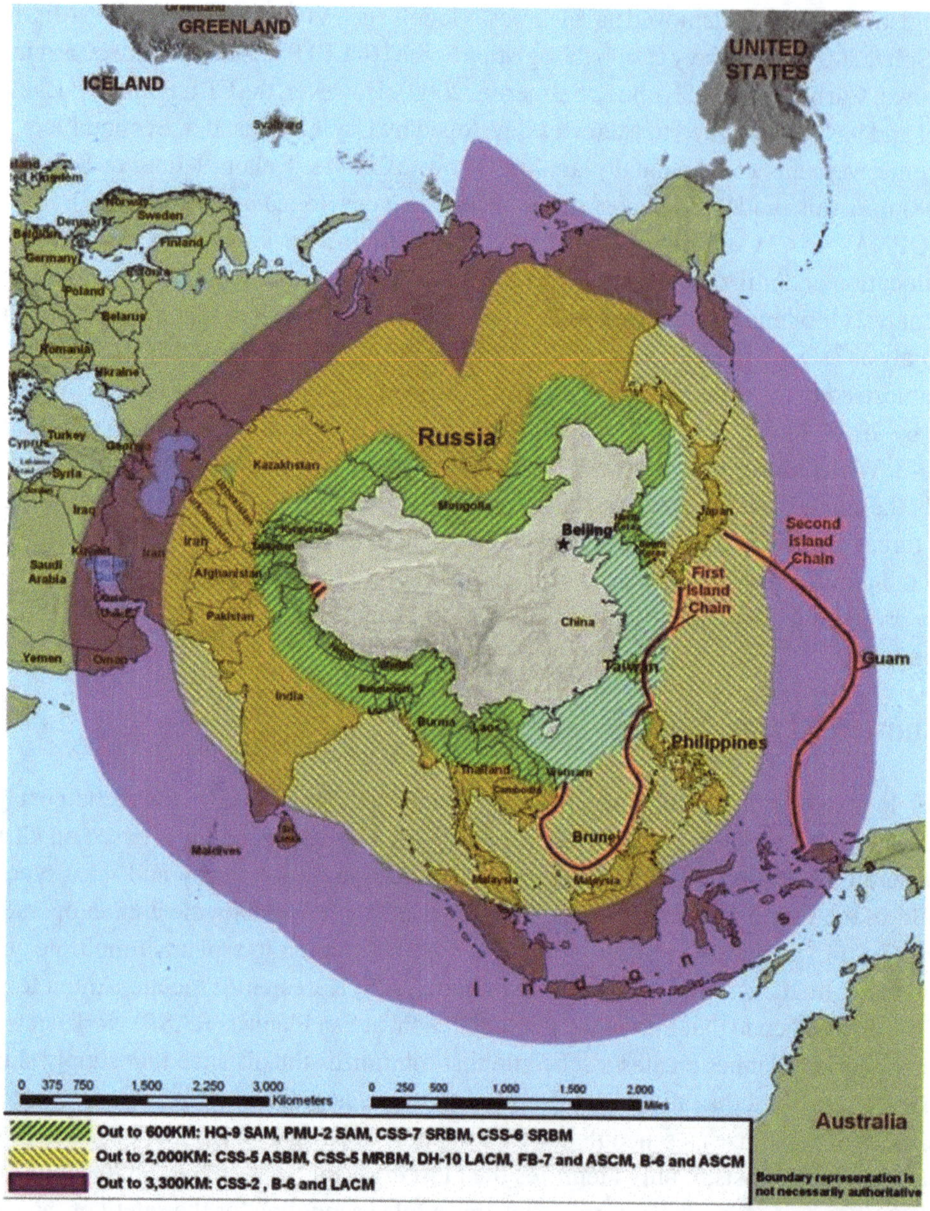

operational parameters for the CLF assets used in our optimization model. While Ross and Harmon (2012) utilize a range of 1500 km for DF-21D, the area cover range images for the DF-21D as of 2010, depicted in Figure 1, allow for a range of 2000 km (shaded in yellow).

More recent information supports the expanded range out to 2000 km (Jane's Strategic Weapons Systems, 2015). This expanded range further limits the operational area of the CLF assets to preserve their safety and further complicates the logistical support problem.

Figure 2. The aftermath of a missile test, published in the Want China Times (2013), reveals the possibility of striking an aircraft carrier flight deck.

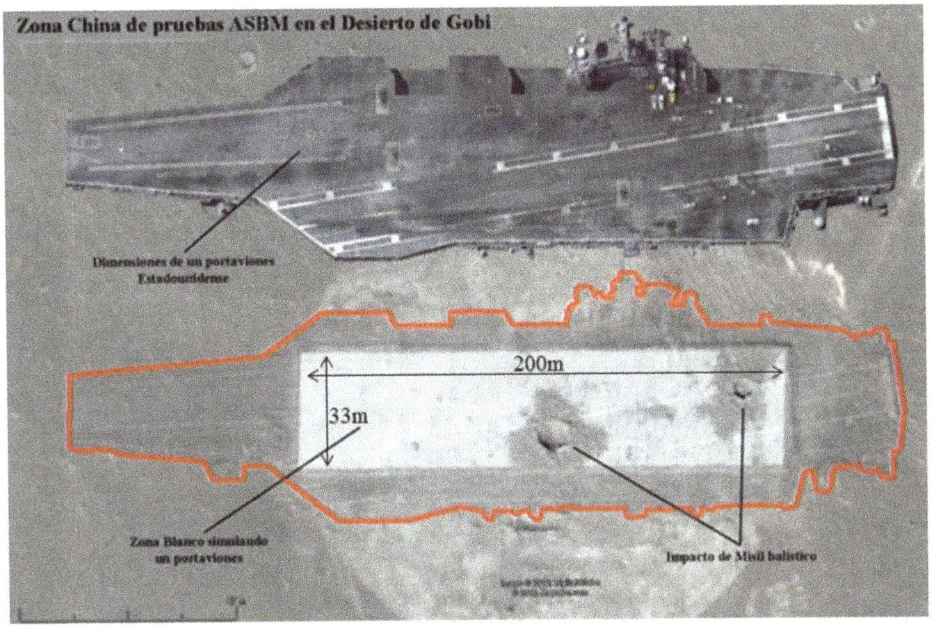

As can be seen in Figure 2, a satellite image published in the Want China Times (2013) reveals two large craters on a 200-meter-long by 33-meter-wide white platform in the Gobi desert, used to simulate the flight deck of an aircraft carrier. The photo was first posted on SAORBATS, an Internet forum based in Argentina. Military analysts at the time believed the craters were created by China's DF-21D anti-ship missile, dubbed the "carrier killer" (Want China Times, 2013).

This realized threat has raised the question, how should the US Navy protect its highly valued CLF assets while providing logistic support for deployed assets, given that current CLF ships are approximately the dimensions listed in Table 1 underscores this threat by showing the dimensions of the T-AKE and T-AO utilized in the DL-RASM model.

New Navy Fighting Machine

Although the current U.S. naval fleet needs to change to counter the evolving missile threat and the transition of sea warfare to the littorals using missile boats, the logistics aspect of how to support the new fleet has escaped most analyses performed to date. Hughes (2009) published such a paper identifying

Table 1. CLF ship dimensions

	Lewis and Clark (T-AKE)	Henry J. Kaiser (TAO)
Length	210.0 m (689.0 ft)	206.5 m (677.5 ft)
Beam	32.2 m (105.6 ft)	29.7 m (97.4 ft)

From Jane's Fighting Ships, 2015.

Table 2. NNFM Green Water Fleet Composition

NNFM Green Water Fleet	
Ship or Combatant	**Number of Units**
Coastal Combatant	30
Offshore Patrol	160
Fleet Station Ship	12
Inshore Patrol	400
Gunfire Support	12
Fast Mine Warfare (MIW)	12
Anti-Submarine Warfare (ASW)	12
CVL (Green Water Fleet)	8
Coastal Combatant Tender	2
Total	648

From Hughes, 2009.

the concept of the NNFM Green Water Fleet. Table 2 depicts the composition of the proposed fleet to combat the emerging threat of our adversaries in the littorals.

Utilizing current force deployment concepts of 1/3 of the fleet deployed at any point in time, this fleet composition results in approximately 216 ships requiring at-sea support, a considerable increase over today's numbers (Status of the Navy, 2015). In addition to being extremely vulnerable to attack, the current CLF design is limited in the number of ships it can service in a given period of time. Thus, a new shuttle ship can assist in supporting the transformation of the U.S. naval fleet and allow diversification of the fleet while decreasing commodity risk and increasing logistics network resiliency

MODEL AND SCENARIO DEVELOPMENT

This section introduces the Dual Lane Replenishment At Sea Model (DL-RASM). DL-RASM is a discrete-time mixed integer linear program that optimizes the ability of a set of shuttles and CLF ships to provide required liquid fuel logistic support to deployed warships for a given planning horizon. DL-RASM prescribes the movement and refueling schedules of a set of warships, shuttles, and CLFs. It incurs penalties for each barrel of fuel shortage that warships, shuttles or CLF ships experience below prescribed safety stock levels, and it accumulates rewards for every barrel of fuel transferred and transmitted through the network, as well as for every time period a warship spends on a forward operating station (FOS). DL-RASM is a network-based model; its nodes consist of forward operating stations, forward RAS lanes (FRLs), intermediate transit points (ITPs), aft RAS lanes (ARLs), and ports. It derives its name from the fact that there are two lanes for replenishment, forward and aft. A schematic representation of the nodes and arcs within the network is provided in Figure 3, and it indicates which nodes are in the threat area, as well as which nodes and arcs are used by the warships, shuttles and CLF ships.

This work makes the following assumptions in order to produce a realistic mathematical model and accompanying input data:

Figure 3. Schematic representation of the nodes, arcs, and threat area modeled by DL-RASM. Warships only visit forward operating stations and forward RAS lanes, shuttles visit forward RAS lanes, aft RAS lanes, and ports in the threat area and CLFs visit aft RAS lanes and ports outside the threat area

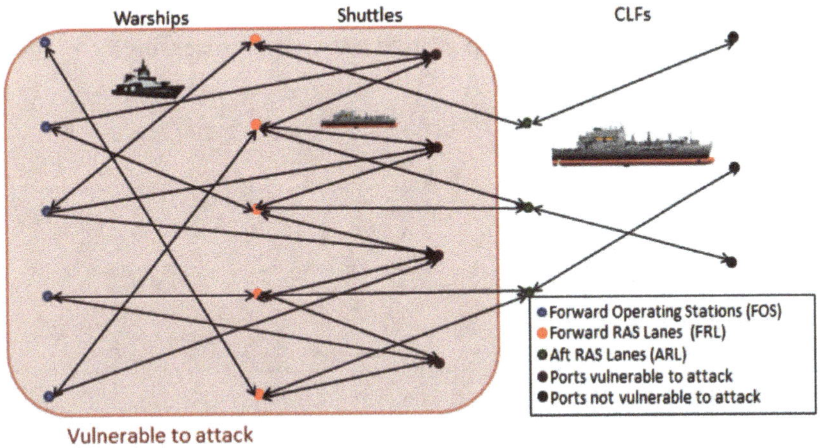

- All ships in a battle group must travel together as a unit, unless a ship encounters a fuel shortage of sufficient severity. If this occurs, that ship may separate from the battle group in order to obtain fuel. To enforce the requirement that all ships in a battle group remain together, the model requires that each ship be collocated with its battle group's carrier unless low on fuel. This may involve creating a fictitious carrier for any battle group that does not contain a carrier.
- Warships have a single fuel tank containing burnable fuel. Shuttles and CLFs have two fuel tanks: one containing burnable fuel, and one containing deliverable fuel. Each shuttle can transfer fuel from its deliverable fuel tank into the burnable fuel tank of a warship or into its own burnable fuel tank. Each CLF ship can transfer fuel from its deliverable fuel tank into the burnable fuel tank of a shuttle, the deliverable fuel tank of a shuttle or into its own burnable fuel tank.
- For a RAS event to occur, the relevant ships must be collocated at a FRL or ARL for a sufficiently long time.
- For modeling purposes, FOSs and FRLs are collocated; however, when a warship is performing RAS operations, it occupies the FRL node and is not rewarded for operational time at the FOS.
- During peace-time operations, RAS events only occur during daylight time periods. This is representative of actual operational restrictions currently practiced during peacetime operations.
- The combat fleet, shuttles and CLF fleet do not suffer any losses during the scenario. That is, all ships remain available throughout the planning horizon.
- Each warship, shuttle, and CLF ship consumes fuel at the same rate at each period in time, unless it is in port.
- Once a ship enters port, it must remain in port for a prescribed number of time periods.
- Requirements outside of the AOR and scenario do not impact the availability of assets.

Dual Lane, Replenishment At Sea Model Formulation

Indices and Sets [Approximate Cardinality]

$bg \subseteq WxG$: warship *w* belongs to battle group *g* (used only in preprocessing)
$c \in C$: CFL ships [2]
$f, f' \in F \subseteq N$: forward operating station nodes [7]
$g \in G$: battle groups [2]
$n, i, j \in N$: nodes [~30]
$p, p' \in P \subseteq N$: port nodes [6]
$r, r' \in R \subseteq N$: RAS lane nodes [10]
$s, s' \in S$: shuttle ships [~3-8]
$t, t', t'' \in T$: time periods [~20]
$t \in D \subseteq T$: daylight time periods
$w, w' \in W$: warships [~10]
$(i, j) \in ac \subseteq NxN$: arc *(i,j)* can be traversed by CLF ships
$(i, j) \in as \subseteq NxN$: arc *(i,j)* can be traversed by shuttle ships
$(i, j) \in aw \subseteq NxN$: arc *(i,j)* can be traversed by warships
$(w, w') \in carrier \subseteq WxW$: warships *w's* carrier is warship *w'*
$(t, t', c, i, n) \in dC \subseteq TxTxCxNxN$: if CLF ship *c* departs node *i* at time *t*, it will arrive at node *n* at time *t'*
$(t, t', s, i, n) \in dS \subseteq TxTxSxNxN$: if shuttle ship *s* departs node *i* at time *t*, it will arrive at node *n* at time *t'*
$(t, t', w, i, n) \in dW \subseteq TxTxWxNxN$: if warship *w* departs node *i* at time *t*, it will arrive at node *n* at time *t'*
$(w, t, n) \in wayp \subseteq WxTxN$: waypoint schedule: battle group *g* must be at location *n* at time *t*

Parameters [Units]

$berths_p$: berths at port *p* [berths]
$burnDFMc_c$: capacity of burnable fuel tank for CLF ship *c* [barrels]
$burnDFMs_s$: capacity of burnable fuel tank for shuttle *s* [barrels]
$burnDFMw_w$: capacity of burnable fuel tank for warship *w* [barrels]
$burnSafetyc_c$: percentage of $burnDFMc_c$ below which CLF ship *c* incurs a penalty [percentage]
$burnSafetys_s$: percentage of $burnDFMs_s$ below which shuttle *s* incurs a penalty [percentage]
$burnSafetyw_w$: percentage of $burnDFMw_w$ below which warship *w* incurs a penalty [percentage]
$capDFMc_c$: capacity of deliverable fuel tank for CLF ship *c* [barrels]
$capDFMs_s$: capacity of deliverable fuel tank for shuttle *s* [barrels]
$\mathcal{E}$: small reward for fuel transfers; intended to mitigate end-of-horizon effects [reward/barrel]
FC_c: fuel burn per time period by CLF ship *c* [barrels]
FS_s: fuel burn per time period by shuttle *s* [barrels]
FW_w: fuel burn per time period by warship *w* [barrels]
$inportc_c$: time periods required for CLF ship *c* to remain in port on each port visit

inports$_s$: time periods required for shuttle *s* to remain in port on each port visit
penaltyw$_w$: per-unit- fuel shortage penalty for warship *w* [penalty/barrel]
inportw$_w$: time periods required for warship *w* to remain in port on each port visit
nrpp: number of ships that can begin RASing per time period, for each shuttle and CLF [ships]
penaltyc$_c$: per-unit fuel shortage penalty for CLF ship c [penalty/barrel]
penaltys$_s$: per-unit fuel shortage penalty for shuttle *s* [penalty/barrel]
reward: reward for on-station time [reward/(ship*period)]
safetyDFMc$_c$: percentage of *capDFMs$_s$* below which shuttle *s* incurs a penalty [percentage]
safetyDFMs$_s$: percentage of *capDFMs$_s$* below which CLF ship *c* incurs a penalty [percentage]
simult_c: number of shuttles that can simultaneously RAS from a single CLF [ships]
simult_s: number of warships that can simultaneously RAS from a single shuttle [ships]
$TC_{c,i,j}$: time periods required for CLF ship *c* to transit from location *i* to location *j* [periods]
$TS_{s,i,j}$: time periods required for shuttle *s* to transit from location *i* to location *j* [periods]
$TW_{w,i,j}$: time periods required for warship *w* to transit from location *i* to location *j* [periods]
TR_S$_s$: time periods required to RAS shuttle *s* [periods]
TR_W$_w$: time periods required to RAS warship *w* [periods]

Decision Variables

Binary Decision Variables

$FDIPc^t_{c,p}$: 1 if CLF ship *c* enters port *p* at time *t* [binary]
$FDIPs^t_{s,p}$: 1 if shuttle *s* enters port *p* at time t [binary]
$FDIPw^t_{w,p}$: 1 if warship *w* enters port *p* at time *t* [binary]
$LOWF^t_w$: 1 if warship *w's* fuel is below safety stock at time *t* [binary]
$REVUc^t_{s,c,r}$: 1 if at start time *t* shuttle *s* collocated with CLF ship *c* in RAS lane *r* long enough to fill shuttle *s* [binary]
$REVUw^t_{w,s,r}$: 1 if at start time *t* warship *w* is collocated with shuttle *s* in RAS lane *r* long enough to fill warship *s* [binary]
$YC^t_{c,i,j}$: 1 if at time *t* CLF ship *c* departs location *i* bound for location *j* [binary]
$YS^t_{s,i,j}$: 1 if at time *t* shuttle *s* departs location *i* bound for location *j* [binary]
$YW^t_{w,i,j}$: 1 if at time *t* warship w departs location i bound for location j [binary}

Nonnegative Decision Variables

$HOLDc^t_c$: deliverable fuel inventory at start of time period *t* for CLF ship *c* [barrels]
$HOLDs^t_s$: deliverable fuel inventory at start of time period *t* for shuttle *s* [barrels]
$HOLD\_INVc^t_c$: fuel transferred from deliverable tank to burnable tank in CLF ship *c* during time *t* [barrels]
$HOLD\_INVs^t_s$: fuel transferred from deliverable tank to burnable tank in shuttle *s* during time *t* [barrels]
$INPc^t_c$: burnable fuel replenished in port during time period *t* for CLF ship *c* [barrels]
$INPs^t_s$: burnable fuel replenished in port during time period *t* for shuttle *s* [barrels]
$INPw^t_w$: burnable fuel replenished in port during time period *t* for warship *w* [barrels]
$INPhc^t_c$: deliverable fuel replenished in port at time *t* for CLF ship *c* [barrels]
$INPhs^t_s$: deliverable fuel replenished in port at time *t* for shuttle *s* [barrels]

$RAS\_burn\_s^t_{s,c}$: burnable fuel transferred to shuttle s from CLF c at time t [barrels]
$RAS\_hold\_s^t_{s,c}$: deliverable fuel transferred to shuttle s from CLF c at time t [barrels]
$RASw^t_{w,s}$: fuel transferred to warship w from shuttle s at time t [barrels]
$SHORTBURNc^t_c$: barrels below safety stock of burnable fuel at time period t for CLF ship c [barrels]
$SHORTBURNs^t_s$: barrels below safety stock of burnable fuel at time period t for shuttle s [barrels]
$SHORTBURNw^t_w$: barrels below safety stock of deliverable fuel at time period t for CLF ship c [barrels]
$SHORTc^t_c$: barrels below safety stock of deliverable fuel at time period t for CLF ship c [barrels]
$SHORTs^t_s$: barrels below safety stock of deliverable fuel at time period t for shuttle s [barrels]

Free Variables

$INVc_c^t$: burnable fuel inventory at start of time period t for CLF ship c [barrels]
$INVs_s^t$: burnable fuel inventory at start of time period t for shuttle s [barrels]
$INVw^t_w$: burnable fuel inventory at start of time period t for warship w [barrels]

Formulation

$$\begin{aligned} Max \sum_{t,w,(n,f)\in aw} reward * YW^t_{w,n,f} - \sum_{w,t} penaltyw_w * SHORTBURNw^t_w \\ -\sum_{s,t} penaltys_s * (SHORTBURNs^t_s + SHORTs^t_s) \\ -\sum_{c,t} penaltyc_c * (SHORTBURNc^t_c + SHORTc^t_c) \\ +\varepsilon * \sum_{t\in D,s}\left(\sum_w RASw^t_{w,s} + \sum_c (RAS\_burn\_s^t_{s,c} + RAS\_hold\_s^t_{s,c})\right) \\ +\varepsilon * \sum_t \left(\sum_w INPw^t_w + \sum_c (INPc^t_c + INPhc^t_c) + \sum_s (INPs^t_s + INPhs^t_s)\right) \end{aligned} \tag{1}$$

$$\sum_{j:(n,j)\in aw} YW^t_{w,n,j} = \sum_{t',i:(t',t,w,i,n)\in dW} YW^{t'}_{w,i,n} \quad \forall w,n,t : t > 1; \exists i,j : (i,n) \in aw \text{ and } (n,j) \in aw \tag{2}$$

$$\sum_{j:(n,j)\in as} YS^t_{s,n,j} = \sum_{t',i:(t',t,s,i,n)\in dS} YS^t_{s,i,n} \quad \forall s,n,t : t > 1; \exists i,j : (i,n) \in as \ and \ (n,j) \in as \tag{3}$$

$$\sum_{j:(n,j)\in ac} YC^t_{c,n,j} = \sum_{t',i:(t',t,c,i,n)\in dC} YC^t_{c,i,n} \quad \forall c,n,t : t > 1; \exists i,j : (i,n) \in ac \text{ and } (n,j) \in aw \tag{4}$$

$$\sum_{(n,j)\in aw} YW_{w,n,j}^{t=1} = 1 \qquad \forall w \tag{5}$$

$$\sum_{(n,j)\in as} YS_{s,n,j}^{t=1} = 1 \qquad \forall s \tag{6}$$

$$\sum_{(n,j)\in ac} YC_{c,n,j}^{t=1} = 1 \qquad \forall c \tag{7}$$

$$\sum_{j:(n,j)\in aw} YW_{w,n,j}^{t=1} = 1 \qquad \forall w,n:(w,1,n)\in wayp \tag{8}$$

$$\sum_{t',j:(t',t,w,j,n)\in dW} YW_{w,\mathrm{j},\mathrm{n}}^{t'} = 1 \qquad \forall (w,t,n)\in wayp : t>1 \tag{9}$$

$$\sum_{t',w,n:(t',t,w,j,n)\in dW} YW_{w,n,p}^{t'} + \sum_{t',s,n:(t',t,s,n,p)\in dS} YS_{s,n,p}^{t'} + \sum_{t',c,n:(t',t,c,n,p)\in dC} YC_{c,n,p}^{t'} \le berths_p \qquad \forall p,t \tag{10}$$

$$INVw_w^t = INVw_w^{t-1} + INPw_w^{t-1} - FW_w * \left(1 - \sum_{t',n,p:(t',\mathrm{w},n,p)\in dW} YW_{\mathrm{w},n,p}^{t'}\right) + \sum_s RASw_{w,s}^{t-1} \qquad \forall w,t>1 \tag{11}$$

$$INVs_s^t = INVs_s^{t-1} + HOLD\_INVs_s^{t-1} - FS_s * \left(1 - \sum_{t',n,p:(t',t,s,n,p)\in \mathrm{dS}} YS_{s,n,p}^{t'}\right) + INPs_s^{t-1} \qquad \forall s,t>1 \tag{12}$$

$$INVc_c^t = INVc_c^{t-1} + HOLD\_INVc_c^{t-1} - FC_c * \left(1 - \sum_{t',n,p:(t',c,n,p)\in dC} YC_{c,n,p}^{t'}\right) + INPc_c^{t-1} \qquad \forall c,t>1 \tag{13}$$

$$INPw_w^t \le burnDFMw_w \sum_{t',n,p:(t',t,w,n,p)\in dW} YW_{w,n,p}^{t'} \qquad \forall w,t \tag{14}$$

$$INPs_s^t \le burnDFMs_s \sum_{t',n,p:(t',t,s,n,p)\in dS} YS_{s,n,p}^{t'} \qquad \forall s,t \tag{15}$$

$$INPc_c^t \le burnDFMc_c \sum_{t',n,p:(t',c,n,p)\in dC} YC_{c,n,p}^{t'} \qquad \forall c,t \tag{16}$$

$$INPhs_s^t \le capDFMs_s \sum_{t',n,p:(t',t,s,n,p)\in dS} YS_{s,n,p}^{t'} \qquad \forall s,t \tag{17}$$

$$INPhc_c^t \le capDFMc_c \sum_{t',n,p:(t',t,c,n,p)\in dC} YC_{c,n,p}^{t'} \qquad \forall c,t \tag{18}$$

$$HOLDs_s^t = HOLDs_s^{t-1} - HOLD\_INVs_s^{t-1} + INPhs_s^{t-1} - \sum_w RASw_{w,s}^{t-1} + \sum_c RAS\_hold\_s_{s,c}^{t-1} \qquad \forall s,t>1 \tag{19}$$

$$\begin{aligned} HOLDc_c^t &= HOLDc_c^{t-1} - HOLD\_INVc_c^{t-1} + INPhc_c^{t-1} \\ &-\sum_w RAS\_burn\_s_{s,c}^{t-1} - \sum_s RAS\_hold\_s_{s,c}^{t-1} \qquad \forall c,t>1 \end{aligned} \tag{20}$$

$$\sum_w REWUw_{w,s,r}^t \le nrpp \qquad \forall t \in D,s,r \tag{21}$$

$$\sum_s REVUw_{w,s,r}^t \le nrpp \qquad \forall t \in D,s,r \tag{22}$$

$$\sum_{w,t',\in D:t-TR\_W_w<t'\le t} REVUw_{w,s,r}^{t'} \le simult\_s \sum_{t',n:(t',t,s,n,r)\in dS} YS_{s,n,r}^{t'} \quad \forall s,r,t \in D \tag{23}$$

$$\sum_s REVUc_{s,c,r}^t \le nrpp \qquad \forall t \in D,c,r \tag{24}$$

$$\sum_{c} REVUc_{s,c,r}^{t} \leq nrpp \qquad \forall t \in D, s, r \tag{25}$$

$$\sum_{s,t' \in D: t-TR\_S_s < t' \leq t} REVUc_{s,c,r}^{t'} \leq simult\_c \sum_{t',n:(t',t,c,n,r) \in dC} YC_{c,n,r}^{t'} \qquad \forall c, r, t \in D \tag{26}$$

$$REWUw_{w,s,r}^{t} \leq \sum_{t'',n:(t'',t',n,r) \in dW} YW_{w,n,r}^{t''} \qquad \forall w, s, r, t \in D, t' \in D : t \leq t' < t + TR\_W_w \tag{27}$$

$$REWUw_{w,s,r}^{t} \leq \sum_{t'',n:(t'',t',s,n,r) \in dS} YS_{s,n,r}^{t''} \qquad \forall w, s, r, t \in D, t' \in D : t \leq t' < t + TR\_W_w \tag{28}$$

$$REWUc_{s,c,r}^{t} \leq \sum_{t'',n:(t'',t',s,n,r) \in dS} YS_{s,n,r}^{t''} \qquad \forall s, c, r, t \in D, t' \in D : t \leq t' < t + TR\_S_s \tag{29}$$

$$REWUc_{s,c,r}^{t} \leq \sum_{t'',n:(t'',t',c,n,r) \in dC} YC_{c,n,r}^{t''} \qquad \forall s, c, r, t \in D, t' \in D : t \leq t' < t + TR\_S_s \tag{30}$$

$$RASw_{w,s}^{t} \leq \sum_{r} (\min(capDFMs_s, burnDFMw_w) * REVUw_{w,s,r}^{t-TR\_W_w+1}) \qquad \forall s, w, t \in D \tag{31}$$

$$\begin{aligned} &RAS\_burn\_s_{s,c}^{t} + RAS\_hold\_s_{s,c}^{t} \\ &\leq \sum_{r} (\min(capDFMc_c, capDFMs_s + burnDFMs_s) * REVUc_{s,c,r}^{t-TR\_S_s+1}) \qquad \forall c, s, t \in D \end{aligned} \tag{32}$$

$$SHORTBURNw_{w}^{t} \geq burnSafetyw_w * burnDFMw_w - INVw_{w}^{t} \qquad \forall w, t \tag{33}$$

$$SHORTBURNs_{s}^{t} \geq burnSafetys_s * burnDFMs_s - INVs_{s}^{t} \qquad \forall s, t \tag{34}$$

$$SHORTBURNc_{c}^{t} \geq burnSafetyc_c * burnDFMc_c - INVc_{c}^{t} \qquad \forall c, t \tag{35}$$

$$SHORTs_s^t \geq safetyDFMs_s * capDFMs_s - HOLDs_s^t \qquad \forall s,t \tag{36}$$

$$SHORTc_c^t \geq safetyDFMc_c * capDFMc_c - HOLDc_c^t \qquad \forall c,t \tag{37}$$

$$LOWF_w^t \leq 1 - \frac{(INVw_w^t - safetyDFMw_w * burnDFMw_w)}{(burnDFMw_w - safetyDFMw_w * burnDFMw_w)} \qquad \forall w,t \tag{38}$$

$$YW_{w,j,n}^t \geq YW_{w',j,n}^t - LOWF_w^t \qquad \forall w,w',t,j,n:(w,w') \in carrier,(j,n) \in aw \tag{39}$$

$$FDIPw_{w,p}^t = \sum_{(t',i):i \neq p,(t',t,w,i,p) \in dW} YW_{w,i,p}^{t'} \qquad \forall w,p,t:\exists i:(i,p) \in aw \tag{40}$$

$$FDIPs_{s,p}^t = \sum_{(t',i):i \neq p,(t',t,s,i,p) \in dS} YS_{s,i,p}^{t'} \qquad \forall s,p,t:\exists i:(i,p) \in as \tag{41}$$

$$FDIPc_{c,p}^t = \sum_{(t',i):i \neq p,(t',t,c,i,p) \in dC} YC_{c,i,p}^{t'} \qquad \forall c,p:\exists i:(i,p) \in ac \tag{42}$$

$$YW_{w,p,p}^{t'} \geq FDIPw_{w,p}^t \qquad \forall w,p,t,t':t<t'<t+inportw_w, \exists i(i,p) \in aw \tag{43}$$

$$YS_{s,p,p}^{t'} \geq FDIPs_{s,p}^t \qquad \forall s,p,t,t':t<t'<t+inports_s, \exists i:(i,p) \in as \tag{44}$$

$$YC_{c,p,p}^{t'} \geq FDIPc_{c,p}^t \qquad \forall c,p,t,t':t<t'<t+inportc_c, \exists i:(i,p) \in ac \tag{45}$$

$$YW_{w,i,j}^t \in \{0,1\} \ \forall t,w,i,j \tag{46}$$

$$YS^{t}_{s,i,j} \in \{0,1\} \ \forall t,s,i,j \tag{47}$$

$$YC^{t}_{c,i,j} \in \{0,1\} \ \forall t,c,i,j \tag{48}$$

$$REVUw^{t}_{w,s,r} \in \{0,1\} \ \forall t,w,s,r \tag{49}$$

$$REVUc^{t}_{s,c,r} \in \{0,1\} \ \forall t,s,c,r \tag{50}$$

$$LOWF^{t}_{w} \in \{0,1\} \ \forall w,t \tag{51}$$

$$FDIPw^{t}_{w,p} \in \{0,1\} \ \forall w,p,t \tag{52}$$

$$FDIPs^{t}_{s,p} \in \{0,1\} \ \forall s,p,t \tag{53}$$

$$FDIPc^{t}_{c,p} \in \{0,1\} \ \forall c,p,t \tag{54}$$

$$0 \leq HOLDs^{t}_{s} \leq capDFMs_{s} \ \forall t,s \tag{55}$$

$$0 \leq HOLDc^{t}_{c} \leq capDFMc_{c} \ \forall t,\mathrm{c} \tag{56}$$

$$0 \leq HOLD\_INVs^{t}_{s} \leq burnDFMs_{s} \ \forall t,s \tag{57}$$

$$0 \leq HOLD\_INVc^{t}_{c} \leq burnDFMc_{c} \ \forall t,\mathrm{c} \tag{58}$$

$$0 \leq RASw^{t}_{w,s} \leq burnDFMw_{w} \ \ \forall t, w, s \tag{59}$$

$$0 \leq RAS\_burn\_s^{t}_{s,c} \leq burnDFMs_{s} \ \ \forall t, s, c \tag{60}$$

$$0 \leq RAS\_hold\_s^{t}_{s,c} \leq capDFMs_{s} \ \ \forall t, s, c \tag{61}$$

$$0 \leq INPw^{t}_{w} \leq burnDFMw_{w} \ \ \forall t, w \tag{62}$$

$$0 \leq INOs^{t}_{s} \leq burnDFMs_{s} \ \ \forall t, s \tag{63}$$

$$0 \leq INPc^{t}_{c} \leq burnDFMc_{c} \ \ \forall t, c \tag{64}$$

$$0 \leq INPhs^{t}_{s} \leq capDFMs_{s} \ \ \forall t, s \tag{65}$$

$$0 \leq INPhc^{t}_{c} \leq capDFMc_{c} \ \ \forall t, c \tag{66}$$

$$0 \leq SHORTBURNw^{t}_{w} \leq burnSafetyw_{w} * burnDFMw_{w} \ \ \forall t, w \tag{67}$$

$$0 \leq SHORTBURNs^{t}_{s} \leq burnSafetys_{s} * burnDFMs_{s} \ \ \forall t, s \tag{68}$$

$$0 \leq SHORTBURNc^{t}_{c} \leq burnSafetyc_{c} * burnDFMc_{c} \ \ \forall t, c \tag{69}$$

$$0 \leq SHORTs^{t}_{s} \leq safetyDFMs_{s} * capDFMs_{s} \ \ \forall t, s \tag{70}$$

$$0 \leq SHORTc_c^t \leq safetyDFMc_c * capDFMc_c \ \forall t, c \tag{71}$$

$$INVw_w^t \leq burnDFMw_w \ \forall t, w \tag{72}$$

$$INVs_s^t \leq burnDFMs_s \ \forall t, s \tag{73}$$

$$INVc_c^t \leq burnDFMc_c \ \forall t, c \tag{74}$$

The objective function (1) rewards hours spent on station and penalizes any fuel shortages. It also includes a small reward for all fuel transfers; this reward helps to mitigate end-of-horizon effects in which ships' inventories are depleted. Constraint sets (2), (3) and (4) require that each warship, shuttle and *CLF* ship, respectively, must leave a node in a period if and only if it entered node in the same period; in other words, these constraints enforce balance of flow in the ship routing network. Constraint sets (5), (6) and (7) require that each ship begin in a single location. Constraint set (8) ensures that each warship starts at its initial waypoint, if one is given. Constraint set (9) ensures that warships visit their assigned waypoints in all subsequent time periods. Constraint set (10) ensures that the number of ships visiting a port in any time period does not exceed that port's available berths. Constraint sets (11), (12) and (13) calculate the burnable fuel inventory for warships, shuttles, and *CLFs*, respectively, based on each ship's prior inventory, current consumption and any internal transfers, *RAS* transfers and port transfers that occur. Constraint sets (14), (15), (16), (17) and (18) require that each ship be located at a port node in order to take on burnable or deliverable fuel from a port. Constraint sets (19) and (20) calculate shuttle and *CLF* deliverable fuel inventories, respectively, based on each ship's prior inventory and any internal transfers, *RAS* transfers and port transfers that occur. Constraint sets (21), (22), (24) and (25) limit the number of *RAS* events that can be initiated at each ship in a single time period, while constraint sets (23) and (26) limit the total number of *RAS* events ongoing in a single time period for each shuttle and *CLF*. Constraint sets (27), (28), (29) and (30) ensure that all relevant ships are present at a *RAS* lane in order for a *RAS* event to occur. For instance, in order for a shuttle to replenish a warship, both the shuttle and the warship must be located at the same *RAS* lane during the time periods in which the *RAS* event is to occur, and likewise if a *CLF* is to replenish a shuttle. Constraint sets (31) and (32) ensure that fuel is only transferred between ships if the ships have successfully rendezvoused according to the binary *REVUw* and *REVUc* decision variables set in constraint sets (27), (28), (29) and (30). Constraint sets (33), (34) and (35) calculate burnable fuel shortages, while constraint sets (36) and (37) calculate deliverable fuel shortages. Constraint set (38) determines the value of a binary decision that indicates whether a warship's burnable fuel level is below its safety fuel level; this decision variable is used in constraint set (39) to allow the ship to depart from its battle group if it is low on fuel, and otherwise to require it to remain with its carrier. (If a battle group does not contain a carrier, an "imaginary ship" can be defined to serve this role. It is unadvisable to designate a different ship as the carrier, as this ship may run low on fuel but will never be able to separate from the battle group unless the remaining ships are also low on fuel.)

Constraint sets (40), (41) and (42) determine the value of a binary decision variable indicating whether each warship, shuttle and CLF ship, respectively, begins a port visit in period *t*. Following the beginning of a port visit, constraint sets (43), (44) and (45) require that each ship to remain in port for its required number of time periods. Constraint sets (46) to (74) define decision variable domains and bounds.

Scenario Development

This section exercises *DL-RASM* to analyze the transportation and throughput of liquid commodities to deployed vessels as a function of: 1) the type of ship within the network, 2) the operations being conducted by the ship, and 3) the duration of those operations. Our notional, unclassified scenario is based on naval operations in the 7th Fleet AOR. This AOR is important due to its proximity to the South China Sea and the highly contested Spratly Islands as well as numerous U.S. allies (Morse, 2008). As Morse so accurately stated: This chain of geographic features is important to the surrounding nations due to its location along the world's second busiest international sea lane, the abundance of natural gas, oil and other resources, and the desire for nations surrounding the islands to increase their claims of territorial seas and archipelagic waters…many of the nations in the region are trying to control the resources in the Spratly Islands as well as the main trade route through which oil from the Middle East and Africa is delivered (2008). A representation of the status of claims and disputed territory of the Spratly Islands with some minor surrounding geography is given in Figure 4.

In our scenario, tensions have increased with country red (representing the aggressor), thus driving us to evaluate a dual RAS lane logistic support method. Our supply network contains 26 nodes, of which three are ARL nodes, seven are FOS nodes, seven are FRL nodes, three are ITP nodes and six are port nodes. This network is depicted in Figures 5–8; Figure 5 is overlain with our notional threat environment.

Figure 4. Contested Spratly Islands claims and surrounding geography
From Campbell, 2014.

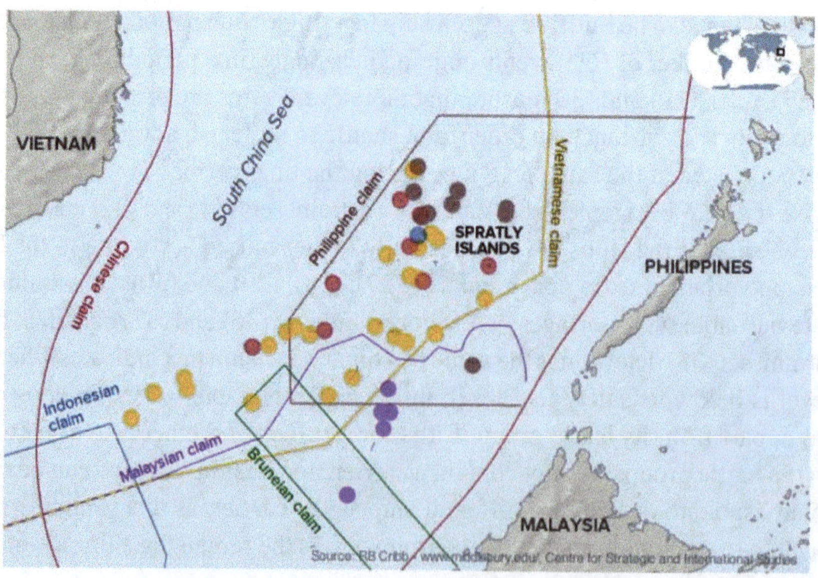

Figure 5. 7th Fleet network nodes and A2AD threat area

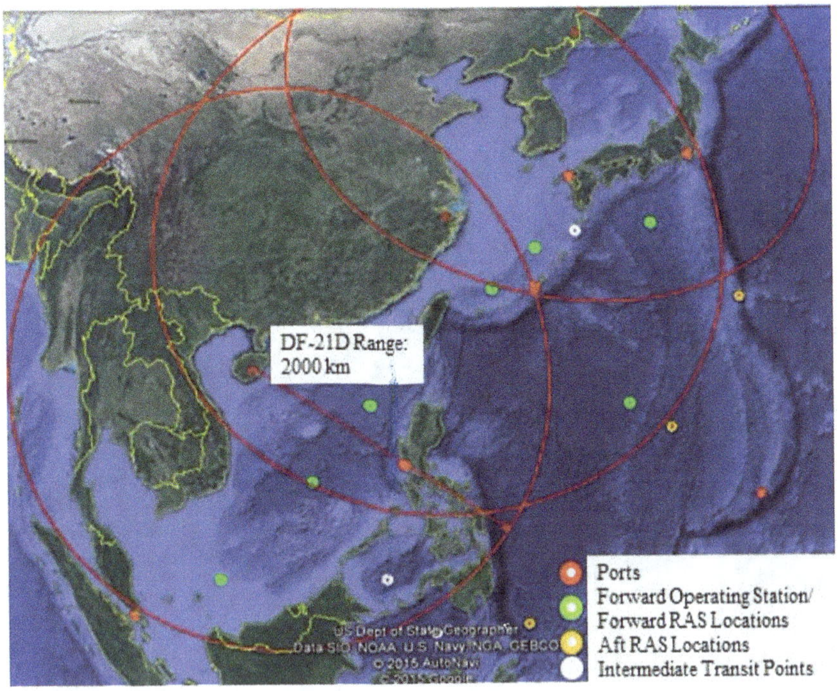

Figure 6. Warship node and arc network

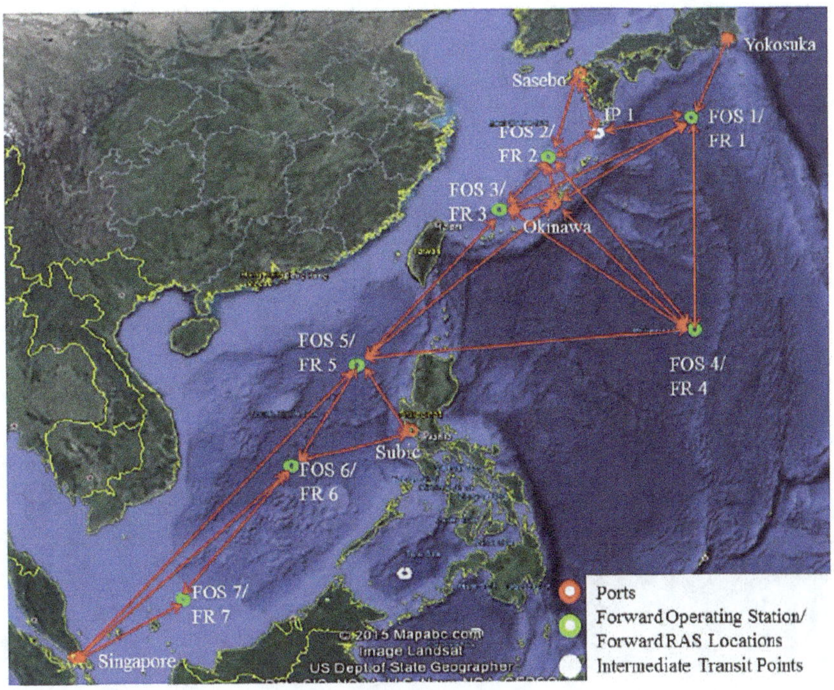

Figure 7. Shuttle node and arc network

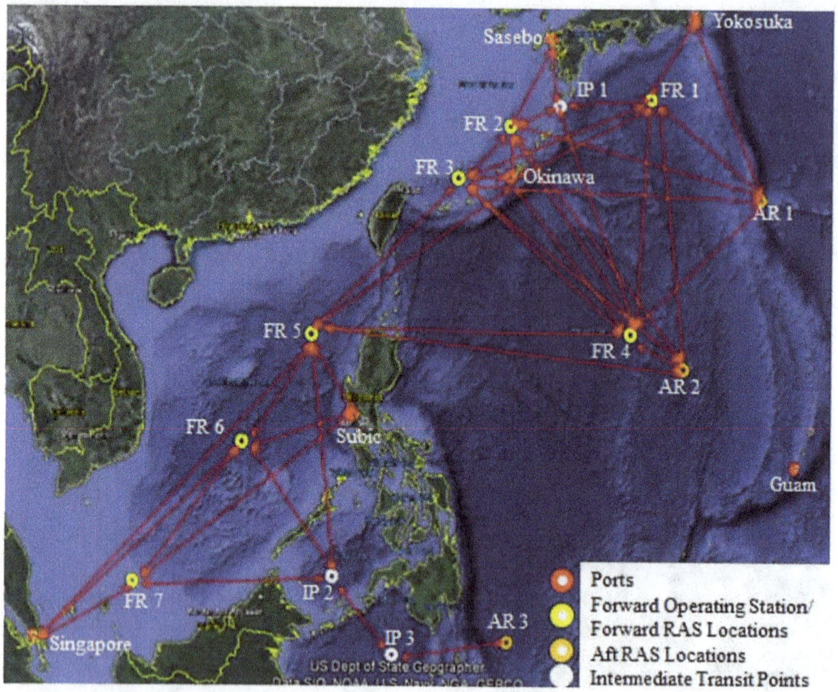

Figure 8. CLF node and arc network

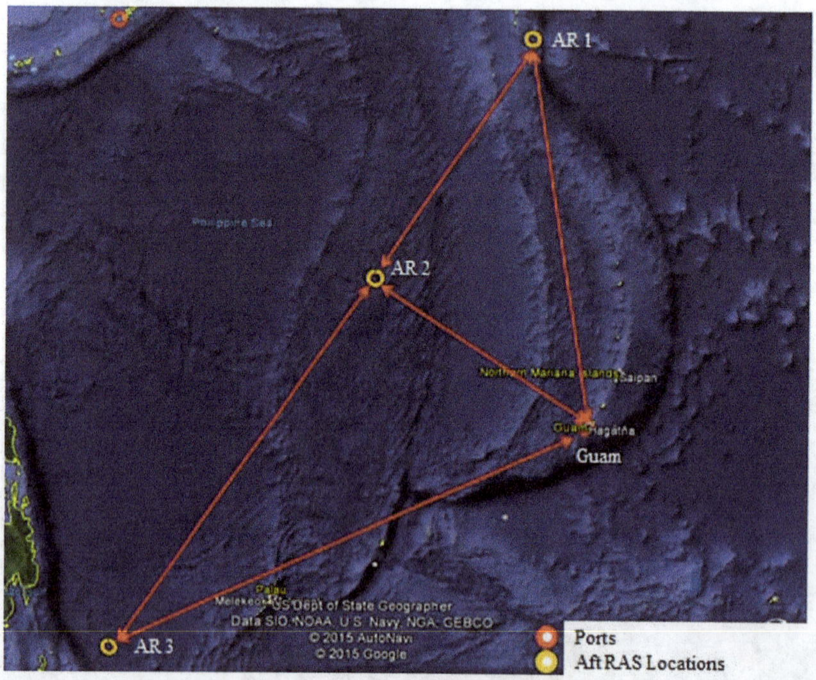

Given this network structure, data requirements for the implemented scenario include the duration of the planning horizon, the number of hours per time period and the number of daylight hours per day available for RAS events during sustained peacetime operations. For each warship, shuttle and CLF ship, daily consumption factors for fuel, transit speeds, safety fuel levels and required days in port, are required. Finally, for each CTF/CRUDES SAG, a group composition and a waypoint schedule are required. The following assumptions facilitate development of this data:

- The CLF fleet consists of only T-AO and T-AKE class ships (one each), as the future of the TAOE is uncertain. This also represents the typical CLF support package assigned to a deployed CTF/CRUDES SAG.
- All warships begin the scenario with fuel at 85% of capacity, and all shuttle and CLF ships begin the scenario at 80% capacity for both burnable and deliverable fuel. This represents a fleet already performing sustained operations.
- All warships and CLF ships transit through the network at 15 knots (kts). This induces the requirement for a moderate but substantial fuel consumption rate to initiate and maintain a commodity demand.
- Our scenarios consider four distinct shuttle speeds. This allows for sensitivity analysis on impact of shuttle speed and ability to support the deployed CTF/CRUDES SAG.
- Shuttle deliverable fuel capacity varies among four distinct levels. This allows for sensitivity analysis on impact of shuttle capacity and ability to support the deployed CTF/CRUDES SAG.
- Warships, shuttles and CLF ships occupy nodes as shown in Figure 6, Figure 7 and Figure 8.

Fleet Composition

Afloat combat forces consist of two battle groups: one CTF and one CRUDES SAG. The CTF consists of one CVN, two guided missile cruisers (CGs), and three guided missile destroyers (DDGs), while the CRUDES SAG consists of two CGs and three DDGs. Our logistics assets consist of one T-AO and one T-AKE, and our sensitivity analysis varies the number of shuttles from three to eight.

CTF/CRUDES SAG Schedule

Both battle groups initialize in their assigned operating area and execute "show of force" operations reflected in their waypoint schedules, shown in Table 3. The time periods represent the first time period the battle group is required to be on station for the assigned exercise to commence in the following time periods.

Planning Factors

The logistics planning factors used in our scenarios come directly from CLF Planner (2008), and were retrieved from NWP 4-01.2 (CNO, 2007) and a more recent study by CNA Corporation (Trickey, 2014). These factors are used to determine the daily consumption rates, fuel capacities, required days in port and speed for the warships, shuttles and CLF ships.

Table 3. Waypoint schedule for battle groups

Time Period (*t*)	Battle Group (*g*)	Location (*i*)
1-47	CTF	FOS_4
1-35	CRUDES SAG	FOS_5
42-71	CRUDES SAG	FOS_6
60-100	CTF	FOS_2
78	CRUDES SAG	FOS_5
90-155	CRUDES SAG	FOS_3
103	CTF	FOS_3
120-140	CTF	FOS_5
148-180	CTF	FOS_6
168-180	CRUDES SAG	FOS_1

Warships

The data used to produce our warship planning factors are shown in Figure 9, Figure 10 and Figure 11. This study utilizes the values circled in red; the values circled or highlighted in yellow were also considered but not adopted. In all cases, the values used were consistent with the most conservative or current values employed. Conservative values were adopted for fuel consumption rates and were taken from the CLF Planner planning factors, while current values were adopted for capacities and were taken from the CNA study. The planning factors for the warships actually applied to the scenarios for DL-RASM are depicted in Table 4.

The *bg*, and *safetyDFM* planning factors are based on the author's observations during multiple assignments to CVN's and DDG's. Note that *FWw* is calculated based on the fuel burn given in Table 4.

CLF

In Figure 12, Figure13 and Figure 14 can be found the data used to produce the CLF planning factors. As with the warship planning factors, red circles indicate the values used, while the values circled or high-

Figure 9. CLF Planner planning factors for the fuel capacity and burn rates of the warships
From Brown & Carlyle, 2008.

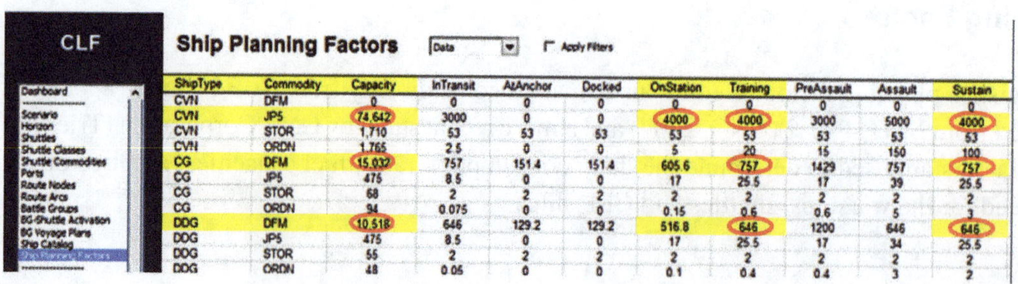

Ship Planning Factors

ShipType	Commodity	Capacity	InTransit	AtAnchor	Docked	OnStation	Training	PreAssault	Assault	Sustain
CVN	DFM	0	0	0	0	0	0	0	0	0
CVN	JP5	74,642	3000	0	0	4000	4000	3000	5000	4000
CVN	STOR	1,710	53	53	53	53	53	53	53	53
CVN	ORDN	1,765	2.5	0	0	5	20	15	150	100
CG	DFM	15,032	757	151.4	151.4	605.6	757	1429	757	757
CG	JP5	475	8.5	0	0	17	25.5	17	39	25.5
CG	STOR	68	2	2	2	2	2	2	2	2
CG	ORDN	94	0.075	0	0	0.15	0.6	0.6	5	3
DDG	DFM	10,518	646	129.2	129.2	516.8	646	1200	646	646
DDG	JP5	475	8.5	0	0	17	25.5	17	34	25.5
DDG	STOR	55	2	2	2	2	2	2	2	2
DDG	ORDN	48	0.05	0	0	0.1	0.4	0.4	3	2

Figure 10. CNA Corporation: Navy Logistics Resiliency Model Description fuel capacity and burn rate warship planning factors
From Trickey, 2014.

Ship	Mission area name	Arrival day	Quantity	Priority	Departure day	Ship number	Mission Area	DFM capacity (bbls)	Assault DFM consumption (bbls/0.1 day)	Pre-assault DFM Consumption (bbls/0.1 day)	Ordnance capacity (stons)	Assault ordnance consumption rate (stons/0.1 day)	Pre-assault ordnance consumption rate (stons/0.1 day)	Ship	Starting DFM (bbls)	Starting ordnance (stons)
AMPH01 LHD	AMPHIB	0	1	1	150	1	2	43,091	107	200	391	3.3	0.66	1	38,782	352
AMPH02 LPD-17	AMPHIB	0	1	2	150	2	2	23,750	107	114	513	0.6	0.12	2	21,375	462
AMPH03 LSD	AMPHIB	0	1	3	150	3	2	19,150	35	73	35	0.2	0.04	3	17,235	32
CG01	CG	0	1	2	150	4	3	15,032	76	143	94	0.5	0.1	4	13,529	85
CG02	CG	0	1	3	150	5	3	15,032	76	143	94	0.5	0.1	5	13,529	75
CG03	CG	0	1	2	150	6	3	15,032	76	143	94	0.5	0.1	6	13,529	66
CG04	CG	0	1	3	150	7	3	15,032	76	143	94	0.5	0.1	7	13,529	56
CVN01	CVN	0	1	3	150	8	4	1	0	0	1,765	15	3	8	1	1,589
CVN02	CVN	0	1	3	150	9	4	1	0	0	1,765	15	3	9	1	1,589
DDG01	DDG	0	1	2	150	10	5	10,518	65	120	48	0.3	0.06	10	9,466	43
DDG02	DDG	0	1	3	150	11	5	10,518	65	120	48	0.3	0.06	11	9,466	38
DDG03	DDG	0	1	2	150	12	5	10,518	65	120	48	0.3	0.06	12	9,466	34

Figure 11. CLF Planner planning factor for minimum in-port duration
From Brown & Carlyle, 2008.

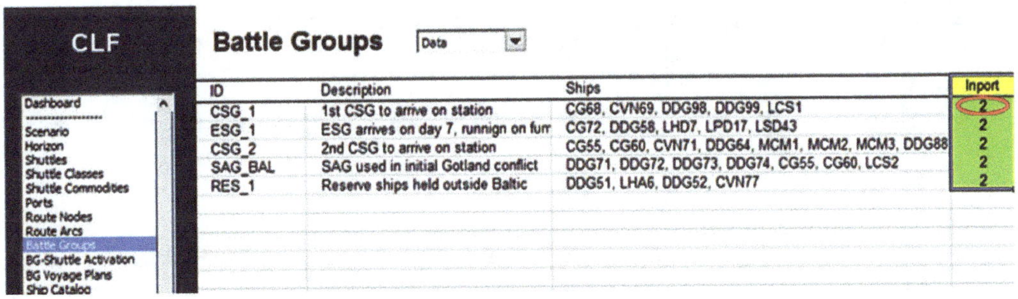

ID	Description	Ships	Inport
CSG_1	1st CSG to arrive on station	CG68, CVN69, DDG98, DDG99, LCS1	2
ESG_1	ESG arrives on day 7, runnign on fum	CG72, DDG58, LHD7, LPD17, LSD43	2
CSG_2	2nd CSG to arrive on station	CG55, CG60, CVN71, DDG64, MCM1, MCM2, MCM3, DDG88	2
SAG_BAL	SAG used in initial Gotland conflict	DDG71, DDG72, DDG73, DDG74, CG55, CG60, LCS2	2
RES_1	Reserve ships held outside Baltic	DDG51, LHA6, DDG52, CVN77	2

lighted in yellow were considered but not adopted. Again, actual values were used where possible, and conservative values were used otherwise. Conservative values were adopted for fuel consumption rates, and current values were adopted for capacities. Our planning factors for CLF ships are shown in Table 5.

Shuttle Ships

The shuttle ship planning factors were modeled after the planning factors for a DDG 51 class ship power plant. Assuming the shuttle design is to be smaller in dimension than the target in Figure 2 and possibly has increased speed capabilities, the DDG 51 class ship power plant provides conservative design characteristics to accomplish both goals. Table 6 is an exhibit of the planning factors for a nominal design

Table 4. DL-RASM warship planning factors

ID (*w*)	*Bg* (Battle Group)	*Speed*$_w$ (kts)	*Inport*$_w$ (days)	*burnDFM*$_w$ (bbls)	*safetyDFM*$_w$ (portion of *burnDFMw*$_w$)	Fuel Burn (%/day)
DDG_1	CTF	15	2	10518	0.7	6.14
DDG_2	CTF	15	2	10518	0.7	6.14
DDG_3	CTF	15	2	10518	0.7	6.14
DDG_4	CRUDES SAG	15	2	10518	0.7	6.14
DDG_5	CRUDES SAG	15	2	10518	0.7	6.14
DDG_6	CRUDES SAG	15	2	10518	0.7	6.14
CG_1	CTF	15	2	15032	0.7	5.04
CG_2	CTF	15	2	15032	0.7	5.04
CG_3	CRUDES SAG	15	2	15032	0.7	5.04
CG_4	CRUDES SAG	15	2	15032	0.7	5.04
CVN_1	CTF	15	3	74642	0.7	5.36

After Brown & Carlyle, 2008; Trickey, 2014.

Figure 12. CLF Planner planning factors for the CLF ships
From Brown & Carlyle, 2008.

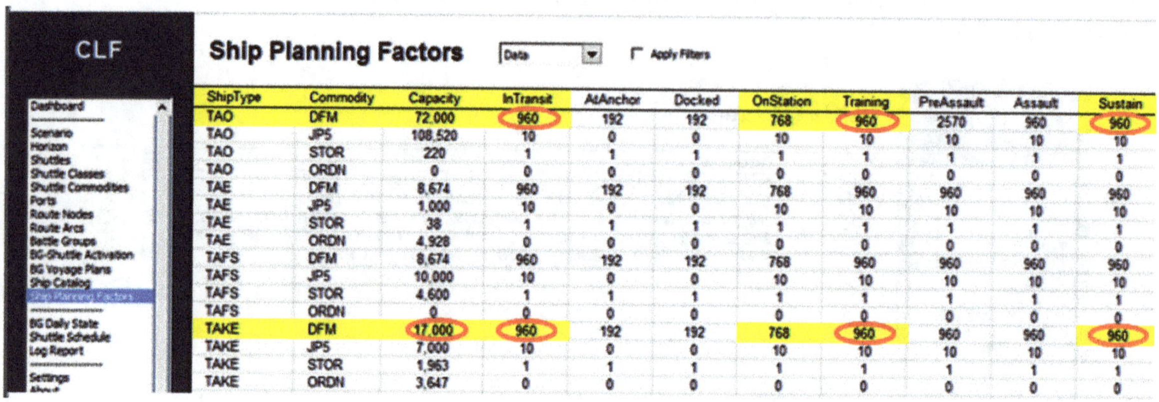

ShipType	Commodity	Capacity	InTransit	AtAnchor	Docked	OnStation	Training	PreAssault	Assault	Sustain
TAO	DFM	72,000	960	192	192	768	960	2570	960	960
TAO	JP5	108,520	10	0	0	10	10	10	10	10
TAO	STOR	220	1	1	1	1	1	1	1	1
TAO	ORDN	0	0	0	0	0	0	0	0	0
TAE	DFM	8,674	960	192	192	768	960	960	960	960
TAE	JP5	1,000	10	0	0	10	10	10	10	10
TAE	STOR	38	1	1	1	1	1	1	1	1
TAE	ORDN	4,928	0	0	0	0	0	0	0	0
TAFS	DFM	8,674	960	192	192	768	960	960	960	960
TAFS	JP5	10,000	10	0	0	10	10	10	10	10
TAFS	STOR	4,600	1	1	1	1	1	1	1	1
TAFS	ORDN	0	0	0	0	0	0	0	0	0
TAKE	DFM	17,000	960	192	192	768	960	960	960	960
TAKE	JP5	7,000	10	0	0	10	10	10	10	10
TAKE	STOR	1,963	1	1	1	1	1	1	1	1
TAKE	ORDN	3,647	0	0	0	0	0	0	0	0

of the shuttle ship, and in Table 7 and Table 8 are shown the planning factor ranges, where applicable, for the shuttle ships. In addition to varying the planning factors, the number of shuttles available varies from three to eight.

Ports

This study conservatively assumes that two berths are available at each port; that is, at most two ships can occupy each port in each time step. Actual available berths for refueling may vary.

Figure 13. CNA Corporation: Navy Logistics Resiliency Model Description CLF ship planning factors
From Trickey, 2014

Ship type name	Ship number	Ship type	Station or Shuttle	Arrival day	Quantity	Priority	Ship number	Speed	Ship's fuel DFM capacity (bbls)	DFM consumption (bbls/day)	DFM capacity (bbls)	JP5 capacity (bbls)	Stores capacity (stons)	Ordnance capacity (stons)
TAO	1	1	1	0	1	2	1	17	14,453	505	90,000	90,000	220	0
TAO	2	1	1	0	1	1	2	17	14,453	505	90,000	90,000	220	0
TAO	3	1	1	0	1	3	3	17	14,453	505	90,000	90,000	220	0
TAO	4	1	2	0	1	2	4	17	14,453	505	90,000	90,000	220	0
TAOE	5	3	2	0	1	2	5	26	31,750	914	62,400	93,600	952	2,016
TAKE	6	2	2	0	1	2	6	18	31,494	593	17,000	7,000	1,300	4,900
TAKE	7	2	1	0	1	2	7	18	31,494	593	17,000	7,000	1,300	4,900
TAKE	8	2	1	0	1	3	8	18	31,494	593	17,000	7,000	1,300	4,900
TAKE	9	2	1	30	1	2	9	18	31,494	593	17,000	7,000	1,300	4,900
TAKE	10	2	1	30	1	3	10	18	31,494	593	17,000	7,000	1,300	4,900

Figure 14. CLF Planner planning factor for minimum in-port duration
From Brown & Carlyle, 2008.

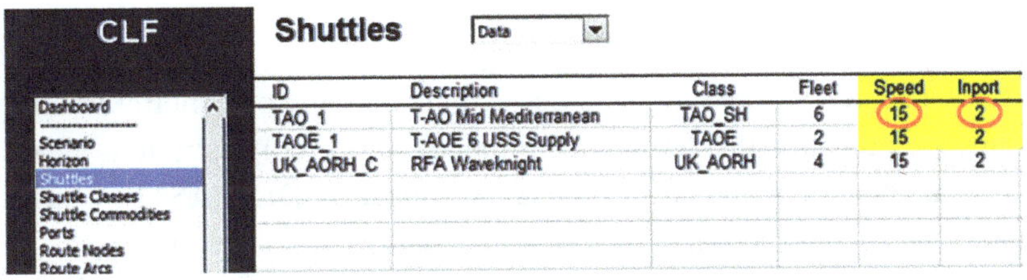

Table 5. DL-RASM CLF planning factors

ID(c)	*Speed$_c$* (kts)	*Inport$_c$* (days)	*burnDFM$_c$* (bbls)	*capDFM$_c$* (bbls)	*burnsafetyDFM$_c$* (portion of *burnDFMc$_c$*)	*safetyDFM$_c$* (portion of *capDFMc$_c$*)	Fuel Burn (%day)
TAO_1	15	2	14453	90000	0.5	0.3	6.64
TAKE_1	15	2	31494	17000	0.5	0.3	3.05

After Brown & Carlyle, 2008; Trickey, 2014.

Table 6. Baseline shuttle planning factors derived from DDG 51 Class power plant consumption rates

Nominal *Speeed$_s$* (kts)	*Inport$_s$* (days)	*burnDFMs$_s$* (bbls)	Nominal *capDFMs$_s$* (bbls)	*burnsafetyDFM$_s$* (portion of *burnDFMs$_s$*)	*safetyDFM$_s$* (portion of *capDFMs$_s$*)	Nominal Fuel Burn (%/day)
15	1	10518	23810	0.5	0.3	6.14

After Brown & Carlyle, 2008; Trickey, 2014.

Table 7. Unclassified planning factors for sensitivity analysis pertaining to shuttle ship speed and corresponding fuel burn percentage based on DDG 51 class daily consumption planning factors

$Speed_s$ (kts)	Fuel Burn (%/day)
13	5.4
15	6.14
17	7.2
19	8.5

After Trickey & Grenwald, 2014.

Table 8. Deliverable fuel capacity levels for sensitivity analysis pertaining to shuttle ship design

Deliverable Fuel Capacity (gal)	$capDFMs_s$ (bbls)
750,000	17,857
1,000,000	23,810
1,250,000	29,762
1,500,000	35,714

Penalties and Rewards

Our penalty and reward parameters are listed in Table 9. For simplicity, this study varies penalties by ship type (i.e., warship, shuttle or CLF) but not by individual ship.

ANALYSIS

The section exercises DL_RASM to study the performance of various shuttle configurations. It first considers peacetime operations. In this setting, a number of shuttle configurations are evaluated in order to determine those characteristics that have the largest impact on DL_RASM's objective value. Next, this analysis considers a wartime scenario in which the shuttles are denied port access and must refuel at the ARLs. The study again evaluates the performance of various shuttle configurations. A 30-day time horizon applies to each computational experiment. Rather than optimize over the entire 30 days simultaneously, this study utilizes a rolling horizon approach. In the first iteration, DL_RASM optimizes over a planning horizon of 20 time periods (80 operational hours). Then the binary decision variables are fixed for the first five time periods, and the planning horizon is expanded by five periods. This process continues until the planning horizon encompasses the entire 30 days. This approach reduces DL_RASM's computational burden while accurately reflecting actual operations. Each design point as applied to DL-RASM was computed using a Dell, Precision T7500 computer with two Intel® Xenon® CPU E5606 @2.13GHz processors, 48 GB of RAM installed and running the Windows 7 Professional operating system. The program software used for the optimization is GAMS 24.0.1 utilizing CPLEX 12.5 solver. With these conditions the model contains approximately 111,530 constraints and 192,913 decision variables in the first iteration, of which181,874 decision variables are integer. In the last itera-

Table 9. Penalty and reward parameter values

Reward (reward/ship*period)	ε (reward/barrel)	$penaltyw_w$ (penalty/barrel)	$penaltys_s$ (penalty/barrel)	$penaltyc_c$ (penalty/barrel)
1000	.00001	1000	100	10

tion of the rolling time horizon implementation, the model contains approximately 747,810 constraints and 555,831 decision variables, of which 55,640 decision variables are integer. Solution times vary by iteration but are typically in the range of 4.96 to 19.75 seconds.

Peacetime Operations

Three design parameters are most crucial to this study: speed, deliverable fuel capacity, and available number of shuttles. In order to efficiently explore this design space, this study applies a three-factor custom nearly orthogonal Latin hypercube (NOLH) design of experiments (DOE). The purpose of the custom NOLH DOE is to efficiently reduce the amount of overall design points evaluated while maintaining near orthogonality across all factors and filling enough design space to determine higher order effects between the factors (Cioppa & Lucas, 2007). Our custom NOLH DOE was derived from the DesignCreatorv2 spreadsheet produced by MacCalman (2012). Table 17 in the Appendix contains our design points as well as DL-RASM's optimal objective value for each design point. These results appear graphically in Figures 15, 16 and 17. In Figure 15, it appears that a speed of 19 kts produces higher optimal objective values; however, at a capacity of 1.5M gallons, all speeds perform well. Moreover, Figure 16 shows that increasing the number of shuttles has some positive effect on the optimal objective value. But again, the 1.5M gallon deliverable fuel capacity consistently results in the higher objective values. Finally, Figure 17 shows a direct positive impact by the increase of deliverable fuel and generally with an increase in the number of shuttles. From the graphs one is able to deduce that in all cases, regardless of speed and with little respect to the number of shuttles, a capacity of 1.5M gallons provides robust good performance. Moreover, this parameter value is determined at the time of ship construction and is not subject to perturbation in the operational environment, as speed and number of shuttles are.

This study now assumes a conservative speed of 15 kts and performs a more detailed sensitivity analysis on a number of shuttles, as this is a value that may change unexpectedly in a wartime scenario. It also examines the impact of shuttle capacity in more detail.

Figure 15. Optimal objective value as a function of shuttle deliverable fuel capacity and speed

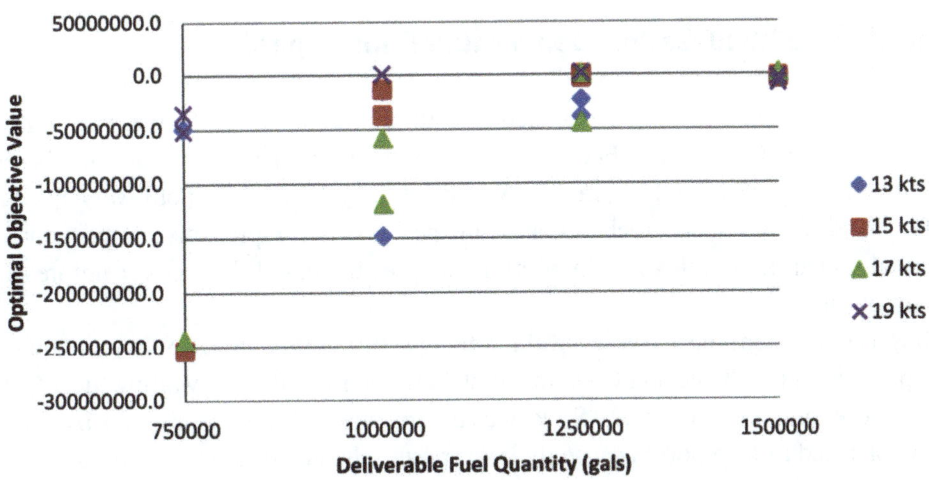

Figure 16. Optimal objective value as a function of #of shuttles and deliverable fuel capacity

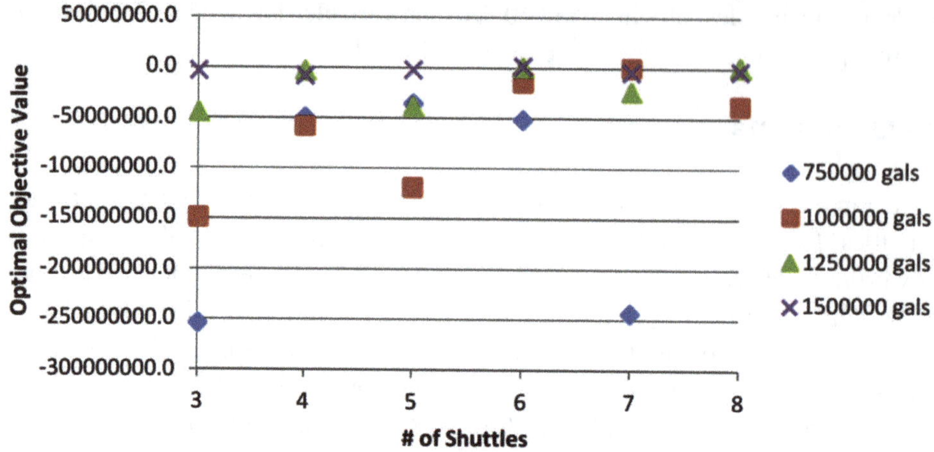

Figure 17. Optimal objective value as a function of deliverable fuel capacity and number of shuttles

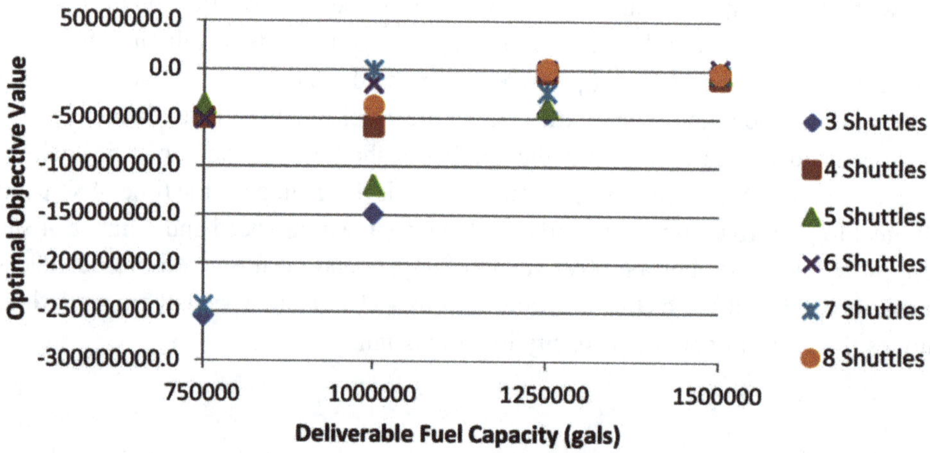

Three Shuttles, 750,000-Gallon Deliverable Fuel Capacity

First, consider a "pessimistic" scenario in which only three shuttles are available, each with a deliverable fuel capacity of 750,000 gallons. Fuel inventories of the warships and shuttles in this scenario are shown in Figure 18, Figure 19 and Figure 20. As indicated in the figures, both warships and shuttles encounter fuel levels below safety stock in many time periods. This represents an undesirable outcome for our nominal scenario, as well as an inability to support increased demands if unforeseen adverse events were to occur.

Table 10 is a summary of the activities of the warships during this scenario. In particular, each warship's time spent on station (Time on FOS), transiting between locations of awaiting fuel (Transit/Wait), and time spent receiving fuel (Active RAS) each as a percentage of the planning horizon are depicted. Warships do not spend time in port because their waypoint schedule does not permit it.

Figure 18. Warship fuel inventory levels during a 30-day time horizon utilizing three shuttles with a 750,000 gallon deliverable fuel quantity and ports available

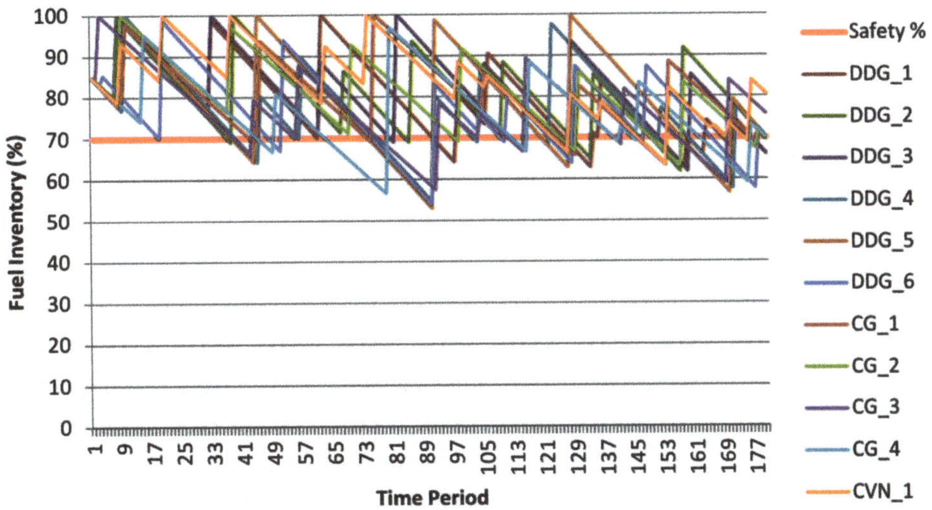

Figure 19. Shuttle burnable fuel inventory levels during a 30-day time horizon utilizing three shuttles with a 750,000 gallon deliverable fuel quantity and ports available

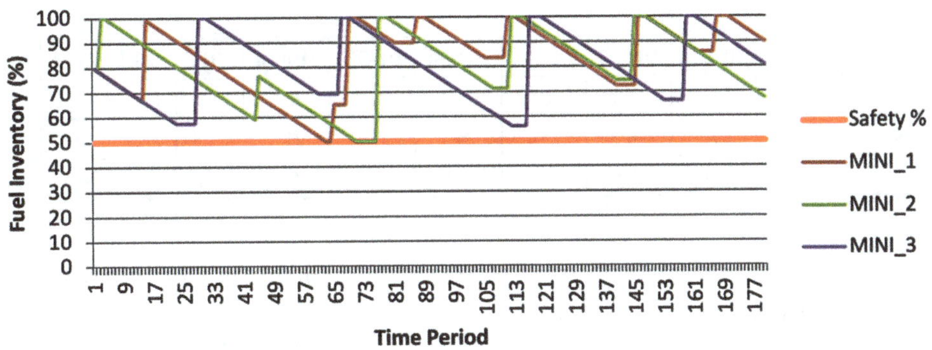

Figure 20. Shuttle deliverable fuel inventory levels during a 30-day time horizon utilizing three shuttles with a 750,000 gallon deliverable fuel quantity and ports available

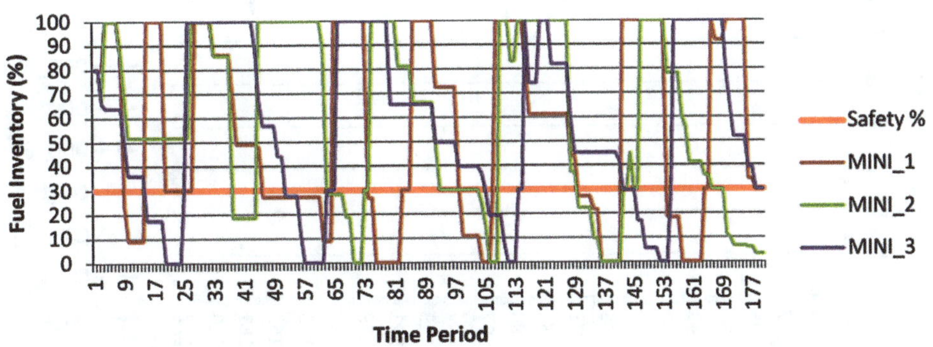

Recall that a warship can only depart from its battle group if its fuel inventory falls below its prescribed safety stock level, as indicated by Equations (38) and (39). Thus, the Transit/Wait times in Table 10 reflect some situations in which an entire battle group visits a FRL so that one or more of its ships can refuel before reaching safety stock level. In this case, the entire battle group is taken off station during the duration of the RAS event.

Six Shuttles, 1.5M-Gallon Deliverable Fuel Capacity

Now consider a more optimistic scenario in which six shuttles are available, each with a deliverable fuel capacity of 1.5M gallons. The fuel quantities of the warships and shuttles in this scenario are depicted in Figure 21, Figure 22 and Figure 23. Note that excursions below safety stock are quite minimal at this design point.

The percentage of time spent on station, transiting or waiting, and actively refueling at sea, for each warship is displayed in Table 11. Note that the 6-shuttle, 1.5M gal design point results in improved on-station times relative to the 3-shuttle, 0.75M gal design point; this improvement is quantified in Table 12.

As indicated in Table 12, there is an average increase in Time on FOS of 3.6%, equating to almost 26 hours per warship during the 30-day planning horizon for our scenario. The warships also enjoy an

Table 10. Warship employment when three shuttles are available, each with a 750,000 gallon deliverable fuel capacity

Percentage	DDG 1	DDG 2	DDG 3	DDG 4	DDG 5	DDG 6	CG 1	CG 2	CG 3	CG 4	CVN 1
Time on FOS	68.9	67.8	70.0	74.4	74.4	73.9	67.8	69.4	74.4	73.9	69.4
Transit/Wait Time	25.0	28.3	25.6	21.1	22.8	21.8	27.8	26.7	21.1	22.2	23.3
Active RAS	6.1	3.9	4.4	4.4	2.8	5.0	4.4	3.9	4.4	73.9	7.2

Figure 21. Warship fuel inventory levels during a 30-day horizon utilizing six shuttles with a 1.5 M-gallon deliverable fuel quantity and ports available

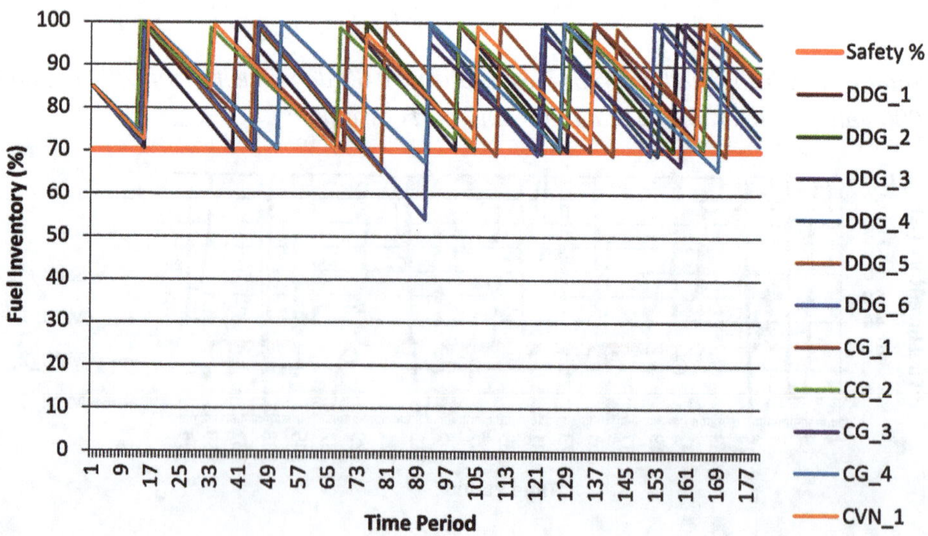

Figure 22. Shutlle burnable fuel inventory levels during a 30-day time horizon utilizing six shuttles with 1.5 M-gallon deliverable fuel quantity and ports available

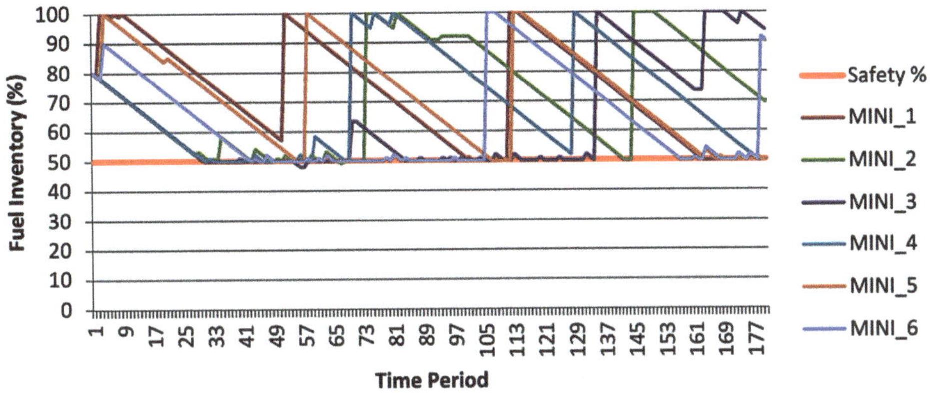

Figure 23. Shuttle deliverable fuel inventory levels during a-30 day time horizon utilizing six shuttles with 1.5 M-gallon deliverable fuel quantity and ports available

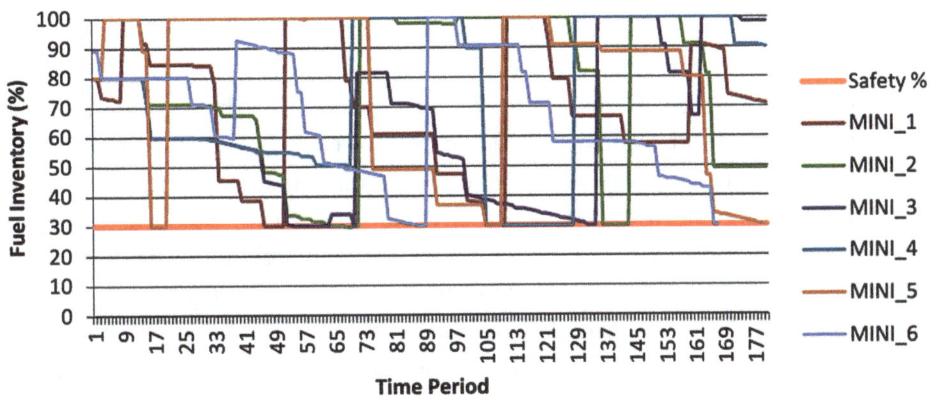

average decrease in Transit/Wait time of 2.5% of the 30-day horizon and a decrease in Active RAS time of 1.1% of the 30-day horizon, corresponding to a decrease of 18 hours and 8 hours per warship for these activities, respectively.

Wartime Operations: Denied Port Access

In the peacetime scenario, the shuttles make heavy use of the ports of Sasebo and Okinawa, Japan, and Subic Bay, Philippines for refueling, while only making limited use of the ARLs and CLFs. While this heavy usage of ports is completely reasonable given the distances involved, it does invite the potential for disruption due to the fact that these ports lie within the A2AD environment. Thus, this study now considers a situation in which access is denied for all ports except Guam. In this situation, the shuttles must refuel at the ARLs, while the CLFs are able to refuel at Guam.

Table 11. Warship employment when six shuttles are available, each with a 1.5 M-gallon deliverable fuel capacity

Percentage	DDG 1	DDG 2	DDG 3	DDG 4	DDG 5	DDG 6	CG 1	CG 2	CG 3	CG 4	CVN 1
Time on FOS	73.3	73.3	73.3	77.2	76.1	76.7	73.3	73.3	77.2	77.2	73.3
Transit/Wait Time	21.7	23.3	23.3	19.4	20.6	20.0	23.3	23.3	20.0	20.0	22.2
Active RAS	5.0	3.3	3.3	3.3	3.3	3.3	3.3	3.3	2.8	2.8	4.4

Table 12. Change in warship employment from the three-shuttle design point to the six-shuttle design point, as a percentage of the 30-day time horizon

Percentage	DDG 1	DDG 2	DDG 3	DDG 4	DDG 5	DDG 6	CG 1	CG 2	CG 3	CG 4	CVN 1
Time on FOS	4.4	5.6	3.3	2.8	1.7	2.8	5.6	3.9	2.8	3.3	3.9
Transit/Wait Time	-3.3	-5.0	-2.2	-1.7	-2.2	-1.1	-4.4	-3.3	-1.1	-2.2	-1.1
Active RAS	-1.1	-0.6	-1.1	-1.1	0.6	-1.7	-1.1	-0.6	-1.7	-1.1	-2.8

Six Shuttles, 1.5M-Gallon Deliverable Fuel Capacity

In Figure 24, Figure 25 and Figure 26 are depicted the fuel quantities of the warships and shuttles when six shuttles are available, each with a deliverable fuel capacity of 1.5M gallons. As indicated, this design point is no longer adequate to support warship requirements in a heightened threat environment. Because warship inventories are prioritized in the objective function, they are the least affected. However, the warship inventories go below safety stock in a number of time periods, particularly near the end of the 30-day horizon. Moreover, the situation is worse for shuttles: shuttle inventories drop below safety stock beginning early in the 30-day horizon, and by the halfway point at least one shuttle's inventory is below safety stock in every time period, for both burnable and deliverable fuel. In fact, shuttle burnable fuel inventories even drop below zero, an operational impossibility that is allowed in DL_RASM only to ensure that a feasible solution exists.

Table 13 is the model output generated Time on FOS, Transit/Wait, and Active RAS times as a percentage of the planning horizon for this scenario.

The change in percentages for this design point in the heightened threat environment with no ports available is found in Table 14. The results indicate an average decrease of 6.3% Time on FOS, equating to almost 45.4 hours per warship during the 30-day planning horizon. They also indicate an average increase of 4.4% Transit/Wait Time and 1.9% Active RAS time, resulting in an increase of 31.7 hours and 13.7 hours per warship for the given activity, respectively.

Eight Shuttles, 1.5M-Gallon Deliverable Fuel Capacity

To mitigate the performance degradation when port access is denied, this variation increases the number of available shuttles from six to eight, each with a 1.5M gallon deliverable fuel capacity. In Figure 27, Figure 28, and Figure 29 are displayed the warship and shuttle inventories for this design point; note the substantial decrease in time spent below safety stock, relative to the six-shuttle design point. There

Figure 24. Warship fuel inventory levels during a 30-day time horizon utilizing six shuttles with a 1.5 M-gallon deliverable fuel quantity and no ports available

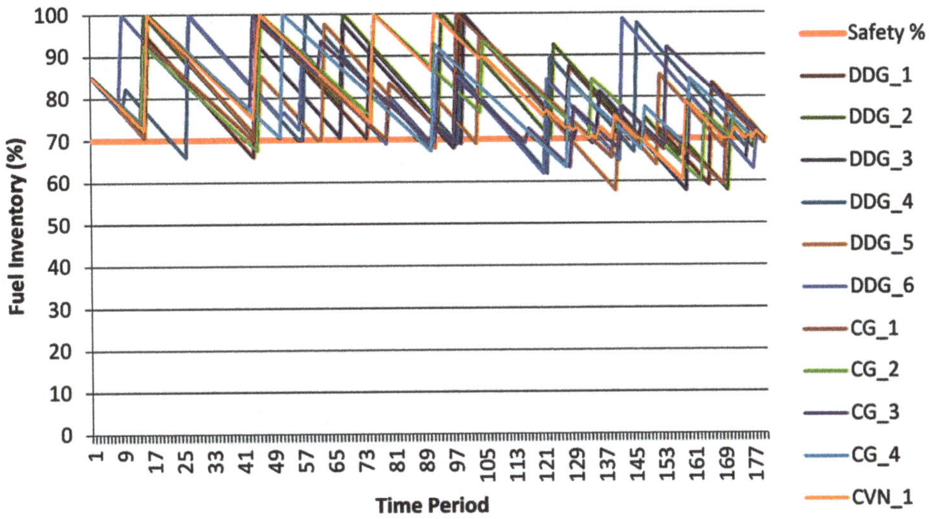

Figure 25. Shuttle burnable fuel inventory levels during a 30-day time horizon utilizing six shuttles with 1.5 M-gallon deliverable fuel quantity and no ports available

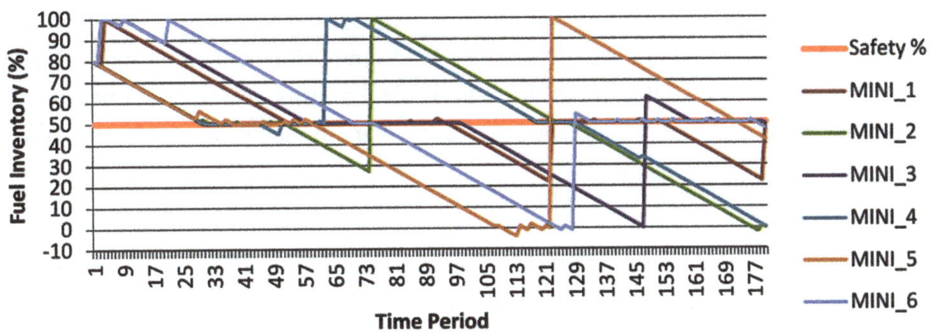

Figure 26. Shuttle deliverable fuel inventory levels during a 30-day time horizon utilizing six shuttles with 1.5 M-gallon deliverable fuel quantity and no ports available

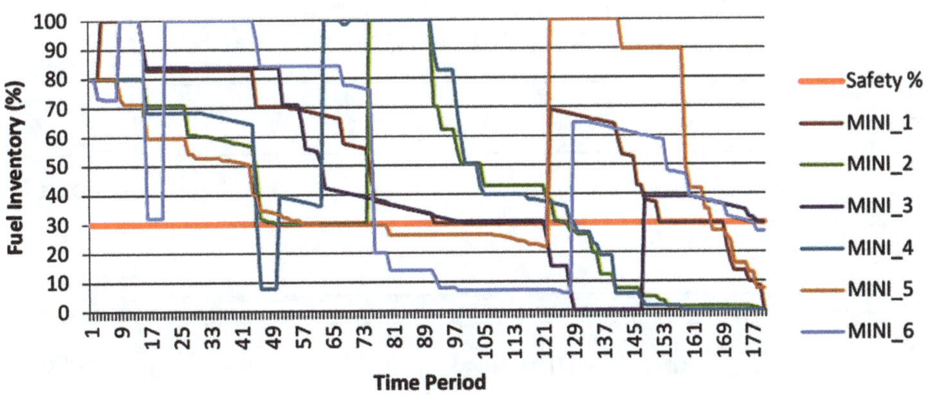

Table 13. Warship employment as a function of percentage of time per given activity for six shuttles and 1.5 M-gallon of deliverable fuel quantity each

Percentage	DDG 1	DDG 2	DDG 3	DDG 4	DDG 5	DDG 6	CG 1	CG 2	CG 3	CG 4	CVN 1
Time on FOS	65.0	66.1	64.4	71.7	71.7	72.2	66.1	66.1	71.7	73.3	66.7
Transit/Wait Time	30.6	28.3	30.0	22.8	23.3	22.2	29.4	30.0	23.3	22.8	22.8
Active RAS	4.4	5.6	5.6	5.6	5.0	5.6	4.4	3.9	5.0	3.9	10.6

Table 14. Percentage of time change per given activity from the optimal design of six shuttles and 1.5 M-gallon deliverable fuel with ports available for shuttle replenishment and the optimal design in the heightened threat scenario with no ports

Percentage	DDG 1	DDG 2	DDG 3	DDG 4	DDG 5	DDG 6	CG 1	CG 2	CG 3	CG 4	CVN 1
Time on FOS	-8.3	-7.2	-8.9	-5.6	-4.4	-4.4	-7.2	-7.2	-5.6	-3.9	-6.7
Transit/Wait Time	8.9	-5.0	6.7	3.3	2.8	2.2	6.1	6.7	3.3	2.8	0.6
Active RAS	-0.6	2.2	2.2	2.2	1.7	2.2	1.1	0.6	2.2	1.1	6.1

are more instances of low fuel inventories toward the end of the 30-day horizon; this can result in part from the end-of-horizon effects and does not necessarily indicate an inability to sustain performance.

The percentage of time spent on station, transiting or waiting, and actively refueling, for each warship are given in Table 15. Although there is a small degradation in performance relative to the design point of six shuttles with no ports available (quantified in Table 16), this design point nearly recovers that level of performance and may be considered adequate for a wartime scenario, assuming no ships are lost.

Figure 27. Warship fuel inventory levels during a 30-day time horizon utilizing eight shuttles with a 1.5 M-gallon deliverable fuel quantity and no ports available

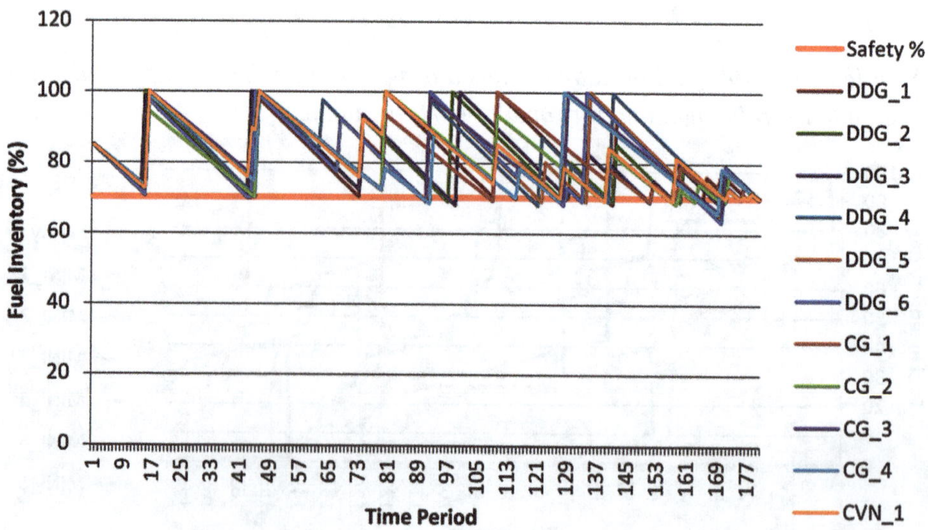

Figure 28. Shuttle burnable fuel inventory levels during a 30-day time horizon utilizing eight shuttles with 1.5 M-gallon deliverable fuel quantity no ports available

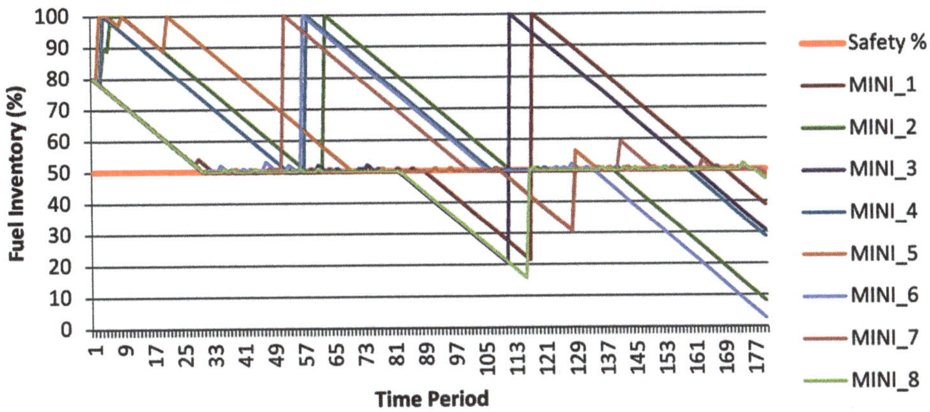

Figure 29. Shuttle deliverable fuel inventory levels during a 30-day time horizon utilizing eight shuttles with 1.5 M-gallon deliverable fuel quantity and no ports available

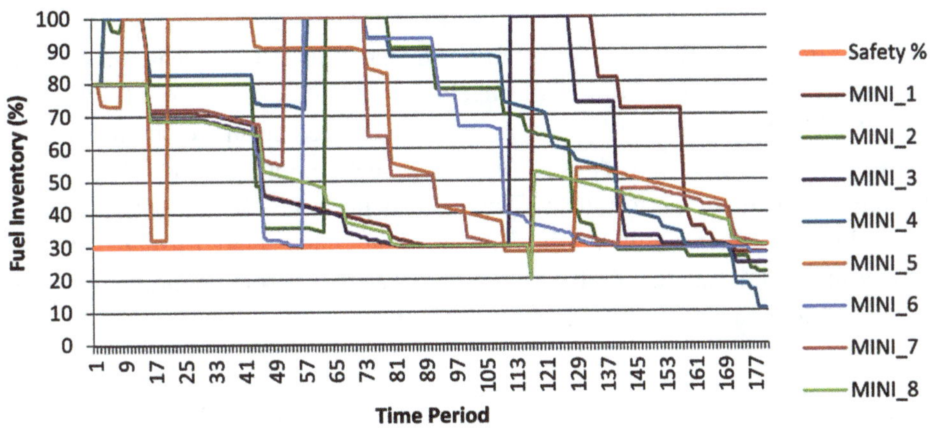

Table 15. Warship employment when eight shuttles are available, each with a 1.5 M-gallon deliverable fuel capacity, and port access is restricted

Percentage	DDG 1	DDG 2	DDG 3	DDG 4	DDG 5	DDG 6	CG 1	CG 2	CG 3	CG 4	CVN 1
Time on FOS	71.1	71.1	71.1	74.4	73.3	73.9	71.7	72.8	73.9	73.3	72.2
Transit/Wait Time	23.3	22.8	21.7	21.7	22.2	21.1	22.8	22.2	22.2	22.8	21.1
Active RAS	5.6	6.1	7.2	3.9	4.4	5.0	5.6	5.0	3.9	3.9	6.7

Table 16. Percentage of time change per given activity from six shuttles and 1.5 M-gallon deliverable fuel each with ports not available to replenish shuttles to eight shuttles and 1.5 M-gallon of deliverable fuel quantity each with no ports

Percentage	DDG 1	DDG 2	DDG 3	DDG 4	DDG 5	DDG 6	CG 1	CG 2	CG 3	CG 4	CVN 1
Time on FOS	6.1	5.0	6.7	2.8	1.7	1.7	5.6	6.7	2.2	0.0	5.6
Transit/Wait Time	-7.2	-5.6	-8.3	-1.1	-1.1	-1.1	-6.7	-7.8	-1.1	0.0	-1.7
Active RAS	1.1	0.6	1.7	-1.7	-0.6	-0.6	1.1	1.1	-1.1	0.0	-3.9

An average increase of 4.0% Time on FOS, equating to 28.8 hours per warship during the 30-day planning horizon is shown in Table 16. It also indicates an average decrease of 3.8% Transit/Wait Time and 0.2% Active RAS time, resulting in a decrease of 27.3 hours and 1.5 hours per warship for the given activity, respectively.

As displayed in Table 17, the change in percentages from the design point of six shuttles and 1.5 M gallons of deliverable fuel with ports available for shuttle replenishment and the design point of eight shuttles and 1.5 M gallons of deliverable fuel with no ports available for shuttle replenishment

The loss of port availability causes an average decrease of 2.3% Time on FOS, equating to a decrease of 16.7 hours per warship during the 30 days planning horizon. There is also an average increase of 0.6% Transit/Wait Time and 1.7% Active RAS time, corresponding to an increase of 4.3 hours and 12.4 hours per warship for the given activities, respectively. The addition of the two shuttles shifts warship employment back towards the six-shuttle scenario when ports were available, though it does not close the gap entirely. Although DL-RASM does not explicitly seek to optimize fuel efficiency, it is still of interest to examine this aspect of performance. Thus, we calculate the quantity of fuel burned per gallon of fuel delivered in each scenario utilizing the following equations, where $Burned_s$ is the total amount of fuel burned by shuttle *s* during the 30-day horizon, $Burned_c$ is the corresponding quantity for CLF *c*, and $Delivered_w$ is the total amount of fuel delivered to warship *w*:

$$\sum_s Burned_s = \sum_s \left(\left(INV_s^1 - INV_s^{180} \right) + \sum_t \left(HOLD\_INV_s^t + INPs_s^t + \sum_c RAS\_burn\_s_{s,c}^t \right) \right)$$

$$\sum_c Burned_c = \sum_c \left(\left(INV_c^1 - INV_c^{180} \right) + \sum_t \left(HOLD\_INV_c^t + INPc_c^t \right) \right)$$

Table 17. Change in warship employment from the six shuttle design point with ports available to the eight-shuttle design point with ports unavailable, as a percentage of the 30-day time horizon

Percentage	DDG 1	DDG 2	DDG 3	DDG 4	DDG 5	DDG 6	CG 1	CG 2	CG 3	CG 4	CVN 1
Time on FOS	-2.2	-2.2	-2.2	-2.8	-2.8	-2.8	-1.7	-0.6	-3.3	-3.9	-1.1
Transit/Wait Time	1.7	-0.6	-1.7	2.2	1.7	1.1	-0.6	-1.1	2.2	2.8	-1.1
Active RAS	0.6	2.8	3.9	0.6	1.1	1.7	2.2	1.7	1.1	1.1	2.2

$$\sum_{w} Delivered_{w} = \sum_{t,w,s} RASw_{w,s}^{t} + \sum_{t,w} INPw_{w}^{t}$$

$$Ratio = \left(\sum_{s} Burned_{s} + \sum_{c} Burned_{c} \right) / \sum_{w} Delivered_{w}$$

When ports are available and there are six shuttles with 1.5M gallons capacity, *Ratio*=0.46. That is, for every gallon of fuel delivered to a warship, the CLF and shuttle ships burn 0.46 gallons transporting it. In the heightened threat scenario (no ports available) with the same shuttle design point, this figure increases to 0.54 gallons. Finally, in the heightened threat scenario with eight shuttle ships, *Ratio*=0.66. This increase reflects the additional burden of transiting to the ARLs vice nearby ports, as well as the impact of having more shuttle ships present and burning fuel.

Recall that the LCS/JHSV has been suggested as a possible candidate for a shuttle platform.

LCS 2 and JHSV have planning factor fuel capacities of 2,663 bbls (111,846 gals) and 3,745 bbls (157,290 gals), respectively (Trickey & Grenwald, 2014). Clearly, these values do not come close to our identified design point of 1.5M gallons deliverable fuel capacity. Even with a proposed 20% increase in fuel capacity as possibly provided by an installed logistics module, their configuration and capabilities do not meet the design requirement for this scenario. Therefore, LCS and JHSV appear to lack the ability to adequately support deployed CTF/CRUDES SAG assets in the 7th Fleet A2AD environment.

Overall, six shuttle ships with 1.5M gallons of deliverable fuel can adequately fill the role as a "delivery boy" in a 7th Fleet A2AD environment, assuming ports are available to replenish the shuttles. In a heightened threat environment when no ports are available to replenish shuttles in our scenario, eight such shuttles are required for sustainable support.

CONCLUSION

The most important finding in our study is the effectiveness of DL-RASM to provide decision support to the decision maker in determining theater level logistics and CLF shuttle requirements. DL-RASM is flexible enough to model a variety of scenarios, and it solves quickly enough to enable detailed sensitivity analysis. Thus, it can provide senior decision makers with an improved understanding of fleet requirements for logistic support, and it can inform future CLF force structure decisions. In the peacetime scenario, this study demonstrates how the deliverable fuel quantity and number of shuttles affects the shuttles ability to support the deployed forces. Our analysis does not indicate a strong correlation between increased shuttle speed and an improved ability to support the warships. While speed does not play a decisive role in our scenario, increased shuttle speed may have other positive influences in other scenarios. For instance, it may allow for improved evasive maneuvering or other defensive measures. Our analysis indicates that a moderate number of shuttles is required to support a peacetime scenario and can allow some resiliency should a wartime scenario ensue. A larger number of shuttles are required to provide sustainable support in a wartime scenario. Indeed, perhaps our most interesting findings result from eliminating the vulnerable ports as a replenishment option for the shuttles. This study finds that the required number of shuttles during peacetime cannot support the warships without port access. Additionally, the efficiency of delivering fuel is dependent upon the number of shuttles available and that efficiency and effectiveness are competing objectives that must be balanced

For our developed scenario in the 7th Fleet AOR, and analysis completed during both peace and wartime situations, this study recommends continued research into developing a CLF shuttle ship with a deliverable fuel capacity of approximately 1.5M gallons. Such a vessel could be employed in a fleet composition of six and eight shuttles in a peace and wartime scenario, respectively, for a deployed warship fleet configuration of one CTF and one CRUDES SAG.

Recommendations for follow-on work are as follows:

- **Expansion to Multi-commodity Flow Model:** For simplicity, this study models liquid fuel as a single commodity combining the DFM and JP5 requirements of surface ships and a deployed airwing on a carrier. To improve DL-RASM's accuracy, the researchers recommend expanding it to a multi-commodity flow model.
- **User Interface Development:** The researchers recommend development of a user interface or scenario developer for DLRASM. This would make DL-RASM more user-friendly and could ultimately increase its level of employment.
- **Mini-CLF Development Cost Analysis:** Development of new platforms requires not only dimension analysis and engineering, but also cost analysis. A cost analysis of the proposed CLF shuttle would provide further insight into the feasibility of adopting and employing the dual-RAS lane concept of operations.
- **Scenario Diversification:** As this study has only considered the use of DL-RASM in the 7th Fleet AOR, the researchers recommend implementation of additional scenarios, to include alternate regions and evaluation of a Battle of the Littorals scenario. Such analysis can lend further credibility to the research and findings of this study and provide additional value to the adoption of the concept of using CLF shuttles.

REFERENCES

Borden, K. D. (2001). *Optimizing the number and employment of combat logistics force shuttle ships, with a case study of the T-AKE ship* (Master's thesis). Naval Postgraduate School, Monterey, CA.

Brown, G. G., & Carlyle, W. M. (2008). Optimizing the U.S. Navy's combat logistics force. *Naval Research Logistics, 55*(8), 800–810.

Brown, G. G., Carlyle, W. M., & Burson, P. (2010). *Replenishment at Sea Planner (RASP) model.* Monterey, CA: Operations Research Department, Naval Postgraduate School.

Brown, G. G., Carlyle, W. M., Kelton, D., Kline, J., & Salmerón, J. (2009). Operational planning tools for U.S. Navy Maritime Commanders. In *International Conference on Harbor, Maritime and Multimodal Logistics Modeling and Simulation*. Universidad de La Laguna.

Cambell, E. (2014, June 10). Spratly Islands: Foreign correspondent visits remote reef flash point where Filipino marines hold out against Chinese navy. *ABC News*. Retrieved from http://www.abc.net.au/news/2014-05-20/oil-key-in-spratlyislands- dispute-south-china sea/5463080

Chief of Naval Operations (CNO). (2007). *Sustainment at Sea* (Navy Warfare Publication 4-01.2). Washington, DC: United States Navy. Retrieved from https://ndls.nwdc.navy.mil

Cioppa, T. M., & Lucas, T. W. (2007). Efficient nearly orthogonal and space-filling Latin hypercubes. *Technometrics*, *49*(1), 45–55. doi:10.1198/004017006000000453

Collins, G., & Erickson, A. (2010, December 26). China deploys first long-range, landbased carrier killer: DF-21D Anti-Ship Ballistic Missile (ASBM) reaches "Initial Operation Capability (IOC). *China Signpost*. Retrieved from http://www.andrewerickson.com/2010/12/china-deploys-world%E2%80%99sfirst-

Commander in Chief, Atlantic Fleet (CINCLANTFLT) (2001), *Fleet analysis of the combat logistics force*. Commander in Chief, Atlantic Fleet. This document is FOR OFFICIAL USE ONLY

Doyle, D. E. (2006). *Evaluation of fleet ownership versus global allocation of ships in the combat logistics force* (Master's thesis). Naval Postgraduate School, Monterey, CA.

Ellis, D. R. (2013). *A distributed logistics concept for flotilla surface combatant support, joint campaign analysis*. Monterey, CA: Naval Postgraduate School.

Givens, R. D. (2002). *A comparison of operational potential and capability of two combat logistics force alternatives* (Master's thesis). Naval Postgraduate School, Monterey, CA.

Hughes, W. P. Jr. (2009). *The new navy fighting machine: A study of the connections between contemporary policy, strategy, sea power, naval operations, and the composition of the United States fleet*. Monterey, CA: Naval Postgraduate School.

Jane's Fighting Ships. (2015, March 26). *Lewis and Clark class (Dry Cargo/Ammunition Ships) (AKEH)*. Retrieved from https://janes.ihs.com.libproxy.nps.edu/CustomPages/Janes/DisplayPage.aspx?DocType=Reference&ItemId=+++1355951&Pubabbrev=JFS_

Jane's Fighting Ships. (2015, March 26). *Henry J Kaiser class (Oilers) (AOH)*. Retrieved from https://janes.ihs.com.libproxy.nps.edu/CustomPages/Janes/DisplayPage.aspx?DocType=Reference&ItemId=+++1355951&Pubabbrev=JFS_

Jane's Strategic Weapons Systems. (2015, April 3). *DF-21 (CSS-5)*. Retrieved from https://janes.ihs.com/CustomPages/Janes/DisplayPage.aspx?DocType=Reference&ItemId=+++1316682&Pubabbrev=JSWS

MacCalman, A. D. (2012). *DesignCreatorv2 spreadsheet*. Retrieved from http://harvest.nps.edu

Morse, T. C. (2008). *Optimization of combat logistics force required to support major combat operations* (Master's thesis). Naval Postgraduate School, Monterey, CA.

Ross, D. B., & Harmon, J. A. (2012). *New navy fighting machine in the South China Sea* (Master's thesis). Naval Postgraduate School, Monterey, CA.

Rowden, T., Gumataotao, P., & Fanta, P. (2015). Distributed Lethality. *Proceedings, 141*(1), 18–23.

Spitz, G. (2007). *Mission resources allocation in the Gulf of Guinea* (Master's thesis). Naval Postgraduate School, Monterey, CA.

Status of the Navy. (n.d.). Retrieved April 26, 2015, from http://www.navy.mil/navydata/nav_legacy.asp?id=146

Stewart, K. A. (2013). *NPS-Developed replenishment at sea program could save millions*. Office of Naval Postgraduate School Public Affairs. Retrieved from http://www.navy.mil/submit/display.asp?story_id=72111

Trickey, W. R. (2014). *Navy Logistics Resiliency Model Description*. Arlington, VA: CNA Corporation.

Trickey, W. R., & Grenwald, M. L. (2014). Navy Logistics Resiliency In A WESTPAC A2AD Scenario. CNA Corporation. (This document is classified SECRET)

Want China Times. (2013, January 23). PLA 'sinks' US carrier in DF-21D missile test in Gobi. *Want China Times*. Retrieved from http://www.wantchinatimes.com/newssubclass-cnt.aspx?id=20130123000112&cid=1101

APPENDIX

Table 18. Nearly Orthogonal Latin Hypercube (NOLH) Design of Experiments (DOE)

Design Point	Speed (kts)	Deliverable Fuel Qty.(gals)	#of Shuttles	Optimal Objective Value
1	13	750000	4	-49701107.7
2	13	1000000	6	-14389238.2
3	19	750000	6	-51560310.6
4	17	750000	7	-243709000.0
5	17	1250000	3	-44309779.1
6	15	1500000	5	-2116539.6
7	15	1000000	8	-37235850.5
8	17	1000000	4	-58387615.4
9	15	1500000	7	-4115407.8
10	13	1250000	7	-2322577038
11	17	1000000	5	-119006000.0
12	19	1250000	8	1259889.9
13	15	1000000	6	-13723848.3
14	15	1250000	6	311896.4
15	17	1500000	6	1483891.5
16	19	1000000	7	179154.4
17	19	750000	5	-357261192.6
18	13	1250000	5	-38715229.2
19	13	1500000	8	-2927663.1
20	15	1250000	4	-3085296.6
21	17	1250000	8	707366.4
22	13	1000000	3	-148675000.0
23	19	1500000	4	-8166286.8
24	15	750000	3	-253934000.0
25	19	1500000	3	-3391122.4

Chapter 12
Optimization of NAS Lemoore Scheduling to Support a Growing Aircraft Population

Manuel Rosas
US Navy, USA

Emily Craparo
Naval Postgraduate School, USA

ABSTRACT

The manual process for flight scheduling at Naval Air Station Lemoore accommodates the needs of 16 fighter resident squadrons as well as constraints imposed by limited military operating area availability. Given the complexity of this problem, attempting to additionally avoid periods of high activity and resultant congestion would challenge the manual process. However, congestion leads to long wait times for flight-line services. Refueling operations are particularly costly when operational time is lost, and resources are backlogged. The problem of avoiding inefficient periods of high demand is complicated by the two types of refueling available: hot refueling, which occurs when the aircraft's engine is running, and cold refueling, which occurs when the aircraft is shut down. Achieving a balance between the two refueling methods is key to maximizing operational effectiveness. The authors discuss an optimization model designed to determine the best daily flight schedules based on the squadrons' flying and training requirements, the refueling infrastructure, and range availability.

INTRODUCTION

Background

Refueling is a critical and costly part of aircraft operation and maintenance. There are two methods to conduct land-based aircraft refueling: hot and cold. In hot refueling, the aircraft's engines are still operating, which means that the aircraft must be equipped with a closed-circuit refueling receiver and single-point pressure refueling receiver that incorporates an automatic fuel shutoff capability. In cold

DOI: 10.4018/978-1-5225-5513-1.ch012

refueling, the aircraft must shut down its engines, turn off all switches, and conduct a turnaround inspection prior to the refueling. The Naval Air Systems Command (NAVAIR) aircraft refueling manual 00-80T-109 provides definitions and guidelines for hot and cold refueling (NAVAIR, 2002). Although cold refueling is more fuel efficient, it is also more time consuming; hot refueling provides a much quicker turnaround. It would be easy to assume that the preferred method is hot refueling because it is faster than cold refueling; however, because fuel is consumed while refueling or waiting to refuel, hot refueling is costly. A squadron requests and schedules an aircraft to receive hot refueling if the aircraft needs to immediately fly another mission. Thus, achieving a balance between these two refueling methods and avoiding unmanageable peak demands are keys to maximizing the effectiveness of a refueling facility.

Naval Air Station (NAS) Lemoore is the home base of 16 fighter squadrons. A number of squadrons are deployed away from NAS Lemoore at any given time. Each of the remaining at-home squadron generates a flight schedule with its respective events a day prior to the actual flying date. Each event includes, among other information, the takeoff and landing times, the number of aircraft, the range where the aircraft will go, and the estimated fuel that will be burned in that event. The events of all squadrons are independently scheduled with the exception of joint missions in which multiple events from one or more squadrons will fly to the same range at the same time.

Because the squadrons do not communicate their schedules among themselves, and because the times when aircraft take off and land are not taken into account as a whole, there are times when the number of aircraft landing exceeds the capacity of the refueling system. The large number of aircraft coinciding in their landing time, in addition to the first-come, first-served distribution of the refueling resources, creates a backlog during some peak hours while leaving refueling resources underused in other times.

Additionally, NAS Lemoore's aircraft population will grow in the coming years, particularly in 2017 and 2018. This growth will be due to the introduction of the Joint Strike Fighters (JSF) F-35 and relocation of one or more F/A-18 E/F squadrons. Given the current bottleneck experienced by the aircraft waiting to be refueled during peak operational periods, the additional future demand will likely increase wait times for refueling if a new approach to scheduling is not developed. Optimizing the flying and refueling schedules in order to operate the refueling system more efficiently will improve the available flying time of the aircraft.

Current Flight Scheduling Process at NAS Lemoore

Every Tuesday, the training officer (TRAINO) of each of NAS Lemoore's homebased fighter squadrons meets with Command Strike Fighter Wing Pacific (CSFWP) staff to request and allocate Lemoore's military operating area (MOA) and field carrier landing practice (FCLP) hours for the following week. Additionally, each squadron's TRAINO requests hours at MOAs and ranges from other bases from the respective base MOA/range coordinators; this occurs a week prior to flight. Figure 1 shows a small example of a report indicating when a particular MOA/range is assigned to squadrons during one week. The numbers in the cells represent the squadrons assigned each day and time, indicated by column and row respectively. Actual reports are more complex than this example, as each location usually includes many MOA/ranges, and it also covers more hours during the day.

The following week, each day, each TRAINO utilizes the MOAs assigned to his or her squadron along with the squadron's training requirements, and, without knowing the other squadrons' flight plans, creates his or her squadron's flight schedule. Figure 2 illustrates the process each squadron's TRAINO follows in order to create a daily flight schedule. Although these squadrons are geographically located at

Figure 1. Example of MOA/Range Assignments

	M		T		W		TH		F	
1000/1015			22	22					86	86
1030/1045		122	122	122	22	22	122	122		
1100/1115	122	122	122	94	22	22			86	86
1130/1145	94/22	94/22	94	94			122	122		
1200/1215	94/22	122	122	122			86	86	86	86
1230/1245	122	122	122		122	122	122	122		
1300/1315	94	94	94	94			86	86		
1330/1345	86	86	94				122	122		

the same base, NAS Lemoore, they do not communicate their flight schedules among themselves; hence, the daily flight plan generated by each squadron is independently scheduled from the other squadrons. The only exception to this lack of communication occurs when planning a joint mission.

Current Refueling Process at Nas Lemoore

The Naval Supply Systems Command (NAVSUP) Fleet Logistics Center (FLC) San Diego provides aircraft refueling services at NAS Lemoore. The current refueling demand at NAS Lemoore is generated by its 16 home-based fighter squadrons and the sporadic arrival of a C-40 or other logistics aircraft. NAS Lemoore has limited refueling resources, including fuel skids or hot pits for conducting hot refueling,

Figure 2. Squadrons' Flight Scheduling Process

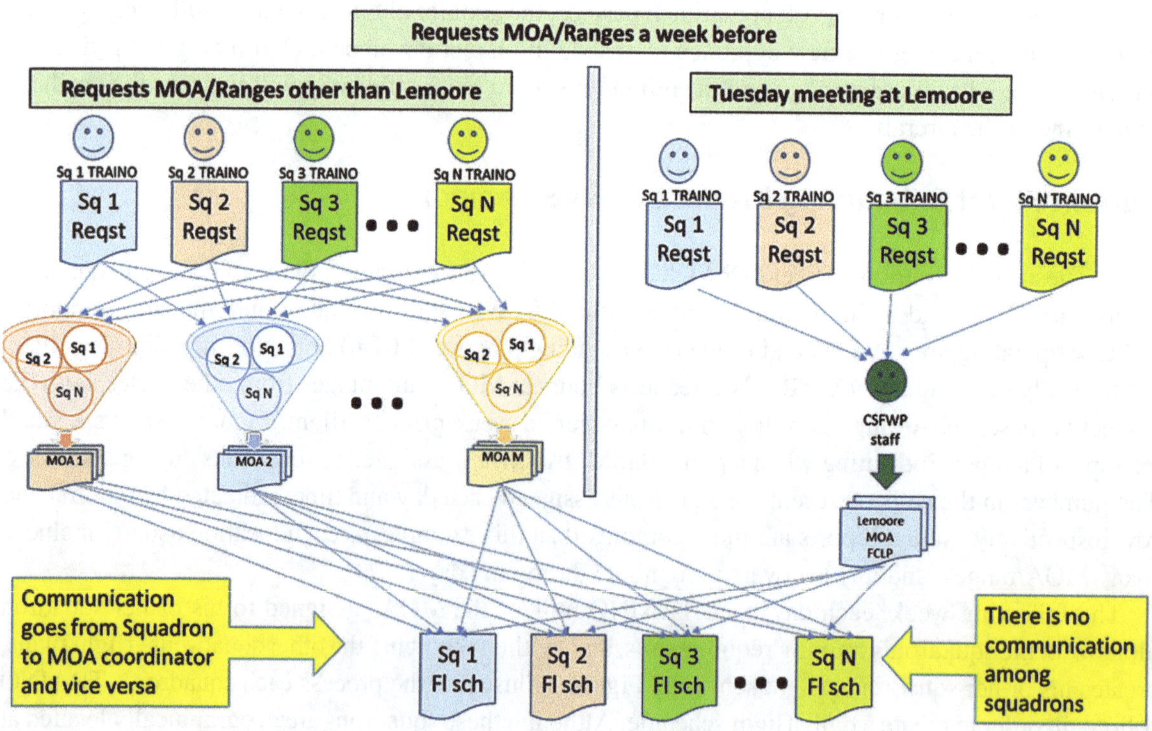

personnel for conducting both types of refueling, and fuel trucks for conducting cold refueling and as a means of transportation for refueling personnel to conduct hot refueling.

The fuel skids are available during specific hours, and each fuel skid is preassigned to specific squadrons due to its proximity to the squadrons' maintenance shop. Cold refueling is conducted at the ramp, which is also close to the squadrons' maintenance shop. The number of available personnel varies during the day. To avoid confusion with other personnel, we use the term "driver" as the designation for the refueling personnel and "maintainer" for personnel from the squadron's maintenance shop. The number of available fuel trucks may also vary due to routine or emergency maintenance requirements.

To conduct cold refueling, the following resources are needed: one driver, one fuel truck, and one maintainer. To perform hot refueling, the following resources are needed: one driver, one fuel truck, and two maintainers. Each fuel skid can refuel two aircraft simultaneously using its right and left hot pits. If the fuel skid is refueling two aircraft simultaneously, then the total needed resources are one driver, one fuel truck, and four maintainers.

Each squadron's maintenance shop submits a daily flight schedule to the refueling team indicating which aircraft require hot and cold refueling. The assignment of the refueling resources to each aircraft is generally done in a first-come, first-served manner. If an aircraft needs cold refueling, it will go to its pre-assigned ramp, shut down, and wait for a driver and a truck to be available. Only one driver with a truck goes to each ramp; another driver with a truck will assist cold refueling in the same ramp only if there are three or more aircraft waiting in queue in that ramp. If an aircraft needs hot refueling and the two sides of its pre-assigned fuel skid are already taken, the aircraft waits until the fuel skid becomes available. Hot refueling also occurs in a first-come, first-served manner.

NAS Lemoore Fuels Division records every refueling event. This information is used to track the percentage of hot refueling events and ensure that the percentage does not exceed the monthly allowed percentage.

Refueling an aircraft involves the waiting time to receive fuel and the actual time consumed while receiving fuel. Although both cold and hot refueling should ideally be done without much waiting, hot refueling must be done expeditiously for two reasons: the cost of burning fuel and the need to ready the aircraft for its next scheduled event, which is what necessitated hot refueling in the first place. In general, it is beneficial to complete both types of refueling efficiently in order to increase flying time availability.

Literature Review

We begin by defining one of our key figures of merit: the concept of operational effectiveness. The Defense Acquisition University (DAU, 2015), in its website Glossary of Defense Acquisition Acronyms and Terms, describes operational effectiveness as "the overall ability of a system to accomplish a mission." Operational effectiveness can be improved by better resource utilization, which often requires changing operating policies.

A number of studies at the Naval Postgraduate School have sought ways to decrease the amount of time aircraft wait to refuel. Geiser (2012) recommends a set of solutions to reduce delays when refueling at NAS Oceana by modifying policy and materiel resources. By using simulation and data analysis, and focusing on improving communication to coordinate and dispatch refueling services, Geiser approaches the fuel demand problem at NAS Oceana. He emphasizes the communication flow between the squadrons and the fuel truck operators when the squadron's maintenance shop calls for refueling services. Moreover, Geiser's thesis explores techniques to help forecast the request for aircraft refueling and minimize the number of times fuel trucks need to refill when conducting aircraft refueling, which consequently decreases aircraft refueling time.

Gerber and Clark (2013) use simulation and data analysis to study refueling operations at NAS Lemoore. Specifically, they vary the aircraft arrival rate and determine the costs and benefits associated with the various policies, with aircraft fuel consumption and delay as their primary figures of merit. They first establish a baseline level of fuel consumed in the current practice; then consider the impact of future transitions to newer aircraft models in the coming years. Based on their analysis, Gerber and Clark recommend that planners "decrease variation in aircraft arrivals during peak periods by establishing a culture of squadron collaboration at the type-wing level through slot management" (p. 113).

Another study was conducted in winter 2015 by a group of 11 students, led by CDR Peter Ward, in the OA4611 Joint Logistics Models class at the Naval Postgraduate School (Ward, 2016). This team modeled and assessed the ability of refueling personnel, processes, and infrastructure at NAS Lemoore to meet future refueling demand around January 2017. They conclude that a range of policy changes, such as smoothing the refueling demand over the course of the operational day, would likely lead to the greatest reduction in refueling wait times.

Notably, prior research has focused on modeling and simulation of airfield operations. In contrast, our work leverages the understanding of flight-line processes developed in prior work to apply the technique of mathematical optimization. Thus, we take a step towards the implementation of slot management as recommended by Gerber and Clark, although ours is a soft approach that values and maintains the autonomy of squadron inputs.

Contributions and Outline

Our goal is to study the factors impacting the land-based refueling wait time at NAS Lemoore, and to determine whether the wait time can be reduced by modifying existing policies or through improved decision support. We develop the Multiple Squadron Input Schedule Enhancer (MSISCHE), an optimization-based decision support tool designed to evaluate the squadrons' daily flight requests and the available refueling resources to create an optimal flying and refueling schedule. The current scheduling practice results in long wait times for refueling during periods of peak demand. These wait times are avoidable. MSISCHE makes small adjustments to scheduled takeoff times that are feasible given airspace scheduling constraints and deferential to squadron inputs. MSISCHE's adjustments smooth demand for limited resources such as refueling, making operations at NAS Lemoore more efficient.

Through optimization and sensitivity analysis, we identify critical elements that impair and constrain the efficiency of land-based aircraft refueling service at NAS Lemoore. The most significant outcome in our study is that, with the current squadrons and refueling infrastructure at NAS Lemoore, MSISCHE is able to find a combined optimal solution that minimizes the aircraft refueling wait time while making only small adjustments to the squadrons' flight schedules.

With the current aircraft refueling demand at NAS Lemoore, the refueling resources are at the edge of sustainability. In order to maintain a balanced system, the refueling demand cannot surpass the available refueling resources. Decreasing the refueling resources or increasing the number of squadrons will force the cancellation of some flying events.

We highly recommend the use of MSISCHE to improve the efficiency of the scheduling process at NAS Lemoore while minimizing changes to the squadrons' flight schedules and deriving the benefits of reduced refueling wait times.

METHODOLOGY

This study was conducted by gathering data from CSFWP and the NAS Lemoore Fuel Division, reviewing previous studies, setting up constraints, limitations, and assumptions, developing the optimization model, creating an interface, and generating scenarios needed to test the model and perform sensitivity analysis.

Objectives

Minimizing the time aircraft spend waiting for refueling service is of paramount importance. However, when refueling demands are concentrated during specific times of the day, there is a natural tension between the customer for refueling who expect immediate service even during periods of peak demand, and the refueling service provider team for whom it would be inefficient to be resourced to support peak demand with idle capability at all other times. Providing additional resources to avoid refueling delays during even peak demand periods might be the best solution from the customer's perspective, but this is treating a symptom rather than addressing a root cause and does not address other effects of congestion on the flight-line or in the airspace. Also, resourcing to meet peak demand is just a temporary solution; eventually, the backlog will return either as the number of aircraft increases or the times of peak demand shift. A deeper analysis is needed to address the root cause of the problem, and consequently develop a robust solution.

The peak demand hours and resulting congestion occur because aircraft participating in one or more events, from one or more squadrons, coincide in their landing time and thus in the time at which they require refueling. This issue can be mitigated by coordination of the squadrons' flight schedules. Because of the complexity involved, reconciling the schedule requests from individual squadrons with an overall objective of avoiding peak demands would be tedious and time consuming to accomplish manually. In addition to the scheduling process, a further area of exploration is the policy of pre-assigning fuel skids to specific squadrons. This policy might be advantageous because the hot skids are close to the squadron's maintenance facility, but it is also a disadvantage because this assignment constrains the squadron's aircraft to wait for refueling if their assigned hot skid is occupied while other hot skids are available. Having aircraft from 16 squadrons creates a high demand, but with the proper coordination of the squadrons and the refueling team, landing and refueling of aircraft can be executed more efficiently than is currently observed. Reducing the aircraft's refueling waiting time increases the aircraft's available flying time.

Our study is based on data provided by CSFWP staff and NAS Lemoore Fuels Division. We use the General Algebraic Modeling System (GAMS) (GAMS, 2015) to implement the optimization model MSISCHE, which is a discrete-time mixed integer linear mathematical optimization model based on series of penalties and rewards. MISISCHE uses as inputs the daily flight schedules developed by the squadrons present at NAS Lemoore, the estimated fuel to be consumed by each flight, the MOA ranges assigned to squadrons, the priority assigned to each squadron, the number of refueling personnel in the various shifts, the number of refueling trucks, the capacity and safety levels of the refueling trucks, the percentage of aircraft allowed to conduct hot refueling in a day, and the times to start and stop hot refueling determined by the refueling team.

A central precept of MSISCHE is to respect the daily flight schedules developed by the individual squadrons to the greatest extent possible. By making only small adjustments to the timing of events proposed by the squadrons, MSISCHE finds the optimal solution to the flight schedules at NAS Lemoore where optimal is defined as a balance between the minimization of refueling waiting time (and consequently maximization of the aircraft's flying time availability) and minimization of deviation from the squadron' proposed schedules. MSISCHE identifies and makes these small adjustments with the greatest positive impact on operational efficiency while preserving squadrons' authority and accountability.

To demonstrate the benefit of using the optimization model, the study assesses the operational availability of the aircraft based on the in-land refueling in both optimized and legacy operating protocols. In addition to modeling the current situation, we evaluate the impact of growing the aircraft population, and we study the impact of modifying some uncertain parameters such as the availability of the fuel trucks and the number of refueling personnel.

Scope, Constraints, Limitations, and Model Assumptions

Scope

This work provides a model to optimize the current daily flight schedules generated by NAS Lemoore's fighter squadrons for the following day in order to provide an alternate and optimal combined schedule that minimizes aircraft refueling waiting time, and accordingly maximizes aircraft available flying time. NAS Lemoore's combined schedule is generated by utilizing all the data, in comma separated values (CSV) files, from the squadrons' daily flight schedules, and data regarding the available refueling resources. By using the optimization model MSISCHE, we show that it is possible to have a collective schedule that, with only small adjustments to the squadrons' requested schedules, guarantees the balance between minimizing schedule changes and minimizing aircraft's refueling wait time.

Constraints

- CSFWP can modify only the pre-assigned times of events going to Lemoore MOAs and FCLPs.
- Times of events going to MOAs and ranges other than NAS Lemoore are pre-assigned a week prior to the event and cannot be modified on short notice.

Limitations

- The model minimizes the refueling wait time by shifting events in small increments before or after the requested takeoff time given by the respective squadron. This self-imposed limitation is set in order to preserve the authority and accountability of individual squadrons.
- Events represent a group of one or more aircraft going to a specific range, and at a specific time, to complete a mission. Missions are represented by a code. The study was not able to gather information linking the mission code to a range.
- The model accounts only for the refueling services provided to NAS Lemoore's squadrons. If other aircraft are visiting NAS Lemoore and require refueling, these requirements must be added to the model in order to take full advantage of the optimization.
- Besides refueling, the fuel truck operators have other duties on base; nonetheless, there is not enough data to represent these other duties, consequently they are not taken into consideration.

Assumptions

We make a number of assumptions in our study. Many of these assumptions can be modified by changing their corresponding parameters in MSISCHE. Because MSISCHE produces daily flight schedules, when an assumption says that a parameter remains constant it actually means that this parameter remains constant across a 24-hour period.

- All events, including refueling of aircraft and trucks, are based on time periods of 15-minute increments.
- Drivers work following their schedule, and they are available for the entire shift.
- Fuel trucks have perfect reliability. The number of fuel trucks remains constant, unless one breaks down or needs to be refueled.
- The safety level percentage of the fuel trucks, which indicates when the truck needs to be refueled, remains constant.
- Available trucks are used if there are enough drivers.
- Trucks' fuel capacity remains constant.
- Fuel trucks are filled to their maximum capacity before starting the day.
- Hot refueling is allowed only during the hours set by NAS Lemoore Fuels Division.
- Cold refueling is available 24 hours per day.
- In a 15-minute period, an aircraft can be refueled up to 1,800 gallons.
- Aircraft requesting hot refuel could do so at any hot pit.
- All fuel skids are operational and available.
- CSFWP provides the priorities for each squadron, giving VFA-122 the highest priority and squadrons preparing to deploy the next highest priority.
- CSFWP provides the penalty incurred for each squadron for a shift in one 15-minute time interval to an event's requested takeoff time.
- CSFWP provides the Lemoore MOAs available hours for the day.
- Aircraft could fly to any of the MOA/ranges included in the NAS Lemoore Inflight Guide 2016 (CSFWP, 2016).
- Transit time to MOA/ranges is assumed as one time period (15 minutes) for most MOA/ranges, and zero time periods for Lemoore, Hunter, FCLP, R_2508, and Ferry.
- All squadrons' flight schedules used as inputs are valid for the following day, and they are provided using time periods of 15-minute increments.
- Some events do not land back at NAS Lemoore Airport (KNLC), and this situation is noted in the flight schedule.
- Events going to joint missions are annotated in the flight schedule.
- The estimated fuel burned in each event is annotated in the flight schedule.
- More than one not-joint event could go to certain missions at the same time (FDMO, Ferry, and missions going to range R_2508).
- Because events in the squadrons' flight schedules do not include aircraft's designation, aircraft in the model are not utilized in more than one event; however, aircraft that require hot refueling effectively model an aircraft that is going to perform multiple events.
- To include the possibility of more constraints, Lemoore MOA is not available all day; instead, it is available during the hours of 0930–1130, 1300–1500, and 1630–2400.

- Events scheduled for Lemoore MOA, FCLP, FDMO, R_2508, and Ferry can be modified on a short notice.
- The monthly percentage of refueling events allowed to conduct hot refueling applies equally to any day.
- There is no restriction on the number of aircraft that can take off at the same time.
- There is no restriction on the number of aircraft that can land at the same time.

THE MSISCHE OPTIMIZATION MODEL

MSISCHE has three components: a Microsoft Excel interface (Microsoft, 2013), a discrete-time linear optimization model implemented in GAMS (GAMS, 2015), and a body of Visual Basic for Applications (VBA) code that links the two. The user interacts with MSISCHE via the Excel-based interface, while the VBA and GAMS code remain invisible to the user.

Interface

Figure 3 through 5 present a few screenshots of the interface. Figure 3 shows the Dashboard, the first and main sheet of MSISCHE's interface. It contains some parameters that the user can modify, a usage guide, and a number of buttons that allow the user to utilize various functions.

Figure 4 shows how, by using the squadrons' worksheets, the user enters the data corresponding to the events of the daily flight schedules. The worksheet includes a few notes to guide the user on the proper formatting when entering the data.

Figure 3. Screenshot of the Dashboard Worksheet, MSISCHE's Interface

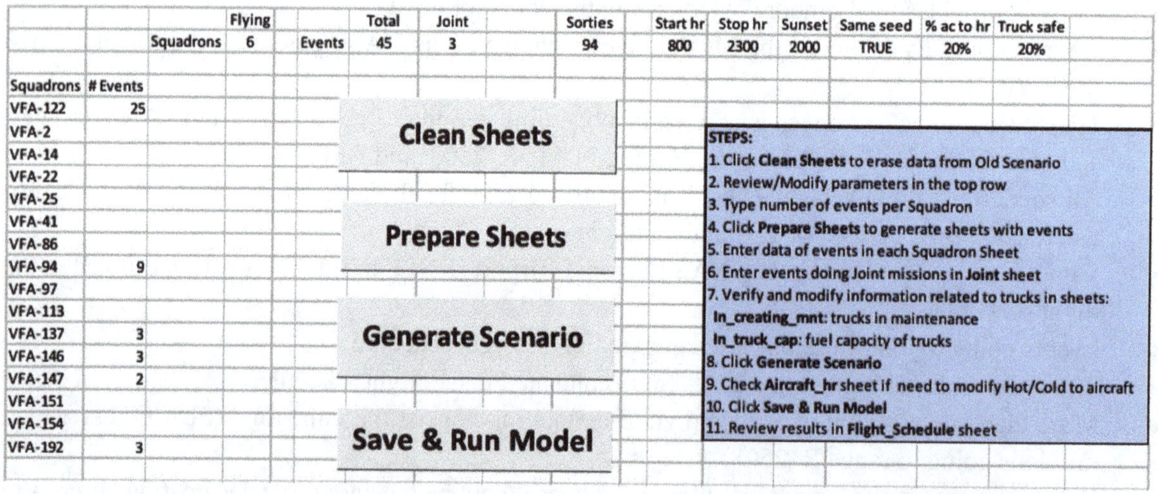

Figure 4. Screenshot of a Squadron Input Worksheet, MSISCHE's Interface

	take-off	land	fuel	joint	KNLC	Range	Sorties	Hot/Cold
V122-e1	1000	1130	12	0	1	R_2508	1	1
V122-e2	1200	1330	17	0	1	R_2508	1	0
V122-e3	1200	1315	17	0	1	R_2508	2	0
V122-e4	1200	1315	17	0	1	Hunter_High	2	1
V122-e5	1215	1345	17	0	1	R_2508	1	0
V122-e6	1245	1415	17	0	1	R_2508	1	0
V122-e7	1315	1445	17	0	1	Superior_Valley	4	0
V122-e8	1315	1400	13	0	1	FCLP	4	0
V122-e9	1400	1515	17	0	1	R_2508	2	1
V122-e10	1400	1515	17	0	1	Hunter_High	2	0
V122-e11	1445	1530	13	0	1	FCLP	4	0
V122-e12	1445	1530	13	0	1	R_2508	1	0
V122-e13	1600	1715	17	0	1	Hunter_High	2	1

Notes:
Take-off: time in periods of 15 minutes
Land: time in periods of 15 minutes
Fuel: estimated in pounds
Joint: number of events going with event at the same mission
KNLC: 1 if event will finish landing at KNLC, 0 otherwise
Range: where the event is going
Sorties: number of aircraft of the event
Hot/Cold: 1 if required hot refueling, 0 if cold

After generating the scenario, saving it, and running the model, the interface presents the results through the Flight Schedule worksheet. Figure 5 illustrates part of the results.

The interface workbook has approximately 50 worksheets, but many of them contain intermediate steps and are hidden because they should not be modified by users.

Model Formulation

We now describe MSISCHE's mathematical formulation.

Figure 5. Screenshot of Flight Schedule Worksheet Results, MSISCHE's Interface

Flight Schedule								
Squadron	VFA-122		Desired	Time	Computed	Time	Start	
Event	Aircraft	Range	T-Off	Land	T-Off	Land	Refuel	Hot/Cold
V122-e1	V122-a1	R_2508	815	1000	815	1000	1000	Hot
V122-e1	V122-a2	R_2508	815	1000	815	1000	1000	Hot
V122-e2	V122-a3	Ferry	1515	1545	1515	1545	1545	Cold
V122-e3	V122-a4	Fallon_N2	930	1030	930	1030	1030	Hot
V122-e3	V122-a5	Fallon_N2	930	1030	930	1030	1030	Hot
V122-e4	V122-a6	R_2508	945	1045	945	1045	1045	Hot
V122-e5	V122-a8	R_2508	830	915	830	915	915	Hot
V122-e5	V122-a9	R_2508	830	915	830	915	915	Hot
V122-e6	V122-a12	Superior_Valley	1100	1200	1100	1200	1200	Hot
Squadron	VFA-2		Desired	Time	Computed	Time	Start	
Event	Aircraft	Range	T-Off	Land	T-Off	Land	Refuel	Hot/Cold
V2-e1	V2-a1	Lemoore_A	1345	1430	1345	1430	1430	Hot
V2-e1	V2-a2	Lemoore_A	1345	1430	1345	1430	1430	Hot
V2-e2	V2-a4	R_2508	1830	1900	1830	1900	1900	Hot
V2-e2	V2-a6	R_2508	1830	1900	1830	1900	1900	Hot

Indices and Sets (Approximate Cardinality)

$i, i' \in I$	Squadrons	$[16]$
$j, j' \in J$	Events	$[\sim 500]$
$k, k' \in K$	Aircraft	$[\sim 500]$
$r \in R$	Ranges/MOAs	$[\sim 114]$
$t \in T = \{t1,..,t11\}$	Refuling trucks	$[11]$
$q, q', q'' \in Q$	Time periods	$[128]$

$E \subseteq I \times J$: Mapping of tuples *(i,j)*: squadron *i* requests event *j*
$A \subseteq J \times K$: Mapping of tuples *(j,k)*: aircraft *k* performs event *j*
$ER \subseteq J \times K$: Mapping of tuples *(j,r)*: event *j* requires range *r*
$EJ \subseteq J \times J$: Mapping of tuples *(j,j')*: events *j* and *j'* belong to a joint mission
$M \subseteq I \times R \times Q$: Mapping of tuples *(i,r,q)*: squadron *i* can occupy range *r* at time *q*
$M' \subseteq I \times R \times Q$: Mapping of tuples *(i,r,q)*: squadron *i* going to range *r* can take off at time *q*
$N \subseteq J \times K \times R \times Q \times Q$: Mapping of tuples *(j,k,r,q,q')*: aircraft *k* of event *j* can go to range *r* taking off at time *q* and refueling at time *q'*
$TRUCKN \subseteq J \times K \times T \times Q$: Mapping of tuples *(j,k,t,q)*: aircraft *k* of event *j* can be refueled by truck *t* at time *q*
$Q^H \subseteq Q$: Set of time periods when hot refueling is available
$Q^{NH} \subseteq Q$: Set of time periods when hot refueling is not available
$R^1 \subseteq R$: Set of ranges that may have more than one event at the same time
$J^1 \subseteq J$: Set of events that will land and refuel in Lemoore (KNLC)
$K^1 \subseteq K$: Set of aircraft assigned to do hot refuel

Parameters (Units)

to_j : Requested takeoff time of event *j* [time period]
la_j : Requested landing time of event *j* [time period]
ff_j : Estimated fuel consumed by event *j* [1000lbs]
$joint_j$: Number of events going with event *j* to a joint mission
tr_r : Number of time periods to reach range *r* [timeperiods]
pr_i : Priority assigned to squadron *i* [reward]
pc_i : Penalty for changing takeoff time of events of squadron *i* [penalty]
pw : Penalty for an aircraft waiting to refuel [penalty]
pt : Penalty per thousand gallon fuel shortage of truck below safety level [penalty/1000 gallons]
$capt_t$: Fuel capacity of truck *t* [1000 gallons]
$safe$: Percentage of fuel capacity below which trucks incur a penalty [percentage]

np_q : Number of fuel personnel assigned to time period q [drivers]
nhp : Number of hot-fuel skids [skids]
ncp : Number of cold pits [pits]
$mnt_{t,q}$: 1 if truck t is in maintenance in time period q, 0 otherwise [binary]
$ptog$: Factor to convert pounds of fuel to gallons [gal/lb]
$hrpercent$: Percentage of aircraft allowed to do hot refueling [percentage]

Derived Data [Units]

et_j : Estimated flying of event j [time period] $et_j = la_j - to_j, \ \forall i, j : (i, j) \in E$
$etfa_j$: Estimated time to refuel aircraft of event j [time periods]

$$etfa_j = \left\lceil \frac{ptog * ff_j}{1.8} \right\rceil, \forall i, j : (i, j) \in E$$

$pec_{j,q}$: Penalty to an aircraft of event j for changing its takeoff time to time period q [penalty]

$$pec_{j,q} = 0.05 * pr_i * pc_i * \left| \frac{q - to_j}{96} \right|, \forall i, j, q : (i, j) \in E$$

$pew^h_{j,q,q'}$: Penalty to an aircraft of event j for waiting hot refueling between landing time q and refueling time [penalty]

$$pew^h_{j,q,q'} = 0.75 * pr_i * pc_i * pw * \left(\frac{q' - (q + et_j)}{96} \right), \forall j, k, r, q, q' : (j, k, r, q, q') \in N$$

$pew^c_{j,q,q'}$: Penalty to an aircraft of event j for waiting for cold refueling between landing time q and refueling time q' [penalty]

$$pew^c_{j,q,q'} = 0.05 * pr_i * pc_i * pw * \left(\frac{q' - (q + et_j)}{96} \right), \forall j, k, r, q, q' : (j, k, r, q, q') \in N$$

Variables (Units)

$Y_{j,r,q}$: 1 if event j arrives at range r at time q [binary]
$X^h_{j,k,r,q,q'}$: 1 if aircraft k of event j going to range r takes off at time q and starts hot refueling at time q' [binary]

$X^c_{j,k,r,q,q'}$: 1 if aircraft k of event j going to range r takes off time q and starts cold refueling at time q' [binary]

$W^h_{j,k,r,q}$: 1 if aircraft k of event j uses truck t to start hot refueling at time q [binary]

$W^c_{j,k,t,q}$: 1 if aircraft k of event j uses truck t to start cold refueling at time q [binary]

$REFUEL_{t,q}$: 1 if truck t is getting fuel at time q [binary]

$SHORT_{t,q}$: Amount by which truck $t's$ fuel is below safety level at time q [1000 gal]

$HOLD_{t,q}$: Fuel inventory of truck t at start of time period q [1000 gal]

$FT_{t,q}$: Fuel received by truck t at time q [1000 gal]

$FA_{j,k,t,q}$: Fuel received by aircraft k of event j from truck t at time q [1000 gal]

Formulation

$$\underset{\substack{Y,X^h,X^c,W^h,W^c,\\ REFUEL,SHORT,\\ HOLD,FT,FA}}{Max} \sum_{\substack{j,k,r,q,q':\exists i:\\ (i,j)\in E,k\in K^1\\ (j,k,r,q,q')\in N,\\ q'\in Q^H}} (pr_i - pec_{j,q} - pew^h_{j,q,q'})X^h_{j,k,r,q,q'}$$

$$+ \sum_{\substack{j,k,r,q,q':\exists i:\\ (i,j)\in E,\\ k\notin K^1,\\ (j,k,r,q,q')\in N}} (pr_i - pec_{j,q} - pew^c_{j,q,q'})X^c_{j,k,r,q,q'} - \sum_{t,q} pt * SHORT_{t,q}$$

s.t.

$$\sum_{\substack{r,q,q':\\ (j,k,r,q,q')\in N,\\ k\in K^1\\ q'\in Q^H,\\ (j,r)\in ER}} X^h_{j,k,r,q,q'} + \sum_{\substack{r,q,q':\\ (j,k,r,q,q')\in N,\\ k\notin K^1,\\ (j,r)\in ER}} X^c_{j,k,r,q,q'} \le 1; \forall j,k:(j,k)\in A \tag{1}$$

$$\sum_{\substack{j,k,r,q,q':\\ (j,k,r,q,q')\in N,\\ k\in K^1,\\ j\in J^1,\\ q'\in Q^H}} X^h_{j,k,r,q,q'} \le hrpercent * \left(\sum_{\substack{j,k,r,q,q':\\ (j,k,r,q,q')\in N,\\ k\in K^1,\\ j\in J^1,\\ q'\in Q^H}} X^h_{j,k,r,q,q'} + \sum_{\substack{j,k,r,q,q':\\ (j,k,r,q,q')\in N,\\ k\in K^1,\\ j\notin J^1}} X^c_{j,k,r,q,q'} \right) \tag{2}$$

$$\sum_{\substack{j,q':\exists i:\\ (i,j)\in E\\ (i.r,q)\in M\\ (i,r,q')\in M\\ (q-et_j+tr_r+1\le q'\le q)}} \frac{Y_{j,r,q'}}{(1+joint_j)} \le 1; \forall q,r: r\notin R^1 \tag{3}$$

$$\sum_{\substack{q':\\(j',k',r,q,q')\in N,\\k'\in K^1,\\q'\in Q^H}} X^h_{j',k',r,q,q'} + \sum_{\substack{q':\\(j',k',r,q,q')\in N,\\k'\notin K^1}} X^c_{j',k',r,q,q'} = \sum_{\substack{q':\\(j,k',r,q,q')\in N,\\k\in K^1,\\q'\in Q^H}} X^h_{j,k,r,q,q'} + \sum_{\substack{q':\\(j,k,r,q,q')\in N,\\k\notin K^1}} X^c_{j,k,r,q,q'}$$

$$\begin{aligned}&\forall j,k,r,q,j',k' : \exists i,i' : (j,j') \in EJ \text{ and } (i,j) \in E \text{ and } (i',j') \in E \text{ and}\\&(j \neq j') \text{ and } (joint(j) > 0) \text{ and } (joint(j') > 0) \text{ and}\\&(j,r) \in ER \text{ and } (j',r) \in ER \text{ and } (j,k) \in A \text{ and}\\&(j',k') \in A \text{ and } (i,r,q) \in M' \text{ and } (i',r,q) \in M'\end{aligned} \tag{4}$$

$$\sum_{\substack{q':\\(j,k,r,q,q')\in N,\\k\in K^1,\\q'\in Q^H}} X^h_{j,k,r,q,q'} + \sum_{\substack{q':\\(j,k,r,q,q')\in N,\\k\notin K^1}} X^c_{j,k,r,q,q'} = Y_{j,r,q+tr(r)}$$

$$\forall j,k,r,q : \exists i : (i,j) \in E \text{ and } (j,k) \in A \text{ and } (j,r) \in ER \text{ and } (i,r,q) \in M' \tag{5}$$

$$\sum_{\substack{t:\\(j,k.t.q')\in TRUCKN}} W^h_{j,k,t,q'} = \sum_{\substack{q:\\(j,k,r,q,q')\in N}} X^h_{j,k,r,q,q'}$$

$$\forall j,k,r,q' : j \in J^1 \text{ and } (j,k) \in A \text{ and } k \in K^1 \text{ and } (j,r) \in ER \text{ and } q' \in Q^H \tag{6}$$

$$\sum_{\substack{t:\\(j,k,t,q')\in TRUCKN}} W^c_{j,k,t,q'} = \sum_{\substack{q:\\(j,k,r,q,q')\in N}} X^c_{j,k,r,q,q'}$$

$$\forall j,k,r,q' : j \in J^1 \text{ and } (j,k) \in A \text{ and } k \notin K^1 \text{ and } (j,r) \in ER \tag{7}$$

$$REFUEL_{t,q} + \frac{1}{2} \sum_{\substack{j,k,q':\\(j,k,t,q)\in TRUCKN,\\q'\in QH,\\k\in K^1,\\q-etfa_j+1\le q'\le q}} W^h_{j,k,t,q'} + \sum_{\substack{j,k,q':\\(j,k,t,q')\in TRUCKN,\\k\notin K^1,\\q-etfa_j+1\le q'\le q}} W^c_{j,k,t,q'} \le 1 - mnt_{t,q}; \forall t,q \tag{8}$$

$$\sum_{\substack{j,k,t,q':\\(j,k,t,q')\in TRUCKN,\\k\notin K^1,\\q-etfa_j+1\le q'\le q}} W^c_{j,k,t,q'} \le ncp; \forall q \tag{9}$$

$$\frac{1}{2}\sum_{\substack{j,k,t,q': \\ (j,k,t,q')\in TRUCKN, \\ q'\in QH, \\ k\in K^1, \\ q-etfa_j+1\le q'\le q}} W^h_{j,k,t,q'} \le nhp; \forall q \tag{10}$$

$$\sum_t \left(REFUEL_{t,q} + \frac{1}{2}\sum_{\substack{j,k,q': \\ (j,k,t,q')\in TRUCKN, \\ q'\in QH, \\ k\in K^1, \\ q-etfa_j+\le q'\le q}} W^h_{j,k,t,q'} + \sum_{\substack{j,k,q': \\ (j,k,t,q')\in TRUCKN, \\ k\notin K^1, \\ q-etfa_j+1\le q'\le q}} W^c_{j,k,t,q'} \right) \le np_q; \forall q \tag{11}$$

$$HOLD_{t,q1} = capt_t; \forall t \tag{12}$$

$$HOLD_{t,q} \le capt_t; \forall t,q \tag{13}$$

$$FT_{t,q} \le capt_t * (REFUEL_{t,q} + mnt_{t,q}); \forall t,q \tag{14}$$

$$FA_{j,k,t,q} \equiv \frac{ptog * ff_j}{etfa_j} * \sum_{\substack{q': \\ (j,k,t,q')\in TRUCKN, \\ q-etfa_j+1\le q'\le q}} W^c_{j,k,t,q'}; \forall j,k,t,q : (j,k,t,q) \in TRUCKN \text{ and } k \notin K^1 \tag{15}$$

$$HOLD_{t,q} = HOLD_{t,q-1} + FT_{t,q-1} - \sum_{\substack{j,k: \\ (j,k,t,q-1)\in TRUCKN, \\ k\notin K^1}} FA_{j,k,t,q-1}; \forall t,q : (q > q1) \tag{16}$$

$$SHORT_{t,q} \ge safe * capt_t - HOLD_{t,q}; \forall t,q \tag{17}$$

$$Y_{j,r,q} \in \{0,1\}; \forall j,r,q : \exists i : (i,j) \in E \text{ and } (i,r,q) \in M \tag{18}$$

$$X^{h}_{j,k,r,q,q'} \in \{0,1\}; \forall (j,k,r,q,q') \in N \quad (19)$$

$$X^{c}_{j,k,r,q,q'} \in \{0,1\}; \forall (j,k,r,q,q') \in N \quad (20$$

$$W^{h}_{j,k,t,q} \in \{0,1\}; \forall (j,k,t,q) \in TRUCKN \quad (21)$$

$$W^{c}_{j,k,t,q} \in \{0,1\}; \forall (j,k,t,q) \in TRUCKN \quad (22)$$

$$FA_{j,k,t,q} \geq 0; \forall (j,k,t,q) \in TRUCKN \quad (23)$$

$$REFUEL_{t,q} \in \{0,1\}; \forall t,q \quad (24)$$

$$SHORT_{t,q} \geq 0; \forall t,q \quad (25)$$

$$HOLD_{t,q} \geq 0; \forall t,q \quad (26)$$

$$FT_{t,q} \geq 0; \forall t,q \quad (27)$$

The objective function maximizes a series of rewards while it minimizes penalties that reflect the quality of NAS Lemoore's combined daily flight schedule. The user can modify the rewards and penalties through MSISCHE's interface. Some time-based penalties are also computed within GAMS (GAMS, 2015). The first and second parts of the objective function include the rewards and penalties for aircraft conducting hot and cold refueling, respectively. These portions reward those aircraft which belong to events that were requested in the input flight schedules and are still included in the optimal solution, and they penalize aircraft for taking off at a different time than desired, and also aircraft that waited for refueling after landing. The last part of the objective function reflects the penalty for trucks being below their safety level of fuel. Constraint set (1) guarantees the time uniqueness of the events by requiring that each aircraft *k* of event *j* going to range *r* does not get more than one time period *q* to take off and one time period q′ to start refueling, either hot or cold. Constraint set (2) ensures that the total number of aircraft conducting hot refueling does not exceed the percentage of hot refueling allowed. Constraint set (3) ensures that each range or MOA *r*, except those in set R^1, is not assigned to more than one event *j* or joint mission in time period *q*. Constraint set (4) pertains to joint missions; it makes sure that all aircraft going to a joint mission take off at the same time period *q* and fly to the same range *r*. Constraint set (5) ensures that all aircraft of event *j* take off at the same time period *q* and go to the same range *r*, but they could refuel at different time period q′. Constraint sets (6) and (7) assign truck *t* to aircraft *k* at time q′ to start refueling, hot and cold respectively. Constraint set (8) ensures that each truck *t* performs at most one activity in each time period. Constraint set (9) limits, at every time period *q*, the number of cold refueling aircraft to the capacity of cold refueling stations. Constraint set (10) prevents the number of hot refueling aircraft from exceeding the capacity of hot refueling stations in any time period. Constraint set (11) ensures that the number of trucks being used, at any time period *q*, do not exceed the number of available drivers. Constraint set (12) fixes a truck's starting fuel to be equal to its fuel capacity. Constraint set (13) prevents truck *t* from exceeding its fuel capacity at any time period *q*. Constraint set (14) allows truck *t,* at any time period *q,* to get fuel without exceeding its fuel capacity; a truck can also receive fuel before coming back from maintenance. Constraint set (15) calculates how much fuel truck *t* transfers to aircraft *k* during cold refueling at time period *q*. Constraint set (16) computes the amount of fuel that truck *t* holds at any time period *q*. Constraint set (17) models the fuel safety levels of the trucks. Constraint sets (18) to (27) define decision variable domains.

ANALYSIS AND RESULTS

We now exercise MSISCHE to study its performance in various scenarios, each with a unique configuration of squadrons and their respective daily flight schedules. In order to conduct sensitivity analysis, 400 scenarios were generated to simulate aircraft landing and refueling at NAS Lemoore.

Data

All data was provided by CSFWP and NAS Lemoore Fuels Division. NAS Lemoore has a total of 5 double fuel skids (10 hot pits) for conducting hot refueling, drivers for conducting both types of refueling, and 11 fuel trucks for conducting cold refueling and as a means of transportation for refueling personnel to conduct hot refueling.

Squadrons

NAS Lemoore currently has 16 fighter squadrons: VFA-122, VFA-2, VFA-14, VFA-22, VFA-25, VFA-41, VFA-86, VFA-94, VFA-97, VFA-113, VFA-137, VFA-146, VFA-147, VFA-151, VFA-154, and VFA-192. To conduct sensitivity analysis, a symbolic squadron VFA-999 is added to this list.

Moa/Ranges

Aircraft conducting events departing from NAS Lemoore may be scheduled to fly to any MOA/range included in the NAS Lemoore inflight guide (CSFWP, 2016).

Fuel Skids

The fuel skids, used to conduct hot refueling, are available during specific hours set by NAS Lemoore Fuels Division. Hot refueling usually goes from 0800 to 2300 Monday through Thursday and from 0800 to 1600 on Friday. Each fuel skid is preassigned to specific squadrons based on proximity to the squadrons' maintenance facility. Cold refueling is conducted at the ramp, which is also close to the squadron's maintenance facility. Table 1 shows these assignments, and Figure 6 displays the locations of hot pits and ramps for hot and cold refueling respectively.

Drivers

The drivers' work schedule is distributed in four shifts provided by NAS Lemoore Fuels Division; thus, the number of drivers varies during the day. Even though refueling is the drivers' primary duty, they have other tasks to complete such as cargo handling and sweeping runways, taxiways, and ramps. Figure 7 displays the distribution of drivers per hour, and Table 2 summarizes the number of drivers per shift assignment.

Trucks

The number of trucks, currently 11, remains constant until one becomes unavailable because of breaking down or refueling its own tank. The fuel capacities of the trucks are 10,000 (10 trucks) or 8,000 (1 truck) gallons. The safety level to be refueled, currently 20%, is set by NAS Lemoore Fuels Division and remains constant. All available trucks can be used if there are enough drivers.

Penalties and Rewards

MSISCHE uses penalties and rewards to find the combined flight schedule's optimal solution. Penalties and rewards are assigned to squadrons based on their priority or proximity to being deployed. The penalty

Table 1. Ramps and Fuel Skids Assignments

Ramp	1	2	3	4	5
Hot Pits	1&2	3&4	5&6	7&8	9&10
Squadrons	VFA-122	VFA-14	VFA-146	VFA-2	VFA-137
		VFA-97	VFA-147	VFA-22	VFA-86
		VFA-25	VFA-192	VFA-154	
		VFA-113	VFA-151	VFA-41	
			VFA-94		

Figure 6. Ramp and Hot Pit Locations

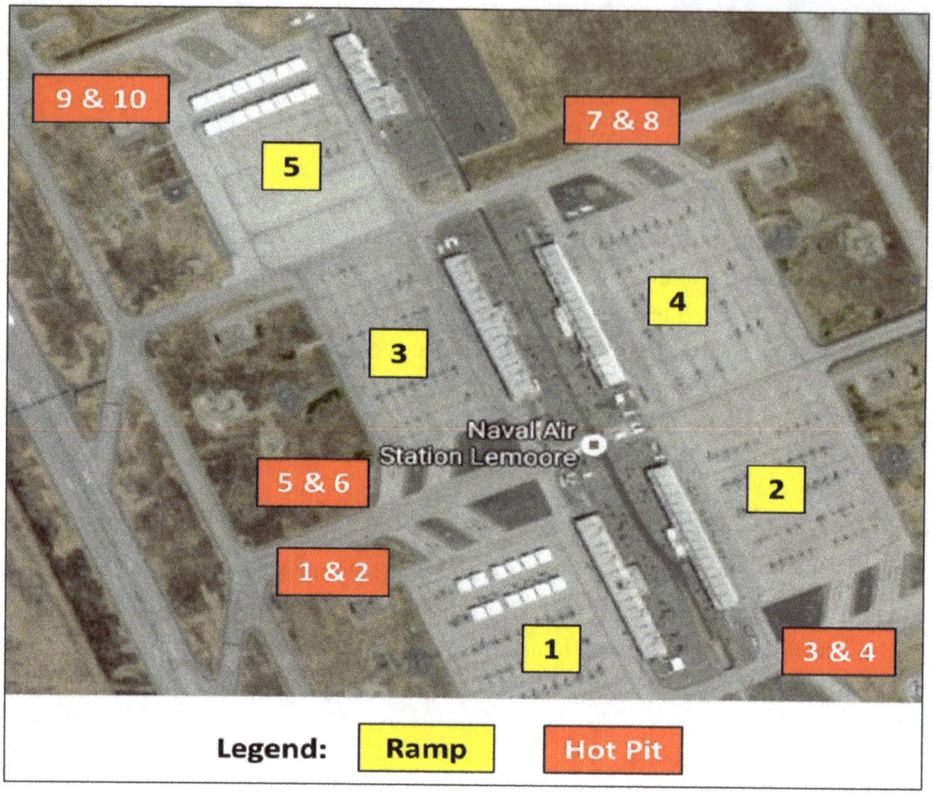

Figure 7. Distribution of Drivers per Hour

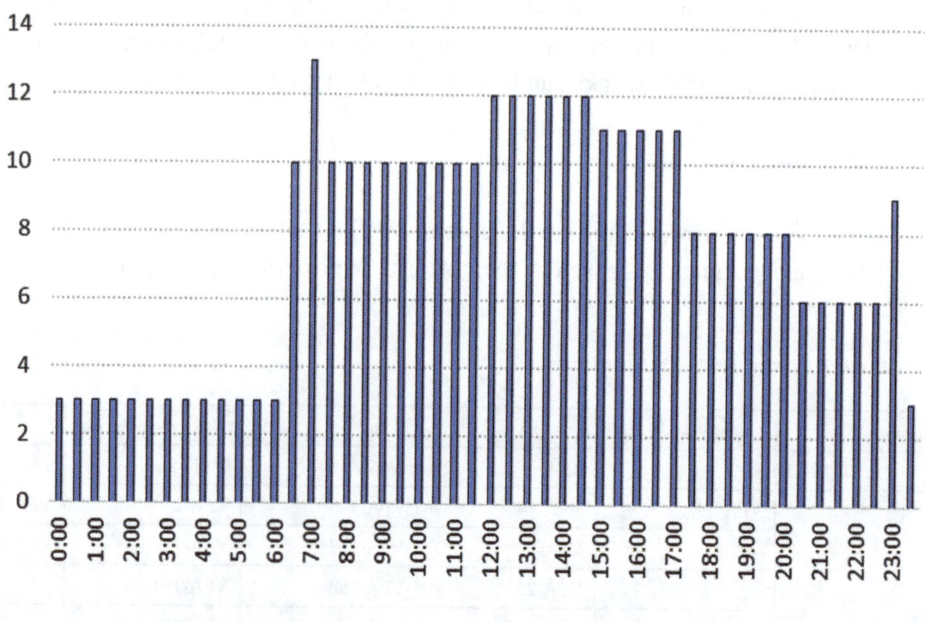

Table 2. Assignment of Drivers per Shift

Shift	1st Shift (0630-1500)	1st Shift Lake (0700-1530)	2nd Shift (1200-2030)	3rd Shift (1500-2330)	4th Shift (2300-0730)
Drivers	7	3	2	6	3

is used to weight the aircraft refueling wait time and the modification of the events' takeoff time. Table 3 displays the values assumed for priorities and penalties, in which VFA-122 has the highest values as the training fighter squadron.

Scenario Development

The scenarios are simulated using Microsoft Excel VBA code (Microsoft, 2013). The VBA scenario developer will be referenced as "Generator."

Assumptions

Besides the assumptions previously described in Section II.B.4, Generator assumes the following in order to produce a realistic scenario that provides the input data to the optimization model and heuristics:

- Some of NAS Lemoore fighter squadrons are deployed. To simulate this, the user defines the minimum of squadrons to fly, and Generator randomly selects a number of flying squadron between that minimum and the total of squadrons at NAS Lemoore.
- The training squadron, VFA-122, gets between 5 and 30 flight events daily. Other squadrons, selected to fly, have between 2 and 8 flight events. The number of flight events for each squadron is selected uniformly at random using these lower and upper bounds.
- The number of daily joint missions is approximately 5% of the total of flying events for that date.
- The number of events going to a joint mission could be 2 or 3, with 75 and 25 percent probability respectively.
- As in MSISCHE, squadrons receive deterministic priorities that are set by CSFWP.
- Assignment of MOA/ranges is based on the probability. Range R_2508 gets 50% of the events.

Table 3. Priorities and Penalties Assigned to Squadrons

Squadron	VFA-122	VFA-2	VFA-14	VFA-22	VFA-25	VFA-41	VFA-86	VFA-94
Priority	20	19	18	17	16	1	1	1
Penalty	10	5	5	5	5	1	1	1
Squadron	**VFA-97**	**VFA-113**	**VFA-137**	**VFA-146**	**VFA-147**	**VFA-151**	**VFA-154**	**VFA-192**
Priority	1	1	1	1	1	1	1	1
Penalty	1	1	1	1	1	1	1	1

- The length of an event, from its takeoff time to its landing time, follows a discrete uniform distribution of 15-minute increments and could be as little as 30 minutes or up to 2 hours.
- The number of aircraft, or sorties, going to an event follows a discrete uniform distribution and could be between 1 and 5.
- The probability of an aircraft going to fly again after being refueled is based on its landing time. The landing times are separated in six groups: before 1500, between 1500 and 1700, between 1700 and 1900, between 1900 and 2100, between 2100 and 2200, and after 2200. The probabilities for each group are 0.9, 0.7, 0.5, 0.3, 0.1, and 0, respectively. Whether or not an aircraft flies again is independent from aircraft to aircraft.
- Considering the monthly allowed hot refueling percentage, and using the assumed probabilities that an aircraft could be flying again after refueling, Generator designates if the aircraft will need hot or cold refueling.
- At most, one truck can be out of service and in maintenance on any day. If the scenario includes a breakdown, the truck is randomly selected out of the 11 trucks.
- Events' takeoff times are generated using a discrete uniform distribution of 15-minute increments from the time KNLC is open to an hour before KNLC is closed.

Scenario Generator

The workbook Generator creates each scenario by filling out the spreadsheets, exporting the spreadsheets as CSV files, executing MSISCHE's GAMS portion using the CSV files as input, and importing the results into a spreadsheet of the same workbook. Displayed in Figure 8 is the main screen of Generator. Besides allowing modification of parameter values to configure the scenarios, Generate also offers five choices to produce and run the scenarios: Generate Scenarios and Run MSISCHE (N), Generate Scenarios and Run MSISCHE (All), Generate Scenarios and Run Modified, Generate Scenarios & Run Current, and Generate Scenarios and Run all 3.

The first choice generates all the scenarios annotated under the Scenarios cell, but it only runs MSISCHE using the data of the last scenario. This option is useful to review the details of each event and each aircraft of the combined flight schedule solution. Figure 9 depicts a portion of the results of a scenario after using this choice. G T-Off and G Land are the takeoff and landing times generated-requested in the squadron's flight schedule, while M T-Off and M Land correspond to the takeoff and landing times recommended by MSISCHE. The Refuel cell indicates when the aircraft will refuel, having the minimum wait time after its landing.

The second choice creates all the scenarios requested under the Scenarios cell and runs MSISCHE using the data of each scenario. This option is useful to review the summary of the combined flight schedule solution. Figure 10 depicts the results after selecting this choice with several scenarios. The report includes number of events and aircraft given, flying, and landing in KNLC; it also summarizes, separately for hot and cold refueling, the total number of aircraft conducting each type of refueling, the number of aircraft waiting to be refueled, the total refueling wait time by all aircraft, and the maximum time an aircraft has to wait to be refueled. Additionally, this report presents other values related to MSISCHE's performance such as the percentage of hot refueling aircraft obtained, the model and solver status, the upper bound and found solutions, and the execution elapsed time.

Figure 8. Screenshot of Generate, the Main Sheet of Generator

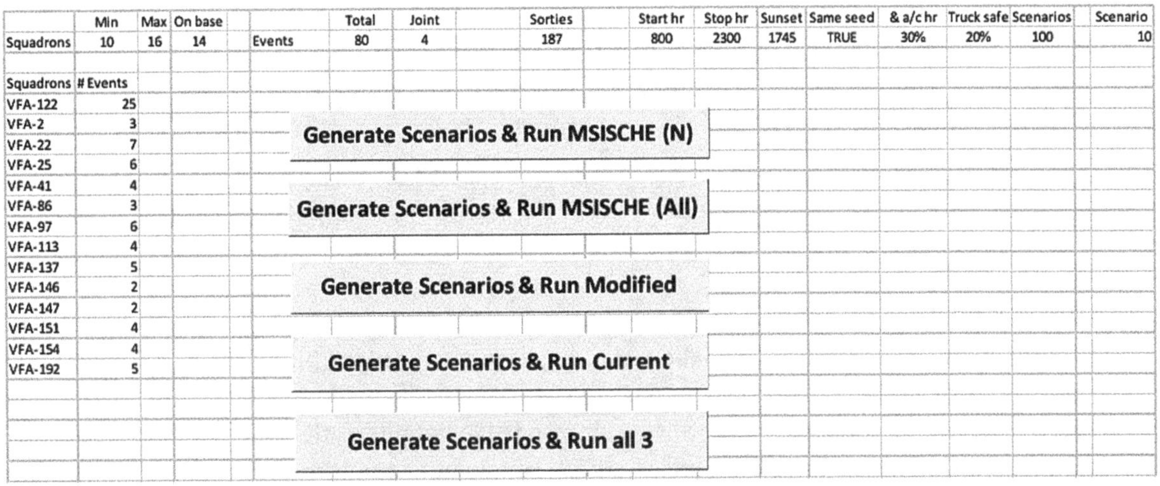

Figure 9. Screenshot of a Portion of the Scenario_Results Sheet from Generator

Event	Aircraft	Range	G T-Off	G Land	M T-Off	M Land	Refuel	Hot/Cold
V122-e1	V122-a1	R_2508	830	930	830	930	930	Hot
V122-e1	V122-a2	R_2508	830	930	830	930	930	Cold
V122-e2	V122-a7	FCLP	845	915	845	915	915	Cold
V122-e3	V122-a8	R_2508	800	845	800	845	845	Hot
V122-e4	V122-a9	FCLP	800	845	800	845	845	Cold
V122-e4	V122-a10	FCLP	800	845	800	845	845	Hot
V122-e5	V122-a11	W_283	815	900	815	900	915	Cold
V122-e5	V122-a12	W_283	815	900	815	900	900	Hot
V14-e1	V14-a1	Fallon_B17	945	1030	945	1030	1045	Hot
V14-e1	V14-a2	Fallon_B17	945	1030	945	1030	1145	Cold
V14-e2	V14-a3	Lemoore_A	930	1015	945	1030	1030	Hot
V14-e2	V14-a4	Lemoore_A	930	1015	945	1030	1115	Cold
V22-e2	V22-a5	R_2508	1445	1515	1445	1515	1515	Hot
V22-e2	V22-a6	R_2508	1445	1515	1445	1515	1700	Cold
V22-e3	V22-a7	FERRY	2215	2300	2230	2315	2330	Cold
V22-e3	V22-a8	FERRY	2215	2300	2230	2315	2330	Cold
V25-e1	V25-a2	FCLP	1245	1315	1245	1315	1315	Hot
V25-e1	V25-a3	FCLP	1245	1315	1245	1315	1345	Cold
V25-e2	V25-a4	Superior_Valley	1315	1400	1315	1400	1415	Cold
V25-e2	V25-a5	Superior_Valley	1315	1400	1315	1400	1430	Cold
V25-e3	V25-a7	W_291	1315	1400	1315	1400	1445	Cold

The module also implements two heuristic approaches: the "current refueling policy" and the "modified refueling policy." The current refueling policy is designed to mimic the current practice at NAS Lemoore, while the modified policy models a slight relaxation of the current policy which MSISCHE also models. Both heuristics handle takeoff requests in order of priority. The highest priority event is examined first given its requested takeoff time. As lower priority events are examined, they may be asked to wait for takeoff if they conflict with higher-priority events that are already assigned to the same MOA/range. Once the takeoff schedule is determined, the heuristics determine when refuelings will occur.

Figure 10. Screenshot of Summarized Results after Using the Second Choice of Generator

Scenario	Events Given	Events Flying	Events la KNLC	Sorties Given	Sorties Flying	Sorties la KNLC	Hot Ref Total AC	Hot Ref AC wait	Hot Ref Total Wait	Hot Ref Max AC Wait	Cold Ref Total AC	Cold Ref AC wait	Cold Ref Total Wait	Cold Ref Max AC Wait	hot ref %	Model Status	Solver Status	Best Possible	MIP Solution	Relative Gap	Elapsed Time
1	86	86	83	210	210	204	61	0	0	0	143	56	3000	240	29.90%	8	3	1946.95	1946.87	0.00%	3623.91
2	68	68	62	158	158	147	44	0	0	0	103	31	1800	180	29.93%	1	1	1271.95	1271.95	0.00%	49.53
3	52	52	51	121	121	120	36	0	0	0	84	10	195	30	30.00%	1	1	1045.98	1045.98	0.00%	29.83
4	57	57	56	112	112	109	32	0	0	0	77	21	720	90	29.36%	1	1	1177.81	1177.81	0.02%	25.79
5	63	63	60	136	136	129	38	0	0	0	91	25	570	45	29.46%	1	1	974	974	0.00%	51.53
6	71	71	67	156	156	144	43	0	0	0	101	32	1905	150	29.86%	1	1	1350.82	1350.82	0.01%	67.3
7	54	54	54	122	122	122	36	0	0	0	86	32	2085	165	29.51%	1	1	875.51	875.51	0.06%	79.75
8	78	78	77	160	160	156	46	0	0	0	110	31	1080	105	29.49%	1	1	1484.9	1484.9	0.01%	139.47
9	76	76	71	167	167	158	47	0	0	0	111	43	3945	255	29.75%	1	1	1359.17	1359.17	0.01%	108.77
10	80	80	76	187	187	178	53	0	0	0	125	55	2430	195	29.78%	8	3	1813.59	1813.59	0.02%	3619.5

Although takeoffs are assigned according to priority, landings are done on a first-come, first-served basis. In the current policy, each aircraft is only allowed to use its preassigned hot pit; whereas in the modified policy, all aircraft are allowed to use any hot pit, as in MSISCHE. Under both policies, aircraft arriving for refueling enter a queue and are assigned to their respective hot pits as the pits become available. A pit may be unavailable because it is currently occupied by another aircraft or because no truck is available to service it. Both heuristics model the amount of fuel available in the trucks as in MSISCHE. For the current policy, each hot skid has its own queue. For the modified policy, there is only a single queue for all hot pits. The third and fourth choices of Generator produce all the scenarios annotated under the Scenarios cell and run the VBA code that simulates the modified and current aircraft refueling process at NAS Lemoore. Figures 11 and 12 show screenshots with the results for heuristics after selecting the third and fourth choices respectively.

The fifth choice calls choices two, three, and four; therefore, after generating all the scenarios requested, this choice executes and summarizes the results of using MSISCHE, the VBA code with modified refueling policy, and the VBA code with current refueling policy.

Sensitivity Analysis

To conduct sensitivity analysis, we evaluated the results of 400 scenarios with the fifth choice of Generator; thus, under the three main executions of Generator: using MSISCHE, using the VBA code to simulate the current refueling policy, and using the VBA to simulate the modified refueling policy. These 400 scenarios were distributed as follows: 100 scenarios use the current setup of squadrons and refueling

Figure 11. Screenshot of Summarized Results for a Heuristic Solution after Using the Third Choice of Generator

Scenario	Events Given	Events la KNLC	Sorties Given	Sorties la KNLC	Hot Ref Total AC	Hot Ref AC waiting	Hot Ref Total Wait (m)	Hot Ref Max AC Wait (m)	Cold Ref Total AC	Cold Ref AC waiting	Cold Ref Total Wait (m)	Cold Ref Max AC Wait (m)	hot ref %
1	86	82	209	195	58	0	0	0	137	70	2205	75	29.74%
2	72	72	172	172	51	7	105	15	121	61	1950	75	29.65%
3	79	75	187	175	52	9	165	30	123	37	1155	60	29.71%
4	56	52	126	117	35	0	0	0	82	17	330	30	29.91%
5	65	61	161	152	45	2	30	15	107	33	570	30	29.61%
6	65	61	146	136	40	6	90	15	96	68	3030	75	29.41%
7	63	61	142	138	41	6	90	15	97	51	2355	75	29.71%
8	59	58	134	133	39	0	0	0	94	10	180	30	29.32%
9	65	63	152	147	44	0	0	0	103	10	210	30	29.93%
10	77	74	181	171	51	2	30	15	120	21	390	30	29.82%
11	68	65	164	160	48	7	105	15	112	70	2280	60	30.00%
12	81	79	203	199	59	2	30	15	140	91	3525	75	29.65%

Figure 12. Screenshot of Summarized Results for a Heuristic Solution after Using the Fourth Choice of Generator

Scenario	Events Given	Events la KNLC	Sorties Given	Sorties la KNLC	Hot Ref Total AC	Hot Ref AC waiting	Hot Ref Total Wait (m)	Hot Ref Max AC Wait (m)	Cold Ref Total AC	Cold Ref AC waiting	Cold Ref Total Wait (m)	Cold Ref Max AC Wait (m)	hot ref %
1	86	82	209	195	58	23	1185	105	137	118	6765	165	29.74%
2	72	72	172	172	51	14	450	60	121	92	7080	180	29.65%
3	79	75	187	175	52	11	390	60	123	107	7980	225	29.71%
4	56	52	126	117	35	13	435	60	82	58	2445	135	29.91%
5	65	61	161	152	45	10	240	60	107	76	3420	105	29.61%
6	65	61	146	136	40	15	675	75	96	73	6345	195	29.41%
7	63	61	142	138	41	8	210	45	97	72	4410	135	29.71%
8	59	58	134	133	39	4	75	30	94	59	2565	135	29.32%
9	65	63	152	147	44	10	345	60	103	57	1830	180	29.93%
10	77	74	181	171	51	18	420	45	120	87	4905	240	29.82%
11	68	65	164	160	48	13	420	60	112	89	8445	240	30.00%
12	81	79	203	199	59	15	330	45	140	122	9990	225	29.65%

resources, and assumptions previously stated in Section II.B.4; 100 scenarios utilize the same setup but add a squadron; 100 scenarios use the original setup but with one less truck; and 100 scenarios utilize the same basic setup but with one less driver. Even though an optimization model provides the value of its objective function, the sensitivity analysis was focused on figures of merit such as maximum and average aircraft refueling wait time, number of aircraft waiting to be refueled, and frequency of refueling wait time, all measured during both hot and cold refueling. Each of our scenarios applied to MSISCHE was generated and computed using a Dell, Precision T7910 computer with two Intel® Xenon® CPU E5-2699 @2.3GHz processors, 128 GB of RAM installed and running the Windows 7 Professional operating system (Microsoft, 2009). The program software used for the optimization is GAMS 24.4.2 utilizing CPLEX solver (GAMS, 2015). With these conditions our model typically contains approximately from 45,397 to 113,674 constraints and from 122,911 to 346,881 decision variables, of which 89,085 to 923,222 decision variables are integer. Solution times vary for each scenario; Figure 13 depicts a histogram with the time used to solve a scenario using MSISCHE with five and zero percent optimality gaps. Figure 13 and subsequent figures are developed using Microsoft Excel (Microsoft, 2013).

Using Current Squadrons and Refueling Resources

Figure 14 presents the comparisons, in regards to hot refueling wait time, among the three executions of Generator after running the 100 scenarios produced using the current squadrons, current refueling resources, and assumptions. Figure 15 displays a comparison of histograms with the frequency of waiting time that aircraft experience before receiving hot refueling over all scenarios.

The following histogram compares the hot refueling wait time when aircraft follow the three alternatives of flight and refueling schedules: flight and refueling schedules obtained from MSISCHE, flight schedule generated-requested by Generator and refueling schedule per current refueling policy, and flight schedule generated-requested by Generator and refueling schedule per modified refueling policy.

The previous figures indicate a large difference in values during hot refueling between aircraft that land per the generated-requested flight schedules and refuel according to the current and modified refueling policies against aircraft that land and refuel according to MSISCHE's schedule. Additionally, these graphs show that, using the generated-requested flight schedules, their values improve by just using the modified instead of the current refueling policy; however, those values improve even more when aircraft

Figure 13. Frequency of Solution Times Used by MSISCHE over All Scenarios with 0% and 5% Optimality Gaps

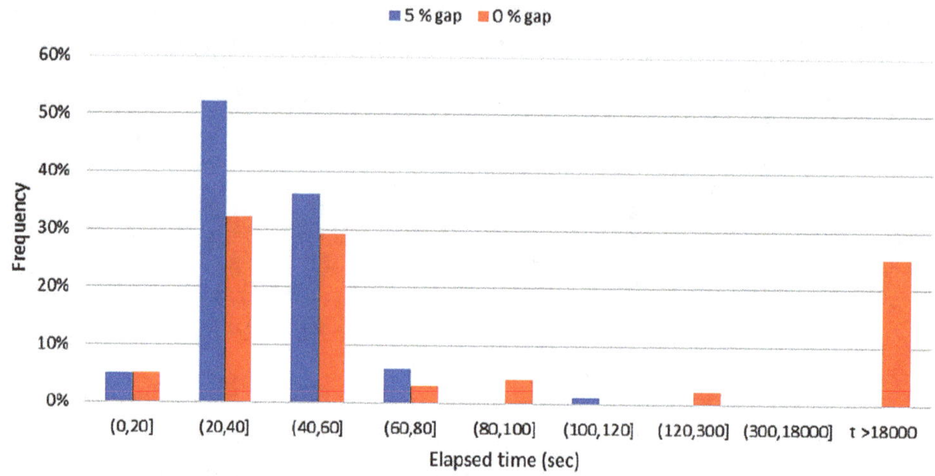

follow MSISCHE's schedule. The values of maximum aircraft refueling wait time, average aircraft refueling wait time, and number of aircraft waiting for hot refueling is small in most of the scenarios when following MSISCHE's solution. On average, these values were reduced by 96.9%, 99.5%, and 99.1%, respectively, when using MSISCHE instead of the current refueling policy. Figure 14 indicates that, in some scenarios, the maximum aircraft refueling wait time following MSISCHE might be higher than when using the modified refueling policy, but the average wait time and number of aircraft waiting are lower using MSISCHE. This happens because the objective value of MSISCHE minimizes the waiting time over all aircraft, not just for one. Since hot refueling wait time is more expensive than cold, the main objective of MSISCHE is to minimize the hot refueling wait time. A secondary objective of MSISCHE is to minimize the cold refueling wait time without jeopardizing the hot refueling wait time. Similar results for cold refueling appear in Figures 16 and 17.

The following histogram compares the cold refueling wait time when aircraft follow the three variants of flight and refueling schedules.

The previous two figures indicate the differences in values during cold refueling between aircraft that land based on the generated-requested flight schedules and refuel according to the current and modified refueling policies against aircraft that land and refuel according to MSISCHE's schedule. Additionally, and following the same behavior with the hot refueling comparison, these graphs show that their values improve simply by using the modified instead of the current refueling policy; however, those values improve even more when following MSISCHE's schedule. On average, the values of maximum aircraft refueling wait time, average aircraft refueling wait time, and number of aircraft waiting for cold refueling improved by 62.7%, 90.1%, and 82.2%, respectively, when using MSISCHE rather than the current refueling policy. The values of average aircraft refueling wait time and number of aircraft waiting for cold refueling are lower in all the scenarios by following MSISCHE; however, the maximum aircraft refueling wait time, in some scenarios, might appear higher when using MSISCHE instead of the modified policy. This is explained by the objective value of MSISCHE which minimizes the waiting time over all aircraft. Figure 18, from Gerber and Clark (2013), illustrates the number of aircraft arriving by day of the week and by time of the day. The peaks during certain hours, which can represent more than

Figure 14. Maximum Wait Time, Average Wait Time, and Number of Aircraft Waiting to Conduct Hot Refueling in 100 Scenarios with Current Squadrons and Refueling Resources

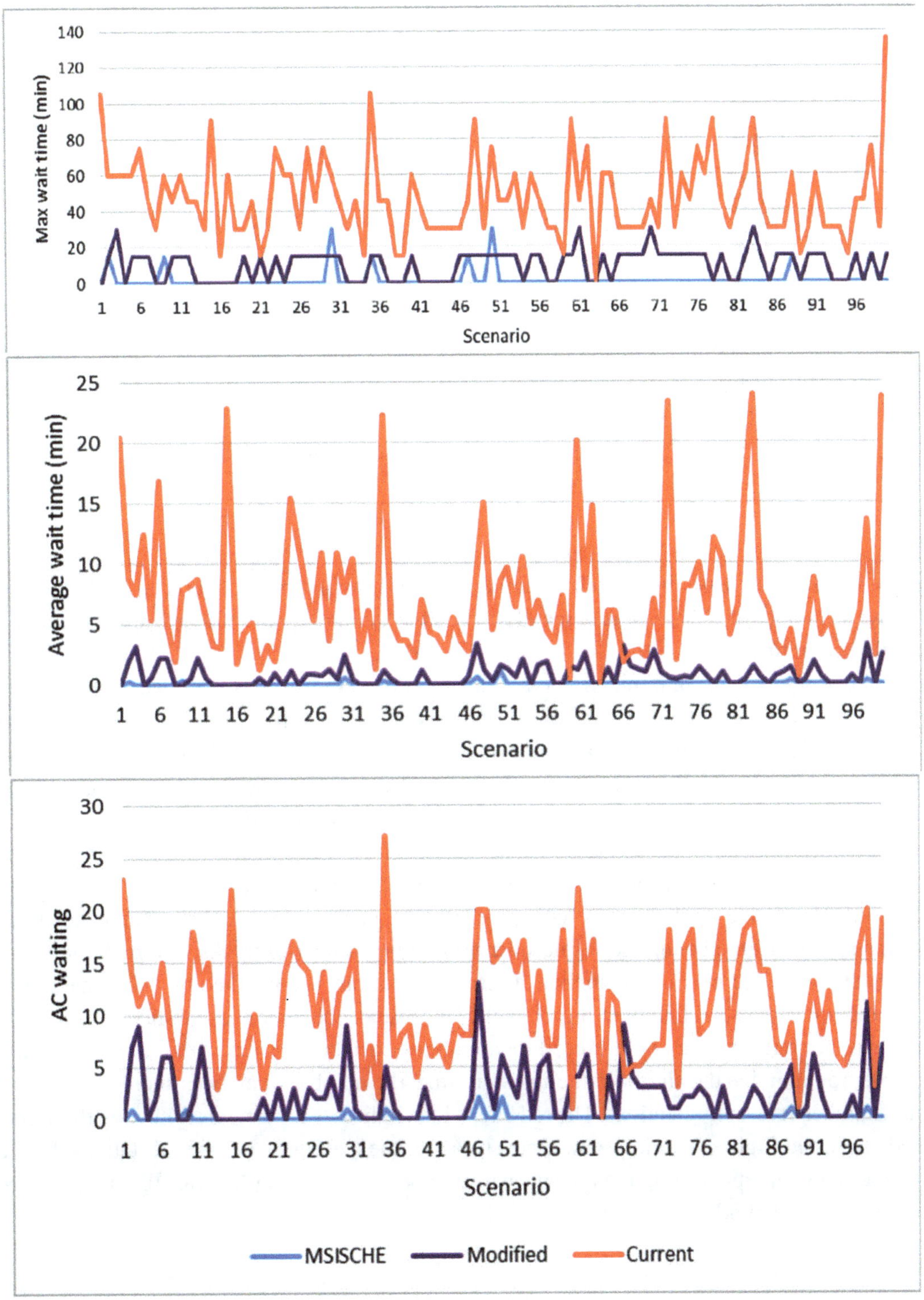

Figure 15. Frequency of Hot Refueling Wait Time over 100 Scenarios with Current Squadrons and Refueling Resources

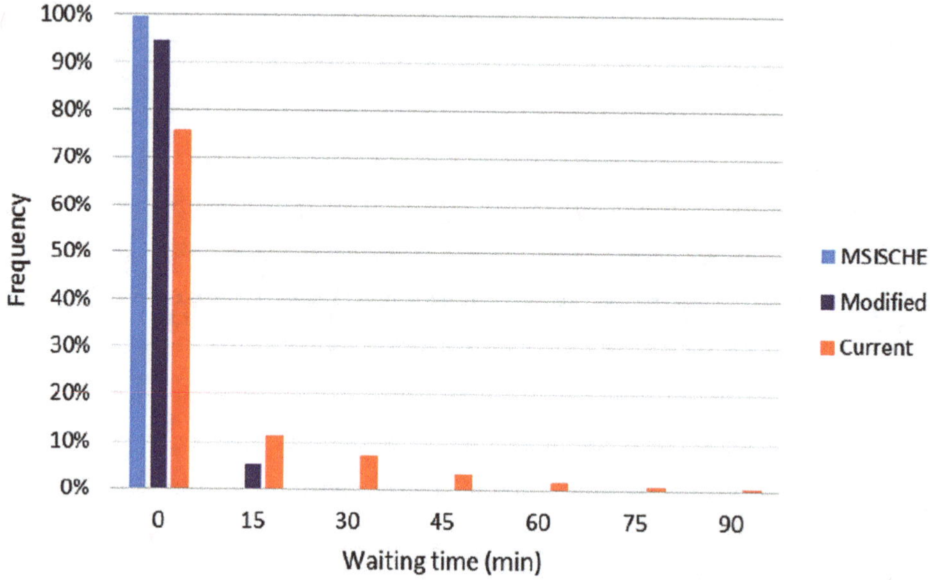

15 arrivals, certainly increase the aircraft refueling demand. Having multiple arrivals at the same time is fine; however, sometimes these peak demands exceed the availability of refueling resources and cause high refueling wait time.

MSISCHE's objective of minimizing the aircraft refueling wait time is obtained by leveraging the number of aircraft arrivals throughout the day. Comparisons of the number and time of aircraft arrivals are displayed in Figures 19 and 20. For a randomly selected scenario, Figure 19 shows the difference of aircraft arriving when they follow the generated-requested flight schedules provided from the scenario against aircraft arriving per the flight schedules obtained from MSISCHE.

For a randomly selected scenario, Figure 20 displays a radar-type comparison of the number and time of aircraft landing when aircraft follow the flight schedules from the scenario and the flight schedules obtained from MSISCHE against the refueling resources per time period. This graph illustrates the aircraft arrivals against the boundaries of number of drivers, number of trucks, and maximum number of refueling stations. The maximum number of refueling stations is computed by using the number of drivers, number of trucks, and number of hot skids. Because of the double capability of each hot skid, the number of aircraft arriving could exceed the number of drivers or trucks at any time period; however, in order to minimize the aircraft refueling wait time, the number of arrivals should not exceed the maximum number of refueling stations. The red line, which indicates the arrivals following the scenario's generated-requested flight schedules, sometimes surpasses the maximum number of stations' boundary. The blue dashed line, representing the arrivals per time period following MSISCHE's flight schedule, does not exceed this boundary.

Figure 16. Maximum Wait Time, Average Wait Time, and Number of Aircraft Waiting to Conduct Cold Refueling in 100 Scenarios with Current Squadrons and Refueling Resources

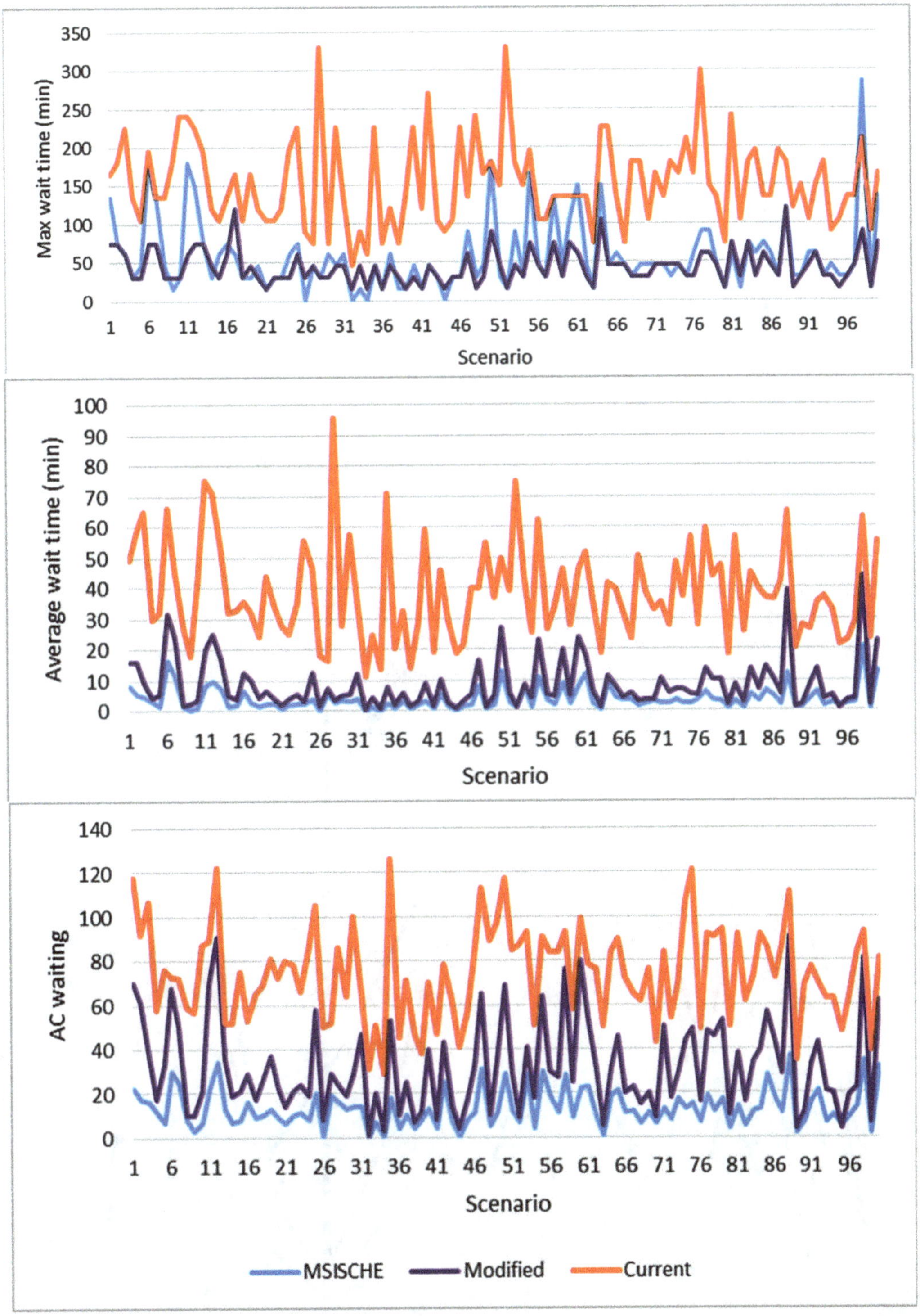

Figure 17. Frequency of Cold Refueling Wait Time over 100 Scenarios with Current Squadrons and Refueling Resources

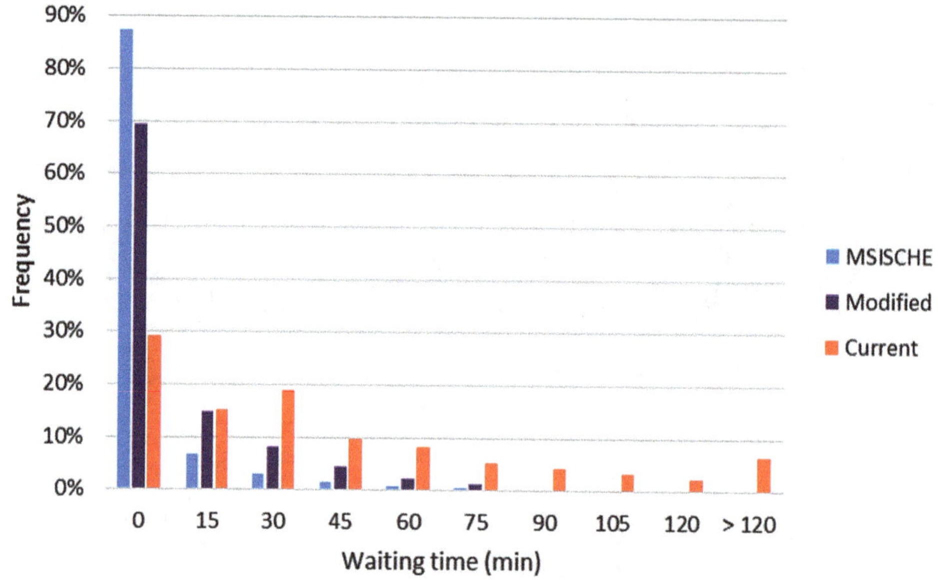

Figure 18. Daily Aircraft Arrival Patterns
Source: Gerber & Clark (2013).

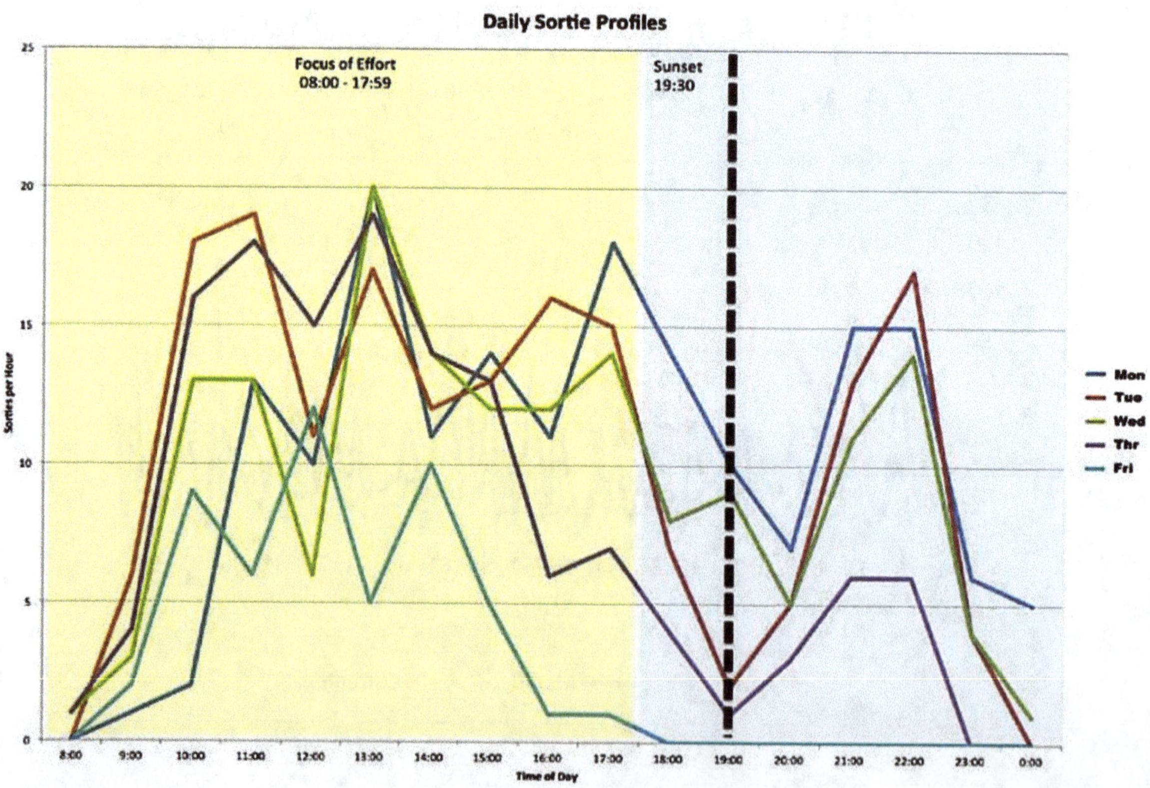

Figure 19. Number of Aircraft Landing during a Scenario when Aircraft Take Off at the Requested Time and Take Off as Recommended by MSISCHE

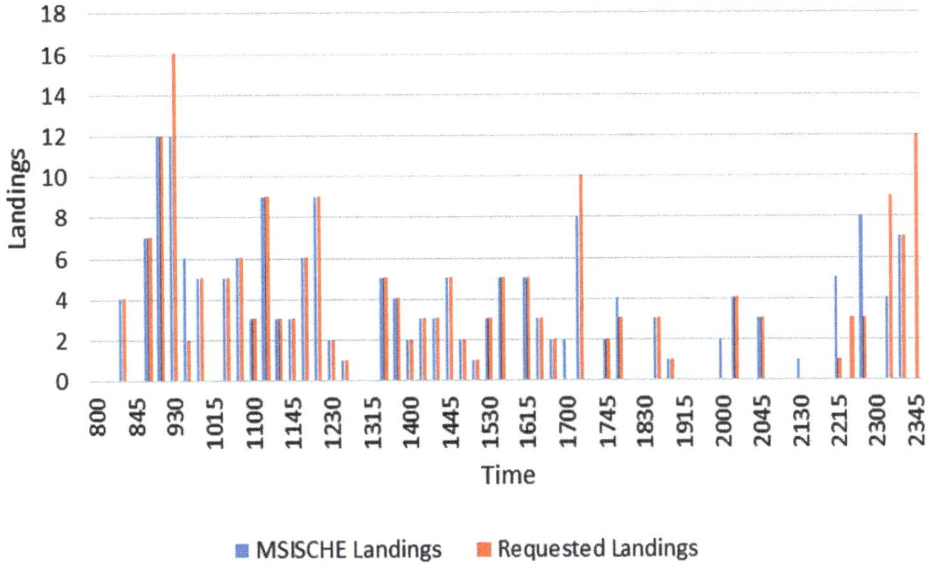

In order to increase the aircraft flying time available while preserving squadrons' authority and accountability, MSISCHE's optimal solution corresponds to a balance between the minimization of refueling wait time and minimization of deviation from the squadrons' proposed schedules. Figure 21 displays the percentage of events that changed their takeoff time in each scenario. On average, 97% of the squadron's requested flying times are followed. When MSISCHE suggests a change from the request, it is with the knowledge that the change is feasible based on MOA/range availability, the change is minimal, and the requests from the highest priority squadrons are protected.

Adding One Squadron to the Original Setup

To simulate the arrival of the JSF to NAS Lemoore, a new squadron VFA-999 was added to the list of Lemoore home-based fighter squadrons. Per Lemoore Fuels Division, even though a JSF holds more fuel than an F-18, the average rate of fuel transfer to the JSF is about the same as that to an F-18; thus, refueling a JSF is expected to take longer. Figures 22 and 23 present the comparisons, in regards to hot refueling wait time, among the three executions of Generator with the 100 scenarios produced after adding the JSF squadron. The UnA column in Figure 23 represents the 0.1% of unassigned sorties in MSISCHE's solution. Unassigned sorties occur because refueling demand cannot be adequately smoothed using at most a two-hour schedule shift in either direction to avoid refueling delays. These unassigned sorties reflect instances in which CSFWP cannot achieve the goal of short refueling wait times while also preserving a large degree of squadron-level autonomy. The two heuristic approaches do not result in unassigned aircraft because they allow arbitrarily long delays. However, in practice, refueling delays can cause cancellations if the aircraft is scheduled for another event before refueling can be completed; this is not reflected in our model. Figure 22 only displays data pertaining to the aircraft that actually fly

Figure 20. Aircraft Landings When Aircraft Follow Takeoff from the Requested Flight Schedule and Takeoff Recommended by MSISCHE against Drivers, Trucks, and Maximum Refueling Stations per Time Period

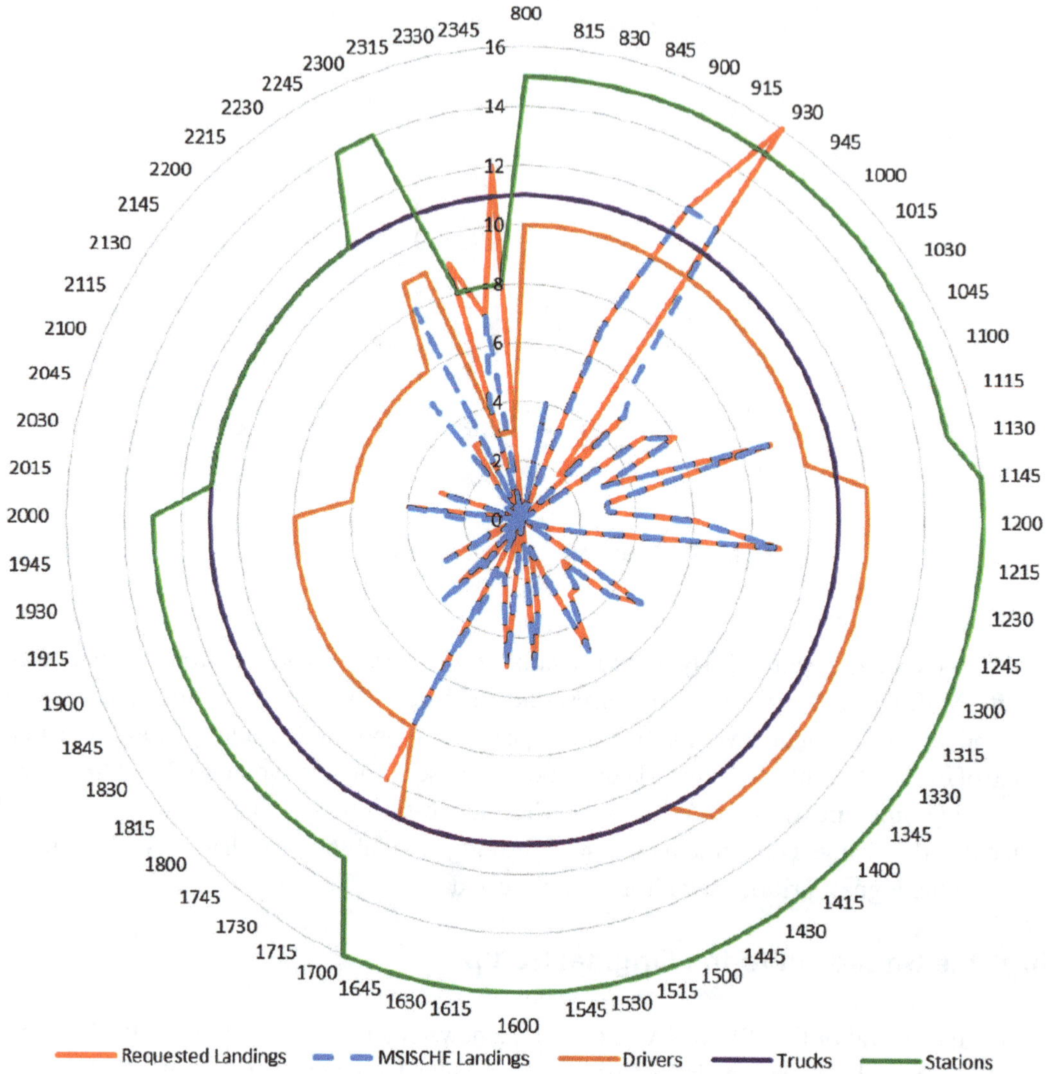

in each of the three executions (MSISCHE, modified refueling policy, and current refueling policy) and do not reflect the flights that are unassigned by MSISCHE.

The previous two figures indicate that the results of conducting hot refueling with the added squadron generally mirror those from conducting hot refueling with the baseline. On average, the value of average aircraft refueling wait time increased by 5% after adding the squadron; however, by using MSISCHE, this value is reduced by 99.2% for those events scheduled by MSISCHE. On average, the values of maximum aircraft refueling wait time and number of aircraft waiting for hot refueling, for those events scheduled by MSISCHE, improved by 95.8% and 98.8%, respectively, when following MSISCHE instead of the current refueling policy. Similar results for cold refueling appear in Figures 24 and 25. Again, the UnA column represents the 0.1% of unassigned sorties or aircraft not included in MSISCHE's solution.

Figure 21. Percentage of Events Changed per Scenario

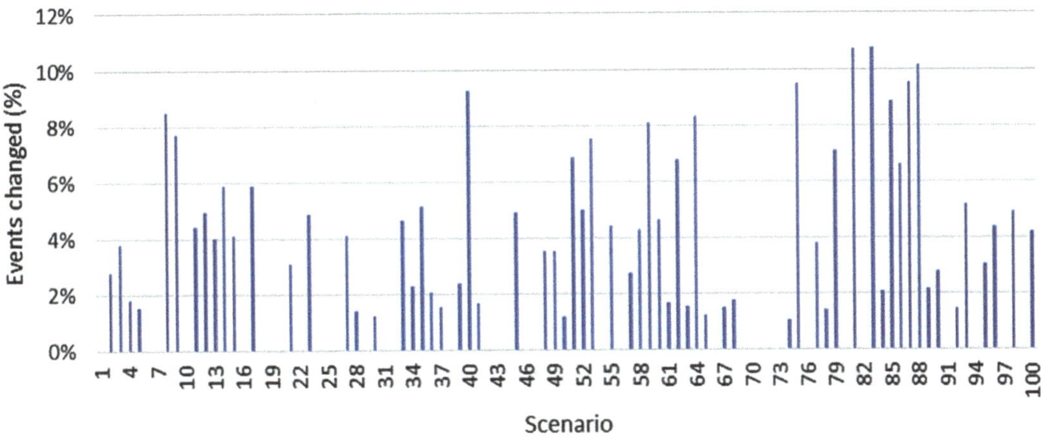

Figure 24 only displays data pertaining to the aircraft that actually fly in each of the three executions (MSISCHE, modified refueling policy, and current refueling policy).

Figure 24 and Figure 25 show that the results of conducting cold refueling with the added squadron in general mirror those from conducting cold refueling with the baseline. On average, the values of maximum aircraft refueling wait time, average aircraft refueling wait time, and number of aircraft waiting for cold refueling, for those events scheduled by MSISCHE, improved by 62.7%, 91.2%, and 85.1%, respectively, when following MSISCHE in place of the current refueling policy. About four percent of the 100 scenarios did not include every requested event in MSISCHE's optimal solution, resulting on 0.1% of requested sorties not being scheduled over all scenarios. Therefore, to keep a balanced solution, in which high peak demands do not supersede the available refueling resources, adding a squadron requires the addition of more refueling resources or decrease the number of flying events per squadron.

Adding Truck Breakdowns to the Original Setup

In order to conduct sensitivity analysis of having a truck breakdown, Generator randomly selected a truck to be out of service and in maintenance for a specific length of time. The times to start and finish maintenance are randomly chosen between 0000 to 0800 and 1800 to 2400, respectively by using a discrete uniform distribution of 15- minute increments. Only one truck can be in maintenance during a day, and it could refuel before going back to service. Figures 26 and 27 present the comparisons, in regards to hot refueling wait time, among the three executions of Generator with the 100 scenarios produced while having a truck in maintenance. Again, the UnA column represents the 0.1% of unassigned sorties in MSISCHE's solution. Referring to our previous statement in Section IV.C.2, unassigned sorties occur because refueling demand cannot be adequately smoothed using at most a two-hour schedule shift in either direction to avoid refueling delays. Figure 26 only displays data pertaining to the aircraft that actually fly in each of the three executions (MSISCHE, modified refueling policy, and current refueling policy) and do not reflect the flights that are unassigned by MSISCHE.

Figure 22. Maximum Wait Time, Average Wait Time, and Number of Aircraft Waiting to Conduct Hot Refueling in 100 Scenarios When Adding a Squadron to the Original Setup

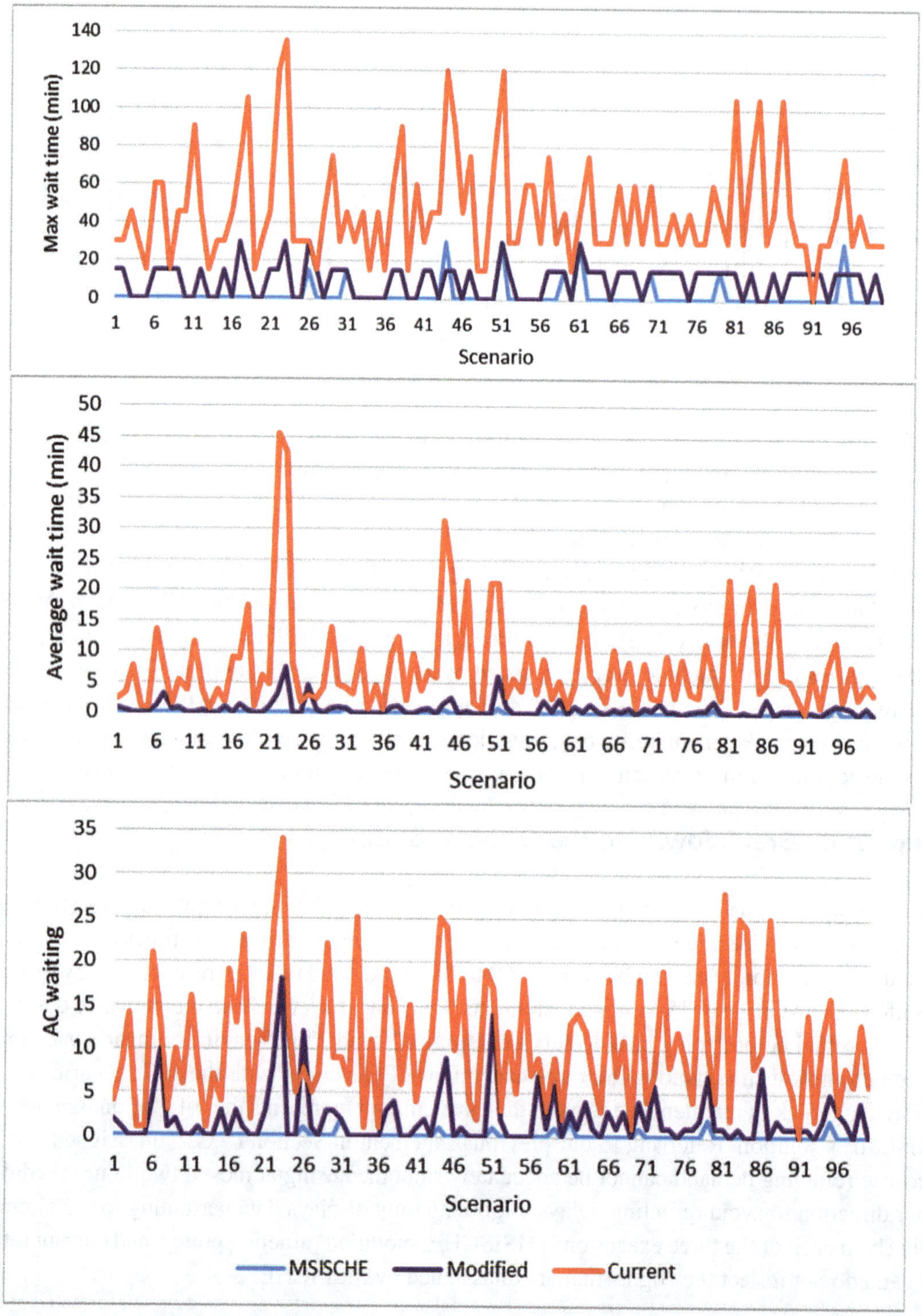

Figure 23. Frequency of Hot Refueling Wait Time over 100 Scenarios When Adding a Squadron to the Original Setup (The far-right column represents the percentage of unassigned aircraft)

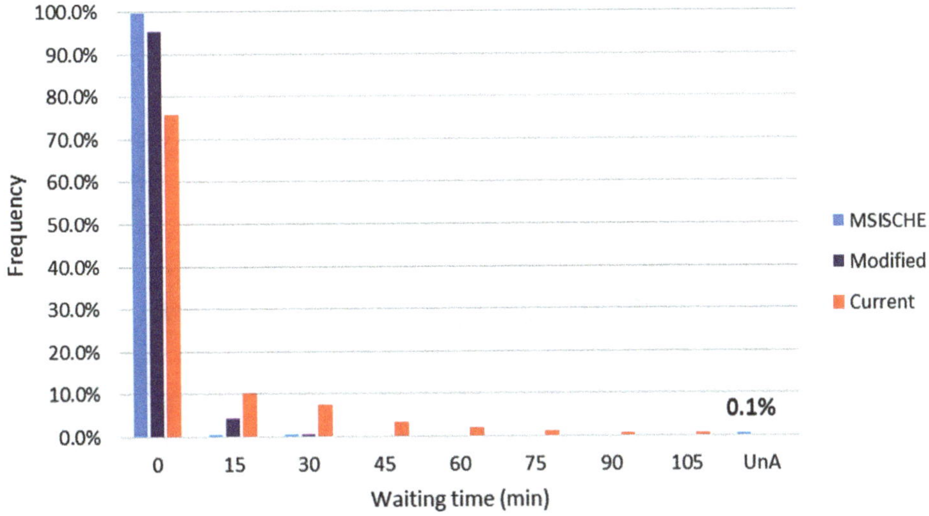

The previous two figures indicate that the results of conducting hot refueling while including a truck breakdown generally behave like those from conducting hot refueling with the baseline. On average, the value of average aircraft refueling wait time increased by 6% after including the truck breakdown; however, by using MSISCHE, this value is reduced by 99.2% for those events scheduled by MSISCHE. On average, the values of maximum aircraft refueling wait time and number of aircraft waiting for hot refueling, for those events scheduled by MSISCHE, improved by 95.1% and 98.6%, respectively, when following MSISCHE instead of the current refueling policy. Similar results for cold refueling appear in Figures 28 and 29. The UnA column represents the 0.1% of unassigned sorties or aircraft not included in MSISCHE's solution. Figure 28 only displays data pertaining to the aircraft that actually fly in each of the three executions (MSISCHE, modified refueling policy, and current refueling policy).

The previous two figures show that the results of conducting cold refueling while including a truck breakdown in general behave like those from conducting cold refueling with the baseline. On average, the values of maximum aircraft refueling wait time, average aircraft refueling wait time, and number of aircraft waiting for cold refueling, for those events scheduled by MSISCHE, improved by 61.1%, 88.9%, and 81.6%, respectively, when following MSISCHE rather than the current refueling policy. About three percent of the 100 scenarios did not include every requested event in MSISCHE's optimal solution, resulting in 0.1% of requested sorties not being scheduled over all scenarios. Therefore, we do not recommend decreasing the number of fuel trucks; in other words, a fuel truck must be replaced before it goes to maintenance. Otherwise, to keep a balanced solution, in which high peak demands do not supersede the available refueling resources, it will be necessary to decrease the number of flying events per squadron.

Figure 24. Maximum Wait Time, Average Wait Time, and Number of Aircraft Waiting to Conduct Cold Refueling in 100 Scenarios When Adding a Squadron to the Original Setup

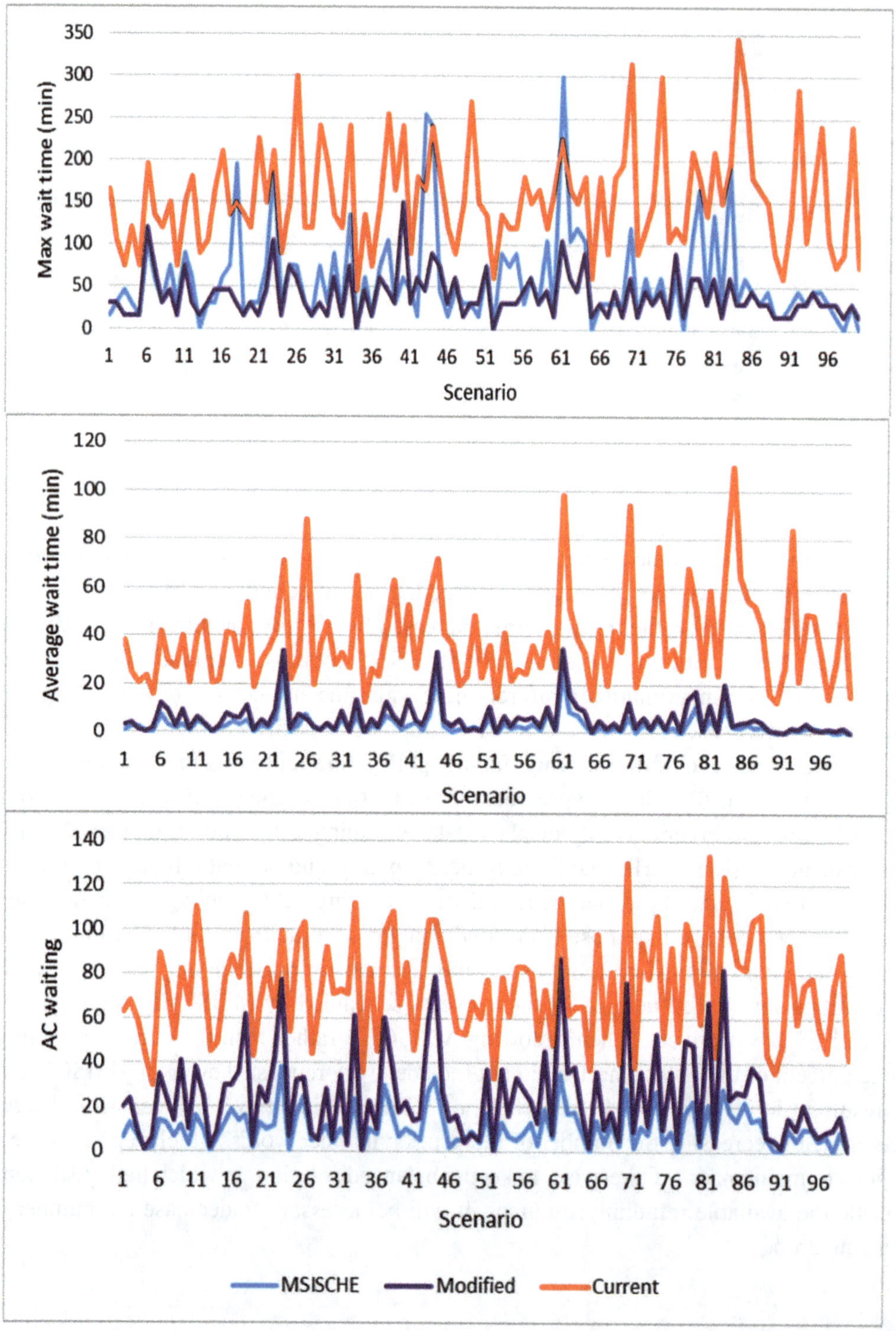

Figure 25. Frequency of Cold Refueling Wait Time over 100 Scenarios When Adding a Squadron to the Original Setup (The far-right column represents the percentage of unassigned aircraft)

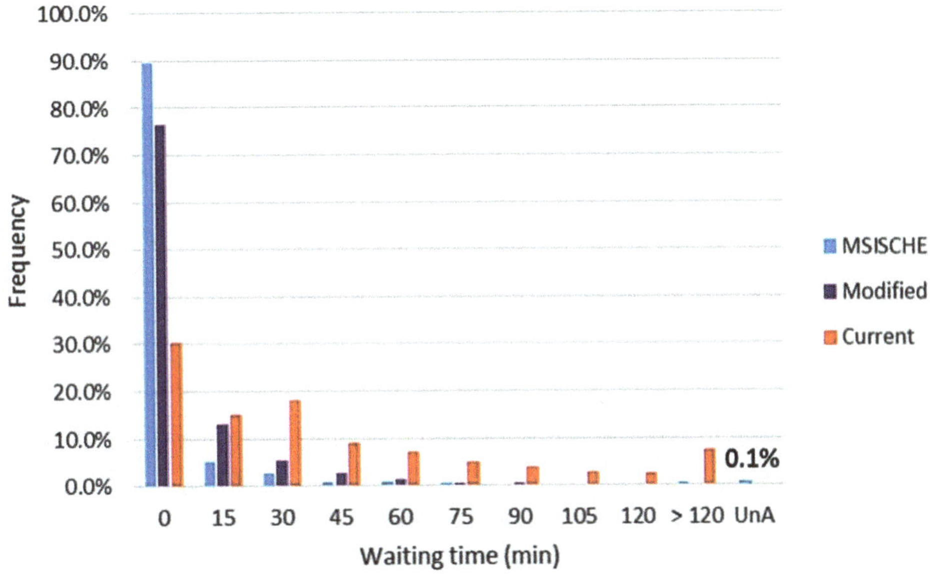

Including Reduction in Fuel Truck Driver Workforce to the Original Setup

Sensitivity analysis of reducing the fuel truck driver workforce by one driver was done by randomly selecting one of the four drivers' shifts to have one fewer driver. Each of the 100 scenarios had its shift selected independently.

Figures 30 and 31 present the comparisons, in regards to hot refueling wait time, among the three executions of Generator with the 100 scenarios produced while removing one driver from the workforce. The UnA column represents the 0.4% of unassigned sorties in MSISCHE's solution. As previously mentioned in Section IV.C.2, unassigned sorties occur because refueling demand cannot be adequately smoothed using at most a two-hour schedule shift in either direction to avoid refueling delays. Figure 30 only displays data pertaining to the aircraft that actually fly in each of the three executions (MSISCHE, modified refueling policy, and current refueling policy) and do not reflect the flights that are unassigned by MSISCHE.

The previous two figures indicate that the results of conducting hot refueling while including a reduction on the fuel truck driver workforce generally mirror those from conducting hot refueling with the baseline. On average, the value of average aircraft refueling wait time increased by 1% after including the workforce's reduction; however, by using MSISCHE, this value is reduced by 99.7% for those events scheduled by MSISCHE. On average, the values of maximum aircraft refueling wait time and number of aircraft waiting for hot refueling, for those events scheduled by MSISCHE, improved by 98.4% and 99.4%, respectively, when following MSISCHE instead of the current refueling policy. Similar results

Figure 26. Maximum Wait Time, Average Wait Time, and Number of Aircraft Waiting to Conduct Hot Refueling in 100 Scenarios When Adding a Truck Breakdown to the Original Setup

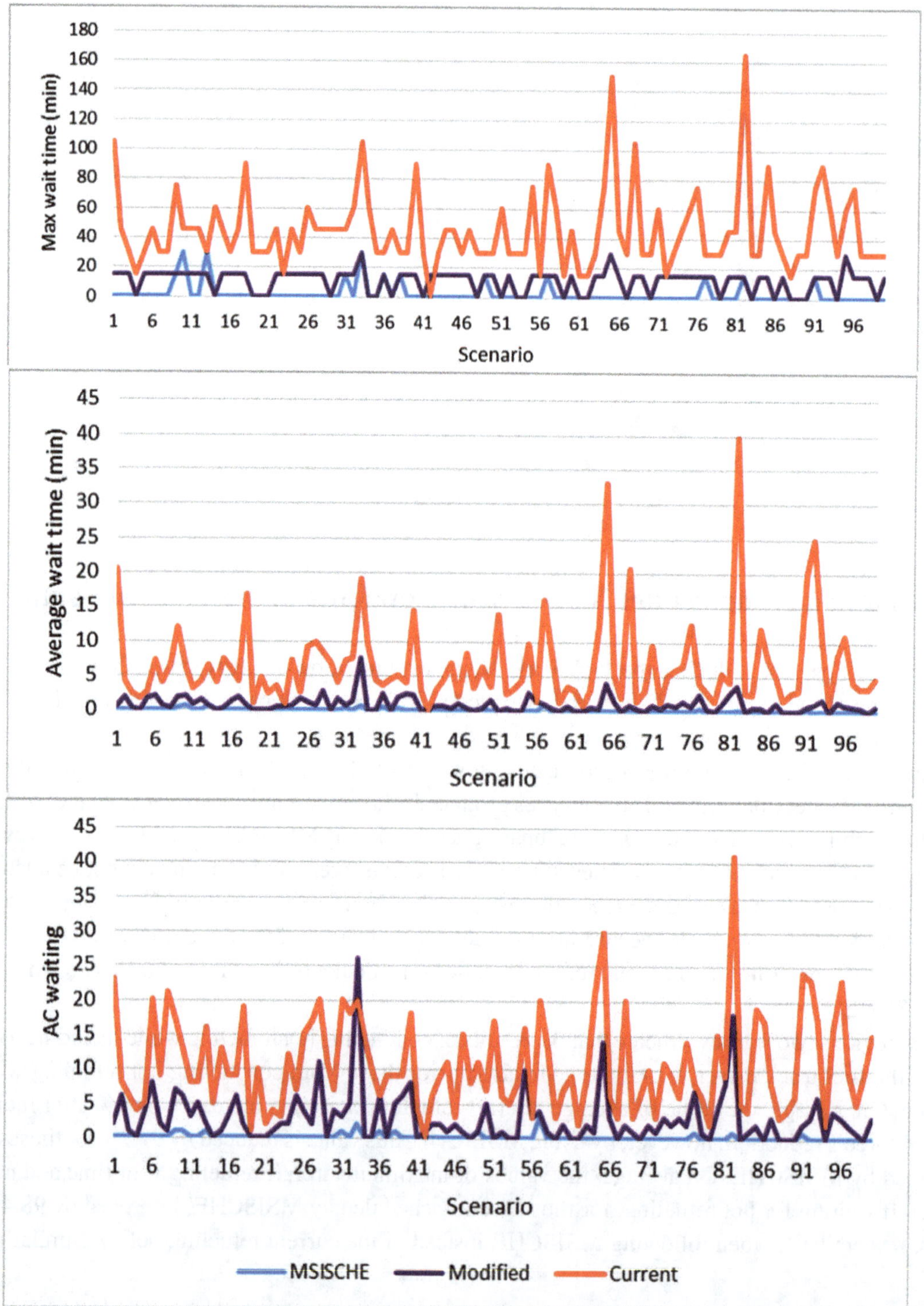

Figure 27. Frequency of Hot Refueling Wait Time Over 100 Scenarios When Adding a Truck Breakdown to the Original Setup (The far-right column represents the percentage of unassigned aircraft)

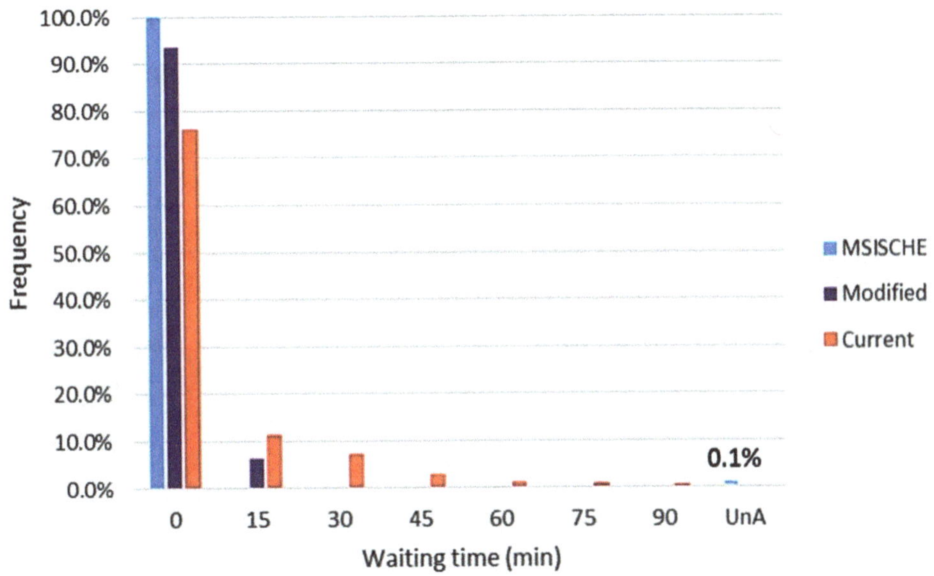

for cold refueling appear in Figures 32 and 33. The UnA column represents the 0.1% of unassigned sorties or aircraft not included in MSISCHE's solution. Figure 32 only displays data pertaining to the aircraft that actually fly in each of the three executions (MSISCHE, modified refueling policy, and current refueling policy).

The previous two figures show that the results of conducting cold refueling while including a reduction on the fuel truck driver workforce in general mirror those from conducting cold refueling with the baseline. On average, the values of maximum aircraft refueling wait time, average aircraft refueling wait time, and number of aircraft waiting for cold refueling, for those events scheduled by MSISCHE, improved by 68.8%, 92.5%, and 84.2%, respectively, when following MSISCHE rather than the current refueling policy. About six percent of the 100 scenarios did not include every requested event in MSISCHE's optimal solution, resulting on 0.2% of requested sorties not being scheduled over all scenarios. Therefore, we do not recommend decreasing the number of truck drivers. If a truck driver does not show for work and in order to keep a balanced solution, in which high peak demands do not supersede the available refueling resources, it might be necessary to decrease the number of flying events per squadron.

Figure 28. Maximum Wait Time, Average Wait Time, and Number of Aircraft Waiting to Conduct Cold Refueling in 100 Scenarios When Adding a Truck Breakdown to the Original Setup

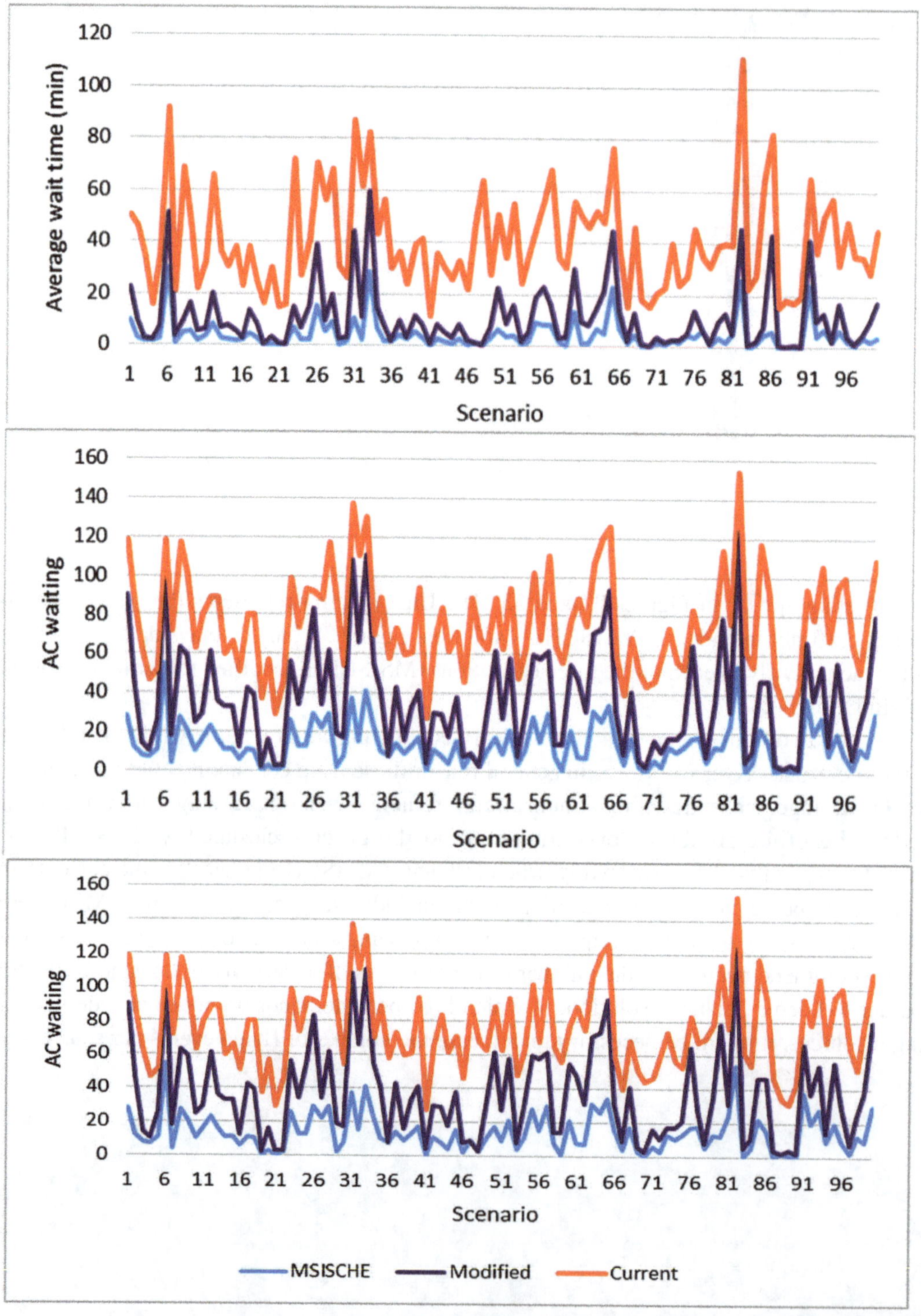

Figure 29. Frequency of Cold Refueling Wait Time over 100 Scenarios When Adding a Truck Breakdown to the Original Setup (The far-right column represents the percentage of unassigned aircraft)

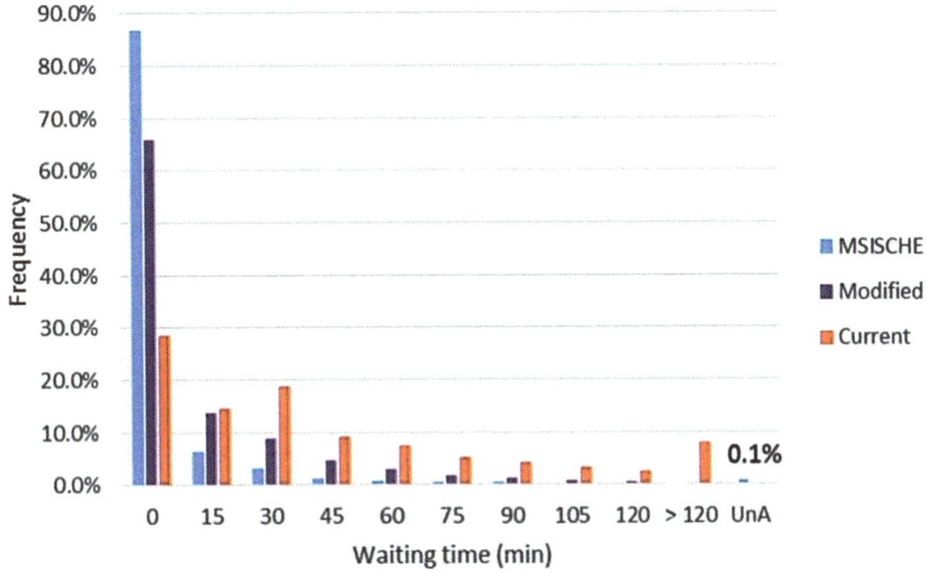

CONCLUSION

The most important finding in our study is that, with the current squadrons and refueling infrastructure at NAS Lemoore, it is possible to significantly reduce aircraft refueling wait time with only small adjustments to the squadrons' flight schedules; thus, the squadrons maintain their autonomy when creating their flight schedules and the aircraft increase their available flying time. MSISCHE has the ability to produce an optimized combined flying and refueling schedule, a balanced solution in which high peak demands do not supersede the available refueling resources. By using penalties and rewards, MSISCHE allows CSFWP and NAVSUP FLC San Diego to emphasize the priorities of the squadrons and their refueling service. MSISCHE is fast and flexible, and can be adjusted if the refueling infrastructure changes. Nevertheless, if the refueling resources decrease or the number of squadrons increases, there is a possibility that some flying events will be unscheduled because MSISCHE does not allow aircraft to wait beyond two hours. In practice, flying events do get cancelled due to delays in refueling, and such cancellations would become more frequent as aircraft number increases or refueling infrastructure declines. NAS Lemoore leadership must carefully consider the cost of burning fuel while aircraft wait at hot skids and the balance between minimizing the aircraft refueling wait time and minimizing the changes to the squadrons' flight schedules when using MSISCHE or any scheduling tool. Our findings also indicate that the aircraft refueling wait time can also be decreased simply by changing the refueling policy from only allowing aircraft to conduct hot refueling at the pre-assigned fuel skid to allowing aircraft

Figure 30. Maximum Wait Time, Average Wait Time, and Number of Aircraft Waiting to Conduct Hot Refueling in 100 Scenarios with One Fewer Driver from the Original Setup

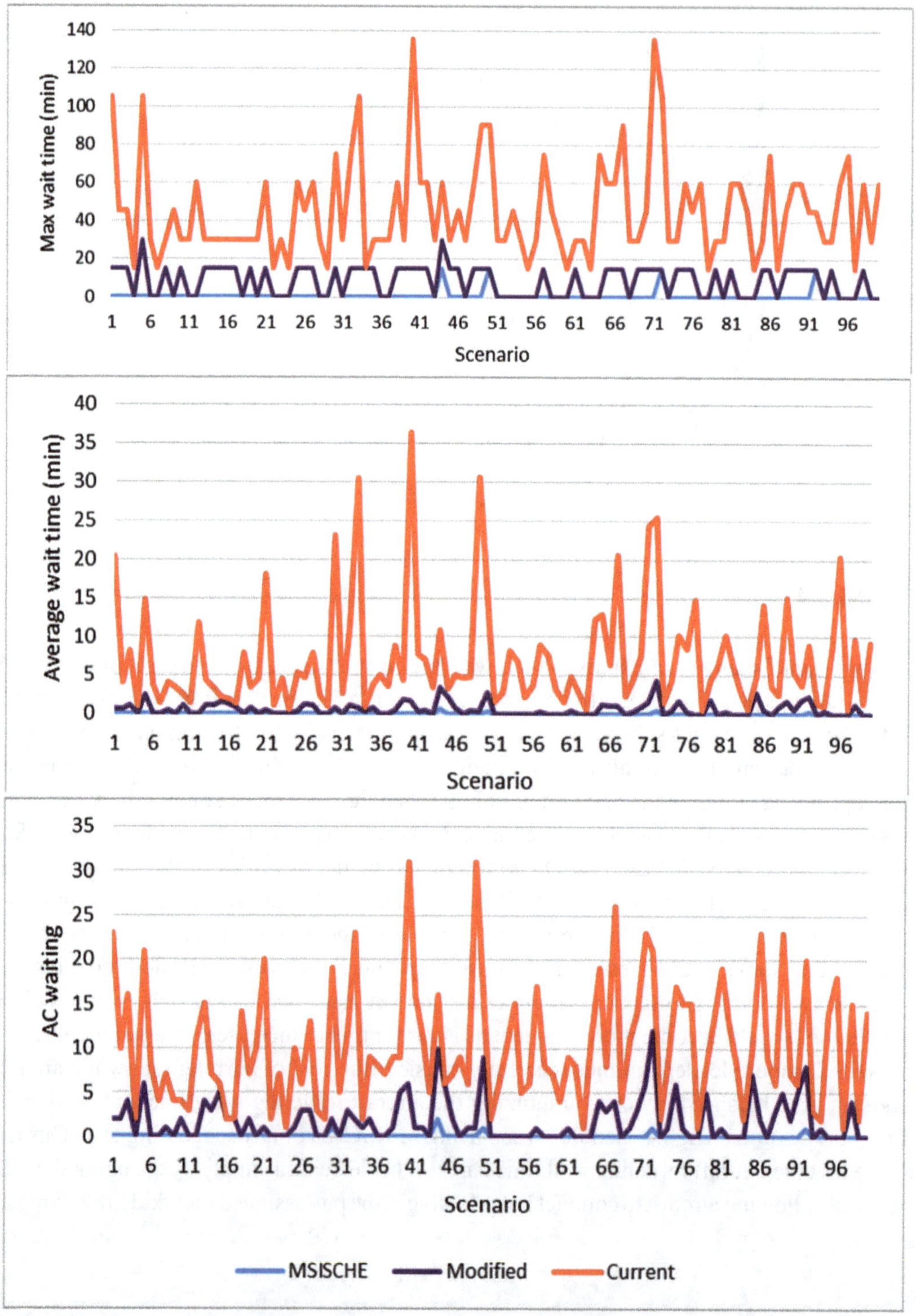

Figure 31. Frequency of Hot Refueling Wait Time over 100 Scenarios with One Fewer Driver from the Original Setup (The far-right column represents the percentage of unassigned aircraft)

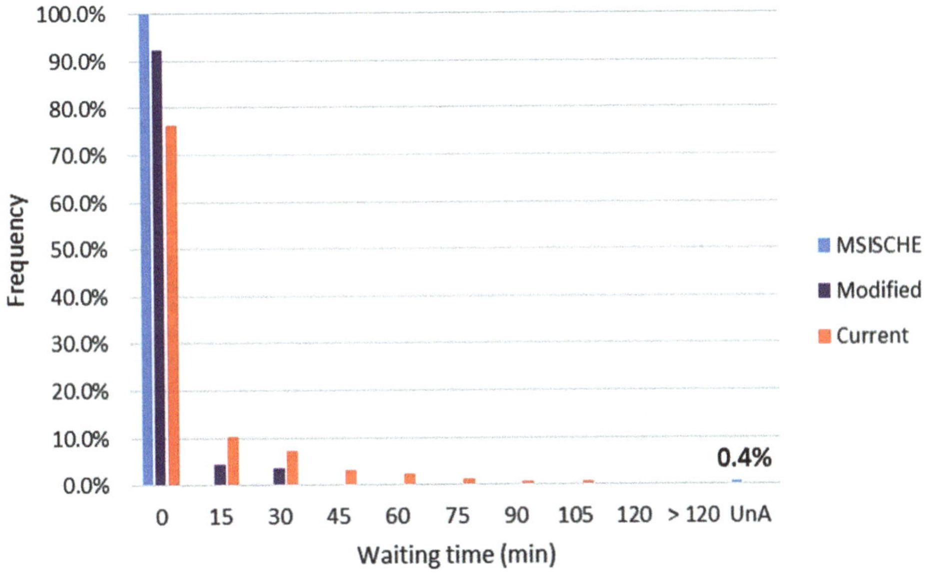

to receive hot refueling at any fuel skid. Our simulation results indicate that this modification results in lower wait times than the current policy, but not as low as when aircraft follow MSISCHE's schedule.

The following recommendations can improve MSISCHE's design and performance.

1. **Week-Prior Inputs:** Because the squadrons' flight schedules are provided only a day prior to the flying date, the possibilities of modifying events are very low. To allow the possibility of modifying events of aircraft flying to additional MOA/ranges, we recommend providing the flight requests a week before the flying date.
2. **Hot Refueling Vehicle:** Fuel trucks are currently used for both cold and hot refueling; however, for hot refueling, fuel trucks are used just as means of transportation for refueling personnel. Having another type of vehicle for hot refueling services could improve cold refueling wait time.
3. **Personnel Breaks:** For simplicity, we do not model breaks during drivers' work shifts. To improve MSISCHE's accuracy, we recommend that breaks be modeled.
4. **Time Horizon:** MSISCHE is a discrete-time model, and our study used 15-minute time increments in order to coincide with the squadrons' flight schedules as provided. In order to improve MSISCHE's accuracy while refueling aircraft, time increments could be shortened.
5. **Aircraft Assignment:** Our study did not track individual aircraft in order to ensure that they have completed refueling prior to being assigned to additional events. Future work could add this level of fidelity.

Figure 32. Maximum Wait Time, Average Wait Time, and Number of Aircraft Waiting to Conduct Cold Refueling in 100 Scenarios with One Fewer Driver from the Original Setup

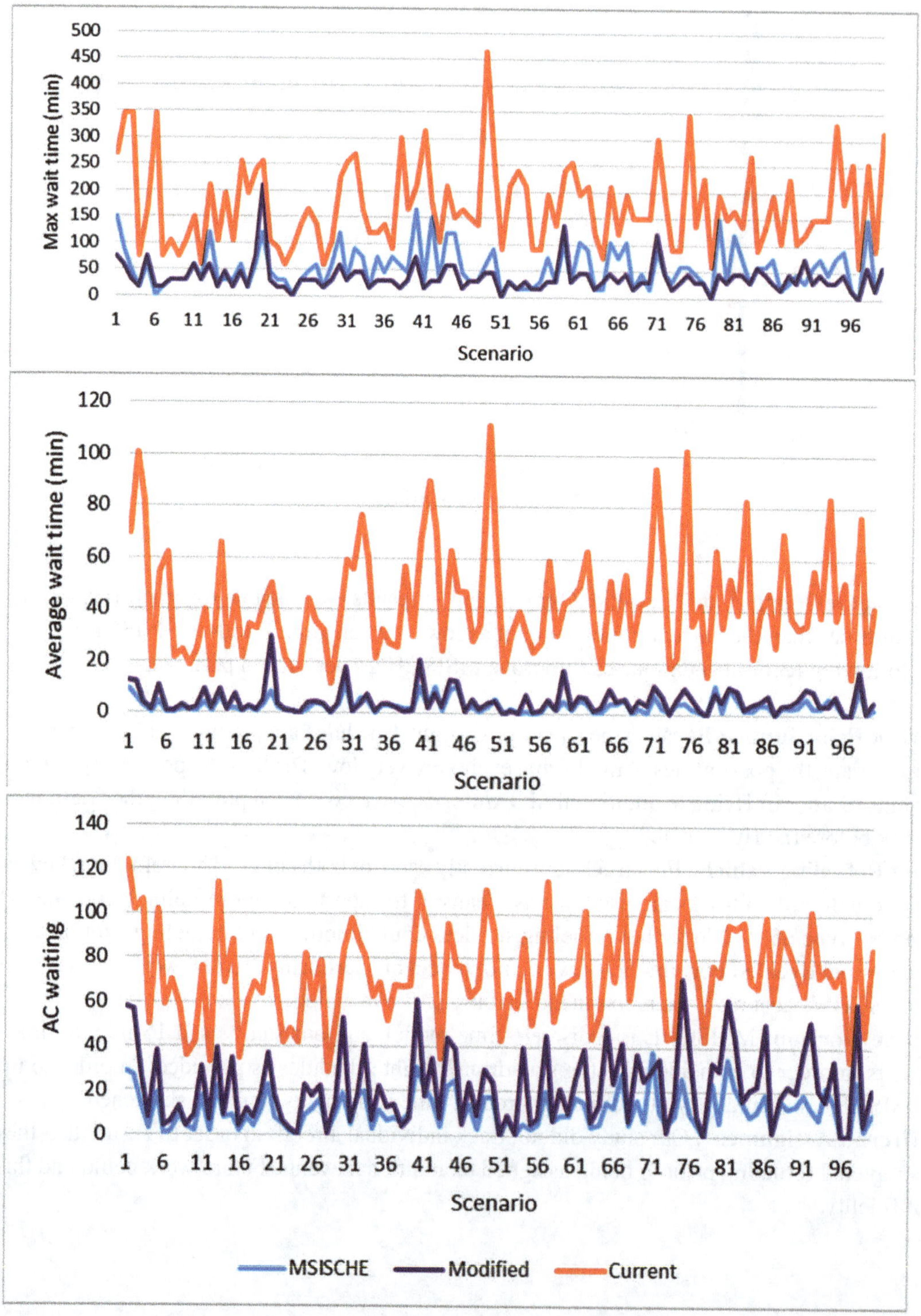

Figure 33. Frequency of Cold Refueling Wait Time Over 100 Scenarios with One Fewer Driver from the Original Setup (The far-right column represents the percentage of unassigned aircraft)

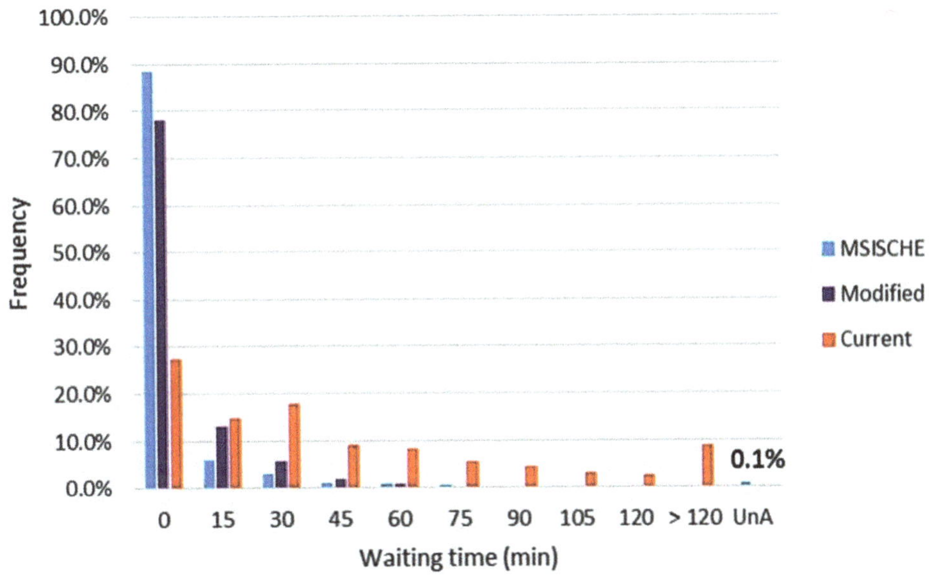

6. **Compute Fuel:** For simplicity, we assumed that the estimated fuel burned in an event is annotated in the flight schedule. If more information about the events is provided, the fuel burned for each event could be computed using the type of mission and duration of the flight.
7. **Include Aircraft Defueling:** Along with aircraft refueling, NAS Lemoore Fuels Division also performs aircraft defueling. Fuel trucks are not an issue because, besides the 11 trucks used for refueling, another truck exists for defueling purposes; however, one of the available drivers will be occupied with that duty. We recommend taking aircraft defueling into consideration.

REFERENCES

Commander Strike Fighter Wing Pacific. (2016). *NAS Lemoore inflight guide*. Lemoore, CA: Author.

Defense Acquisition University. (2015). *Glossary of defense acquisition acronyms and terms*. Retrieved from https://dap.dau.mil/glossary/Pages/Default.aspx

GAMS. (2015). *General Algebraic Modeling System Development Corporation, version 24.4.2 r51415 (64 bit)*. Retrieved from https://www.gams.com

Geiser, M. T. (2012). *Improving aircraft refueling procedures at NAS Oceana* (Master's thesis). Retrieved from Calhoun: http://calhoun.nps.edu/handle/10945/7346

Gerber, C. A., & Clark, J. A. (2013). *More fight–less fuel: Reducing fuel burn through ground process improvement* (Master's thesis). Retrieved from Calhoun: http://calhoun.nps.edu/handle/10945/34667

Microsoft. (2009). *Microsoft Windows 7 Professional (64 bit)*. Retrieved from https://www.nps.edu/Technology/SoftwareLib

Microsoft. (2013). *Microsoft Office Professional Plus 2013: Microsoft Excel version 15.0.4885.1000 (64 bit)*. Retrieved from https://www.nps.edu/Technology/SoftwareLib

Naval Air Systems Command. (2002). *Aircraft refueling manual (NAVAIR 00-80T-109)*. Washington, DC: Department of the Navy.

Ward, P. W. (2016). *NAS Lemoore modeling project. Presented at OA4611 Joint Logistics Models*. Monterey, CA: Naval Postgraduate School.

Chapter 13
Load Optimization for Navy Landing Ship Tank

Mevlut Savas Bilican
Istanbul Technical University, Turkey

ABSTRACT

The success of military operations mainly relies on the proper flow of the logistical supplies such as water, food, ammunition, etc. from source to the operation theater on time. There are special types of transportation vessels regarding the feature of supply. However, when transporting special material like ammunition, most navies usually prefer utilizing their own transportation capabilities since they require special treatment. For this reason, such material is carried in special boxes, called containers. To minimize the transportation cost and time, an efficient container stowage plan is necessary in terms of loading and unloading these containers. This chapter aims to develop a solution methodology to the problem with the focus on military logistics planning. For this purpose, the author develops a mathematical model that attempts to minimize the transportation time by creating proper loading and unloading sequence of containers to military cargo ships.

INTRODUCTION

Over the past two decades there has been a continuous increase in demand for cost efficient containerized transportation. To meet this demand, shipping companies have deployed larger container vessels, which can nowadays transport up to 19,000 TEUs (Twenty-Foot Container Equivalent Units). These vessels sail from port to port loading and unloading thousands of containers.

Minimizing the time, a vessel spends at port involves, among other issues, an efficient stowage plan; a plan that describes where each container should be loaded on the vessel. Note that containers are only accessible when they are on the top of other containers. Unloading and loading movements of the same containers in the same port are called shifts. Shifts arise either when planners want to unload containers destined for current port which however are beneath those destined for subsequent ports, or when planners want to reorder the sequence of containers to prevent more shifts in the future. Usually it is called necessary shifts in the former case and voluntary shifts in the latter. Shifts are time-consuming and

DOI: 10.4018/978-1-5225-5513-1.ch013

money-consuming activities, hence the arrangement of containers on board is crucial to achieve effective operations by reducing the number of shifts. The task of determining the arrangement of containers is called, as mentioned before, stowage or load planning.

In the military domain, the logistic supplies are usually delivered by private shipping companies or military cargo ships such as LST (Landing Ship Tank), LCM (Landing Craft Mechanized), LCU (Landing Craft Utility) and other types of ships, depending on the size of the cargo. But some critical military cargo, like ammunition, must be delivered in special boxes or containers with delicate treatment. This may require the utilization of navy cargo ships like LSTs and LCMs. In peace time, transportation of critical military cargo is made through several military ports by unloading filled boxes and loading empty boxes along with other loads. Particularly in the event of crisis or any operation however, these deliveries should be completed in minimum time interval. This can be achieved only by shortening the time spent at any port. In order to shorten this time, the stowage plan must be made in a fashion that there is no or minimum number of shifts. In this paper, the author proposes an algorithm that minimizes the number of shifts that may occur during loading process.

The purpose of this study is to show the close relation between Military Transportation and Container Stowage Plan and to produce an optimal solution for Military Transportation problems in designated time by utilizing CSP algorithms. Since the duration of Military Transportation is crucial to planners in critical times, it is assumed that proposed model will make contributions to the studies in this field.

LITERATURE REVIEW

An early attempt at this problem explained by Sheilds (1984), uses a combination of simulation using Monte-Carlo techniques and human interaction to provide possible vessel plans.

One of the early works on this problem was the one conducted by Imai and Miki (1989) who considered the maximization of GM and the minimization of the loading-related re-handle when loading containers onto one of the ship holds. For simplicity in the solution process, they formulated the problem as a two-objective assignment problem, employing the estimated number of re-handles in the objective function instead of the exact one. The precise number of re-handles is obtained from the resulting solution.

Imai et al. (2001) followed another approach in order for the exact number of re-handles to be considered in the problem. They formulated the stowage problem as a GM objective assignment problem for identifying a non-inferior solution set in terms of the GM and number of re-handles. Multiple solutions of the assignment problem were enumerated, thus computing the exact number of re-handles based on each enumerated solution. This approach, however, generated enormous multiple solutions for nearly the same range of a non-inferior solution set as the method employed in Imai and Miki (1989). Surprisingly, the former took 6000 times longer computation time than the latter. Imai et al. (2002) modified the problem only for finding non-inferior solutions with acceptable GM. The first attempt to derive some rules for determining good container stowage plans is reported in Ambrosino and Sciomachen (1998), where a constraint satisfaction approach is used for defining and characterizing the space of feasible solutions.

Wilson and Roach (1999, 2000) and Wilson et al. (2001) presented a realistic model, considering all technical restrictions in order to implement a commercial usable decision support system. Their approach was based on decomposing the planning process into two phases. In the first phase, called the strategic process, they created a rough stowage plan, based on grouping the containers with the same characteristics in terms of size, destination, etc., and on assigning them to blocks of stowage space in the ship. Ship stability was kept to an acceptable degree by this assignment process. These calculations were performed by a branch and bound procedure. In the second phase, called the tactical process, individual containers were assigned to specific locations, resulting in a detailed stowage plan. They employed a Tabu search heuristic for the second phase calculation. Their objectives included, among others, the unloading-related re-handle and ship stability; however, no detail of the stability was described in their study. Due to the complexity of their solution methodology, they only show a solution result for a small sample problem; therefore, its efficiency from the practical viewpoint is not shown.

Avriel and Penn (1993) and Avriel et al. (1997) focus on stowage planning considering the problem of minimizing the number of unproductive shifts (temporary unloading and reloading of container at a port before their destination ports in order to access containers below them for unloading) without any consideration for ship's stability. They formulated the problem as a 0–1 Integer Programming and applied it for loading onto a single hold like Imai and Miki (1989). The MBPP (Master Bay Plan Problem) is an NP-complete combinatorial optimization problem (see. e.g. Avriel et al. (2000) that connect the MBPP to the coloring of circle graphs problem). Avriel et al. (2000) developed some characteristics in the relationship between stowage planning and the coloring of circle graphs. The scale of the problem is enormous. For a 2000 TEUs (twenty-foot equivalent unit) containership, the number of possible stowage configurations is approximately 3.3^{105775} and as the size of containerships increases the reliance on manual systems becomes increasingly difficult and costlier.

Todd and Sen (1997) implemented a GA procedure with multiple criteria such as proximity in terms of container location on board and the minimization of unloading-related re-handle. Their study is interesting because it examined the relationship between the re-handle and the ship stability; however, their stability is not well defined as used in practice.

Dubrovsky et al. (2002) use a GA for minimizing the number of container movements in the stowage planning problem, while being able to include appropriate constraints with some ship stability criteria; the authors significantly reduce the search space using a compact and efficient encoding scheme.

The description of the entire problem in the form of an optimization problem is given in Ambrosino et al. (2004). Ambrosino et al. (2004) consider the stowage planning problem while taking a set of structural and operational restrictions into account. The problem is denoted as "Master Bay Plan Problem (MBPP)". The objective is the minimization of the total stowage time. Practical constraints such as different dimensions or weights of containers are considered. A basic 0/1 LP model is proposed. Since this model is not practically useful for large instances, a heuristic is proposed for preprocessing. Thus, pre-stowage rules are applied allowing to relax some constraints of the exact model resulting in solutions of the combinatorial optimization problem. The approach strives to avoid unloading-related re-handles by assigning some ship holds to containers with the same destination. Loading-related re-handles that occur if specific containers are stacked on the yard below others which are to be picked up later are not considered in the approach. The approach is evaluated with real size cases based upon data from a terminal in Genoa (Italy), showing good performance in terms of precision and computational time.

In Ambrosino et al. (2006) the authors present a three-phase algorithm for defining stowage plans. The objective is the minimization of the total loading time. Weight, size, and stability constraints have to be satisfied related to weight distribution. In the first phase, main branching tree splits the ship into different partitions. Containers with different subsets of bays are associated without specifying their actual position. Thus, the search space for the following procedure is limited to portions of ships. The successive solution of the 0/1 LP model results in the optimal plan for loading containers into each partition of the ship. Finally, a local search and exchange procedures are performed for checking and removing possible infeasibilities of the global solution caused by stability constraints. The approach facilitates parallel loading operations of different ship portions and shows good performance in numerical experiments based upon data from Genoa.

Sciomachen and Tanfani (2006) utilize the relation between the MBPP and the three-dimensional bin packing problem. Containers are regarded as items; the only bin is the vessel. The approach aims at optimizing important terminal performance measures, such as crane productivity. Structural and operational constraints of containers and ships are considered.

Imai et al. (2006b) focus on container stowage and loading plans of a ship. Two conflicting criteria are considered: requiring the minimum number of container re-handles (loading-related and unloading-related ones) in yard stacks as well as meeting ship stability. A multi-objective integer programming model is proposed. The weighting method proposed by Cohon (1978) is used to obtain a set of non-inferior solutions. In the weighting method, the problem is defined as a mathematical programming model with a single objective incorporating multiple objectives. Numerical experiments show that the solutions are acceptable for practical purpose if no re-handles occur during unloading. Furthermore, the GA is tested with and without a tournament approach showing that the tournament provides better solutions but needs more computation time.

Álvarez (2006) proposes an approach using TS and multiple initial plans for automated generating of vessel loading plans in reach stacker-based terminals. For judging the quality of alternative solutions, the algorithm takes the impact of container reshuffling, ground travel of reach stackers as well as on-board weight distribution into account. The algorithm's performance is evaluated in comparison with an MIP formulation of the problem. Computational experiments using data from a Spanish terminal show that the algorithm can provide good solutions for test instances with hundreds of containers within a few minutes. The basic MIP formulation failed to solve problem instances with more than 40 containers.

Alternatively, Álvarez (2007) proposes a Lagrangian relaxation approach for the problem. Powerful lower bounds are obtained by the algorithm within a branch-and-bound framework. Sub-gradient optimization is applied to the relaxation sub-problems. Furthermore, important implementation issues for speeding up the algorithm are discussed. Computational results on small test instances (10, 15, 19 containers) are conducted for a comparison between the performance of the proposed algorithm and strengthened MIP formulation. Findings are that the proposed approach is better scalable. Additionally, it is competitive in terms of computing time. Further research will focus on testing the promising Lagrangian approach with larger problem instances as well as on combining the algorithm with a TS approach.

Stahlbock and Voß (2007) provided an expository update of research on operations research methods applied on maritime container terminal operations. The paper is an extension and update of Steenken et al. (2004) providing a comprehensive survey of the state of the art of operations at a container terminal as well as of methods for their optimization.

BACKGROUND

Military Logistics

Army logistics, also known as military logistics, refers to the art and science of the process of planning and making sure that the movement and maintenance of the armed forces is in due accordance with the laid-down plans and procedures. Army logistics deals with procurement, maintenance, distribution and replacement of personnel and material. But, when taken in a comprehensive sense, it deals with:

1. Designing, developing, storage, maintenance and acquisition, disposition and evacuation of material.
2. Hospitalization, evacuation and movement of personnel.
3. Furnishing or acquisition of services.
4. Maintenance, operations, acquisition and disposition of available facilities.

Army logistics is as old as the concept of war. In ancient times, people used to find their own knotted clubs, place of shelter and food during the time of war. Every soldier was himself responsible for obtaining his necessities. With the advances associated with time, soldiers joined larger groups and some men were assigned the task of arranging these items of necessities for their counterparts. These were among the first instances of an army logistics organization.

Logistics cannot be termed as an exact science as it does not follow a defined set of principles or adhere to previously laid guidelines in the same manner each time. It does not even follow a pre-defined sets as in case of science. Military logistics is dynamic in nature. Therefore, a particular solution cannot be prescribed for every similar situation. It is primarily concerned with ensuring services and supplies at the right time, at the right place, when and where needed, in the most optimal and economical way and by making use of the best available options to do the task.

The personnel engaged in army logistics take the decisions regarding any type of problems which are logistical in nature. This is done by them through their ability to analyze, interpret and by using their past experiences. The decision-making ability of these people is not solely dependent upon the professional knowledge imparted to them from the military institutions, but also from tactics, strategy, personnel and training.

Matters related to logistics have frequently been of utmost importance in deciding the outcome of wars. This can be clearly understood by taking as example the failure of the German naval force to sink enough cargo in the 2nd Atlantic Battle of the Second World War, which helped the British troops to continue fighting. It is often believed that the destruction of the enemy's supply lines and guarding own supply lines is a crucial element in deciding the outcome of the war.

Modern developments have led to enhanced army logistics performance due to less use of time and financial resources. This has raised the standard of the armed forces and in a way, has proved to be beneficial for their overall growth and development. Accuracy and speed are the major factors nowadays for the flow of information and materials from its providers to its users.

Proper Logistics Transport: A Must Have for Success

In the current scenario of serious competition in the global business environment, an effective and efficient logistical system is a must for providing and delivering quality products to the point of destination. The organization must be vigilant enough to eliminate any inefficiencies and improve their operational performance on a regular basis. Any organization, whether big or small, can benefit from a proper logistics transport system.

What are the available ways of acquiring transportation? There are three possible ways of acquiring transportation, namely: 1.) Purchase the vehicles for transportation; 2.) Lease or hire the vehicles; 3.) Use a distributor or a courier company for movement of goods.

Logistics involve all steps in the process of providing and delivering raw materials, semi-finished and finished goods from the origin to the destination. One may wonder how the goods can be transported. The following ways should give a clearer view.

- **Freight Forwarders:** They are specialists in the process of consolidation and usually transport goods of various organizations under one container or vehicle. This helps in lowering the transportation costs.
- **Couriers:** They are known for secure and speedy delivery of goods and packages, normally those smaller and lighter in weight. They can deliver both domestically and internationally.
- **Logistics Providers:** They are responsible for managing the complete transportation process on their own, from tracking of goods to organizing documents, and from warehousing to storage.

Logistics and transportation focus on several benefits inherent to various forms of transportation such as railways, metros, ships, ports, buses, ships, trolleys, rickshaws or trucks. The selection of the mode of transportation depends on numerous factors such as quality and quantity of goods, their cost, available time, instructions of customers, availability of vehicles, climatic conditions and past experiences.

MAIN FOCUS OF THE CHAPTER

Container Stowage Plan Problem (CSP Problem)

A container (Figure 1) is a large reusable box which is used in intermodal transportation to carry the goods between destinations in commercial domain. The containers are defined according to their lengths. A standard container is either 20-feet or 40-feet. A 20-feet equivalent container is referred to as 1 TEU (twenty feet equivalent unit). 40- feet containers are called as 2 TEU. It is important to have such standardization in containers since they are used interchangeably by both carriers and companies. Containers are flexible in the sense that they can carry a wide variety of goods such as liquids, manufactured goods, cars and perishable goods. Containerization enables effective transfer between several modes of transport, which is a major requirement for intermodal transportation. Containers are made of durable materials which help to keep the content of the containers safe and stable. Another advantage of the endurance of the containers is that they can easily be stacked. Since the space on the vessels and on the land side of the terminals are limited, the stackability of containers helps manage the storage capacities better.

Containers are placed on the vessels according to a stowage plan. Each container is attributed with a specific bay, row and tier number. Bays are the cross-wise areas of a vessel. Row is the position of the containers in bays while tier is the level on which the container is stacked. These concepts are visualized in Figure 2. The basic structure of a containership and its cross section are shown in Figure 3, which are used to provide an idea of how container stowage takes places. The stowage plan of a vessel induces precedence relations among containers. There are some physical constraints which should be considered while planning the loading and unloading operations. The stowage plan is modified as the vessel proceeds between its destination ports with the addition and removal of containers.

The stowage of a containership is one of the problems that have to be solved on a daily basis by any company which manages a container terminal. In the past, stowage plans for containers were performed by the Captain of the ship; today, as a consequence of containerization, the maritime terminal has to

Figure 1. A view of a container
Source: www.bombshelters.com

decide the stowage of containers following the instructions coming from the maritime company holding the ship and considering different interrelated terminal requirements. Formally, the stowage planning problem consists of determining how to stow a given set C of n containers of different type into a set S of m available locations of a containership, respecting some structural and operational constraints, related to both the containers and the ship; the goal is the minimization of the loading time of all containers. In fact, while the turnaround time of a ship includes the time for berthing, unloading, loading and departure, the major activities affecting the turnaround time are the unloading and loading processes, and, in particular, loading is more difficult and sensitive to the efficiency of the operations.

Figure 2. A view of container positions on a vessel
Source: www.bestshippingnews.com

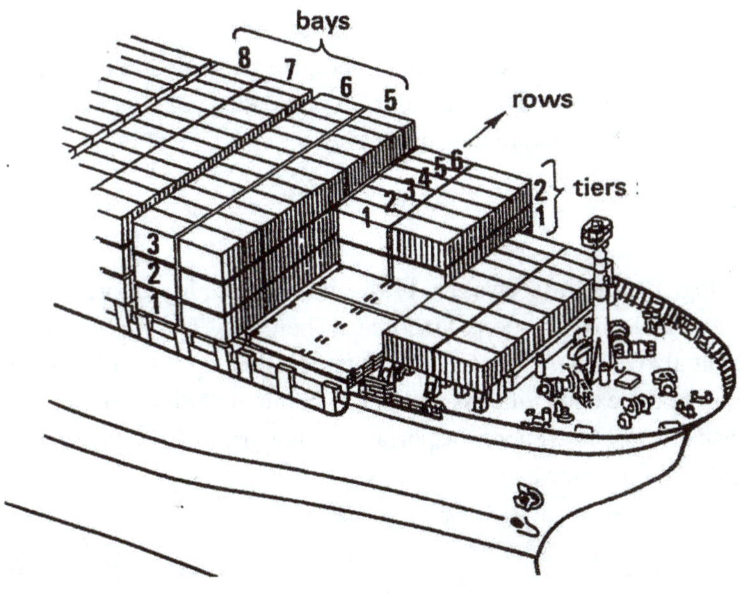

Figure 3. Containership Structure

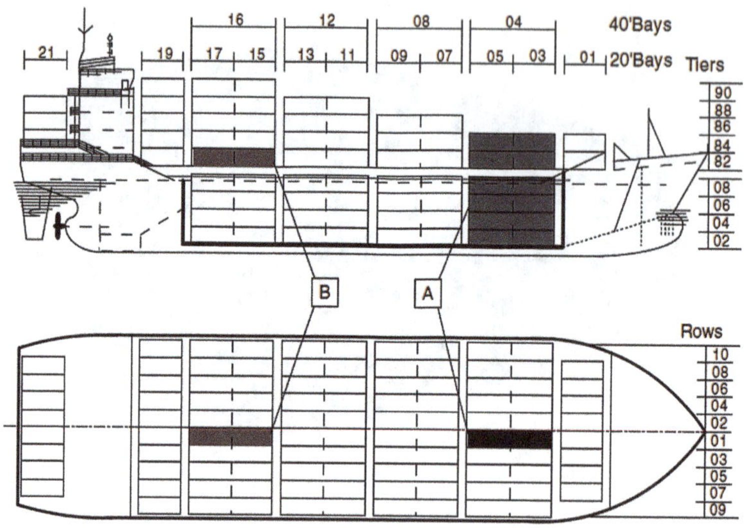

In the CSP each container $c \in C$ must be stowed in a location $l \in S$ of the ship. Note that the I_{th} location is actually addressed by indices i, j, k representing, respectively: the bay (i), that gives its position related to the cross section of the ship, the row (j), that gives its position related to the horizontal section of the corresponding bay (counted from center to outside), and the tier (k) that gives its position related to the vertical section of the corresponding bay (counted from bottom to top). I, J and K denote respectively, the set of bays, rows and tiers of the ship.

Since the main objective is to minimize the time spent at port, Stowage plans, if not prepared well enough, may cause unnecessary handling time, time required for temporary unloading and reloading of containers from/onto the ship, called shifting, and for movements of gantry cranes at the ports. Consequently, port efficiency and ship utilization are largely affected by stowage plans.

In general, shifting is caused by overstowage, which denotes the situation when containers that should be unloaded at the current port are placed under other containers that should go further in the ship's route. In this case, the latter containers should be temporarily unloaded in order or the former containers to be unloaded at the current port. These temporarily unloaded containers, commonly referred to as overstows, must then be re-loaded before the ship leaves.

Problem Definition

As mentioned above, logistics can be seen as the bridge between deployed forces and the industrial base, which produces the materials and weapons for forces. Therefore, The mission is not only to get your forces from A to B, but also to keep them fed and watered as they go. In this context, the art of movement is one of the most complex and vital issue that any commander must master, if he is going to win. One way to accomplish this task is proper transportation which includes the movement of personnel and material into the theater of operations.

In the transportation process, accuracy and speed are the major factors. However, in this paper the main concern is the speed of transportation process. It is also considered that most of the transported items are carried in special boxes called as containers.

In this sense, for providing fast delivery in transportation, one must decrease the loading and unloading time. There are several factors which effect loading and unloading time. Among these factors, the most important one is the proper loading, that is, the proper sequence of loading and unloading of containers onto the ship. In container transportation, notice that containers are only accessible when they are on the top of other containers. Unloading and loading movements of same containers in the same port are called shifting. Shifting arises either when unloading containers destined for current port which however are beneath those destined for subsequent ports, or when reordering the sequence of containers to prevent more shiftings in the future. Shiftings are time-consuming and money-consuming activities, hence the arrangement of containers on board is crucial to achieve efficient operations by reducing the number of shiftings. The task of determining the arrangement of containers is called, as mentioned before, stowage planning. Therefore, minimizing the time, a vessel spends at port involves, among other issues, an efficient stowage plan; a plan that describes where each container should be loaded on the vessel.

It should be noted that military equipment is transported via either commercial container ships or military logistics ships, as mentioned before. Either way, the minimum delivery time is of great importance. Especially if this delivery is conducted in operation then the fast delivery is getting vital for the survivability of the units. Utilizing the resemblance between CSP and military transportation may possibly pave the way for fast delivery. In this case, as for the implementation of CSP in military domain, the problem can be stated as follows:

1. For the transportation of equipment Navy ship LST is used.

At this point it would be very informative to get acquainted with Navy ship LST. LST (Landing Ship Tank) is a type of Naval amphibious vessel designed for transporting and deploying military troops along with their equipment onto either the scene of action or the theater where the amphibious operation is being conducted. The bow of the LST has a large door that would open with a ramp for unloading the vehicles. Although these ships may have different features, the general view of the LST is shown in Figure 4.

Featuring long and flat deck makes these platforms preferable in transporting cargo. Therefore, it is inevitable to utilize these ships for cargo operation for their ability to land on shore. The general outline of the deck that can be used for cargo operation on LST is depicted in Figure 5.

2. There are military supplies and ammunitions which need to be delivered to 5 different ports by navy ship LST.

The transportation matrix is used to show the quantities of cargo which will be transported among 5 different ports. The transportation matrix for the model is indicated in Table 1. If studied carefully, one can easily notice that each cell indicates the quantity of the containers that will be transported between corresponding ports. For example, T_{34} is the number of containers that is transported from port 3 to port 4. In the model, this value is 44.

Figure 4. LST Ship

Figure 5. Deck Structure

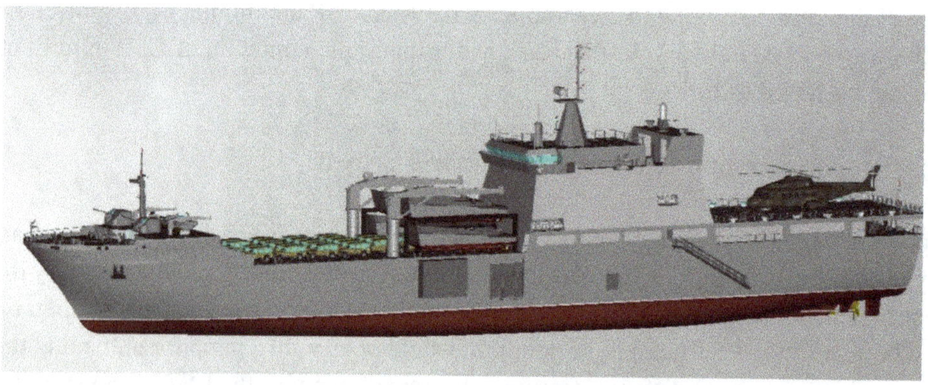

Table 1. Transportation Matrix

Ports	2	3	4	5
1	186	117	58	71
2	0	43	32	111
3	0	0	44	116
4	0	0	0	134

3. These supplies and ammunitions are delicate materials and require special care. So that these supplies are carried in special boxes, that is, containers with standard dimensions.

In the model, it is assumed that all the containers are of the same dimension. Therefore, standardized container dimensions are 5 meters in length, 1.5 meters in width and 1 meter in height.

4. Once the ship arrives any port, it unloads all containers bound for that port and then receives other containers bound for next ports.

Since there are still containers which need to be unloaded at the next ports, loaded containers at the current port should not cause extra container movements at the next ports. That is, loaded containers at the current port should not be located on top of any container destined for any of the next ports which in this case gives rise to shifting. If this has to be done, then it should be kept at the minimum number.

The problem stated above is in fact a Container Stowage Problem (CSP). However, it is needed to make some minor adjustments in order to convert the problem to exact CSP form. The ship`s deck structure comes the first to be tackled with. Normally a container ship consists of bays to carry allocated containers. Whereas the bays are in standard dimensions on containerships, there is no bay plan on any naval vessel, including LST. Since the containers are of the same dimensions, an arrangement is needed on how to load these containers onto LST ship. As mentioned above, the deck on LST is flat and has no bay structure. In this sense, the ship deck is partitioned into 18 sections regarding the dimensions of military containers for the LST ship to have bay structure. These new sections, namely new bays, are enumerated in a way to start from stern of the ship (Figure 6). In the process of creating these fictitious bays, following principles have been taken into consideration.

- A space is left both in the middle of the deck and between successive bays on purpose to allow free movement of both crew and apparatus.
- Containers are to be stowed vertically.

Figure 6. LST Deck adapted to Bay structure

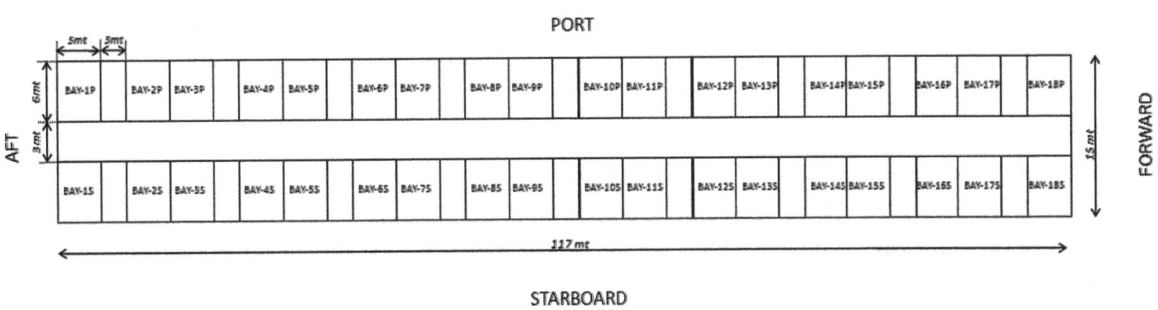

- All bays have the same containers capacity that is 24 containers per bay.
- On each of these bays, only 3 stack heights are allowed.
- Each container has the same weight.
- Containers are to be loaded/unloaded only with palettes (tracks) since there is no crane utility.
- In case of an overstow, the shifted container will be temporarily stacked onboard deck without being unloaded, as in container ships.

The problem with the simulations stated above resembles "Container Stowage Optimization" problem.

SOLUTION AND RECOMMENDATIONS

The main principle of the solution approach is to minimize overstow. In other words, the main aim is to distribute the containers to the bays in such a way to create minimum overstow.

LST ship carries its cargo among 5 ports, calling at each port in a round tour way. This ship starts its tour from first port. At this port (first port), it loads all the containers bound for second, third, fourth and fifth ports. Afterwards, it sails through the second port. At this port (second port) it unloads all containers bound for this port and then load the containers which need to be transported to ports 3,4 and 5. And this process continues in this fashion until the ship arrives the final port. At the final port, the ship discharges all containers, hence completing the tour. During this process, the aim is allocating containers to bays in a way to create minimum overstow. The suggested model puts forth the percentage of bay capacity that should be assigned for any type of container while minimizing overstow. For example, considering bay 1 and the containers transported from port 1 to port 2, the model generates a percentage for bay 1, that is, it tells how much of the bay 1 should be reserved for containers 1-2. Considering the example studied in this chapter, Transportation Matrix is presented in Table 1. Each element in this matrix will be expressed as T_{ij}, the number of container to be transported from port i to port j. For example, $T_{23} = 43$ is the number of containers to be transported from 2 to 4.

Suggested model determines the bay and *i-j* containers that are to be loaded to designated bay. Some assumptions have been made. These assumptions are as follows;

- All containers have the same dimensions. That is, 5 meters in length, 1.5 meters in width and 1 meter in height.
- The container capacity of the ship is 432.
- There is no special container.
- The ship unloads all containers at the last port.

The Model which proposed in this part, is derived and implemented from the model that was proposed by Gumus et al. (2008).

In this phase of the model, the author determines at which percentage of the corresponding bay is reserved for specific (*i-j*) containers. Containers are classified as *i-j* containers, that is, containers being transported from *i* to *j*. We use the notation as follows:

T_{ij} = Number of i - j containers.

X_{ij}^{b} = The percentage of Bay b reserved for i - j containers.

Q_{b} = The capacity of Bay b .

Y_{ij}^{b} = If there is any i-j container loaded on Bay b, then Y_{ij}^{b} = 1, otherwise Y_{ij}^{b} =0.

W_{j}^{b} = Maximum overstow that can occur on Bay b at port j.

For example; if $X_{13}^{3} = 0.4$ then, 40% capacity of Bay 3 is reserved for containers being transported from port 1 to port 3. If the capacity of Bay 3 is 20, then 20×0.4=8 slots are going to be spared for 1-3 containers. If the capacity of Bay 3 is 20, then $Q_{3} = 20$. It is needed to stress the importance of variable W_{j}^{b} at this point since it constitutes the core of the model in this phase. To grasp the importance of this variable, please see Figure 7. In Figure 7, there are *N* ports. Further assume that Bay *b* is being analyzed. Each arc in this network represents the presence of containers being transported from *i* to *j* on Bay *b*. In example on Bay *b*, one can easily see that there are containers already loaded to be transported from 1 to 2, from 1 to 3, from 1 to *N*, from 2 to 4, from 2 to *N* and from 3 to 4.

It can easily be seen from the Figure 7 that maximum overstow that can occur at port 3 is $X_{2N}^{b} + X_{24}^{b}$. There is an arc from 1 to 3 which states that there are containers loaded at port 1 to be discharged at port 3. The ship loads all these 1-3 containers at port 1 and move to port 2. Once it arrives at port 2, it first unloads 1-2 containers and then load 2-4 and 2-*N* containers on top of containers loaded at port 1, namely on top of 1-3 containers. Afterwards, ship sails to port 3. When she arrives port 3, 1-3 containers must be unloaded. But remember that at port 2, 2-4 and 2-N containers have been placed on top of 1-3 containers. In this case to discharge 1-3 containers, 2-4 and 2-N containers have to be unloaded thereby causing extra movements of containers. It is clear that the number of this extra movements is $X_{2N}^{b} + X_{24}^{b}$. But from definition, it is known that X_{2N}^{b} and X_{24}^{b} are the ratios which show the percentage of Bay b, reserved for 2-4 and 2-N containers. If this ratio is multiplied by the capacity of Bay b, the exact number of overstow occurred at this port is obtained. In summary $W_{3}^{b} = X_{2N}^{b} + X_{24}^{b}$ and the number of overstow equals to $W_{3}^{b} \times Q_{b}$. Since the number of overstows to be minimized, then this expression will constitute objective function.

In this case the model will be as follows:

Objective function:

$$\min \sum_{i=1}^{N-1} \sum_{b=1}^{B} (W_{b}^{i} \times Q_{b})$$

Subject to:

Figure 7. Network representation of Bay plan

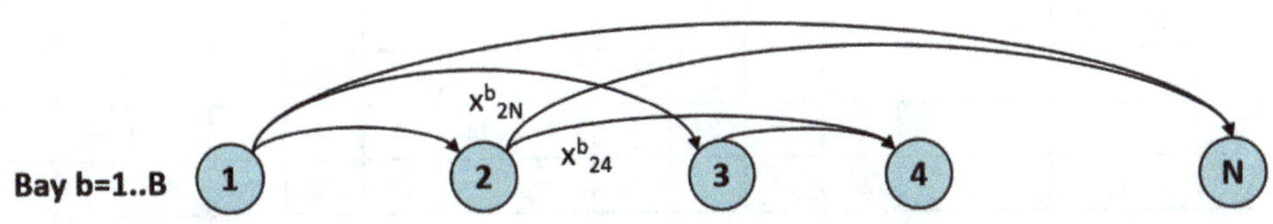

$$\sum_{i_1=1}^{i}\sum_{i_2=i+1}^{N} x_{i_1 i_2}^{b} \le 1, \forall i$$

$$\sum_{b=1}^{B} Q^b x_{ij}^b = T_{ij}, \forall (i,j)$$

$$\sum_{i_1=i+1}^{j-1}\sum_{i_2=j+1}^{N} X_{i_1 i_2}^b + Y_{ij}^b - 1 \le W_j^b$$

$$X_{ij}^b \le Y_{ij}^b, \forall (i,j)$$

Objective function is the total number of overstows that should be minimized. Note that overstow cannot occur at the last port since all the containers are to be unloaded according our assumption. That`s why the final port in the objective function is not included.

The first constraint guarantees the total usage of any bay to be less than %100. The second constraint ensures that all containers are to be transported. The third constraint determines the maximum number of overstow that may occur. That is, it defines the upper limit for overstows.

When the transportation matrix is formulated on this model, it comes up with 535 constraints. This model is solved on GAMS solver and after 161 iterations optimal solution is obtained. Solution obtained from GAMS is indicated in Table 2. Closer look at the result reveals that the overstow that may appear in this matrix is 0. One can easily load these 432 containers with 0 overstow. The loading plan is presented in Table 2. Notice that total transported i-j container numbers are consistent with the values in transportation matrix.

Table 2. Solution obtained from GAMS

i-j	Bay Number																		Total
	1	2	3	4	5	6	7	8	9	10	11	12	13	14	15	16	17	18	
1-2			24	24	24				24			3	24		15		24	24	186
1-3	24	24								24		21		24					117
1-4							24	24							9	1			58
1-5						24					24					23			71
2-3									24			3					16		43
2-4													17		15				32
2-5			24	24	24								7				8	24	111
3-4										24				20					44
3-5	24	24							24			24		4			16		116
4-5							24	24		24			17	20	24	1			134

FUTURE RESEARCH DIRECTIONS

The algorithm allocates containers to bays with minimum overstow. However, algorithm disregards both container weights and stability issues. Considering the weights of containers and main structure of LST ship in this example, the stability issue might pose a concern. As the ship size and the weights of the loads increase, stability becomes a major issue in this kind of problems. In this case, an algorithm which yields an efficient loading plan by considering stability factors is needed. The other issue that should be considered in future research is the crane utilization. If the results are examined carefully, it can be seen that the distribution of the containers transported to same ports is very wide. That is, some containers which are going to same destination are put into bays that are far from each other. This situation obviously requires more crane or palette operation, thereby increasing the loading and unloading time. In this sense, future studies should be intensified on to abolish these drawbacks by acquiring sound and applicable algorithm.

CONCLUSION

As Admiral Raymond A. Spruance said,

A sound logistics plan is the foundation upon which a war operation should be based. If the necessary minimum of logistics support cannot be given to the combatant forces involved, the operation may fail, or at best be only partially successful.

That is, the success of the military operation lies in the heart of logistic support provided by corresponding units. The backbone of this support is the accurate and fast transportation. In this context, most of the military logistic transportation is carried out by navy ships. LST ships are the most favorable container transportation vessel. The main issue for logistic transportation with this kind of ships is to conduct the operation in minimum time interval, hence meeting the main principle of logistics. This can be done only with efficient loading plan.

The theme of the problem studied in this paper is creating an efficient load plan for the fast delivery of delicate materials (like ammunition within standard boxes) to various military ports. For a proper solution, the problem is simulated to Container Stowage Problem with minor modifications and then some parts of the model proposed by Gumus et al. (2008) has been utilized. The solution acquired from the model is satisfactory for a ship in LST size. However, a new holistic algorithm which considers stability issues, should be developed for future prospective.

REFERENCES

Álvarez, J. F. (2006). A heuristic for vessel planning in a reach stacker terminal. *Journal for Maritime Research*, *3*(1), 3–16.

Álvarez, J. F. (2007). *A Lagrangian relaxation approach to the vessel planning problem*. Working paper, Pompeu Fabra University, Barcelona.

Ambrosino, D., & Sciomachen, A. (1998). A constraint satisfaction approach for master bay plans. *Water Stud. Ser.,* 175–184.

Ambrosino, D., Sciomachen, A., & Tanfani, E. (2004). Stowing a containership: The master bay plan problem. *Transport Res. A, 38,* 81–99.

Ambrosino, D., Sciomachen, A., & Tanfani, E. (2006). A decomposition heuristics for the container ship stowage problem. *Journal of Heuristics, 12*(3), 211–233. doi:10.100710732-006-5905-1

Avriel, M., & Penn, M. (1993). Exact and approximate solutions of the container ship stowage problem. *Computers & Industrial Engineering, 25*(1-4), 271–274. doi:10.1016/0360-8352(93)90273-Z

Avriel, M., Penn, M., & Shpirer, N. (2000). Container ship Stowage Problem: Complexity and Connection to the Colouring of Circle Graphs. *Discrete Applied Mathematics, 103*(1-3), 271–279. doi:10.1016/S0166-218X(99)00245-0

Avriel, M., Penn, M., Shpirer, N., & Witteboon, S. (1998). Stowage planning for container ships to reduce the number of shifts. *Annals of Operations Research, 76,* 55–71. doi:10.1023/A:1018956823693

Dubrovsky, O., Levitin, O. G., & Penn, M. (2002). A genetic algorithm with a compact solution encoding for the container ship stowage problem. *Journal ofHeuristics, 8*(6), 585–599. doi:10.1023/A:1020373709350

Gumus, M., Kaminsky, P., Tiemroth, E., & Ayik, M. (2008). A multi-stage decomposition heuristic for the container stowage problem. *Proceedings of the 2008 MSOM Conference.*

Imai, A., Nishimura, E., & Papadimitriou, S. (2005a). The dynamic berth allocation problem for a container port. *Transport Res B, 39*(3), 197. doi:10.1016/j.trb.2004.03.004

Imai, A., Sasaki, K., Nishimura, E., & Papadimitriou, S. (2006b). Multi-objective simultaneous stowage and load planning for a container ship with container rehandle in yard stacks. *European Journal of Operational Research, 171*(2), 373–389. doi:10.1016/j.ejor.2004.07.066

Kim, K. H., & Bae, J. W. (1999). A dispatching method for automated guided vehicles to minimize delay of containership operations. *International Journal of Management Sciences, 5,* 1–25.

Kim, K. H., & Bae, J. W. (2004). A look-ahead dispatching method for automated guided vehicles in automated port container terminals. *Transportation Science, 38*(2), 224–234. doi:10.1287/trsc.1030.0082

Saginaw, D. J., & Perakis, A. N. (1989). A decision support system for containership stowage planning. *Materials Technology, 26,* 47–61.

Sciomachen, A., & Tanfani, E. (2003). The master bay plan problem: A solution method based on its connection to the three-dimensional bin packing problem. *IMA Journal of Management Mathematics, 14*(3), 251–269. doi:10.1093/imaman/14.3.251

Sciomachen, A., & Tanfani, E. (2006). A 3D-BPP approach for optimising stowage plans and terminal productivity. *European Journal of Operational Research.* doi:10.1016/j.ejor

Shields, J. J. (1984). Container-ship stowage: A computer-aided preplanning system. *Materials Technology, 21,* 370–383.

Stahlbock, R., & Voß, S. (2007). Operations research at container terminals: A literature update. *OR-Spektrum, 30*(1), 1–52. doi:10.100700291-007-0100-9

Steenken, D., Voß, S., & Stahlbock, R. (2004). Container terminal operation and operations research – a classification and literature review. *OR-Spektrum, 26*(1), 3–49. doi:10.100700291-003-0157-z

Todd, D. S., & Sen, P. (1997). A multiple criteria genetic algorithm for containership loading. *Proceedings of the Seventh International Conference on Genetic Algorithms*, 674–681.

Wilson, I.D., & Roach, P.A. (1999). Principles of combinatorial optimization applied to container-ship stowage planning. *Journal of Heuristics, 5*(4), 403–418.

Wilson, I. D., & Roach, P. A. (2000). Container stowage planning: A methodology for generating computerized solutions. *The Journal of the Operational Research Society, 51*(11), 1248–1255. doi:10.1057/palgrave.jors.2601022

Chapter 14
Bearing-Only Target Motion Analysis

Alper Aytun
Marmara University, Turkey

Serol Bulkan
Marmara University, Turkey

ABSTRACT

The main goal of the bearing-only target motion analysis (BOTMA) is to determine the target's kinematic parameters such as position, course, and speed by only using the bearings reported by an onboard passive sensor (e.g., a sonar or an ESM [electronic support measures] device). This chapter provides a brief description of the BOTMA problem. Next, it discusses the implementation of the Kalman filtering technique to solve the problem. The authors then discuss the variations of the Kalman filtering (i.e., the extended and unscented Kalman filters). They also propose a genetic algorithm metaheuristic that incorporates a novel search space narrowing technique to solve the BOTMA problem and present numerical results for different noise conditions. They finally highlight the future research directions for modeling and solving the BOTMA problem.

INTRODUCTION

In a naval operation, the detection, classification, range, and localization of a hostile target play a key role while making a decision of an engagement. Platforms executing surface and subsurface warfares generally rely on their active and passive sensors as their primary sources of information. When high emission control policy is applied and the stealthiness is necessary, passive sensors constitute the only source of data. In such cases, the target-tracking algorithms implemented in the Combat Information Center (CIC)'s of a naval vessel rely on the data provided by passive sensors and those algorithms are known as Target Motion Analysis (TMA).

Although TMA is generally used in anti-submarine warfare with passive sonars as the source of data, it is also used for any type of passive sensor, e.g. ESM (Electronic Support Measures) systems, which can provide bearing data. In an anti-submarine warfare, the sonar is the primary information source and

DOI: 10.4018/978-1-5225-5513-1.ch014

it provides bearing information along with the frequency of the sound signals. In case of Electronic Warfare, ESM Systems are the primary passive source and these systems usually provide bearing, frequency, PRI (Pulse Repetition Interval), PRF (Pulse Repetition Frequency), PW (Pulse Width) and antenna scan period information of the target. There are several different implementations of TMA algorithms, which take into account not only the bearing information but also the frequency, range rate or other available information in hand. In this chapter, we restrict our scope to the bearing only case, i.e. Bearing Only Target Motion Analysis (BOTMA), and do not discuss other types of implementations.

Direction finding algorithms used in passive systems such as sonar and ESM systems are error-prone because of the noise present in the propagation channel of the sound and electromagnetic waves. The additive receiver noise on the receiver front end may also generate erroneous bearing information. Moreover, due to the multipath effect, signals bouncing and arriving at the receiver front end from different paths may degrade the signal quality, which makes the direction finding process more complicated.

TMA algorithm provides the means of predicting the target's next position given the previous bearing measurements. In this sense, it can be defined as a tracking method as well as a target trajectory extractor. While working on a general TMA problem, one needs to make two important assumptions. First, it is assumed that target moves with a constant speed in a linear direction without changing its course in Cartesian coordinates (x-y plane). Second, is that the vessel collecting measurements needs to maneuver and change its course at least once in order to make the target observable. Therefore observability plays a key role in the success of TMA algorithm and will be discussed thoroughly in this chapter.

The well-known methods found in the literature usually use iterative methods such as the reputable Extended Kalman Filtering (EKF) to solve the target-tracking problem. There are two possible ways of processing data collected from the passive sensor. Batch processing is the offline processing of the noisy bearing measurements collected for a period of time. That is to say, data cannot be processed in real-time. On the other hand, iterative methods (e.g. EKF) are more suitable and give an opportunity to implement real-time TMA (Farina, 1999).

In addition to iterative methods, there are several papers in the literature solving the target-tracking problem by the use of evolutionary algorithms (Ince, Sezen, Saridogan, & Ince, 2009; Sonmez, Hocaoglu, & Genc, 2017) and they usually consider the batch processing case. The evolutionary algorithm used for TMA uses a set of noisy bearing measurements, simulates the target's movement with different speed, course and location parameters, calculates the bearings based on targets simulation parameters and own-ship parameters and tries to minimize the difference between bearings acquired from the passive sensor and the simulated bearings. As the difference decreases, the algorithm converges to the global minima and outputs a not optimal but a good enough solution. Since our objective is to minimize the difference, the problem becomes an optimization problem. When the maximum sensor detection ranges, different course, and speed combinations are taken into consideration, the solution space for TMA problem becomes very large and finding an optimal solution turns out to be a challenging task.

In this book chapter, target-tracking problem and Target Motion Analysis algorithms used in the literature will be examined in a detailed way. In the following subsections, observability, target-tracking algorithms such as the Extended Kalman Filtering and other iterative methods will be discussed. Following that the implementation of evolutionary algorithms on TMA problem will be considered. An implementation of the Genetic Algorithm (GA) on the Target Motion Analysis (TMA) problem will be given and a brief discussion of the results will be provided.

BACKGROUND

Problem Description

TMA algorithm provides the means of predicting the target's next position given the previous bearing measurements. In this sense, it can be defined as a tracking method as well as a target trajectory extractor. While working on a general TMA problem, one needs to make two important assumptions. First, it is assumed that target moves with a constant speed in a linear direction without changing its course in Cartesian coordinates (x-y plane). Second, is that the vessel collecting measurements needs to maneuver and change its course at least once in order to make the target observable. A clear definition of observability will be given in next section. In batch processing scheme, m angular measurements that are T seconds apart are collected from the on board passive sensor (see Figure 1).

The position of the own-ship in x-y plane is represented with X_O and Y_O while the position of the target is represented as X_T and Y_T in this paper. Farina (1999) and Ince et al. (2009) defined the bearing calculation with respect to x-axis as follows:

$$h(n) = \arctan \frac{Y_T(n) - Y_o(n)}{X_T(n) - X_o(n)} \tag{1}$$

Figure 1. Own-ship and target-ship traces for general TMA problem

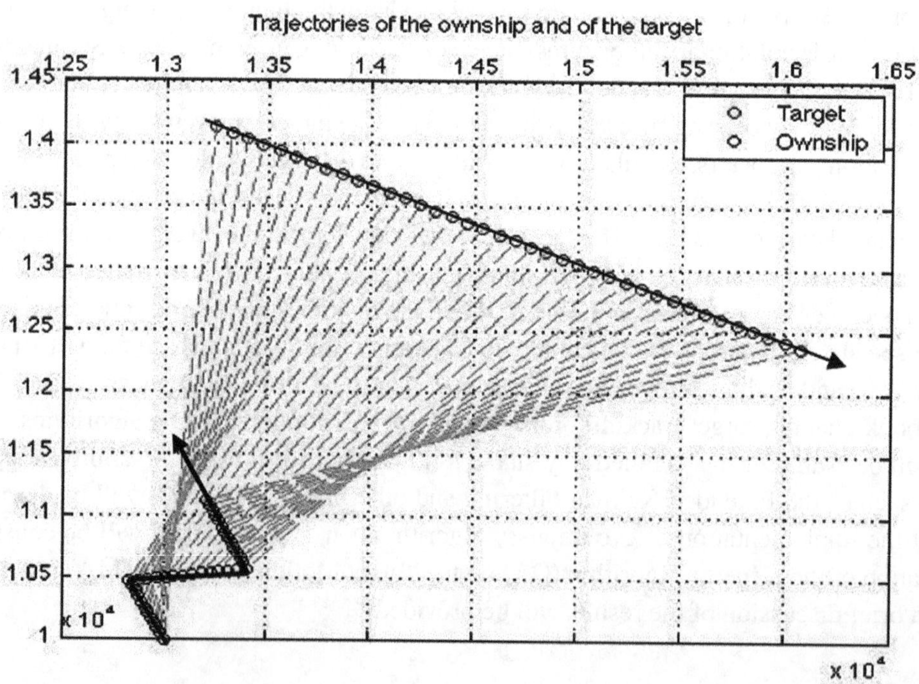

$$\theta(nT) = h(nT) + w(nT), n = 1,2,3,\ldots,m \tag{2}$$

where $\theta(n)$ represents the noisy bearing measurements; $X_T(n)$, $Y_T(n)$ pair is the unknown target coordinates at the nth time; $X_O(k)$, $Y_O(k)$ pair is the known sensor coordinates and $w(n)$ is the measurement error at the nth time. Measurement error is produced from Normal (Gaussian) probability density function where mean $\mu = 0$ and variance $\sigma^2 = 1$ is assumed. Velocities of the own-ship and target along the x- and y-axis can be extracted from the course (C) and speed (S) values:

$$V_X = S_O \cos(C) \tag{3}$$

$$V_Y = S_O \sin(C) \tag{4}$$

$$X_O(n) = X_{O0} + nTV_{Ox} \tag{5}$$

$$Y_O(n) = Y_{O0} + nTV_{Oy} \tag{6}$$

$$X_T(n) = X_{T0} + nTV_{Tx} \tag{7}$$

$$Y_T(n) = Y_{T0} + nTV_{Ty} \tag{8}$$

Thus (2) becomes:

$$\theta(n) = \arctan\left(\frac{(Y_{T0} + nTV_{Ty}) - (Y_{O0} + nTV_{Oy})}{(X_{T0} + nTV_{Tx}) - (X_{O0} + nTV_{Ox})}\right) + w(n), n = 1,2,3,\ldots,m \tag{9}$$

TMA algorithm's main goal is to find the unknown parameters of the target which are X_{TO}, Y_{TO}, V_{Tx}, V_{Ty} and these unknown parameters can be represented as a vector $\mathbf{x} = [X_{TO}, Y_{TO}, V_{Tx}, V_{Ty}]$.

Observability

In order to extract target parameters in TMA problem, either the target or the own-ship must have acceleration during the measurement period. In this study, the own-ship acceleration is assumed. Observability can be thought as finding a unique set of solution parameters using m target bearings. If there is more than one unique solution or if one of the parameters tends to infinity or zero, then it is concluded that target is not observable. Observability requirements for TMA problem are studied rigorously in Nardone and Aidala (1981) and Pillon et al. (2016).

Figure 2 shows the examples of observability. Figure 2 (a, b, c) are all the examples of observable conditions, however in Figure 2 (d) since the own-ship and target move with a constant speed in a linear direction and none of the platforms makes a maneuver and change its course, this case indicates the non-observable condition for TMA.

Maximum Likelihood Estimator (MLE)

One of the ways to estimate the unknown parameters of vector **x** is to use Maximum Likelihood Estimator (MLE). It is obtained from the maximization of the log-likelihood function for the bearing measurements. The maximization problem of MLE can be expressed as a minimization problem. Farina (1999) expressed TMA using MLE as:

$$\arg\min\left\{n * \ln\left(\sigma\sqrt{2\pi}\right) + \left(\frac{1}{2\sigma^2\sum\left(\theta\left(k\right)-\varphi\left(k\right)\right)^2}\right)\right\} \tag{10}$$

From equation (10) it can be seen that TMA problem turns out to be a Least Squares (LS) problem. Details of various estimators and proposed instrumental variable estimator can be found in Doğançay (2005) 's work.

Figure 2. Examples of observability (a) stationary target (upper-left) (b) own-ship maneuvers (upper-right) (c) both ships maneuver (lower-left) (d) both ships move but no maneuver (lower-right)

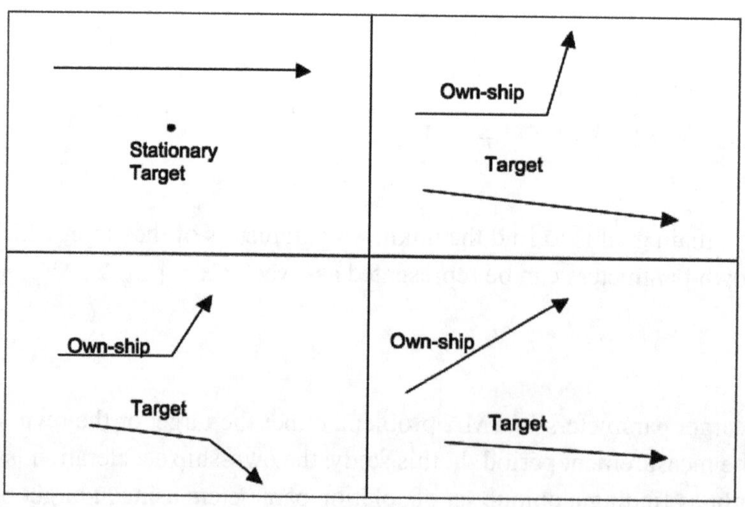

Kalman Filtering

Maybeck (1979) defines Kalman filter as an optimal recursive data processing algorithm. Kalman filter estimates the current state of a linear dynamic system based on the system's previous state and the current measurement or observation made on the system. Kalman filter also accounts for the uncertainties involved in the measurement and any process uncertainties (Grewal & Andrews, 1993). For a single parameter problem uncertainty associated with measurement and estimation are expressed as variance while for a multi-parameter problem covariance matrices are used. The mechanics of the Kalman Filter is given in Figure 3.

In Figure 3, $\hat{X}$ represents the estimation of the state matrix, u represents the control variable matrix, P represents the process covariance matrix, z represents the measurement taken at a particular time, K represents the Kalman gain, Q represents the process noise covariance matrix, R represents the sensor noise covariance matrix.

Kalman Filter works in a recursive manner and first predicts the state of the system based on the previous state of the system and then corrects the prediction upon the receipt of a measurement update. Here the Kalman Gain plays a key role and determines how much weight is put on the estimate and the measurement. If the measurement is noisy and unreliable then the projected estimate is weighted more. However, if the measurements taken are so accurate then more weight is put on the measurement.

Kalman filter was designed for optimally solving the linear estimation problems and is widely used in process control, tracking and navigation. Although Kalman Filter works well on linear dynamic systems, most dynamic systems in the real world are usually nonlinear due to the factors such as noise and inherent randomness involved in the processes. Therefore, the variations of Kalman Filters such as Extended Kalman Filter (EKF) and Unscented Kalman Filter (UKF) have been used in order to estimate the state of a nonlinear dynamic system. Both EKF and UKF use the linearization process and apply the principles of the Kalman Filtering. Although the EKF is a good choice for solving TMA problem, it can also exhibit erratic behaviors under some circumstances. The erratic behavior of EKF is discussed and candidate methods to overcome this problem are proposed by Aidala (1979).

Figure 3. Kalman filtering process

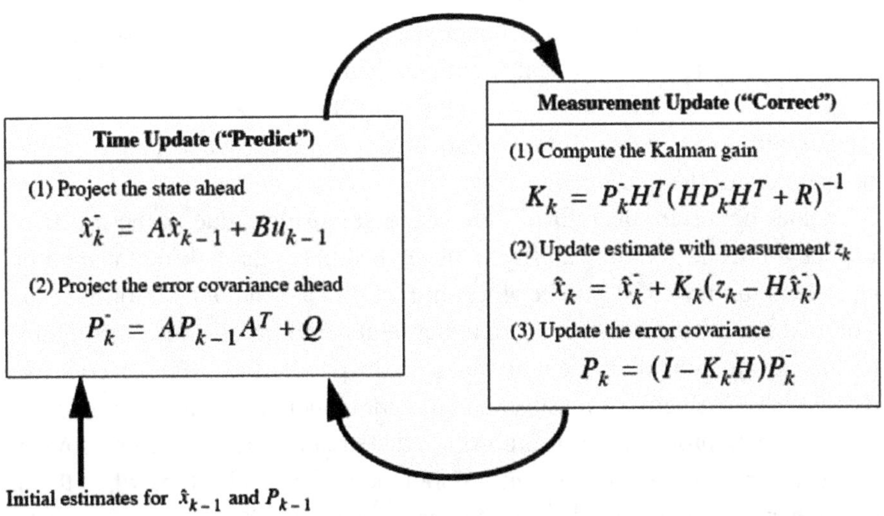

BOTMA problem can be characterized as a nonlinear dynamic system due to the inaccurate noisy bearing measurements and the uncertainties involved in the process. Therefore the application of linear Kalman Filter will produce erroneous results. In order to overcome this problem the widely accepted solution technique is the use of either the Extended Kalman Filter (EKF) or Unscented Kalman Filter (UKF). Gibbs (2011) described the dynamic model of the problem as

$$x_k = \begin{bmatrix} 1 & 0 & t & 0 \\ 0 & 1 & 0 & t \\ 0 & 0 & 1 & 0 \\ 0 & 0 & 0 & 1 \end{bmatrix} * x_{k-1} \text{ where } x_{k-1} = \begin{bmatrix} X_T \\ Y_T \\ V_x \\ V_Y \end{bmatrix}$$

and the measurement is taken according to Equation 9. Several applications of UKF on TMA problem can be found in Santhosh et al. (2015) and Raju et al. (2015).

However, the main focus of this chapter is the implementation of an evolutionary algorithm, namely the Genetic algorithm (GA), and to evaluate the performance of the algorithm on the TMA problem. In the following section Genetic Algorithm will be briefly introduced and then the implementation of BOTMA problem is discussed.

Genetic Algorithm

The genetic algorithm gets its inspiration from nature and it uses the natural phenomenon seen in human genetics, evolution, natural selection and inheritance as its source in the area of machine learning. Genetic Algorithm creates a set of individuals, also called population, where each individual consists of different genetic material, which is analogous to a human DNA. Genetic algorithm starts with these randomly generated chromosomes and individuals in the population exchange genes with its mate and sometimes exposed to mutation after mating. These processes can be thought as an evolution of the population.

In genetic algorithm gene exchange process is called crossover. Crossover operation operates on two individual chromosomes and the selection of chromosomes may vary depending on the implementation. However, most genetic algorithm applications implement the selection of chromosomes according to their fitness values. Roulette-wheel and tournament selection schemes maybe counted as the probabilistic ways of selecting chromosomes where the chromosomes with high fitness values have more chance compared to the chromosomes with low fitness values. Crossover and fitness based selection can be stated as the most important operators of the genetic algorithm.

Mutation is another important operator for the genetic algorithm, due to the fact that it provides a diversity of the population and gives the ability to the algorithm to search different areas of search space without staying at local minima. The genetic algorithm follows a cycle of operations until it reaches the desired output or until it reaches the maximum number of generations allowed (see Figure 4). The cycle starts with the fitness evaluation all of the individuals in the population. After that, a new population is created using fitness-based selection, crossover and mutation operations on the individuals in the population. During the selection process, the chromosomes that are not selected for crossover are discarded. Then, fitness evaluation is performed on the new population. This cycle stops when the desired fitness value achieved or when the maximum number of generations is reached.

Figure 4. The flow of Genetic Algorithm

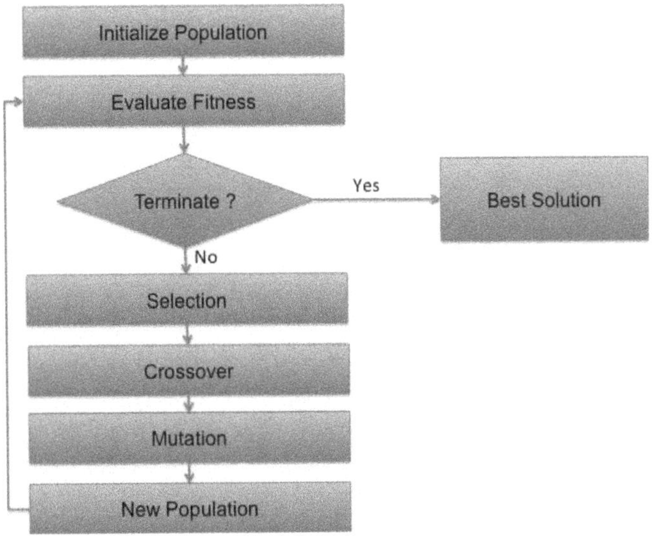

SOLUTIONS AND RECOMMENDATIONS

Chromosome Representation

TMA genetic algorithm uses Cartesian coordinates (*X*, *Y*) as the position vector and the velocities along the *x*- and *y*-axis are derived from the speed and course values using Equations (3) and (4). Each chromosome has $[X_{TO}, Y_{TO}, C, S]$ values where X_{TO} and Y_{TO} are the initial positions of the target; and *C* and *S* are the course and speed values of the target. In this study, binary representation has been used. Therefore, each field in the chromosome is represented by a number of bits, 1s and 0s, which is determined after the search space narrowing process.

Search Space Narrowing

Genetic Algorithm used in this study to solve TMA problem starts with search space narrowing process. The reason for the need of narrowing process is because the *X* and *Y* coordinate values can get really large values and the number of bits required to represent can get quite large which might cause the algorithm to be ineffective. As previously stated generally Genetic Algorithm starts with a randomly generated population. However, in this study, a different type of initialization process has been used. The bearing measurements reported by the sensor indicate that the target should be somewhere on this bearing line or more or less close to the bearing line reported due to the ambient noise. Using this fact, the X_{TO}, Y_{TO} coordinate pairs were generated on the first bearing line reported by the sensor. Distance between each X_{TO}, Y_{TO} pair generated is selected to be 2000 meters. The maximum passive sensor range is assumed to be 80 kilometers (see Figure 5).

After X_{TO}, Y_{TO} pairs generated, course (*C*) and speed (*S*) values were assigned to each X_{TO}, Y_{TO} pairs. Course values were selected in all directions - 0 to 359 degrees - and speed values were selected as 0 to 20 meters per second. Thus, each X_{TO}, Y_{TO} pair was duplicated 7200 times and assigned a different

Figure 5. X, Y pairs on the first bearing reported

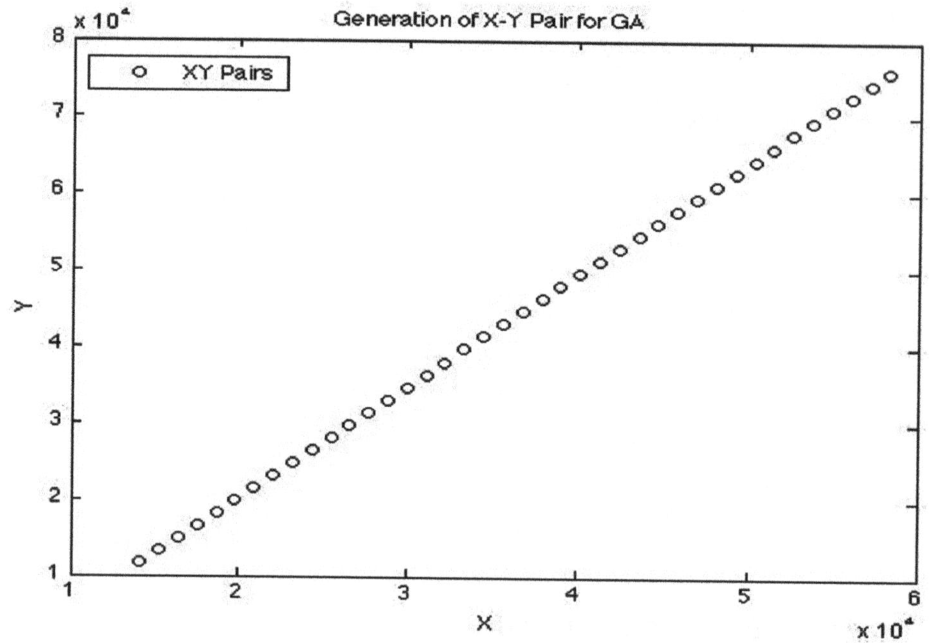

course and speed value for the fitness evaluation, and finally, the initial solution set of $[X_{TO}, Y_{TO}, C, S]$ values were obtained.

As previously mentioned passive sensor on board the own-ship collects *m* angular measurements *T* seconds apart. Each $[X_{TO}, Y_{TO}, C, S]$ element in the solution set is moved according to its course (*C*) and speed (*S*) values starting from the position X_{TO}, Y_{TO}, using Equations (3), (4), (7), (8). This yields *n* target positions on the *x-y* plane *T* seconds apart. Since the own-ship position is known at angular measurement instances, one can easily calculate *m* bearings for each solution set element, $[X_{TO}, Y_{TO}, C, S]$, using Equation (1).

The fitness of each element in this set was evaluated according to Equation (10) aiming to minimize the total angular difference between *m* simulated bearing and *m* reported bearing from the passive sensor. Basically, the differences between the corresponding solution set and reported bearings are calculated and then these difference values were summed up. Since the objective function looks for a minimum value, the smaller the sum, the better the fitness. GA is usually used in maximization sense, so fitness function is defined as reciprocal of the total angular difference value in this study. Upon completion of the fitness evaluation, the solution set consisting of $[X_{TO}, Y_{TO}, C, S]$ elements were sorted in a descending order.

In order to narrow the parameter space, first 1000 fittest $[X_{TO}, Y_{TO}, C, S]$ elements were selected and the rest was discarded. Afterward, minimum, maximum and range values for each parameter are determined based on the selected 1000 fittest elements. These minimum and maximum values of each parameter actually give the boundaries of the search space. Application of this procedure narrows the parameter and search space helps the algorithm work more efficiently and uses less number of bits to represent each parameter in a chromosome.

TMA Genetic Algorithm uses a classical genetic algorithm where each parameter in $[X_{TO}, Y_{TO}, C, S]$ is represented as a binary string. Upon completion of the parameter space narrowing, the fittest 1000 values are used for the initial population, instead of starting with a totally random population. Minimum and maximum values of parameters, found after parameter narrowing, are used to calculate the number of bits to represent decimal values in binary format. Each parameter is then converted to binary representation and then combined together to form a chromosome. Each bit in this chromosome is called a gene and carries important information to the next generations. Next, the fitness values of all chromosomes in the population are calculated and the population is sorted in the descending order with regard their fitness values.

To introduce randomness roulette-wheel selection is applied in this study. Selected chromosomes are paired to generate new offspring by using the crossover operation. The crossover operation is applied to 4 parameters in a chromosome-*X, Y, C, S* values- separately; therefore four-point crossover is used. Next, mutation is applied to introduce diversity and not to stick at a local minima. Probabilities of crossover and mutation were selected 0.99 and 0.1 respectively. After crossover and mutation, a new population with new offspring is produced and this procedure is repeated for a number of generations. Upon the execution of the algorithm for a predetermined number of the generations; the best result found so far is displayed.

Computational Study and Results

In this section, the results of various trials are shown. In all trials, the own-ship parameters were kept constant. It was assumed that the target moved along a straight line with no course change and with a constant speed. Cartesian coordinates were used to represent the position of target and own-ship. MATLAB software was selected for the implementation of TMA genetic algorithm. The parameters of the own-ship are presented in Table 1. In all trials totally 37 bearings were collected. The own-ship makes a zig-zag movement to increase the observability. In the first leg own-ship proceeds to 340 and after 12 were collected it turns to 080 and another 12 bearings were collected. Finally, it turns to 340 again and after 12 bearing collection, the scenario ends. The crossover and mutation probabilities were 0.99 and 0.1, respectively.

In the first two trials the effects of bearing rate were examined. In trial 1, the target was placed (13246, 18127) point on the *x-y* plane with a heading to east with a speed of 4 m/s. TMA algorithm was run 25 times and the standard deviation, σ, of the Gaussian noise was selected as 0.5°. As it can be seen, the genetic algorithm performs fairly well. Even though it generates erroneous results sometimes, on the average it performs well and the error is quite low. In trial 2 the speed of the target is raised to 8 m/s and the location of the target was closer to the own-ship in order to get a higher bearing rate. A higher bearing rate provides better results especially for the course and speed values. The results for the first 2 trials are given in Table 2. The plots of ground and the results of the genetic algorithm are plotted on to the same figure and it is shown in Figure 6.

Table 1. Own-ship parameters used in trials

	X	Y	C	S
Target	13000	10000	340 (Leg1) 80 (Leg 2)	3 m/s

Table 2. Results of trials 1 and 2 (Bearing rate test)

Runs	Trial 1				Trial 2			
	X0	Y0	Course	Speed	X0	Y0	Course	Speed
1	13266	18076	87	4	13258	16078	89	8
2	13237	18214	97	4	13223	16029	89	8
3	13220	17735	83	4	13216	15358	89	7
4	13171	17105	75	4	13240	16090	90	8
5	13176	18434	70	5	13212	15990	89	8
6	13250	17970	88	4	13203	15859	87	8
7	13237	18093	92	4	13221	16069	90	8
8	13310	18321	119	4	13282	16324	92	8
9	13296	18601	103	4	13278	16385	93	8
10	13223	18094	92	4	13238	15448	90	7
11	13256	18167	89	4	13333	17118	93	9
12	13186	17664	84	4	13168	15755	86	8
13	13194	17716	86	4	13203	15944	88	8
14	13219	18093	96	4	13197	15256	88	7
15	13257	18373	100	4	13192	16008	90	8
16	13376	20665	110	5	13287	17555	90	10
17	13144	16568	69	4	13391	18141	97	10
18	13179	17697	89	4	13203	16043	90	8
19	13222	18714	73	5	13284	16861	90	9
20	13270	18411	103	4	13281	16202	91	8
21	13229	18650	70	5	13269	16258	91	8
22	13198	17681	82	4	13206	15987	89	8
23	13215	17824	86	4	13291	16892	91	9
24	13193	17410	80	4	13183	15361	91	7
25	13274	18323	95	4	13313	16461	94	8
Average	13231.92	18103.96	88.72	4.16	13246.88	16218.88	90.28	8.12
Ground	13246	18127	90	4	13246	16127	90	8
Error	14.08	23.04	1.28	0.16	0.88	91.88	0.28	0.12

In Trials 3 and 4 the initial position and speed of the target kept constant, while its course value changed. In these trials, examination and validation of the algorithm's performance were examined by changing the heading values. Table 3 shows that even in the change of course, the genetic algorithm performs well enough. Results can be seen in Table 3 and Figure 7.

In trials 5, 6, 7 and 8 the ambient noise was increased gradually. In this process, the standard deviation value of the Gaussian noise was increased to introduce more noise in bearing measurements. The algorithm ran 10 times and the results were recorded. The results can be seen in Table 4, Table 5, and Figure 8. It clearly shows that when there is no noise the algorithm performs extremely well. However,

Figure 6. Bearing rate test

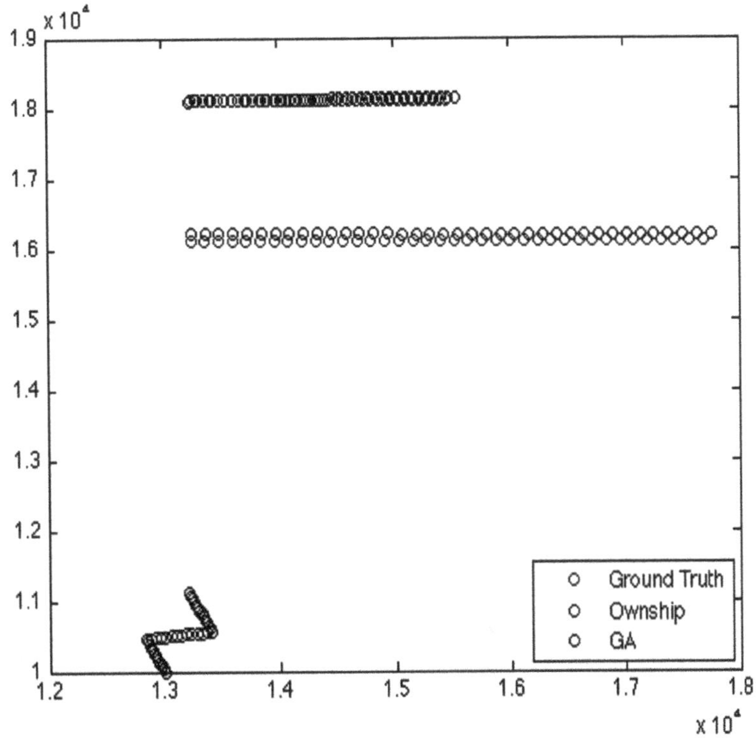

as the standard deviation increases, the performance of the algorithm decreases as expected. When the standard deviation value exceeds 5° we started to get erroneous results from the algorithm.

FUTURE RESEARCH DIRECTIONS

Target Motion Analysis problem is one of the tracking problems studied extensively in the literature. Due to the uncertainties and noise involved in the measurement process of bearings by the passive sensors the nonlinear nature of the problem demonstrates challenges to the solution methodologies. Since the TMA problem is closely related to the target tracking, the first option the researchers used was the application of the Kalman Filter and its variations such Extended Kalman Filter and Unscented Kalman Filter applicable to the nonlinear problems. Therefore researchers mainly focused on these areas and the literature on TMA is established heavily on Kalman Filtering.

Table 3. Results of trials 3 and 4 (Course value change test)

Runs	Trial 3				Trial 4			
	X0	Y0	Course	Speed	X0	Y0	Course	Speed
1	13210	14035	119	9	13174	13694	153	8
2	13816	14021	120	9	13167	13689	153	8
3	13224	14077	121	9	13214	13729	154	8
4	13176	13649	120	8	13211	13727	154	8
5	13319	14619	124	10	13168	13687	153	8
6	13255	14131	122	9	13170	13694	153	8
7	13192	14044	121	9	13176	13698	153	8
8	13205	14043	121	9	13252	14125	155	9
9	13291	14572	123	10	13175	13691	153	8
10	13250	14129	122	9	13172	13690	153	8
11	13193	14044	121	9	13246	14128	155	9
12	13209	14026	120	9	13180	13699	153	8
13	13253	14164	123	9	13169	13691	153	8
14	13275	14560	123	10	13206	13718	154	8
15	13195	13657	120	8	13174	13689	153	8
16	13255	14131	122	9	13177	13695	153	8
17	13182	13646	120	8	13168	13690	153	8
18	13202	14043	121	9	13209	13725	154	8
19	13190	13653	120	8	13211	13717	154	8
20	13202	14042	121	9	13216	13724	154	8
21	13177	13641	120	8	13206	13719	154	8
22	13270	14187	123	9	13214	13730	154	8
23	13198	13664	120	8	13177	13698	153	8
24	13220	13702	121	8	13166	13694	153	8
25	13198	13665	120	8	13293	14540	156	10
Average	13246.28	14005.8	121.12	8.8	13195.64	13771.24	153.6	8.16
Ground	13246	14127	122	9	13246	14127	155	9
Error	0.28	121.2	0.88	0.2	50.36	355.76	1.4	0.84

On the other hand, to the best of authors' knowledge, there are only a few research articles related to the application of evolutionary algorithms or heuristics on Target Motion Analysis. Ince et al. (2009) applied Genetic Algorithm; Genc (2010) proposed Big Bang-Big Crunch (BB-BC) optimization algorithm and recently Sonmez et al. (2017) applied Covariance Matrix Adaptation Evolutionary Strategies (CMA-ES) to the bearing-only TMA problem. Therefore the authors believe that evolutionary algorithms, meta-heuristics, and swarm optimization algorithms can provide a promising research area on the optimal solution of the bearing only TMA problem and different flavors of evolutionary algorithms such as Differential Evolution and Particle Swarm Optimization can be applied for further studies.

Figure 7. Heading change test

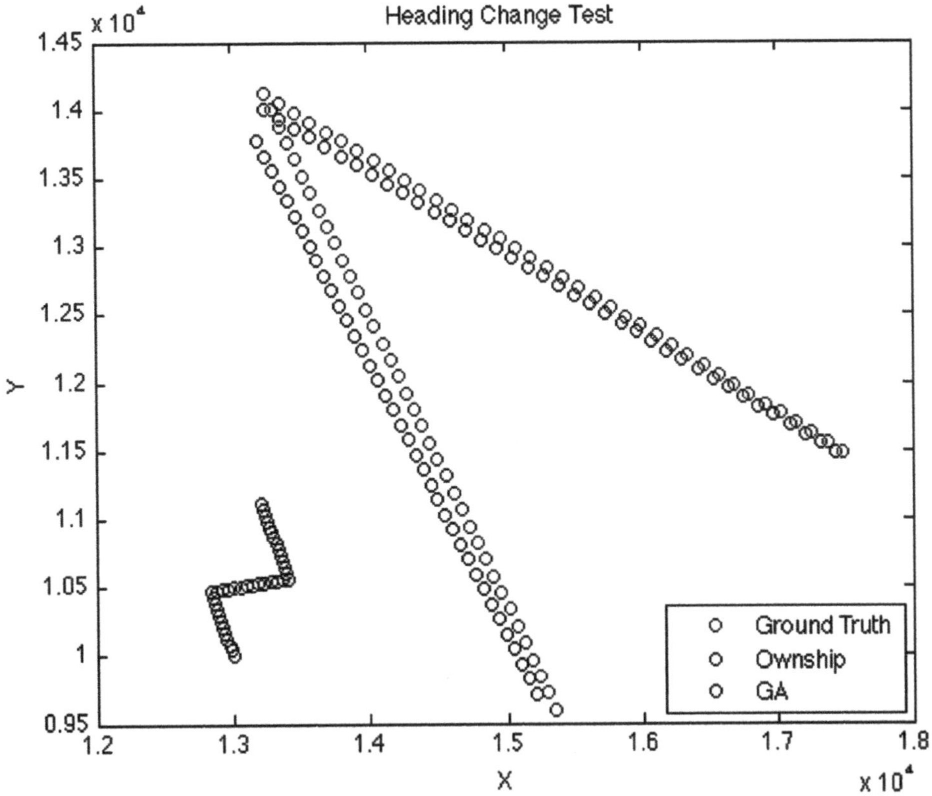

Table 4. Effects of standard deviation change on algorithm (Trials 5 and 6)

Runs	Trial 5 (σ=0)				Trial 6 (σ=0.5)			
	X0	Y0	Course	Speed	X0	Y0	Course	Speed
1	14001	16001	122	9	14003	15992	122	9
2	14000	16000	122	9	13969	15988	122	9
3	13999	15999	122	9	13976	15935	121	9
4	14000	16000	122	9	13843	15293	119	8
5	14001	16001	122	9	14036	16034	122	9
6	13998	16000	122	9	14043	16059	123	9
7	14003	16005	122	9	13956	15935	121	9
8	13999	15999	122	9	14044	16055	123	9
9	14000	16001	122	9	13992	15993	122	9
10	14001	16001	122	9	14024	16053	123	9
Average	14000.20	16000.7	122	9	14003	15933.7	121.8	8.9
Ground	14000	16000	122	9	14000	16000	122	9
Error	0.2	0.7	0	0	3	66.3	0.2	0.1

Table 5. Effects of standard deviation change on algorithm (Trials 5 and 6)

Runs	Trial 7 (σ=1)				Trial 8 (σ=5)			
	X0	Y0	Course	Speed	X0	Y0	Course	Speed
1	14012	15900	120	9	13651	14330	112	7
2	13906	15388	120	8	13485	13360	116	5
3	14344	17863	124	12	13722	15744	109	10
4	14014	16032	122	9	14913	18409	142	12
5	14262	17380	126	11	14566	20104	127	16
6	14077	16588	122	10	14011	16310	118	10
7	14161	16751	124	10	14646	18594	139	12
8	13974	15987	123	9	13990	15423	122	8
9	14012	16013	123	9	13632	14041	118	6
10	14495	18134	128	12	14201	15923	130	8
Average	14125.70	16603.6	123.2	9.9	13651	16223.8	123.3	9.4
Ground	14000	16000	122	9	14000	16000	122	9
Error	125.7	603.6	1.2	0.9	349	223.8	1.3	0.4

Figure 8. Trials 5-8 standard deviation change

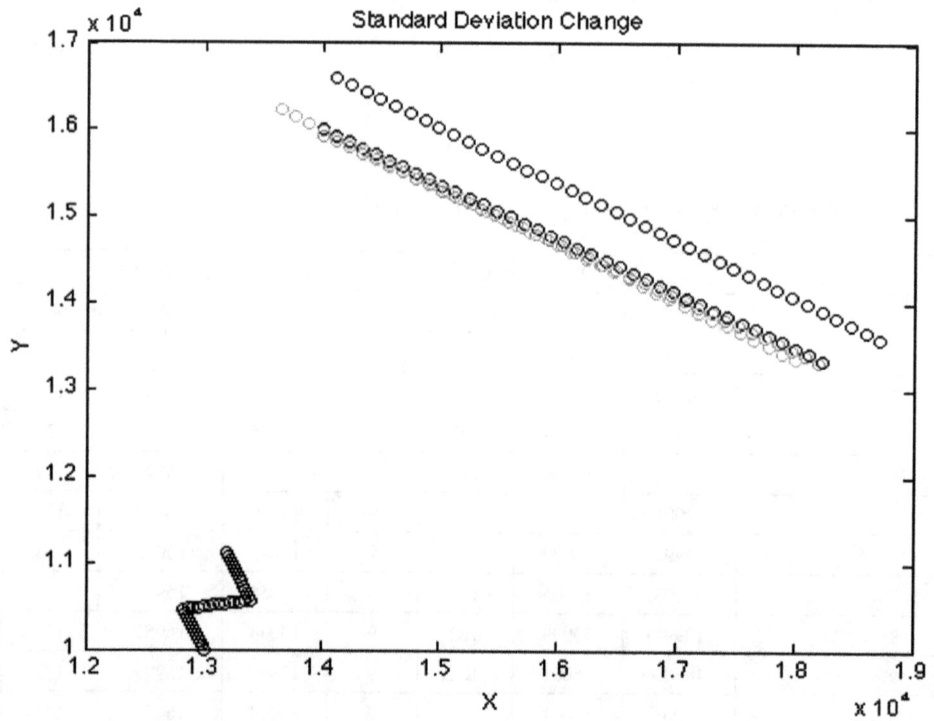

CONCLUSION

In this chapter, authors provide a solution method using Genetic Algorithm on Bearing-only Target Motion Analysis (BOTMA) problem rather than using Extended Kalman Filtering, which is widely proposed in the literature. Although there is a particular research paper based on the Genetic Algorithm proposed by Ince et al. (2009), authors of this chapter use a different search space narrowing technique that eliminates the need of dummy runs and chromosome tuning. The results of the algorithm presented in the previous section reveal that TMA genetic algorithm works well when the bearing rate is high and the noise level is low. The algorithm successfully estimates target motion parameters even when the bearing data is corrupted with Gaussian noise.

REFERENCES

Aidala, V. (1979). Kalman Filter Behavior in Bearings-Only Tracking Applications. *IEEE Transactions on Aerospace and Electronic Systems*, *AES-15*(1), 29–39. doi:10.1109/TAES.1979.308793

Doğançay, K. (2005). On the efficiency of a bearings-only instrumental variable estimator for target motion analysis. *Signal Processing*, *85*(3), 481–490. doi:10.1016/j.sigpro.2004.10.014

Farina, A. (1999). Target tracking with bearings – Only measurements. *Signal Processing*, *78*(1), 61–78. doi:10.1016/S0165-1684(99)00047-X

Genc, H. M. (2010). A new solution approach for the bearing only target tracking problem. *4th International Workshop on Soft Computing Applications*. 10.1109/SOFA.2010.5565618

Gibbs, B. P. (2011). *Advanced Kalman filtering, least-squares and modeling: a practical handbook*. Hoboken, NJ: A John Wiley & Sons.

Grewal, M. S., & Andrews, A. P. (1993). *Kalman filtering: theory and practice*. Englewood Cliffs, NJ: Prentice Hall.

Ince, L., Sezen, B., Saridogan, E., & Ince, H. (2009). An evolutionary computing approach for the target motion analysis (TMA) problem for underwater tracks. *Expert Systems with Applications*, *36*(2), 3866–3879. doi:10.1016/j.eswa.2008.02.046

Maybeck, P. S. (1979). *Stochastic Models, Estimation, and Control: Vol.: 1*. New York: Academic Press.

Nardone, S., & Aidala, V. (1981). Observability Criteria for Bearings-Only Target Motion Analysis. *IEEE Transactions on Aerospace and Electronic Systems*, *AES-17*(2), 162–166. doi:10.1109/TAES.1981.309141

Pillon, D., Perez-Pignol, A., & Jauffret, C. (2016). Observability: Range-only vs. bearings-only target motion analysis for a leg-by-leg observers trajectory. *IEEE Transactions on Aerospace and Electronic Systems*, *52*(4), 1667–1678. doi:10.1109/TAES.2016.150016

Raju, K. L., Rao, S. K., Das, R. P., Santhosh, M. N., & Murthy, A. S. (2015). Passive Target Tracking using Unscented Kalman Filter based on Monte Carlo Simulation. *Indian Journal of Science and Technology*, *8*(29). doi:10.17485/ijst/2015/v8i1/76981

Santhosh, M. N., Rao, S. K., Das, R. P., & Raju, K. L. (2015). Underwater Target Tracking using Unscented Kalman Filter. *Indian Journal of Science and Technology*, *8*(31). doi:10.17485/ijst/2015/v8i31/77054

Sonmez, H. H., Hocaoglu, A. K., & Genc, H. M. (2017). A new solution method for the passive target motion analysis problem using evolutionary strategies. *2017 25th Signal Processing and Communications Applications Conference (SIU)*. doi:10.1109iu.2017.7960401

KEY TERMS AND DEFINITIONS

Evolutionary Algorithm: A branch of evolutionary computation that is based on a population of candidate solutions and evolves through the optimal solution or nearly optimal solution via using the selection, reproduction, and mutation phenomenon seen in the nature.

Genetic Algorithm: A widely used evolutionary algorithm for solving combinatorial optimization problems that is based on probabilistic selection, crossover, and mutation operators.

Kalman Filter: A recursive optimal data processing algorithm runs on a computer that is widely used in target tracking, process control, and navigation.

Noise: Additive random perturbations caused by the medium and the random fluctuations of electrons at the receiver front end that causes difficulty in detecting signals of interest if excessively high.

Observability: The kinematic conditional requirements such as acceleration and wide turns of observing vessel that make the TMA problem solvable.

Passive Sensor: An electronic device, equipment, or system that can measure and extract parameters of a physical phenomenon (e.g. electromagnetic, optical, acoustic) caused by the exciters without actively emitting or transmitting to the medium.

Target Motion Analysis (TMA): Extraction of target kinematic parameters based on the measurements made by the passive sensors such as bearing, frequency, and range.

Chapter 15
Similarity Measure Optimization for Target Detection:
A Case Study for Detection of Keywords in Telephone Conversations

Batuhan Gundogdu
National Defence University, Turkey

Murat Saraclar
Bogazici University, Turkey

ABSTRACT

Today, in the era of data and computing, fast and reliable retrieval of information has become of great importance for security and military applications, and continues to be such, as the amount available digital data increases every second. While the search and retrieval of text data has produced mature products and are today being used in search engines everyday by everyone, the retrieval of spoken content still remains a young research, especially for low resource languages where the available data is scarce to train reliable speech recognition systems. This chapter provides a thorough introduction of a speech retrieval task called "keyword search" and presents a novel similarity measure optimization-based approach. The case study was experimented on telephone conversations in three different languages and thousands of keywords randomly selected from each language were searched in the document. The experiments show that the technique introduced in this chapter offers a new methodology to handle the terms that does not even exist in the vocabulary of the speech recognition systems.

INTRODUCTION

The digitalization of data and the ever-increasing storage capacity has brought people closer to a life like the fictional world in Jorge Luis Borges' *Library of Babel* (Borges, 1998), —an infinite library *whose books contain every possible string of letters and, therefore, somewhere an explanation of why the library exists and how to use it.* Though, it was highly doubted that the librarians would ever find

DOI: 10.4018/978-1-5225-5513-1.ch015

that book amid the miles of nonsense. The situation the society faces today is somewhat similar: When it comes to retrieval of data, and especially the spoken data, the conundrum appears to be the "*Paradox of Abundance*" defined in Haidt (2006): *Quantity undermines the quality of our engagement*. Hence, it is vital now, in the Internet era; to efficiently and correctly retrieve the data of interest to make our lives easier as evidenced by the well-developed text retrieval technology, which is undoubtedly one of the biggest life-changing innovations of the past century. On the other hand, speech retrieval technology is still nascent even though speech is definitely one of the most appealing forms of information content. The retrieval of speech is therefore not only attractive but also necessary for users to efficiently browse across the vast quantities of multimedia content.

Detection of certain words in telephone conversations is also a very important task for security reasons and fight against terrorism. Such a task may require a dynamic list of 'suspicious' keywords, so that there no longer exists a necessity for developing the retrieval system from the beginning once new keywords become of interest.

This chapter focuses on a speech retrieval task called spoken term detection, also known as keyword search (KWS). In KWS, the user provides a query in text form consisting of one or more words. This query is searched in an un-transcribed audio archive and the locations in the audio where the query exits are returned to the user in an ordered list, with respect to a similarity score the KWS system attributes to them. The text counterpart of the task that was just defined is a very common one in everyday life. Nowadays, people rarely spend a day without searching something in Google©. Text browsing also comes in handy and used in text editors with very well-known short-cut button configurations. Despite this wide usage of text retrieval acknowledged by almost every computer user, searching for speech still remains a young research. Searching for audio-visual content in multimedia archives, such as YouTube, only provides videos whose title or meta-data has high similarity scores to the text query. The actual locations where the terms of interest are uttered are not available upon such a search. In KWS, on the other hand, the audio documents are returned to the user along with the time stamps where the query occurs. Such a system maybe desired in situations where the user is interested in finding locations of their queries in lecture videos, broadcast news reports, radio communications, meeting recordings. Furthermore, detecting and locating the utterance of certain keywords in telephone conversations may be very important for security applications for prevention and/or investigation of criminal or terrorist activities. System diagram demonstrating the KWS task is given in Figure 1.

Since the text retrieval task is now a mature product and has been providing reliable results, the first practiced approach to KWS was using an automatic speech recognition (ASR) system on the audio archive and converting the task into a text retrieval task. Once the spoken document is converted into a string of words by the ASR system, the task is now to finding text in text, which is not so hard. However, only depending on the ASR system output limits the retrieval system's performance to the accuracy and the vocabulary size of the ASR systems being used. Furthermore, when circumstances are harder as in the situations when the recording conditions are very noisy and there is little or no available data for the language of interest, which is referred to as a "*low resource language*" in the literature.

This chapter aims to address the problem of KWS especially for low resource languages by presenting the methodologies adopted in literature and introduces a novel approach that conducts a similarity measure optimization and a dynamic search. In the next section the definitions are presented and the conventional techniques are introduced. The problem brought by scarcity of resources is addressed in a thorough manner and the solution methodology is described. The validity of the approach is presented

Figure 1. A brief demonstration of the keyword search task

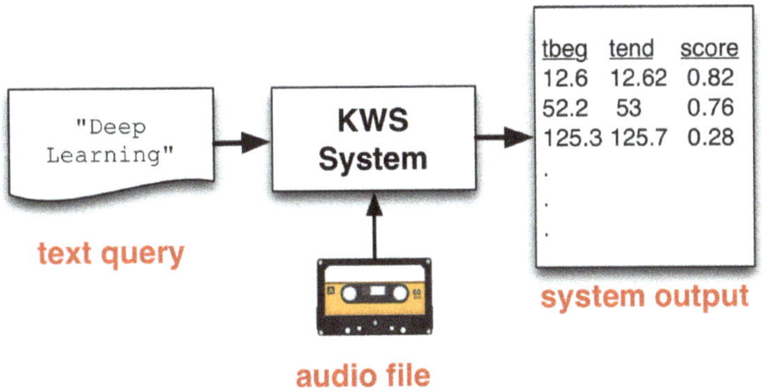

with the experiments conducted on three low resource languages of the IARPA Babel Program (Harper, 2014).

BACKGROUND

This section summarizes the definition of the task in hand and the presents the background information explaining the methodologies adopted in the literature. KWS is defined as the task of obtaining the locations of a user provided text query term in an audio speech archive. The speech archive is referred to as the *"document"* in this book and in literature. The earliest mature works date back to mid-1990s when the ASR and information retrieval (IR) technologies began to achieve satisfactory performances. The Text Retrieval Conference (TREC) evaluations conducted in 1999 and 2000 are considered a milestone in the KWS applications. The application in (Van Thong et al, 2002) was the first KWS system used in web. Since these applications mainly focused on using broadcast news speech, the ASR results had fewer errors due to better recording conditions when compared to conversational speech. The actual and deep endeavor on KWS was motivated with the 2006 NIST Spoken Term Detection (STD) evaluations (Fiscus et al, 2007). Many successful applications of KWS were implemented on broadcast news (Wang, D. et al, 2008), video lectures (Garcia & Gish, 2006) and web-videos (Lee, S. W. et al, 2005) since then. As an extension to the 2006 NIST STD evaluations, which worked on KWS systems in English, Mandarin, and Arabic, the IARPA Babel Program (Harper, 2014), was initiated in 2011 to specifically address low resourced languages.

Large Vocabulary Continuous Speech Recognition (LVCSR) Based KWS Systems

The most convenient approach to approach the KWS problem would be thought of using an ASR system, more specifically a large vocabulary continuous speech recognition (LVCSR) system to obtain a text transcription from the audio and applying text retrieval techniques on the LVCSR output. The next subsection briefly discusses how an LVCSR system operates on a speech document.

A Brief Description of LVCSR Systems

Before proceeding onto the mathematical definitions and operations taking place under the hood, the explanation the task *large vocabulary continuous speech recognition* is needed. "Speech recognition" is a subtask of machine learning and basically defined as the process of a machine to identify words and phrases in spoken language and convert them to into understandable sequences of texts of actions. "Continuous speech recognition" is, in contrast to "isolated speech recognition" or "isolated word recognition" is a speech recognition task where the word boundaries are unknown by the machine. A machine conducting continuous speech recognition will convert the speech into combinations of the words or phrases it was trained to "know". The set of words that the system knows is defined as the "*vocabulary*" of the speech recognition system. The vocabulary size of the speech recognition system depends on the application it is used. For applications such as "command recognition" or "digit recognition" will have certainly a limited vocabulary and the recognition accuracy is expected to be high in these applications. An example of such a system can be the automated systems used by the call centers of banks asking you to explain what you wish to do in a brief sentence. Large vocabulary continuous speech recognition, on the other hand, has a more comprehensive vocabulary and a pronunciation lexicon, which primarily aims to operate on conversational speech. The problem is put in a statistical perspective. Since the goal of LVCSR is stated as obtaining the sequence of words $\boldsymbol{W} = \{w_1, w_2, \ldots, w_N\}$ from the acoustics $\mathbf{X} = \{x_1, x_2, \ldots, x_T\}$. The task is nothing but a maximum likelihood optimization problem on the conditional probability distribution

$$\hat{\boldsymbol{W}} = \arg\max_{\boldsymbol{W} \in \boldsymbol{W}^*} p(\boldsymbol{W} \mid \boldsymbol{X}) \tag{1}$$

where, the maximization is taken over all possible word sequences $\boldsymbol{W}^*$ that can be produced from the words in the vocabulary. There's hardly a way to model this probability distribution for an LVCSR task, yet the feasible means is to obtain the maximum a posteriori optimization equation from Bayes formula i.e.

$$\hat{\boldsymbol{W}} = \arg\max_{\boldsymbol{W} \in \boldsymbol{W}^*} p(\boldsymbol{X} \mid \boldsymbol{W}) p(\boldsymbol{W}) \tag{2}$$

where the term $p(\boldsymbol{X} \mid \boldsymbol{W})$ is called the "*acoustic model*" and $p(\boldsymbol{W})$ is referred to as the "*language model*". The acoustic model is estimated from a dataset of transcribed speech data and presents a probability distribution in the acoustic feature space for each possible word sequence $\boldsymbol{W}$. Since it is practically impossible to obtain enough samples to estimate the distribution on word level, the general approach is to use sub-word units like *phonemes* and model their duration with Hidden Markov Models (HMM). On the other hand, the language model is obtained as n-gram conditional probabilities in word level, from transcriptions in the target language.

Acoustic Modeling With Gaussian Mixture Models

As mentioned above, the acoustic model is a probability distribution over the set of phonetic units in the target language, estimated from the training data. Each sub-word unit is generally approximated by a multivariate Gaussian Mixture Model (GMM).

$$p(\boldsymbol{X}) = \sum_{i=1}^{K} c_i \mathcal{N}(\boldsymbol{X} \mid \mu_i, \Sigma_i) \tag{3}$$

where μ_i and Σ_i are the mean and variance parameters, respectively.

Each sub-word unit is modeled as a mixture of K Gaussians and the set of parameters are estimated using the training set via a expectation-maximization procedure called "Baum-Welch Algorithm". Further discussion on LVCSR would be out of the scope of this book, hence interested readers are referred to the very useful sources such as (Rabiner, 1989) and (Huang et al., 2001).

Acoustic Modeling With Deep Neural Networks

With the invent of high computing power brought by Graphical Processing Units (GPU) and the increasing amounts of available digital data, neural network computation yielded satisfactory results in many fields. Deep neural networks (DNN) are also used in acoustic modeling in speech recognition field (Hinton et al, 2012). DNNs are used in LVCSRs as stochastic structures whose inputs are acoustic feature vectors and outputs are posterior probabilities of the input frame belonging to one of the many classes. In the LVCSR task, these classes are phonetic sub-word units. A simple DNN structure is given in Figure 2. In the training, the weights $w_{i,j}$ that connect the nodes of the network are learned.

The output layer size of a DNN is equal to the number of phonetic sub-word units in the target language. Hence, the output of the network for a given time instant can be considered as a vector depicting a posterior probability mass function along the classes. The acoustic model from this posterior vector can easily be obtained by dividing them by the priors, again estimated from the training data. The most widely used acoustic feature vectors inputted to the neural network are linear prediction cepstral coefficients (LPCC) or mel frequency cepstral coefficients (MFCC). On the other hand, it should be noted

Figure 2. A sample DNN model with two layers

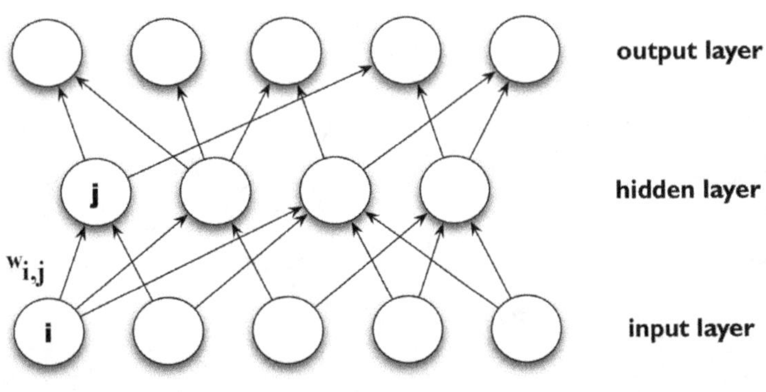

that, with deep enough networks, the gaps of efficiencies of different acoustic feature vectors are closed, due to the representation learning power of DNNs.

KWS on LVCSR Output

Since LVCSR systems may fail to return perfect transcriptions when the available data to estimate the model parameters and the recording conditions are bad, the performance of the KWS systems that use them will be far from efficient as well. In order to address this problem, practical LVCSR based KWS systems use stochastic structures like *lattices* or *confusion networks* obtained from the LVCSR systems instead of the single-best transcriptions. Lattices are graph structures that store weighted hypotheses in a compact form. Indexing ASR lattices, instead of the best ASR hypothesis, is a widely used KWS method that partly compensates for the negative effects of recognition errors (Can & Saraçlar, 2011). In this method, spoken documents are divided into short segments called utterances. For each utterance, an ASR lattice, representing ASR hypotheses, is generated. A sample lattice structure is given in Figure 3.

Once the lattices are created from the LVCSR output, the search is then conducted by means of weighted finite state transducer (WFST) operations (Mohri et al, 2002), (Saraçlar & Sproat, 2004).

Retrieval of Out-of-Vocabulary Terms

As mentioned above, the LVCSR systems will produce lattices based on how similar the acoustics to the sequences of words within its vocabulary. However, if a query term includes words that are not in the vocabulary of the LVCSR system, they will never appear on the lattice. Hence it will be impossible to retrieve them. Such terms are referred to as out-of-vocabulary (OOV) terms, in the literature. Furthermore, users generally tend to search for distinctive words like names and technical terms, not common words like "sun" and "sea". This problem presents a bigger conundrum even when working on low resource languages, since there is already not enough data to estimate parameters for the words in vocabulary. Also, the vocabulary size is going to be consequently small for such languages. In order to cope with this problem, one solution is to create lattices for sub-word units instead of whole words (Karakos & Schwartz, 2014). In this case, the usage of language model is surpassed; making the KWS systems create many false alarms. While sub-word indexing provides a method to retrieve OOV query words given a sub-word level representation for the query, it does not answer how to obtain these rep-

Figure 3. A sample lattice structure

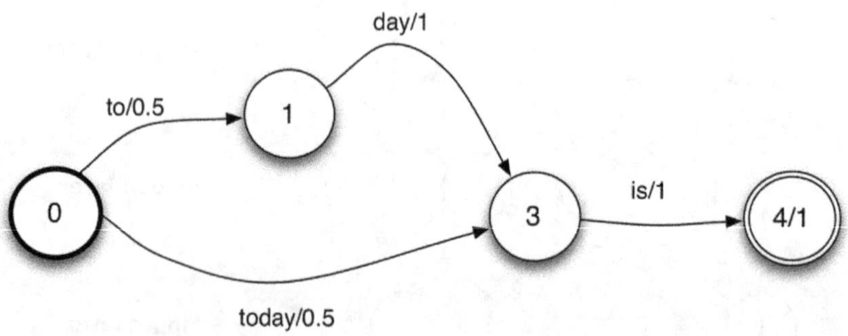

resentations. Most KWS systems tackle this problem at the retrieval stage by employing pronunciation models. However, it is not an easy task to generate proper pronunciations for words that haven't been seen before. For that reason, most KWS systems utilize inexact matching methods to handle the discrepancies between the actual and hypothesized pronunciations. This solution is implemented by searching the web for expansion of the language model to obtain automatically generated pronunciations (Gandhe et al., 2013). A very efficient approach to OOV handling is obtaining a confusion model from the training set and using acoustically similar in-vocabulary (IV) terms, instead of the OOV terms. These terms are called "proxy keywords" and this approach is preferred by many LVCSR based KWS systems due to its efficiency (Chen et al, 2013). The methodology adopted in this chapter follows a very different scheme and conducts the search in the feature space without using the LVCSR based systems. The addressing of the OOV terms in this works methodology is the main strength of the work. Next subsection introduces a pertinent task called query by example (QbE) spoken term detection, for what this chapter does is nothing but extending the success of QbE techniques onto KWS to provide a means of OOV retrieval.

Query by Example Spoken Term Detection

Query by example (QbE) spoken term detection, which is a very pertinent task, aims to search audio in audio. In this task, the user provides an audio snippet, instead of text, as a query and this term is searched in an audio archive (Shen, et al., 2009). Examples of this can be found in applications like Shazaam and SoundHound. Two main approaches proved useful for QbE STD.

1. Symbol-based approaches.
2. Frame-based approaches.

The symbol-based approaches use a decoder system and express the query and document as sequences of symbols from a finite alphabet. An LVCSR system can be thought of such a decoder for these systems. Since the query is also an audio, it will also be modeled as a lattice, in other words a weighted finite state transducer conveying the uncertainty of the decoding. A similarity score depicting the probability that the query exists in the utterance is then obtained by weighted finite state transducer operations.

Frame-based approaches, on the other hand, aims to obtain feature representations for each time window, called *frames,* from the document and the query, and then conduct the similarity search by means of versions dynamic time warping algorithms. In the MediaEval campaign (Rodriguez-Fuentes & Penagarikano, 2013), which focused on QbE spoken term detection on low resource languages, frame-based approaches yielded better performances compared to symbol-based approaches due to their freedom on search. The following subsection describes the dynamic time warping algorithm and their versions to conduct keyword search.

Dynamic Time Warping

Dynamic time warping (DTW) is a dynamic programming algorithm that finds an alignment between two temporal sequences possibly of different length (Albrecht & Muller, 2009). The algorithm also outputs a similarity score for them. The traditional DTW assumes that the two sequences start and end at the same time and they are monotonic. The DTW algorithm has been widely used in speech recognition or verification tasks since speech is a monotonic signal (Sakoe & Chiba, 1978). An illustration of DTW can be seen in Figure 4. Of the two sequences, the red line is faster than the blue line. DTW algorithm fines the alignment of the two lines based on their similarities at any given location.

The dynamic programming algorithm is as follows: Given two sequences (henceforth called utterances) $\boldsymbol{Q}$ and $\boldsymbol{X}$ to be aligned, each of the sequences are represented as a time series of feature vectors, $\boldsymbol{Q} = \{\boldsymbol{q}_1,\ldots,\boldsymbol{q}_M\}$ and $X = \{\mathbf{x}_1,\ldots,\mathbf{x}_N\}$, respectively. The optimal alignment path between $\boldsymbol{Q}$ and $\boldsymbol{X}$ is denoted as $\hat{\Phi}$. An alignment Φ between $\boldsymbol{Q}$ and $\boldsymbol{X}$ is a sequence of pairs $\Phi = (i_k, j_k); k = 1,\ldots,T$ that represents the mapping $(q_{i_1}, x_{j_1}), (q_{i_2}, x_{j_2}), \ldots, (q_{i_T}, x_{j_T})$, where T is the length of the alignment path.

In the traditional DTW, the two utterances are aligned in such a way that their starting and ending frames coincide, i.e. $i_1 = j_1 = 1$, $i_T = M$ and $j_T = N$. The optimal alignment path, $\hat{\Phi}$, is the one that has the minimum total distance, henceforth called as accumulated distance, defined as:

$$D_{\Phi} = \sum_{k=1}^{T} d(\boldsymbol{q}_{i_k}, \boldsymbol{x}_{j_k}) \tag{4}$$

where $d(\boldsymbol{q}_{i_k}, \boldsymbol{x}_{j_k})$ is a distance measure between the two frames on the alignment path. The accumulated distance on the optimal path between the two sequences is found using the following algorithm:

DTW Algorithm

```
for i=1 to M
DTW(i,0) = infinity
for j=1 to N
DTW(0,j) = infinity
DTW(0,0) = 0
for i=1 to M
for j=1 to N
cost = d(q(i), x(j))
```

Figure 4. An illustration of the DTW alignment
Source: (Hundt et al, 2014).

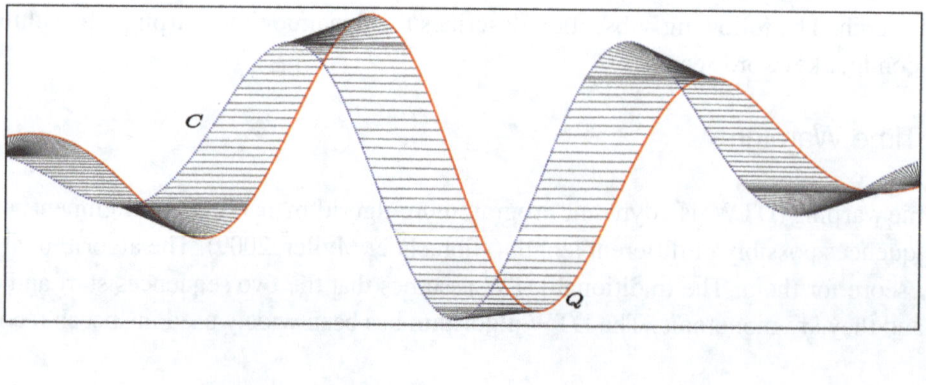

```
DTW(i,j)= cost + min(DTW(i-1,j),
    DTW(i,j-1),
    DTW(i-1,j-1))
return DTW(M,N)
```

However, in QbE spoken term detection, the two sequences do not start and end at the same time as required by DTW, rather one of the sequences (the document) is significantly longer than the other (the query) and the query can be aligned with any subsequence of the document. For this reason, two versions of DTW are used for the search: segmental DTW and subsequence DTW.

Segmental Dynamic Time Warping

In segmental dynamic time warping, short and overlapping segments from the document are obtained and the DTW distance is calculated between the query and each of the segments. The down part of this method is that the selection of the segment size is another optimization parameter that needs to be chosen. The segmental DTW employs the regional search by incorporating global constraints and restricting the allowable shapes that the alignment path can take. Likewise, it is also possible to allow generating multiple alignment paths, in many sub-sequences of the two utterances. It is intuitive to expect that for the two utterances to be aligned, one should not get too far ahead of frames from the other. An illustration of the segmental DTW algorithm can be seen in Figure 5.

Subsequence Dynamic Time Warping

The subsequence dynamic time warping (sDTW) algorithm is another version of the DTW algorithm specifically suitable for the task of interest. While in the traditional DTW algorithm the best alignment between Q and X are found by fixing and aligning the start and end frames of the both utterances, in sDTW, it is a fact that Q is considerably shorter than X and the starting point of Q can be aligned with any point of X. Furthermore the segmentation modification offered by segmental DTW fails to perform well when the durations of the two utterances differ by a larger margin. It has been reported in (Lee et al., 2015) that the audio recordings can be as long as twice the original term especially in broadcast news since the speaker tend to slow down the speaking pace drastically in order to make sure the pronunciations are good. Hence, sDTW is used for finding the best alignment between Q and all subsequences of X with a solution to this problem inherent in segmental DTW. The best alignment yields

Figure 5. An illustration of the segmental DTW.

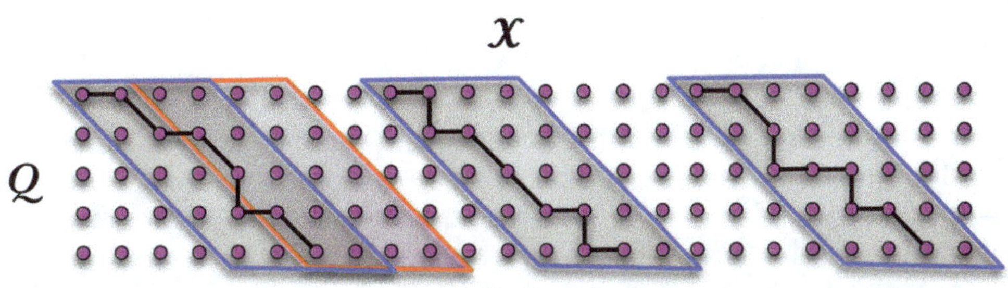

Figure 6. An illustration of the recurrent sDTW Algorithm
Source: (Sarı et al, 2015).

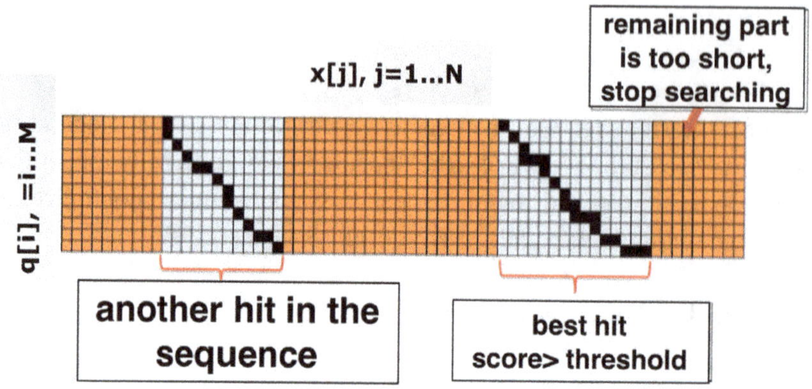

the most similar subsequence of X to Q. The similarity search is done through the accumulated distance scores between Q and all sub-sequences of X averaged by the alignment path lengths T. An illustration of the recursive sDTW algorithm can be seen in Figure 6. Interested readers are referred to (Gundogdu et al, 2016) for more information about the algorithm and the usage in KWS.

Evaluation Metrics

The evaluation methodology is a very important aspect for developing systems. For KWS applications, the success can be defined as the degree of relevance and perfection in detection. Two very pertinent evaluation metrics for this task are *Precision-Recall Rate* and the *F-Measure*, which is very similar to it. Precision (Pr) is the fraction of returned documents from the collection that are relevant to the query, and recall (Rc) is the fraction of relevant documents in the collection that are returned. These values are calculated as follows: Given Q queries, let R(q) be the total number of documents that are relevant to query q, A(q) be the total number of retrieved documents and C(q) be the number of relevant documents correctly retrieved. Then:

$$P_r = \frac{1}{Q}\sum_{q=1}^{Q}\frac{C(q)}{A(q)} \tag{5}$$

$$R_c = \frac{1}{Q}\sum_{q=1}^{Q}\frac{C(q)}{R(q)} \tag{6}$$

$$F = \frac{2 \times P_r \times R_c}{P_r + R_c} \tag{7}$$

Similar metrics derived from precision and recall rates are *Mean Average Precision (MAP), R-Precision, Precision@N* metrics. For the computation of MAP, Pr values are calculated at each of the Rc values (it should be noted that Rc varies between 0 to 1) and the average of the areas under this Pr -Rc curve for all queries are calculated. Precision@N is the Pr value of the top N returned documents (where N = 10 is a common choice). R-precision is similar to Precision@N, only that N varies for each given query q, and is set to the number of relevant documents R(q). Other widely used evaluation metrics are *Receiver Operating Curve (ROC), Figure Of Merit (FOM), Detection Error Trade*-off (DET) Curve, and the *Term Weighted Value (TWV)* metrics. In order to describe these metrics, the methodology for detection of a given term should be introduced. The outputs of the system are ordered with respect to some confidence score. If the score of an output is above some given threshold, then the output is marked "YES" and it is marked "NO" if it is below the threshold. In other words, a binary decision-making is required per the task definition. The evaluation system looks for the overlaps between the reference occurrences and the outputs of the system and marks the outputs as "HIT" if the central point of the outputs labeled as "YES" overlaps with the 0.5 sec extension of the reference and "FA" (referring to False Alarm) otherwise. The missed detections and the detections labeled as "NO" which overlaps with the reference occurrence are marked "MISS". This scoring scheme was introduced in NIST STD 2006 evaluations (Fiscus et al, 2007). The methodology of labeling of the system outputs is visualized in Figure 7.

Given the following definitions:

- *term*: the term being searched,
- *th*: the decision threshold,
- *T:* the total duration of the document,
- $N_{target}(term)$: the number of all correct occurrences of the term in the dataset,
- $N_{HIT}(term, th)$: the number of correctly detected term occurrences for the given *th* value,
- $N_{FA}(term, th)$: the number of false alarms for the term for the given *th* value

With the definitions of the above counts, the following statistics are calculated:

$$p_{HIT}(term, th) = \frac{N_{HIT}(term, th)}{N_{target}(term)} \tag{8}$$

Figure 7. Detection and hit labeling procedure

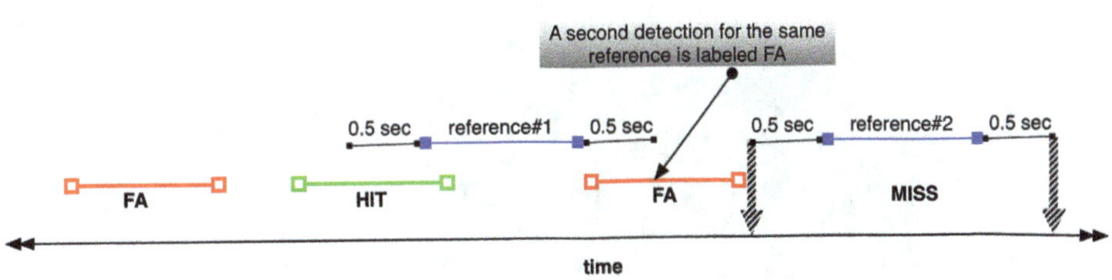

$$p_{MISS}(term, th) = 1 - P_{HIT}(term, th) = 1 - \frac{N_{HIT}(term, th)}{N_{target}(term)} \tag{9}$$

$$p_{FA}(term, th) = \frac{N_{FA}(term, th)}{T - N_{target}(term)} \tag{10}$$

Generally speaking, the performance of an STD system is defined by the trade-off between p_{HIT} and p_{FA}. In response to demands of the specific applications, different measures are considered using these statistics. An *ROC* for one *term* is the *term*'s p_{HIT} *(term; th)* as a function of N_{FA}*(term; th)* per hour. *FOM* is an upper-bound estimate on spoken term detection accuracy averaged over 1 to 10 false alarms per hour. It can be defined as the area under *ROC* from 0 to 10 false alarms per hour. The definitions of these metrics are illustrated in Figure 8.

However, the FOM metric is somewhat dependent on each terms' threshold and does not convey useful information about how the system would perform with regards to a single *th* for all queries. Hence the DET curve, which is a more explanatory metric considering the decision threshold, is introduced. For a given threshold th, the DET curve shows the dependency of term's false alarm probability $p_F A$*(term; th)*(x-axis) to the miss probability p_{MISS}*(term; th)* (y-axis). Contrary to ROC, the DET curve is a plot for all possible values of the threshold this is why it is widely used in the recent KWS applications. An example of a DET curve is illustrated in Figure 9.

In order to be able to compare two systems based on one number, the metric *Term Weighted Value (TWV)* was introduced. Although helpful in monitoring the system performance, DET curves cannot give such a number. The definition of TWV is as follows:

Figure 8. Receiver Operating Curve (ROC) and Figure Of Merit (FOM) evaluation metrics

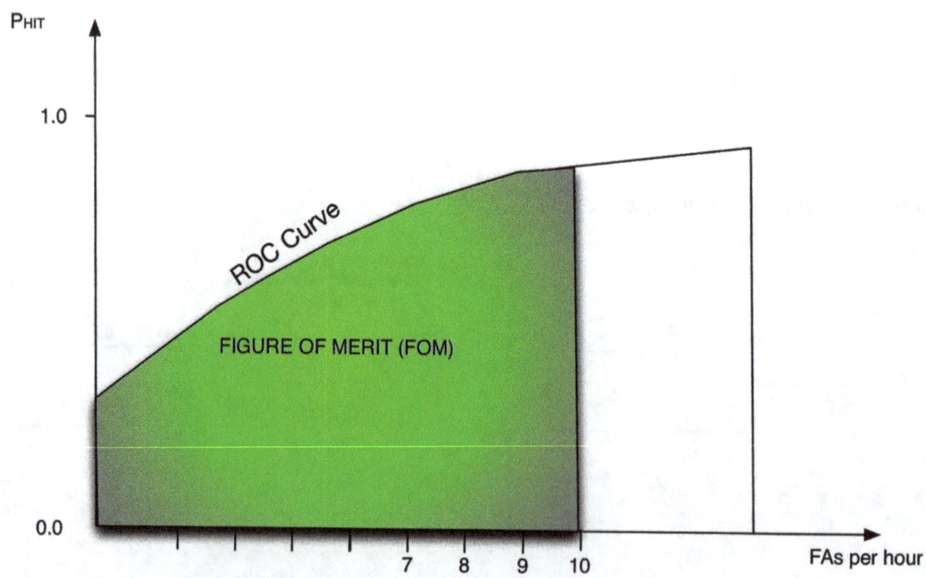

Figure 9. Detection Error Trade-off (DET) curve

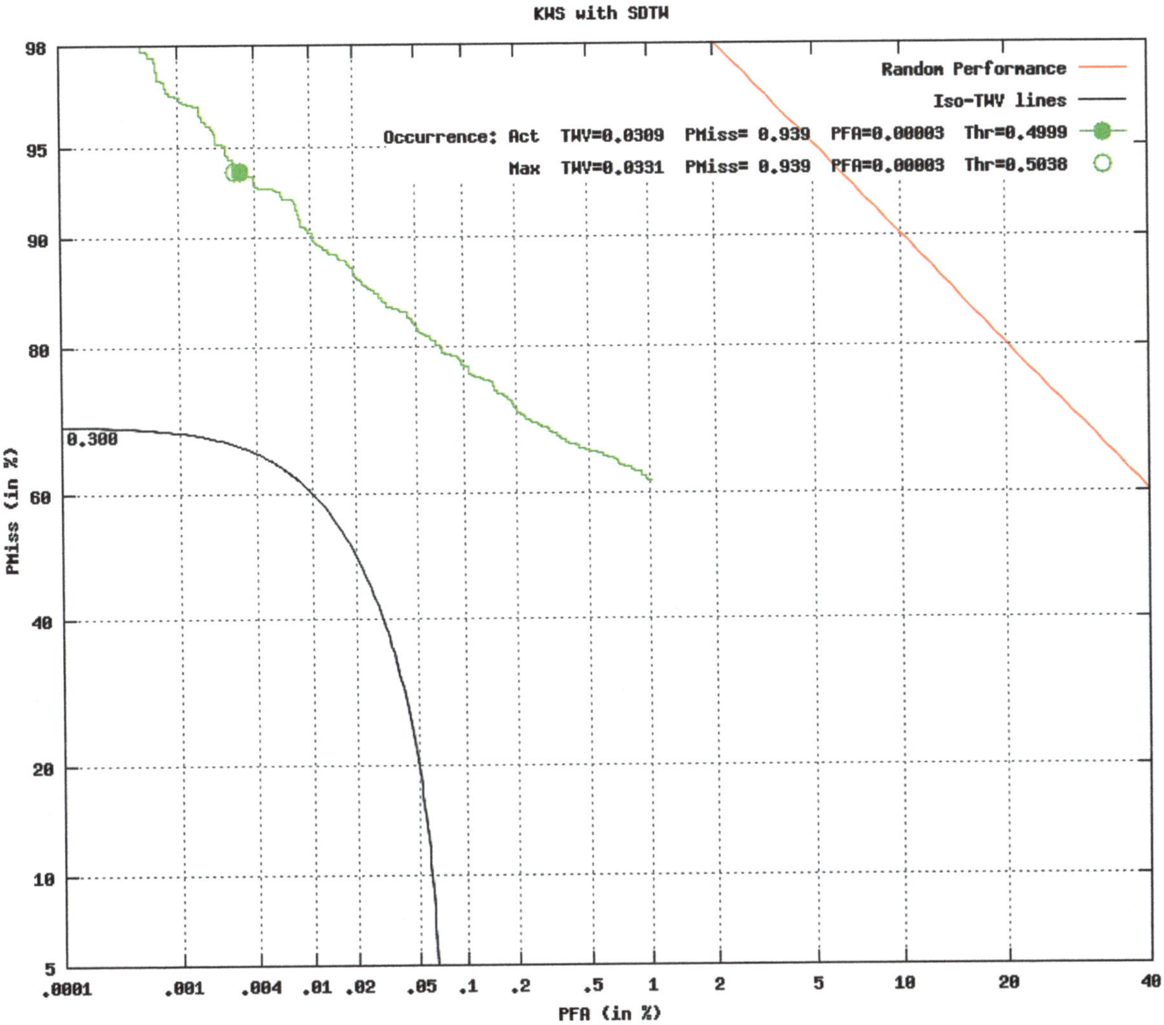

$$TWV(th) = 1 - \frac{1}{Q}\sum_{term}(p_{MISS}(term, th) + \beta p_{FA}(term, th)) \tag{11}$$

where, β is a constant that gives an intuition on how valuable a "HIT" is when compared to many number of false alarms. Hence, TWV gives a number between negative infinity and 1. A system that succeeds to detect all of the occurrences of all terms with no false alarms will have a TWV of 1, and a system that gives no outputs will give a TWV of 0. Hence, it is theoretically possible to do worse than doing nothing at all. Similar to TWV, Actual Term Weighted Value (ATWV) is the TWV using the actual decisions. ATWV represents the system's ability to predict the optimal operating point given the TWV scoring metric. Maximum Term Weighted Value (MTWV) is the maximum value the TWV can get over all *th* values. Optimum Term Weighted Value (OTWV) is calculated using the best *th* value for each query. Supremum Term Weighted Value (STWV) is calculated by only considering the "HIT"s and ignoring the cost of "FA"s. It is simply the ratio of the system's correctly detected outputs to the actual number of occurrences averaged over all terms. Clearly;

$$ATWV \leq MTWV \leq OTWV \leq STWV \quad (12)$$

Score Normalization

As stated in the earlier sections, the KWS systems are expected to return the hit hypotheses based on confidence scores that are assigned to them. The decision of a hypothesis being relevant or irrelevant is made based on this similarity scores. Since the system is developed before the run time, the user can only use one threshold value to decide the relevancy. On the other hand, keywords and utterances vary in length and acoustic ingredient. Because of this reason, the similarity scores for each keyword tend to vary in range, too. A very unlikely hypothesis for a specific keyword may have higher scores than the best (and relevant) hypothesis for some other keyword, for this reason.

In order to address this problem, the hypotheses of the keywords are normalized so that a single threshold value can be used for decision in the run time. Many normalization techniques have been used in the literature. First and the most widely used normalization technique is called the *Sum to One (STO) Normalization.* In STO normalization all hypothesis for a certain keyword are normalized by their sum, so that they add up to one. It is a great means of taking the ranges of hypotheses belonging to different keywords to the same scale by protecting their order. Another very widely used normalization technique is called the *Histogram Equalization (HE),* which rescales the hypotheses' scores so that the best hypothesis has a score of 1 and the worst has a score of 0 for all keywords. This normalization technique becomes useful in that it prevents the scores taking too small numbers, where this is the case for STO.

Another very pertinent score normalization technique that the conventional LVCSR-based KWS systems use called the *Keyword Specific Threshold (KST) Normalization.* In KST, the main idea is to optimize the TWV metric given in (11). However, the actual number of occurrences ($N_{target}(term)$) of a keyword is not known by the system developer. The estimation of $N_{target}(term)$ is done by the summation of each hypothesis. Since in the LVCSR-based approach the lattices yield posterior probabilities of the keyword existing in a certain utterance, the summation of these probabilities, i.e. scores, gives an estimate of $N_{target}(term)$. The optimal threshold for the KST normalization is then obtained using this estimate.

System Fusion

Fusion of heterogeneous systems has been studied and used in many detection and verification applications. The main idea is to utilize superiorities of one system to compensate for the impotencies of others in different areas. The fusion can take place at the search part, using different structures and levels, yet generally the fusion is applied to the system scores of different systems. One general fusion methodology is called the *majority voting,* in which when majority of the systems have high scores for the same hypothesis, the weighted average of their scores is used as a single output of the fusion system. This methodology is very useful for elimination of the false alarms.

MAIN FOCUS OF THE CHAPTER

This chapter aims to address the task of keyword search primarily for the low resource languages in order to compensate for the impotencies of the LVCSR systems trained on little data. As the main search methodology, the fruitful technique of frame-based QbE spoken term detection scheme is applied and it is extended to the keyword search problem. With this, the primary intention is to release the necessity for a high performance decoder and propose a new method for the retrieval of the terms out of the vocabulary of such an LVCSR system. This proposed methodology already carries the problems to be addressed in the task. For one, the frame-based QbE systems, which follow the sDTW recipe, obtains frame representations from the document (utterances) and the spoken queries. The methodology of this chapter also obtains such frame representations from the utterances using *deep neural networks* and uses the *posteriorgram* representations to be used in the DTW search. The query, however, which is given in text form in this task, has to be handled separately. Hence, the fist problem is modeling the text queries as *pseudo posteriorgrams*.

In QbE, the frames of the posteriorgrams are posterior vectors, obtained generally from neural network, hence in sDTW, as the distance metric, versions of *cosine distance* is used. However, in the methodology proposed in this chapter, the queries are modeled artificially and the cosine distance may not be the best similarity measure to be used between posterior vectors of the utterance. Hence, the second problem to be addressed is to learn a similarity measure to be used in accordance with the artificially modeled query frames.

In LVCSR based systems, the output scores of the hypotheses are posterior probabilities and an estimation of the actual occurrence number is feasible making it possible to use a normalization like the KST normalization. Furthermore, the number of hypotheses is small compared to the sDTW based search systems, where it is basically possible to propose a similarity score for any subsequence of the utterance. Because of this reason, a sum-to-one normalization is directly feasible for LVCSR based systems, where in sDTW based search systems; a pruning of unlikely subsequences is necessary to keep the STO-normalized scores in a big enough range. This pruning threshold is another optimization parameter to make the problem harder than it is. The third problem to be addressed is an suitable normalization scheme for the novel approach proposed in this chapter.

The fusion of heterogeneous systems is a non-trivial subject. For the *majority voting* to work, the score ranges of the two systems have to be aligned. However, the scoring methodologies of the LVCSR based systems and the sDTW-based systems are completely different, where the former outputs probability values for each hypothesis and the latter assigns similarity values through accumulated DTW distances for each hypothesis. A generally applied solution is to use STO normalization before combination so that the relative significances of each hypothesis have the same meaning across systems. However, this approach brutally cancels out the benefits of the individual normalization being studied as the third problem. Furthermore, as mentioned before, the number of hypotheses to be normalized in the sDTW based approach is significantly higher than the LVCSR based systems making it impossible to apply a proper STO normalization. Hence, the fourth problem is to work on a suitable fusion methodology to utilize powerful parts of each of the LVCSR and sDTW based systems.

In summary, the problems to be addressed for the methodology proposed in this chapter can be enumerated as follows:

Problem One: Modeling the text keyword into a proper frame level representation suitable for template base search.

Problem Two: Learning an optimal similarity measure to be used in sDTW for these representations.

Problem Three: Finding a proper score normalization technique to work properly with the sDTW based KWS output.

Problem Four: An efficient system fusion methodology for the combination of the sDTW based and LVCSR based systems.

METHODOLOGY

The solution to the first problem follows the recipe introduced in (Gündoğdu et al, 2016). In the beginning, of the system set-up, there is a set of telephone transcribed telephone conversations to train a DNN to obtain the posteriorgrams. After training the network until convergence, this DNN is used to obtain document posteriorgrams from the telephone conversations, on which the desired keyword search is to be conducted. The posteriorgram representations and the phone-frame alignments depicting which frame belongs to which phone are also obtained using this DNN. The flowchart of obtaining the posteriorgrams and the alignments are depicted in Figure 10.

From, the training set alignments, the estimations of statistics for phonemes can be obtained. Two of these statistics are the average phoneme posterior vectors depicting a way of inter-phone confusions introduced by the DNN and the average durations of each phoneme. Using these statistics, the pseudo query models can be models can be obtained with two methods. The text query is fist converted into

Figure 10. The flowchart for obtaining the document posteriorgrams and the training set frame-phone alignments

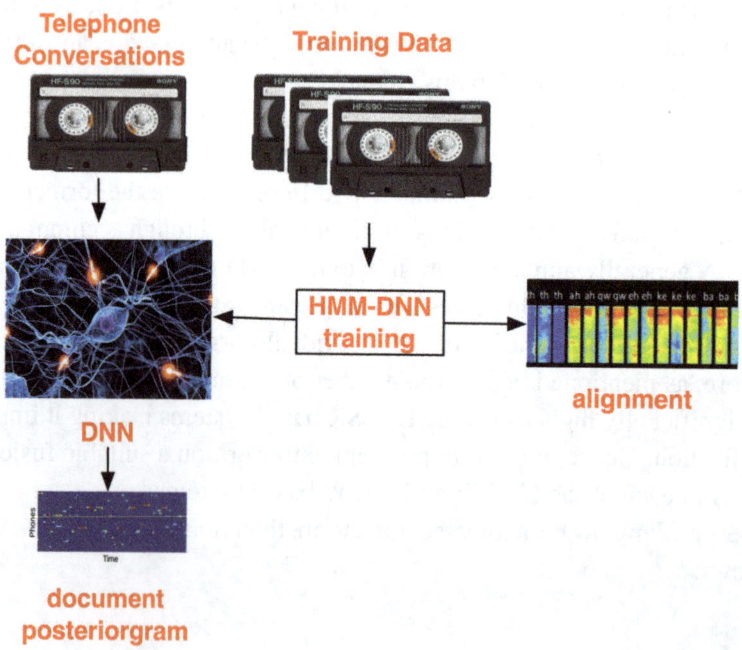

Figure 11. The methodology for obtaining indices from text queries

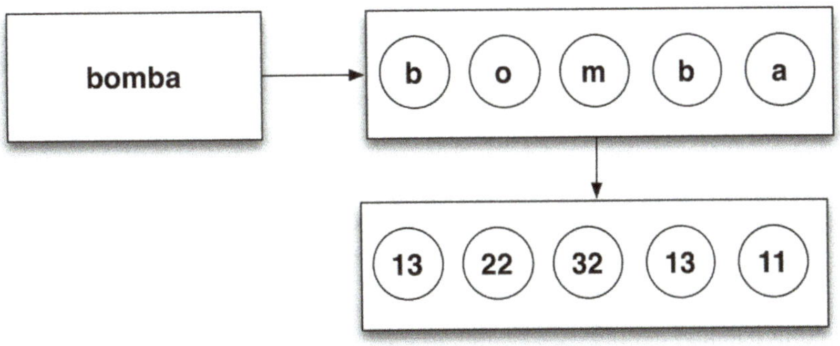

sequences of phonemes depicting its pronunciation using a pronunciation lexicon for IV words, and a grapheme-to-phoneme conversion system for OOV words. Each phoneme in the language will basically have an index number for which the DNN outputs a posterior probability value. The initial step for obtaining index numbers from a keyword is depicted in Figure 11.

From the indices, the statistics obtained from the training alignment are used and they are modeled as pseudo posteriorgrams. In the first *query modeling* methodology, one-hot vectors are used to represent phonemes. A one-hot vector is a vector where only one of the indices is 1 (the one depicted by the phoneme number in the pronunciation) and the other indices are zero. Each phoneme is repeated by the number of their average durations to model the phoneme durations in the artificial query model posteriorgram. This methodology is called *binary query modeling*. The other query modeling technique uses phoneme averages obtained from the training alignment instead of one-hot vectors to model the confusions. The second technique is called *average query modeling*. Sample query models obtained from the keyword *"arma"* using these query modeling techniques are demonstrated in Figure 12.

Once the pseudo posteriorgrams are obtained for the keywords, the sDTW algorithm is used to get similarity scores for every subsequence of the utterance. These similarity scores are calculated from the average accumulated distance through the alignment path:

$$score = 1 - \frac{1}{|\Phi|} \sum_{k=1}^{T} d(\boldsymbol{q}_{i_k}, \boldsymbol{x}_{j_k}) \tag{13}$$

Figure 12. Pseudo Query Models obtained using binary query modeling (left) and average query modeling (right) for the sample keyword "arma"
Source (Gundogdu & Saraclar, 2017b)

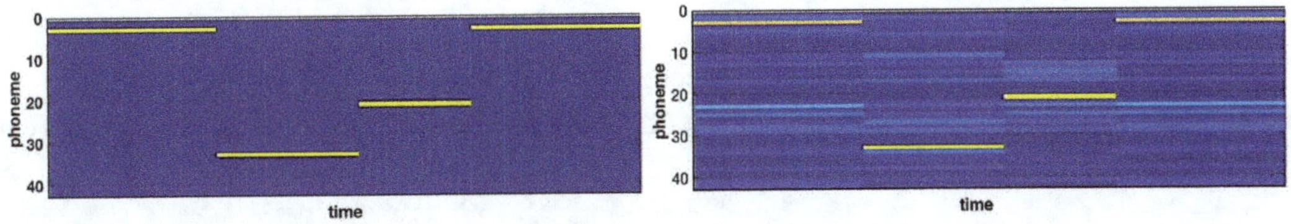

Similarity Measure Optimization

As it is given in (13), system similarity scores for the hypotheses offered by the proposed system is as relevant to the distance measure ($d(\boldsymbol{q}_i, \boldsymbol{x}_j)$) as it is to the alignment algorithm (Φ). Several distance measures can be used for this. In a generic distance measure, denoted $d(\boldsymbol{x}, \boldsymbol{y})$, the distance value between vectors **x** and **y** are desired to be high, if they are "far" or "dissimilar"; and small if they are "close" or "similar". For example, in the Euclidean space where the **x** and **y** values denote location information, the Euclidean distance measure (14) would be proper. On the other hand, for the posterior features being used in QbE, cosine distance (15) and logarithmic cosine distance (16) are more appropriate to use, because of their geometric and probabilistic interpretations.

$$d_{Euc}(\boldsymbol{x}, \boldsymbol{y}) = \sqrt{\|\boldsymbol{x}\|^2 + \|\boldsymbol{y}\|^2 - 2\boldsymbol{x}^T\boldsymbol{y}} \tag{14}$$

$$d_{\cos}(\boldsymbol{x}, \boldsymbol{y}) = 1 - \frac{\boldsymbol{x}^T\boldsymbol{y}}{\|\boldsymbol{x}\| . \|\boldsymbol{y}\|} \tag{15}$$

$$d_{\log}(\boldsymbol{x}, \boldsymbol{y}) = -\log\left(\frac{\boldsymbol{x}^T\boldsymbol{y}}{\|\boldsymbol{x}\| . \|\boldsymbol{y}\|}\right) \tag{16}$$

This methodology, on the other hand, learns models the queries artificially making these distance metrics no longer appropriate for the new model. Hence, the second problem of learning a new similarity measure that would work better than the other ones is going to be addressed here. For this, the following Siamese neural network model is proposed.

Hence, once the parameters are learned the new similarity measure will be:

$$f(\boldsymbol{x}, \boldsymbol{y}) = \sigma\left(\boldsymbol{x}^T\boldsymbol{W}^T\boldsymbol{W}\boldsymbol{y} + b\right) \tag{17}$$

where, $\sigma(z)$ is the sigmoid function defined as

$$\sigma(z) = \frac{1}{1 + e^{-z}} \tag{18}$$

In this model, the parameters to be optimized are as follows:

- **W:** Shared weight matrix, projecting the vectors onto another space.
- **b:** The bias value to take the new inner product to the range where the sigmoid non-linearity works better.

Figure 13. Siamese neural network based similarity measure learning scheme.
Source: (Gundogdu & Saraclar, 2017a)

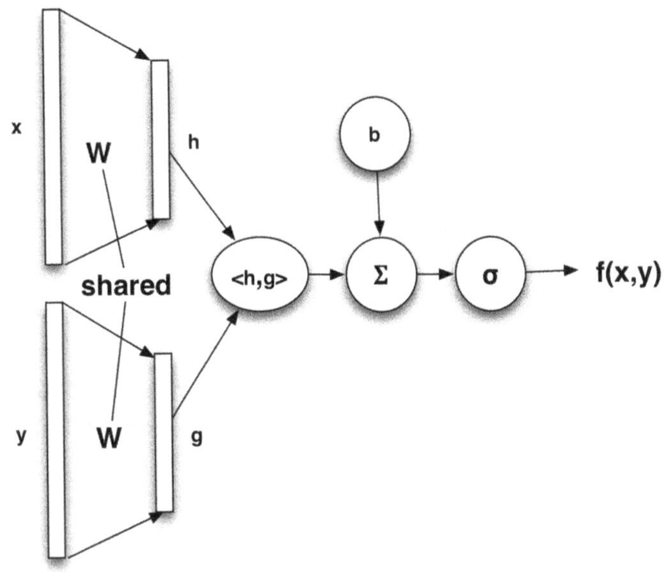

This similarity learning algorithm, first proposed in (Gundogdu & Saraclar, 2017), uses all pairs of frames in the training set alignment and finds the **W** and b values such that the total cross entropy cost is minimum.

$$J\left(\boldsymbol{W},b\right)=\sum_{t}-r_t\log(f_t)-(1-r_t)\log(1-f_t) \quad (19)$$

where f_t is $f(\boldsymbol{x}_t,\boldsymbol{y}_t)=\sigma\left(\boldsymbol{x}_t^T\boldsymbol{W}^T\boldsymbol{W}\boldsymbol{y}_t+b\right)$, the sigmoid similarity between t[th] (x,y) pair in the training set, and r_t is the label or the ground truth of this pairs. Since it is desired that the frames belonging to the same class (*friends*) give high similarity and ones belonging to different classes (*foes*) give low similarity, the following labels are used in the training.

$$r_t=\begin{cases}1, \text{ if } \boldsymbol{x} \text{ and } \boldsymbol{y} \text{ are friends}\\0, \text{ if } \boldsymbol{x} \text{ and } \boldsymbol{y} \text{ are foes}\end{cases} \quad (20)$$

For this cost function, the training was conducted using the online (stochastic) gradient descent. The gradient update rules for the parameters can be found as follows:

$$\begin{aligned}\boldsymbol{W}&=\boldsymbol{W}+\eta(r-f)(\boldsymbol{x}\boldsymbol{y}^T+\boldsymbol{y}\boldsymbol{x}^T)\\b&=b+\mu(r-f)\end{aligned} \quad (21)$$

The similarity measure optimization problem can also be incorporated with the first problem of query modeling. Recently in (Gundogdu & Saraclar, 2017b), authors proposed such a scheme by adding on more layer to the Siamese neural network and having an asymmetric structure.

Score Normalization for the New Approach

The KWS approach being used in this chapter is very similar to the QbE-STD approaches except from the query modeling and similarity learning novelties. Hence, in addition to classical normalization techniques, this new methodology allows for new normalization techniques, as well.

The best performing normalization techniques for the proposed approach have been the versions of b-norm proposed in (Gündoğdu, & Saraçlar, 2017c). The distribution of scores per each keyword approximates a Gaussian distribution with a longer left tail, due to the fact that more number of telephone conversations is irrelevant than relevant. So the number of low scores is inherently higher than the number of high scores (explaining the longer left tail). Furthermore, the score distribution of some other keyword over the same audio archive will have different statistics, but a similar distribution. Hence, it is desired to modify the scores for all keywords so that they have the same relative meaning. This is very important in terms of operation with a single global threshold and fusion with other systems. The unnormalized distribution for 5 different keywords scores using the proposed retrieval approach is shown in Figure 14 (top-left). The sum-to-one and histogram equalization normalizations, along with the b-norm normalized distribution are shown in the same Figure on top-right, bottom-left and bottom-right, respectively

For a specific keyword, If the vector of scores of the hypotheses obtained from the search audio is denoted **s**, the b-norm (and versions of it), proposed in (Gündoğdu, & Saraçlar, 2017c) can be described as seen in (22).

$$\boldsymbol{s}_{norm} = \frac{\boldsymbol{s} - \mathrm{median}(\boldsymbol{s})}{\hat{\sigma}(\boldsymbol{s})} \tag{22}$$

where, $\hat{\sigma}(\mathrm{s})$ is the standard deviation calculated over all scores greater than some chosen vicinity of the median of the distribution.

Fusion of sDTW and LVCSR-Based Systems

The current system fusion methodologies implemented for KWS proved inefficient for the combination of LCVSR and sDTW based systems. The reason is that the LVCSR-based systems use KST normalization while the best practiced normalization technique for the sDTW-based systems is the b-norm (median normalization). The existing system fusion techniques are mainly dependent on taking the score ranges of the two systems to the same values by applying sum-to-one normalizations, while incorporating the information about the number of hits per system and per query. However, the application of sum-to-one normalization on top of the b-norm simply eliminates its efficiency and makes the proposed systems just STO normalized.

In order to utilize the dynamic power of the new normalization techniques, the combination procedure is vectorized and a time efficient algorithm was implemented. Each system output is a list that consists of the locations and scores of each hypotheses per keyword as shown in (23) and visualized in Figure 15.

Figure 14. Effect of score normalization on system scores for 5 different keywords: x-axes denote the score of a detection, where y-axes denote the number of detections with this score. Curves with different colors denote different keywords.
Source: (Gundogdu & Saraclar, 2017c)

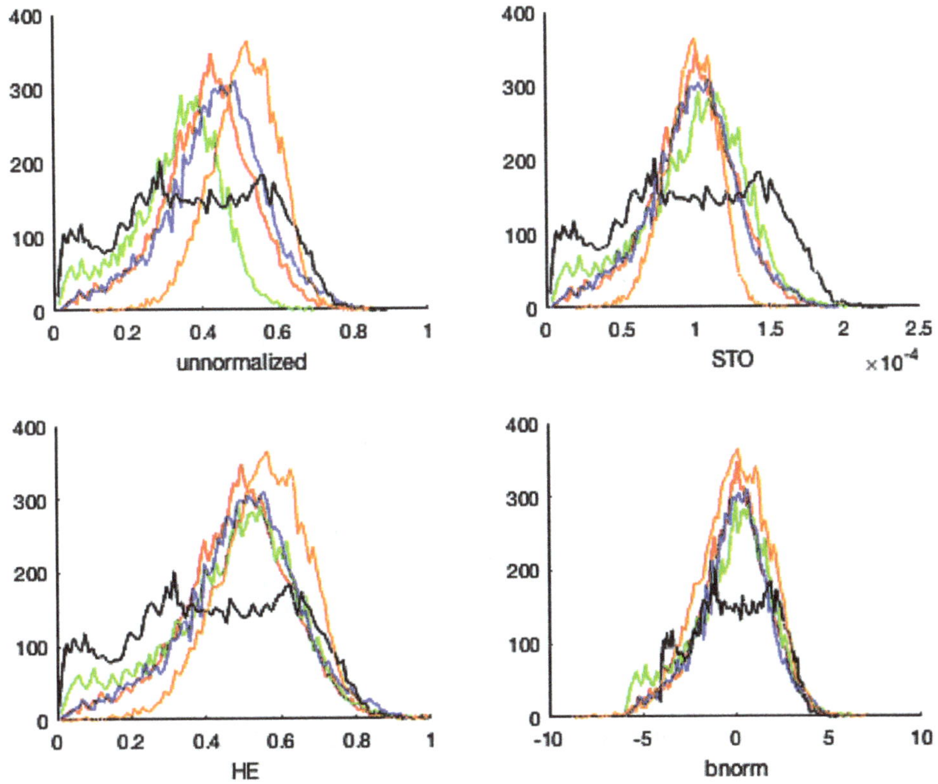

$$sys(i) = \left\{kw(j) : \{file(k), start(k), end(k), score(k)\}\right\} \\ i = 1 \ldots \# systems, j = 1 \ldots \# keywords, k = 1 \ldots \# hits(i, j) \tag{23}$$

The fusion algorithm aims to find the overlaps between score-lists of different systems for the same keyword and combine the scores of the overlapping hypotheses after some normalization. This chapter introduces a time efficient combination scheme for combination of heterogeneous systems. Each system is represented with a sparse 3-D matrix:

$$\mathbf{SYS}\{i\} \in \Re^{K \times U \times F} = \mathbf{SYS}\{i\}(j, file(k), start(k) : end(k)) = score(k) \tag{24}$$

where K is the number of keywords, U is the number if files (utterences), and F is the maximum frame number for which the systems propose a hypothesis. The filling of the sparse matrix is explained in (24) and visualized on Figure 16.

Since the LCVSR based system is KST normalized, their scores range between 0-1 and the most confident hypotheses, which are for in-vocabulary terms, have scores close to 1. On the other hand,

Figure 15. A sample KWS system output scheme for evaluation

	KW_1			
1	FILE_ID	start	end	score
2	FILE_ID	start	end	score
3	FILE_ID	start	end	score
4	FILE_ID	start	end	score
.	FILE_ID	start	end	score
.	FILE_ID	start	end	score
.				
	KW_2			
1	FILE_ID	start	end	score
2	FILE_ID	start	end	score
3	FILE_ID	start	end	score
4	FILE_ID	start	end	score
.	FILE_ID	start	end	score
.	FILE_ID	start	end	score
.				
	KW_3			
1	FILE_ID	start	end	score
2	FILE_ID	start	end	score
3	FILE_ID	start	end	score
4	FILE_ID	start	end	score
.	FILE_ID	start	end	score
.	FILE_ID	start	end	score
.				

Figure 16. Methodology to model the system output as a sparse cubic structure

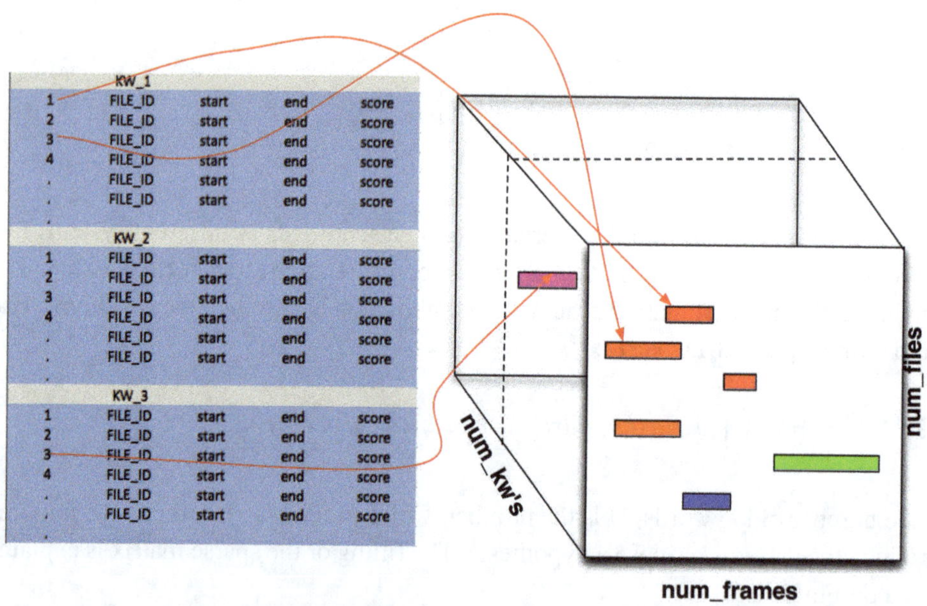

b-normalized sDTW based ssytems will have larger values for confident hits, due to the division by the standard variation. It is desirable to "trust" the LVCSR based systems for their hypotheses for the in-vocabulary words. Hence, the ranges of the two matrices are equalized by simply multiplying the LCVSR based system with the global maximum of the sDTW based system matrix.

$$\mathbf{SYS}_{norm}\{LCVSR\} = \mathbf{SYS}\{LCVSR\} \times \max(\max(\max(\mathbf{SYS}\{sDTW\}))) \tag{25}$$

Now that the whole scheme is vectorized and equalized, the overlap search is simply done via taking the weighted sum of the sparse system matrices (Figure 17) .For the experiments of this chapter, authors took equal weights for the systems. (26).

$$\mathbf{Fusion} = \frac{1}{N}\sum_{i=1}^{N}\mathbf{SYS}\{i\} \tag{26}$$

The fusion system matrix, which is a sum of sparse matrices, will again be sparse matrix. The ultimate list of system hypotheses upon the fusion is then extracted from this matrix by non-zero segments. Due to averaging, non-overlapping hypotheses will be suppressed. These also correspond to hypotheses where the minority rooted for. For other hypotheses, with majority votes, the overlapping part will have a relatively high score. Due to the fact that different systems will report different start and end locations for the same hypothesis, the scores for those will vary along frames on the combination matrix. The colors of the separated areas in Figure 16 correspond to different scores described by (24). Hence the fusion matrix on Figure 17 may have varying range of score values for each separated hypothesis. During the extraction of the hypothesis list from the matrix, their mean is taken for each separated hypothesis in the sparse cube, so that a measure of the overlap ratio is also integrated in the proposed combination methodology. This procedure is visualized in Figure 18.

Figure 17. Procedure to obtain the fusion matrix cube from the two systems

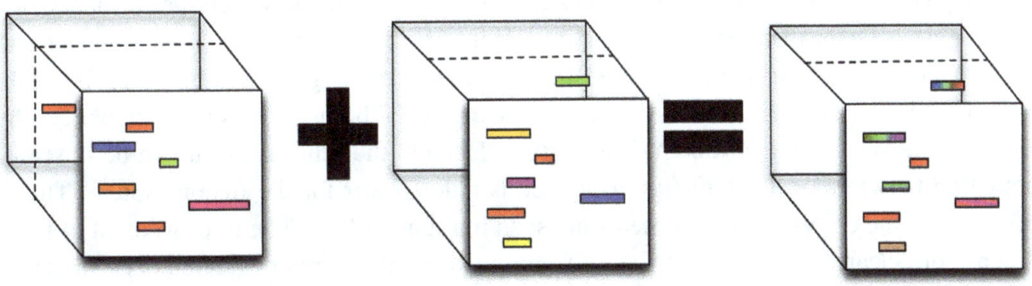

Figure 18. The system output for evaluation is obtained from the fusion cube, by taking the means of the sparse locations recording their addresses

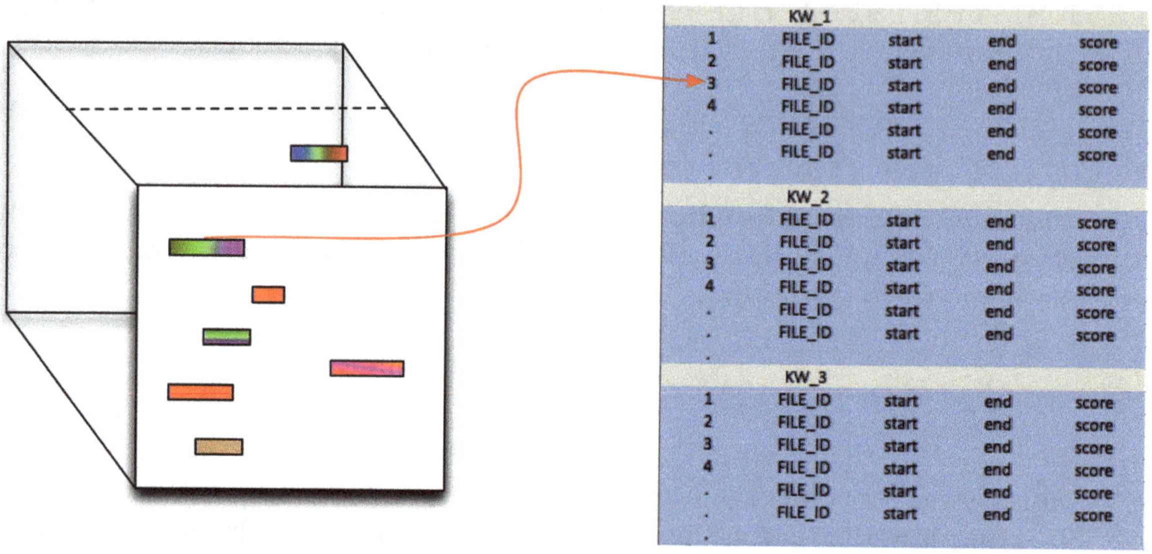

EXPERIMENTAL RESULTS

The techniques proposed in this chapter were tested on varying parts of 45 hours of telephone conversations from three low-resource languages. The keywords of interest were simulated to be taken from random phrases or words from each language. The term weighted value performances were calculated for the LVCSR-based baseline system and the fusion with the sDTW-based system. The initial experiments conducted on each language with the 10-hour telephone conversation data served as system development to learn the optimal operation thresholds for the detection. Hence, for the development experiments MTWV scores were calculated. On the last part of the experiments for each language, with the 5-hour telephone conversation recordings, both the ATWV and MTWV scores were calculated using the optimal decision thresholds learned in the fist part. The list of experiments can be seen in Table 1.

The MTWVs (for development and evaluation experiments) and the ATWVs (for the evaluation experiments), were plotted on Figure 19(a). It can be seen that the fusion system performance (denoted with red) is always better than the LVCSR-based system (denoted with blue). The valleys on the both lines correspond to the experiments with OOV-only experiments (e.g. 2, 5, 6 etc.) and the peaks correspond to IV only experiments (e.g. 3, 6, 9 etc). Obviously, the both systems perform better when the keyword is in the vocabulary of the LVCSR system (IV). On the other hand, it can be observed that the fluctuation between IV and OOV performance is much lower for the fusion system. The system performances for the OOV-only experiments are seen in Figure 19 (b). The improvement of the fusion system can more clearly be seen when the OOV-only experiments are considered. The fusion system performance is almost always more than twice of the baseline system. Although, for the IV-only experi-

Table 1. List of experiments

Exp. No.	Language	Keyword List Size	Telephone Conversation Length	Keywords in the Vocabulary?
1	Turkish	307	10 hours	mixed (IV and OOV)
2	Turkish	88	10 hours	Only OOV
3	Turkish	219	10 hours	Only IV
4	Turkish	3171	5 hours	mixed (IV and OOV)
5	Turkish	1216	5 hours	Only OOV
6	Turkish	1955	5 hours	Only IV
7	Pashto	2065	10 hours	mixed (IV and OOV)
8	Pashto	600	10 hours	Only OOV
9	Pashto	1465	10 hours	Only IV
10	Pashto	4203	5 hours	mixed (IV and OOV)
11	Pashto	971	5 hours	Only OOV
12	Pashto	3232	5 hours	Only IV
13	Zulu	2000	10 hours	mixed (IV and OOV)
14	Zulu	806	10 hours	Only OOV
15	Zulu	1194	10 hours	Only IV
16	Zulu	3310	5 hours	mixed (IV and OOV)
17	Zulu	1112	5 hours	Only OOV
18	Zulu	2198	5 hours	Only IV

ments, the LVCSR based system is expected to perform at a satisfactory level, the experiments show that the fusion with the sDTW-based system provides an improvement on all experiments, as can be seen in Figure 19 (c). The experiments conducted over a mixed set of IV and OOV terms are more akin to realistic, real-world scenarios. The clear performance improvement that can be seen in Figure 19 (d) is the demonstration of this experiment set.

CONCLUSION AND FUTURE RESEARCH DIRECTIONS

In this chapter, a similarity optimization based keyword search technique is introduced and the novel query modeling, score normalization, and system fusion techniques were presented. These techniques were tested on several experimental set-ups on three different languages and various different keyword-lists provided by the IARPA Babel Program. It was shown that, this approach not only provide a new an powerful method for the retrieval of OOV terms, but also improves the conventional LVCSR-based systems' performances over all experiments.

Figure 19. TWV performances evaluated on the experiments conducted: x-axes denote set of experiments, and y-axes denote the MTWV performance in the experiment. (a) All experiments given in Table 1, (b) OOV-only experiments, (c) IV-only experiments, (d) mixed (IV and OOV) experiments.

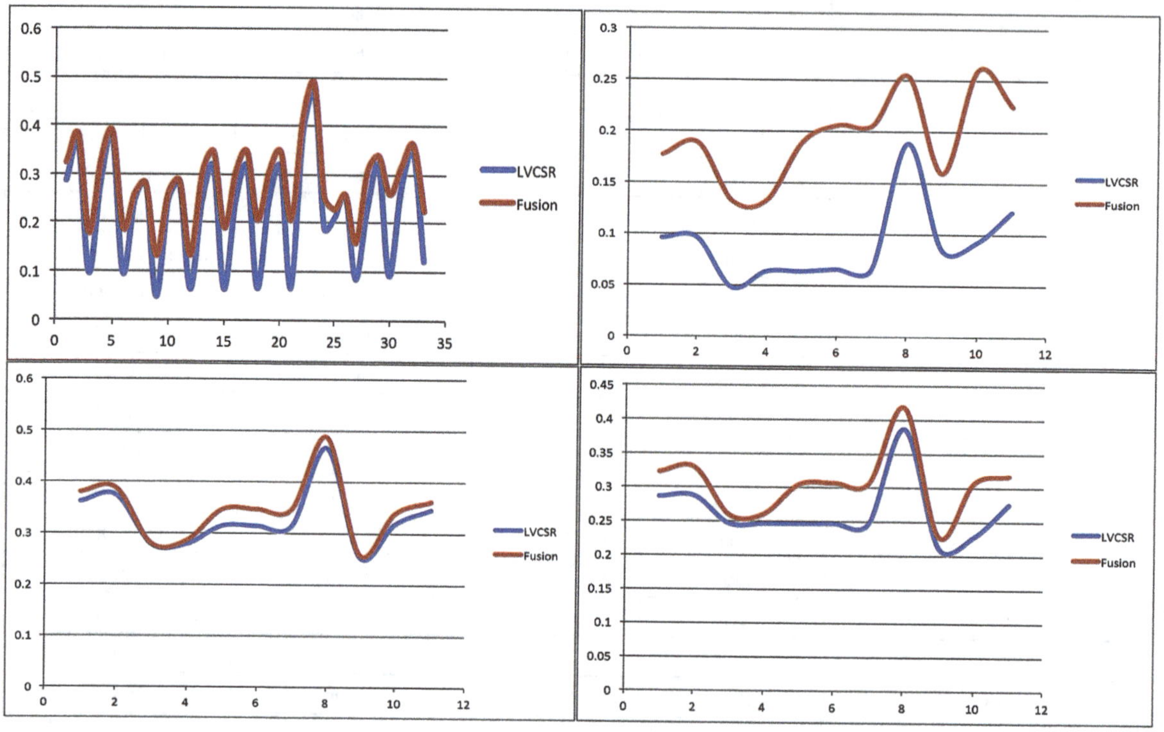

As a future research direction, the authors recommend modeling the query using generative recurrent neural networks. Instead of modeling the phonemes by one-hot vectors (or with fist order statistics) and applying the same phoneme model through its estimated duration, a properly trained generative recurrent neural network would model a more realistic query posteriorgram.

ACKNOWLEDGMENT

This study was supported in part by the Intelligence Advanced Research Projects Activity (IARPA) via Department of Defense U.S. Army Research Laboratory (DoD/ARL) contract number W911NF- 12-C-0012 and the Scientific and Technological Research Council of Turkey (TUBITAK) under Project 116E076. The U.S. Government is authorized to reproduce and distribute reprints for Governmental purposes notwithstanding any copyright annotation thereon. Disclaimer: The views and conclusions contained herein are those of the authors and should not be interpreted as necessarily representing the official policies or endorsements, either expressed or implied, of IARPA, DoD/ARL, or the U.S. Government. The authors would like to thank Mr. Bolaji Yusuf and Mr. Recep Doga Siyli for their invaluable help in conducting the experiments.

REFERENCES

Albrecht, T., & Muller, M. (2009). Dynamic Time Warping (DTW). *Information Retreival for Music and Motion*, 70-83.

Borges, J. L. (1998). *The library of Babel: Collected fictions*. Academic Press.

Can, D., & Saraclar, M. (2011). Lattice indexing for spoken term detection. *IEEE Transactions on Audio, Speech, and Language Processing*, *19*(8), 2338–2347. doi:10.1109/TASL.2011.2134087

Chen, G., Yilmaz, O., Trmal, J., Povey, D., & Khudanpur, S. (2013, December). Using proxies for OOV keywords in the keyword search task. In *Automatic Speech Recognition and Understanding (ASRU), 2013 IEEE Workshop on* (pp. 416-421). IEEE. 10.1109/ASRU.2013.6707766

Fiscus, J. G., Ajot, J., Garofolo, J. S., & Doddingtion, G. (2007). Results of the 2006 spoken term detection evaluation. In Proc. sigir (Vol. 7, pp. 51-57). Academic Press.

Gandhe, A., Qin, L., Metze, F., Rudnicky, A., Lane, I., & Eck, M. (2013, December). Using web text to improve keyword spotting in speech. In *Automatic Speech Recognition and Understanding (ASRU), 2013 IEEE Workshop on* (pp. 428-433). IEEE. 10.1109/ASRU.2013.6707768

Garcia, A., & Gish, H. (2006, May). Keyword spotting of arbitrary words using minimal speech resources. In *Acoustics, Speech and Signal Processing, 2006. ICASSP 2006 Proceedings. 2006 IEEE International Conference on* (Vol. 1, pp. I-I). IEEE. 10.1109/ICASSP.2006.1660179

Gündoğdu, B., & Saraçlar, M. (2017a). Distance Metric Learning for Posteriorgram Based Keyword Search. *International Conference on Acoustics, Speech and Signal Processing (ICASSP)*. 10.1109/ICASSP.2017.7953240

Gündoğdu, B., & Saraçlar, M. (2017b). Similarity Learning Based Query Modeling for Keyword Search. *Proc. Interspeech*, *2017*, 3617–3621. doi:10.21437/Interspeech.2017-1273

Gündoğdu, B., & Saraçlar, M. (2017c). Novel score normalization methods for keyword search. In *Signal Processing and Communications Applications Conference (SIU), 2017 25th* (pp. 1-4). IEEE. 10.1109/SIU.2017.7960262

Gündogdu, B., Sarı, L., Çetinkaya, G., & Saraçlar, M. (2016). Template-Based Keyword Search with Pseudo Posteriorgrams. In *2016 24th Signal Processing and Communications Applications Conference (SIU)* (pp. 973-976). Academic Press. 10.1109/SIU.2016.7495904

Haidt, J. (2006). *The happiness hypothesis: Finding modern truth in ancient wisdom*. Basic Books.

Harper, M. (2014). *IARPA babel program*. Retrieved from https://www. iarpa. gov/index. php/research-programs/babel

Hinton, G., Deng, L., Yu, D., Dahl, G. E., Mohamed, A. R., Jaitly, N., & Kingsbury, B. (2012). Deep neural networks for acoustic modeling in speech recognition: The shared views of four research groups. *IEEE Signal Processing Magazine*, *29*(6), 82–97. doi:10.1109/MSP.2012.2205597

Huang, X., Acero, A., & Hon, H. W. (2001). *Spoken language processing: A guide to theory, algorithm, and system development*. Prentice Hall PTR.

Hundt, C., Schmidt, B., & Schömer, E. (2014, September). Cuda-accelerated alignment of subsequences in streamed time series data. In *Parallel Processing (ICPP), 2014 43rd International Conference on* (pp. 10-19). IEEE. 10.1109/ICPP.2014.10

Karakos, D., & Schwartz, R. (2014). Subword and phonetic search for detecting out-of-vocabulary keywords. *Fifteenth Annual Conference of the International Speech Communication Association.*

Lee, L. S., Glass, J., Lee, H. Y., & Chan, C. A. (2015). Spoken content retrieval—beyond cascading speech recognition with text retrieval. *IEEE/ACM Transactions on Audio, Speech, and Language Processing*, *23*(9), 1389–1420.

Lee, S. W., Tanaka, K., & Itoh, Y. (2005, March). Combining multiple subword representations for open-vocabulary spoken document retrieval. In *Acoustics, Speech, and Signal Processing, 2005. Proceedings. (ICASSP'05). IEEE International Conference on* (Vol. 1, pp. I-505). IEEE.

Mohri, M., Pereira, F., & Riley, M. (2002). Weighted finite-state transducers in speech recognition. *Computer Speech & Language*, *16*(1), 69–88. doi:10.1006/csla.2001.0184

Rabiner, L. R. (1989). A tutorial on hidden Markov models and selected applications in speech recognition. *Proceedings of the IEEE*, *77*(2), 257–286. doi:10.1109/5.18626

Rodriguez-Fuentes, L. J., & Penagarikano, M. (2013). *MediaEval 2013 spoken web search task: system performance measures. n. TR-2013-1.* Department of Electricity and Electronics, University of the Basque Country.

Sakoe, H., & Chiba, S. (1978). Dynamic programming algorithm optimization for spoken word recognition. *IEEE Transactions on Acoustics, Speech, and Signal Processing*, *26*(1), 43–49. doi:10.1109/TASSP.1978.1163055

Saraclar, M., & Sproat, R. (2004). Lattice-based search for spoken utterance retrieval. *Urbana (Caracas, Venezuela)*, *51*, 61801.

Sarı, L., Gündoğdu, B., & Saraçlar, M. (2015). Fusion of LVCSR and Posteriorgram Based Keyword Search. *INTERSPEECH*, *2015*, 824–828.

Shen, W., White, C. M., & Hazen, T. J. (2009). *A comparison of query-by-example methods for spoken term detection*. Massachusetts Inst of Tech Lexington Lincoln LAB.

Van Thong, J. M., Moreno, P. J., Logan, B., Fidler, B., Maffey, K., & Moores, M. (2002). Speechbot: An experimental speech-based search engine for multimedia content on the web. *IEEE Transactions on Multimedia*, *4*(1), 88–96. doi:10.1109/6046.985557

Wang, D., Frankel, J., Tejedor, J., & King, S. (2008, March). A comparison of phone and grapheme-based spoken term detection. In *Acoustics, Speech and Signal Processing, 2008. ICASSP 2008. IEEE International Conference on* (pp. 4969-4972). IEEE. 10.1109/ICASSP.2008.4518773

Chapter 16
Clutter Removal Techniques in Ground Penetrating Radar for Landmine Detection: A Survey

Deniz Kumlu
Istanbul Technical University, Turkey

Isin Erer
Istanbul Technical University, Turkey

ABSTRACT

Ground-penetrating radar (GPR) is a popular technique for landmine detection and widely used by military organizations for landmine clearance purposes. It is well known that GPR is greatly affected by clutter during the landmine detection process. The clutter can be reasoned by soil properties, depth of the buried landmine, different surface types, and ingredient of landmine materials. Thus, the detection of landmine becomes challenging, and clutter removal algorithm must be applied prior to any landmine detection scheme in GPR. In order to remove clutter, various algorithms are proposed, and they can be mainly separated into two groups such subspace-based methods and multiresolution-based methods. This chapter focuses on the performance analysis of these clutter removal algorithms on the simulated dataset that is created by using the gprMax simulation software where it contains four different challenging scenarios.

INTRODUCTION

Landmines are simple yet efficient devices, and they can be easily manufactured with a low cost. They are light, easily transported, simple to bury and difficult to detect. They are widely used by countries to protect their own borders but they can also be used by terrorist organizations to destroy/prevent access to borders, roads, bridges, water sources, and other strategically important regions or points. It is estimated that there are more than 110 million buried landmines in approximately 70 countries and every

DOI: 10.4018/978-1-5225-5513-1.ch016

year nearly 26000 victims die due to landmine explosions (Walsh & Walsh, 2003). Locating the buried landmines is extremely difficult even if they are in marked fields and unfortunately most of the mine-fields are unmarked. In addition, the landmine production technology is still improving, and most of the contemporary landmines are produced solely from the plastic material to reduce the detection probability and increase the false alarm rates. It is known that there are more than 600 types of landmines available and this makes the identification of landmines challenging even if they are detected. The United States currently spends approximately $100 million annually in mine clearance operations (demining) (Furuta & Ishikawa, 2009) and individual removal per mine costs $300-$1000. On the other hand, there is also development in landmine detection technologies (Robledo et al., 2009), and there are several landmine detection techniques available such as;

- **Probing the Ground:** Finding the location of landmines by probing the ground with a stick, and bayonet has been used for many years. For this process, trained soldiers probe the ground however, one small mistake can cost them their lives.
- **Trained Dogs:** Dogs can be trained to locate landmines by sniffing out the vapors from the ingredients of the explosives inside the landmines. However, this is also a slow and dangerous process. Since the vapor from the explosive penetrates the surrounding vegetation by 10 meters, it is impossible for the dogs to find the precise location of the landmines.
- **Metal Detectors:** Early landmines were composed of metal ingredients, and they can be easily detected by metal detectors. Therefore, mine manufacturers have improved the technology in landmine production by including low metal content or making them solely from plastic. Thus, the landmine detection capacity of metal detectors is reduced considerably.

These improvements in landmine production technology caused a decrease in the detection performance, and a new technology called ground-penetrating radar (GPR), is being developed to locate the landmines more accurately, and it emerges as a geophysical non-destructive tool to detect buried objects with low or no metal content (Daniels, 2009). GPR is currently one of the most popular and efficient subsurface sensing devices to investigate buried objects, especially plastic antipersonnel landmines and it has a wide range of other applications such as archeological research, concrete, bridge and road analysis, and geological profiling.

The components of GPR system consist of transmitting and receiving antennas, control and display units as shown in Figure 1. It works by sending pulses of ultra-high frequency radio waves from its transmitting antenna and records the strength, and the time of the reflected energy of the signal from the ground and various buried objects such as roots, stones, gravel and other materials with respect to their dielectric constant and electrical conductivity properties (Daniels, 2005). The received signal is used to create a detailed image of the underground, called a B-scan image or a GPR image. This image contains high amount of clutter caused by several reasons. The major ones are ground-bounce (caused by direct coupling between transmitting and receiving antennas), direct wave (reflection from the ground surface), and scattering from other non-mine objects such as roots, stones and non-uniform terrain conditions.

Generally, landmines are buried close to the surface and clutter dominates the reflected target signal since the ground-bounce is much higher than the target signal. Thus, the clutter degrades the performance of the landmine detectors, and causes an increase in false alarm rates in the non-target region. In addition, there are other factors which can affect the performance of GPR such as soil type, weather conditions, temperature, the depth of the buried target, and varying terrain conditions. Structural parameters in GPR device also affect the detection performance such as antenna design techniques, frequency selection, beamforming techniques (Daniels, 2005), but they are out of the focus of this work.

In order to increase the detection performance of GPR while reducing the false alarm rates, a variety of clutter reduction techniques have been proposed in the literature, and they have obtained certain success on different datasets (Kumlu & Erer, 2017a; Van Der Merwe & Gupta 2000; Verma et al., 2009). Since the terrain preparation and measurement procedures are very time consuming and difficult, there are not many real GPR datasets publicly available containing various challenging scenarios to test the algorithms in all aspects. It is really hard to design a real dataset that contains different scenarios. In addition, it is not easy to make quantitative analysis with real datasets since there is no reference or ground-truth GPR image. The performance comparison of the algorithms for real datasets can only depend on the visual inspection of the authors. To fill this gap in the literature, and to make a fair comparison of algorithms, a new simulated dataset is introduced in this chapter. It is constructed by using the latest version of gprMax simulation software (Warren et al., 2016). gprMax is an open source simulation software that is used for simulation of electromagnetic wave propagation to solve Maxwell's equations in 3D with the Finite-Difference-Time-Domain (FDTD) method, and it is used for modeling GPR. Although gprMax simulation tool is mainly designed for modeling GPR, it can be used to model electromagnetic wave propagation for many other applications (Warren et al., 2016). It has the capability of simulating real commercial antennas used for GPR, and realistic soil types, object materials and various burial depths for the target. Terrain condition parameters can also be easily changed by the user in order to create challenging scenarios. In our simulated dataset, 112 GPR images are created using 7 different soil types, 2 different burial depths, 2 different targets and 4 challenging scenarios. In the first scenario, a flat ground surface is chosen and this is the most basic scenario. In the second scenario, the grass is applied to the flat surface, and the effect of vegetation is tested. In the third scenario, a rough surface is applied and in the last scenario water puddles are added to the rough surface. These challenging and realistic scenarios are used to test the detection performance of GPR clutter removal algorithms.

Figure 1. Block diagram of generic GPR system

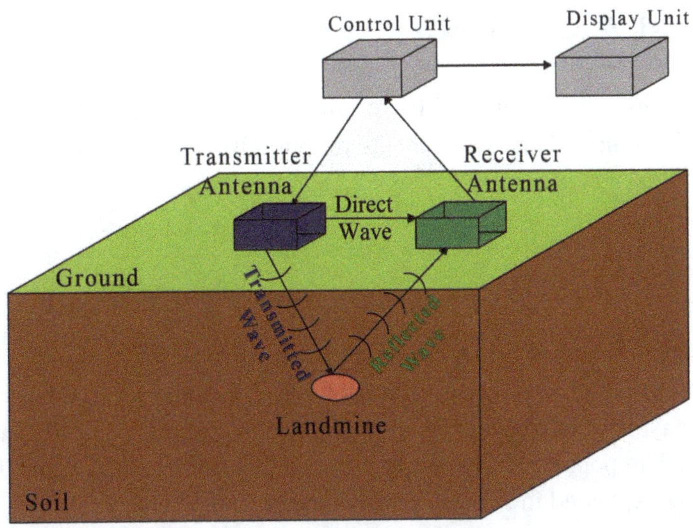

BACKGROUND

Clutter reduction techniques used in the GPR works can be mainly divided into two groups. The first group is based on the subspace-based methods and can be cited as principal component analysis (PCA) (Abujarad & Omar, 2006; Karlsen et al., 2001), singular value decomposition (SVD) (Abujarad et al., 2005; Riaz & Ghafoor, 2012), independent component analysis (ICA) (Abujarad & Omar, 2006), and morphological component analysis (MCA) (Temlioglu & Erer, 2016). The assumption in PCA, SVD, and ICA is quite similar. However each of these algorithms utilizes different constraints for decomposition of the data matrix, which is the GPR image. The clutter removal techniques in these algorithms are based on the assumption that the most significant term after decomposition belongs to the clutter, and the remaining subspaces can be accepted as the target (Verma et al., 2009; Kumlu & Erer, 2017a). The clutter is much stronger than the target in GPR image (Solimene et al., 2014). As a result, GPR image can be decomposed into two separate images as clutter and target images by grouping the subimages during the reconstruction process. The recently proposed MCA method is also in this group. However, it is more complicated in terms of the constraints. This method is based on the sparse representation, and morphological diversity of the GPR image. It takes the image as a linear mixture of several morphological components, and it defines the appropriate dictionary to represent them sparsely. Once the dictionaries are created, the MCA searches for the clutter and the target by making successive iterations by using different thresholds, and decomposes the image into clutter and target components.

The second group is based on multiresolution-multidirectional transforms such as curvelet transform (CT) (Bao et al., 2014; Tzanis, 2015; Terasse et al., 2015), and multiscale directional bilateral filter (MDBF) (Hu & Li, 2012; Kumlu & Erer, 2017b) where each method uses a different scaling, and directional decomposition subbands to express the image as a decomposition of coefficient matrices with different scales and orientations. Then, the information related to different directions is used to separate the target from the clutter. These methods exploit the geometrical property of the clutter and target: the clutter has a horizontal shape, and target has a hyperbolic shape, respectively. In order to remove the clutter in the GPR image, the coefficients that belong to the clutter in the decomposed subspace are equaled to zero, and the inverse transform applied to remaining coefficients to reconstruct the clutter-free target image.

CLUTTER REMOVAL ALGORITHMS

The techniques used for clutter removal in GPR are called as "blind source separation" in signal processing community (Naik & Wang, 2014). The main assumption is that the source signal is a mixture of the clutter and the target signal as given in (1)

$$X = X_{clutter} + X_{\mathrm{target}} \quad (1)$$

where X, $X_{clutter}$, X_{target} are GPR, clutter and target images, respectively.

Various signal processing techniques are proposed to decompose GPR image as clutter and target images without using any prior information in the literature (Verma et al., 2009; Kumlu & Erer, 2017a; Abujarad et al., 2004). The popular and recently proposed clutter removal algorithms available in the literature will be briefly explained in the following sections.

Principal Component Analysis (PCA)

PCA is a popular and widely used mathematical tool from applied linear algebra in statistical data analysis (Wold et al., 1987). It is a simple, non-parametric method for extracting significant information from a mixture of data. PCA works as an orthogonal transformation which converts a mixture of data with possibly correlated variables into a set of linearly uncorrelated variables. To put it simply, it re-expresses the mixture as a linear combination of its basis vectors. If the number of the correlated components is high, the accuracy of the algorithm increases. PCA is used in a variety of applications such as signal separation, image analysis, pattern recognition, and data visualization.

In the PCA based clutter removal method (Abujarad & Omar, 2006; Karlsen et al., 2001), GPR image is represented by a rectangular matrix $X_{i,j}$ where

$$MxN(i=1,2,...,M; j=1,2,...,N)$$

is the dimensions. Here *i* denotes the depth index and *j* denotes the antenna position or down-track index. *N* principal components of the mixture data matrix *X* can be given by (2)

$$P = A^T X \tag{2}$$

where $X = \left[x_1, x_2, x_3, ..., x_N\right]^T$ is the zero-mean input vector, $Y = \left[y_1, y_2, y_3, ..., y_N\right]^T$ is the principal components, and $A = \left[\sigma_1, \sigma_2, \sigma_3, ..., \sigma_N\right]$ is an *MxN* transformation matrix containing eigenvectors in decreasing order, used to transform *X* into *Y*.

After the transformation matrix *A* is found, the principal component matrix *S* can be calculated as (3)

$$S = A^T X \tag{3}$$

After applying PCA on data matrix *X*, we can obtain principal components as the number of A-scans in GPR image where it is defined as *N*. Data matrix *X* can be thought as a mixture of clutter and target images. Therefore, it can be separated into two subspaces, target and clutter such as

$$X = X_{clutter} + X_{\text{target}} = A_1^T S_1 + \sum_{i=2}^{N} A_i^T S_i \tag{4}$$

After GPR image *X* is decomposed into two subspaces as in (4), X_{target} yields the clutter-free target image.

Singular Value Decomposition (SVD)

SVD is used for spectral decomposition in linear algebra (Golub & Reinsch, 1970), and it has many applications in signal processing, image processing and statistics that can be used for many tasks such as pattern analysis, noise reduction, regression, prediction, and compression to find approximate solutions to optimization problems. The main assumption in SVD for data analysis is to use a rank reduced approximation of a matrix to generalize some of the structures.

For clutter removal using SVD (Abujarad et al., 2005; Riaz & Ghafoor, 2012), GPR image is represented by a rectangular matrix $X_{i,j}$, whose dimension are

$$MxN(i = 1,2,...,M; j = 1,2,...,N).$$

Here i denotes the depth index and j denotes the antenna position or down-track index. SVD decomposition of data matrix X can be given by (5)

$$X = USV^T \quad (5)$$

where the U and V are left and right singular vectors with the dimensions *(MxM)* and *(NxN)* respectively. In other words, U and V are the eigenvectors of XX^T and X^TX and $S = diag(\sigma_1, \sigma_2, ..., \sigma_r)$ sorted from maximum to minimum order, the SVD decomposition is given by (6), and (7)

$$X = \sum_{i=1}^{N} \sigma_i u_i v_i^T \quad (6)$$

$$X = M_1 + M_2 + ... + M_N \quad (7)$$

where M_i are matrices of the same size with X, and called as i^{th} eigenimage of X. Data matrix X is a mixture composed of clutter and target matrix. Therefore, it can be separated into two subspaces, target and clutter such as

$$X = X_{clutter} + X_{\text{target}} = \sigma_1 u_1 v_1 + \sum_{i=2}^{N} \sigma_i u_i v_i^T \quad (8)$$

After GPR image X is decomposed into two subspaces as in (8), X_{target} will become the clutter-free target image.

Independent Component Analysis (ICA)

ICA is used to solve the blind source separation problem similarly to PCA and finds a new basis to represent the mixture of data but using different constraints. It is used to separate data matrix into its statistically independent components (Hyvarinen et al., 2004). Therefore, ICA is popular in many applications such as clutter removal, feature extraction and financial data analysis.

In ICA based clutter removal, GPR image is represented by a rectangular matrix $X_{i,j}$, whose dimension are

$MxN(i = 1,2,...,M; j = 1,2,...,N)$.

Here i denotes the depth index and j denotes the antenna position or down-track index. ICA model assumes that the mixture of data matrix X can be generated from source data S through a linear process $X = AS$ where S and A are unknown source and mixing matrices, respectively. Then, ICA model can estimate unknown matrices from data matrix X itself with very few assumptions. The estimation of source matrix can be given as

$$\hat{S} = WX \tag{9}$$

There are various algorithms proposed to solve ICA in literature and the most popular one is FASTICA algorithm which is a fixed point iteration based algorithm to calculate W (Hyvarinen, 1999). After the separating matrix is calculated, the mixing matrix A can be computed by simply taking the inverse of it as $A = W^{-1}$. Once the mixing matrix A is obtained, the related source matrix S can be easily computed.

After applying ICA on the data matrix, the number of independent components equals to the number of A-scans in GPR image. The data matrix X is composed of clutter and target matrix. Therefore, it can be separated into two subspaces, target and clutter such as

$$X = X_{clutter} + X_{\text{target}} = A_1 S_1 + \sum_{i=2}^{N} A_i S_i \tag{10}$$

After GPR image X decomposed into two subspaces as in (10), X_{target} will become the clutter-free target image.

Morphological Component Analysis (MCA)

The recently proposed MCA is a sparse representation based image decomposition method which has found applications in image denoising, image decomposition and source separation (Starck et al., 2005; Bobin et al., 2006). In this approach, the image is assumed to be composed of several morphologically distinct components and each of them is sparsely represented in a specific transform domain. The GPR images consist of the clutter, the target, and system noise which is insignificant compared to the other components. Thus in MCA, it is assumed that the GPR images are composed of two morphological components, namely, the clutter and the target (Temlioglu & Erer, 2016). The clutter component is given by (11)

$$x_c = \phi_c \alpha_c \tag{11}$$

and the target component is given by (12)

$$x_t = \phi_t \alpha_t \tag{12}$$

Here ϕ_c and ϕ_t are redundant dictionaries, α_c and α_t are the sparse coefficients corresponding to the clutter and the target components, respectively. Thus, the optimization problem in MCA can be re-written as in (13)

$$\left\{x_c^{opt},\, x_t^{opt}\right\} = \underset{\{x_c, x_t\}}{\arg\min} \left\|\phi_c^{+} x_c\right\|_1 + \left\|\phi_t^{+} x_t\right\|_1 + \lambda \left\|X - x_c - x_t\right\|_2^2 \tag{13}$$

The MCA decomposition algorithm for the clutter reduction in GPR is given in Algorithm 1 (Temlioğlu & Erer, 2016).

The success of MCA algorithm depends heavily on the selection of dictionaries. Thus, dictionary selection reveals as a major step which has to satisfy the following conditions:

- The dictionaries must be mutually coherent.
- The dictionaries must represent each related component sparsely.

It is well known that isotropic structures can be efficiently represented by the wavelet dictionaries, while the curvelet or ridgelet dictionaries are more suitable candidates for global lines. Similarly, locally oscillating textures can be better represented by a discrete cosine transform (DCT) or brushlet dictionaries. Since clutter components include lines or line-shaped parts, while targets produce isotropic structures, an appropriate choice will be to select wavelet dictionaries for modeling the target, and curvelet dictionaries for modeling the clutter components. The recipe for MCA-based clutter removal is given in Figure 2.

Curvelet Transform (CT)

CT is a high-dimensional generalization of the wavelet transform to represent an image as a multiresolution-multidirection decomposition in the sense of different scales, and orientations (Starck et al., 2002). Although wavelet transform has become popular in image processing, it fails to handle randomly oriented edges. Therefore, CT is proposed to overcome these limitations. It can successfully detect the different orientations of the image edges. When CT was first presented, it was using the concept of discrete ridgelet transform and then, it was improved, and the second generation of CT transform was proposed. The new version of CT is called as fast discrete curvelet transform via wrapping (FDCT) (Candes, et al., 2006). Thus, it is faster, simpler and less redundant than the original CT. The architecture of the FDCT algorithm can be given in the following steps.

1. In order to get Fourier samples, take the 2D FFT of an image

$$\hat{f}\left[m,n\right], -\frac{n}{2} \le m,n < \frac{n}{2},$$

2. For each scale *j* and angle *l*, form the product

$$\tilde{U}_{j,l}\left[m,n\right]\hat{f}\left[m,n\right],$$

Figure 2. MCA-based clutter removal in GPR images

Algorithm 1 MCA-based clutter removal in GPR images

Parameters:
X: GPR image (B-scan data)
$\emptyset_c$ and $\emptyset_t$: clutter and target dictionaries, respectively
N_{iter}: number of iterations
λ_{min}: stopping threshold value

Initialization:
Initial solutions: $x_c^{(0)} = x_t^{(0)} = 0$
Initial residual: $r^{(0)} = X$
Initial threshold: let $k^* = max_k \|\phi_k^+ X\|_\infty$, set $\lambda^{(0)} = max_{k \neq k^*} \|\phi_k^+ X\|_\infty$

Main iteration:
1. Perform N_{iter} times
- Calculate residual $r = X - x_c - x_t$

Part A: Update of x_c assuming x_t is fixed
- Calculate clutter coefficients $\alpha_c = \emptyset_c^+(x_c + r)$
- Update clutter coefficients by thresholding, $\hat{\alpha}_c$
- Reconstruct clutter image x_c by $x_c = \emptyset_c \hat{\alpha}_c$

Part B: Update of x_t assuming x_c is fixed
- Calculate target coefficients $\alpha_t = \emptyset_t^+(x_t + r)$
- Update target coefficients by thresholding, $\hat{\alpha}_t$
- Reconstruct target image x_c by $x_t = \emptyset_t \hat{\alpha}_t$

2. Update the threshold λ
3. If $\lambda \leq \lambda_{min}$ then stop

Output: x_c and x_t clutter and target images, respectively

3. Wrap this product around the origin and obtain

$$\hat{f}_{j,l}[m,n] = W\left(\tilde{U}_{j,l}\hat{f}\right)[m,n],$$

where the range for m, n, and θ is now $0 \leq m < 2^j$, $0 \leq n < 2^{j/2}$ and $-\pi/4 \leq \theta < \pi/4$,

4. Apply IFFT to each to each $\hat{f}_{j,l}$, hence collecting the discrete coefficients $c^D\left(j,l,k_1,k_2\right)$.

The proposed FDCT algorithm is based on the wrapping of Fourier samples. It is an invertible transform and the inverse FDCT algorithm can be computed by reversing all the operations, given in the following steps.

For each scale and angle pair *(j, l)* apply 2D FFT of each array $c^{D,N}\left(j,l,k_1,k_2\right)$, and obtain $W\left(\tilde{U}_{j,l}\hat{f}\right)[m,n]$,

1. For each scale and angle pair *(j, l)* multiply the array by the corresponding $W\left(\tilde{U}_{j,l}\hat{f}\right)\left[m,n\right]$ wrapped curvelet $W\left(\tilde{U}_{j,l}\hat{f}\right)\left[m,n\right]$ that gives

$$W\left(\left|\tilde{U}_{j,l}\right|^{2}\hat{f}\right)\left[m,n\right],$$

2. Unwrap each array $W\left(\left|\tilde{U}_{j,l}\right|^{2}\hat{f}\right)\left[m,n\right]$ on the frequency grid and add them together to recover $\hat{f}\left[m,n\right]$,
3. Take a 2D inverse FFT to reconstruct image.

The code is available for the FDCT and IFDCT algorithm and it is called "Curvelab (version: 2.1.2)" (Candes, et al., 2006). This code is used in our implementation.

The architecture of the FDCT via wrapping algorithm is given in Figure 3.

As it can be seen in Figure 4, The GPR image is decomposed into 5 scales by using the steps in Figure 3. The first scale is called as coarse layer and it contains coefficients of the image, the three middle

Figure 3. CT-based clutter removal in GPR images

Algorithm 2 CT-Based Clutter removal in GPR Images

***A*:** GPR image (B-scan data)
***C*:** Coefficient matrix
***j*:** Number of decomposition levels (scale)
***l*:** Number of decomposition levels (2^l direction)
k_1 and k_2: spatial location of the coefficients
fdct: fast discrete curvelet transform
ifdct: inverse fast discrete curvelet transform

Step 1: Scale and directional decomposition of the image (FDCT)
$c(j,l,k_1,k_2) = fdct(A)$

Step 2: Remove horizontal clutter in detail subbands
  for j = 2,…J and l = 1,…,2^l **do**
    if $C^{j,l}$ == horizontal clutter
      $C^{j,l}$ = zero;
    end if
end for

Step 3: Reconstruction of target image from Coefficient matrix
Arec = ifdct($c(j,l,k_1,k_2)$)

Output :
Arec = reconstructed clutter-free image

Figure 4. CT decomposition of GPR image

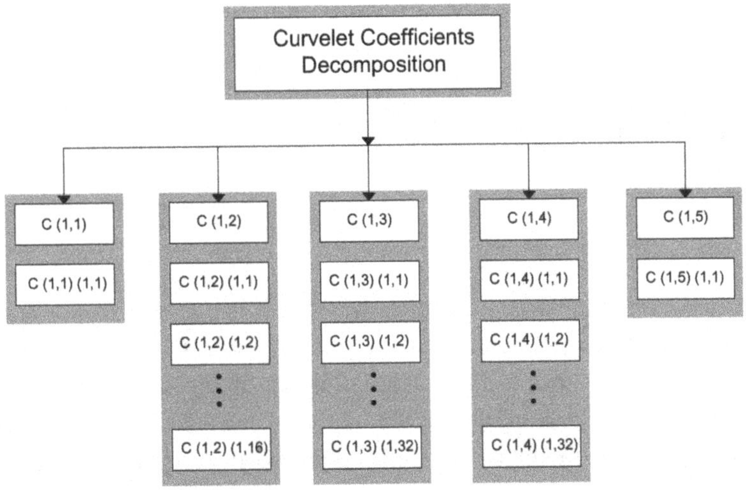

layers and the last layer are detail and fine layers, respectively. Then every detail layer is divided into 16 or 32 directions, called wedges which represent edge orientations at each detail layer.

In order to remove clutter from the GPR image, wedges related to the horizontal details (clutter) are detected in each layer and these wedge matrices are set to zero. Then, the remaining coefficient matrices are used for reconstruction via inverse FDCT transform to get the clutter-free image.

Multiscale Directional Bilateral Filter (MDBF)

The MDBF method is based on bilateral filter which is an edge-preserving smoothing filter (Tomasi & Manduchi, 1998) and it is used in many applications such as image processing, remote sensing and clutter removal in GPR (Kaplan & Erer, 2014; Kaplan & Erer 2012; Kumlu & Erer 2017b). Since clutter and target components have some geometric features, this information can be used to separate them. Generally, multiresolution-based methods in GPR use the horizontal feature of clutter and the hyperbolic feature of the target.

Bilateral filter is an improved version of the classical Gaussian filter. The performance of Gaussian filter fails at edges. To overcome this problem, bilateral filter is proposed where it combines both geometric closeness and photometric similarity of pixel intensity values. Therefore, bilateral filter can be thought as a combination of the spatial filter and the range filter (14)

$$BF\left[A\right]_p = \frac{1}{W_p}\sum_{q\in N} G_{\sigma_s}\left(\left\|p-q\right\|\right)G_{\sigma_r}\left(\left\|I_p - I_q\right\|\right)I_q \tag{14}$$

where W is the normalization parameter and can be given as (15)

$$W_p = \sum_{q\in N} G_{\sigma_s}\left(\left\|p-q\right\|\right)G_{\sigma_r}\left(\left\|I_p - I_q\right\|\right) \tag{15}$$

where A is the input GPR image, p and q are the spatial locations of the GPR image, A_p and A_q are the pixel intensity values of the location p and q respectively. N defines the neighborhood of the pixel value p, G_{σ_s} and G_{σ_r} are called the spatial and range Gaussian kernels. After GPR image is filtered as in (14), it is called approximation subband *BF[A]* and it is substracted from input GPR image A in order to obtain detail subband *D[A]* and it is given by (16)

$$D\left[A\right] = A - BF\left[A\right] \tag{16}$$

This is called one level bilateral filter decomposition and it is not enough to obtain significant geometric information from GPR image since images may contain information at different resolutions and orientations. Therefore, this algorithm can be extended into a multiscale form using *a trous* algorithm (Fattal et al., 2007). Simply, the above equations are applied iteratively to obtain more detail subbands.

After the detail subbands are obtained, the directional information in these detail subbands can be extracted (Hu & Li, 2012), by applying nonsubsampled directional filter bank (NSDFB) to each detail band. (Da Cunha et al., 2006). The size of the resulting multidirectional detail bands, are equal to the size of the input GPR image, thus this transform is shift-invariant.

The recipe of the decomposition of the GPR image is presented in Figure 5 and it provides the MDBF coefficients $c\left(j, l, k_1, k_2\right)$ indexed by a scale j, an orientation l and spatial location parameters k_1 and k_2. Then, the directional subband matrices which contain horizontal information are set to zero and clutter-free image is reconstructed by taking the inverse transform of the resulting coefficient matrix which contains no horizontal information.

The GPR image is decomposed into directional subbands in each scale by using MDBF method given in Figure 5. In this implementation, 5 scale subbands are used where the first subband is the approximation subband and the others are the detail subbands. Then, each detail subband is decomposed into 16 directional subbands by using directional filter bank as shown in Figure 6. During the multiscale decomposition, the critical Gaussian kernel parameters are chosen as $\sigma_s = 2.8$ and $\sigma_r = 0.1$ since the best ROC curve results are obtained with these setups.

EXPERIMENTAL SETUP OF THE SIMULATED DATASET

During the construction of the simulated dataset, the actual challenges of the landmine detection and the current properties of commercial systems such as antennas are taken into account (Warren et al., 2016). In all simulations, Geophysical Survey Systems, Inc. (GSSI) 1.5 GHz (Model 5100) antenna which is common in commercial GPR systems, is used. The constructed simulated dataset has 4 scenarios and each scenario contains 28 GPR images, thus the total number of GPR images in the simulated dataset becomes 112. Each scenario contains 7 different soil types, 2 different objects, 2 different burial depths as shown in Table 1. The two most important parameters of the material that can affect the GPR performance are the dielectric constant, and electrical conductivity (Daniels, 2005). Therefore, soil type is one of the important factors for GPR performance since a more conductive soil (one that is wetter or has a greater clay content) will absorb more energy compared to a low conductive one such as dry sand, and this phenomenon will constrain the maximum depth penetration of the electromagnetic wave (Tretya-

Figure 5. MDBF-based clutter removal in GPR images

Algorithm 3 MDBF-Based Clutter Removal in GPR images

Parameters :
A : GPR image (B-scan data)
C **:** Coefficient matrix
j **:** Number of decomposition levels (scale)
l **:** Number of decomposition levels (2^l direction)
σ_s **and** σ_r **:** Standard deviation of Gaussian filters
nsdfb : directional filter bank function for decomposition
nsdfbrec : directional filter bank function for reconstruction

Step 1: Scale Decomposition of GPR image (MBF)
$A_n^1 = A_n$
for *j = 1, ..., J* ***do***
$A_p^{j+1} = \frac{1}{W_p} \sum_{q \in S} w^j(p-q) G_{\sigma_s}(A_p^j - A_q^j) A_q^j$
$C^j = A^{j+1} - A^j$
end for

Step 2: DFB decomposition of GPR image (MDBF)
for j = 2,…,J **do**
$C^{j,l} = nsdfb(C^j, l)$
end for

Step 3: Remove horizontal clutter in detail subbands
  for j = 2,…J and l = 1,…,2^l **do**
    if $C^{j,l}$ == horizontal clutter
      $C^{j,l}$ = zero;
    end if
end for

Step 4: Reconstruction of target image from Coefficient matrix
$C^j = nsdfbrec(C^{j,l}, l)$
for j = 2,…,J **do**
Arec = Arec + C^j
end for

Output :
Arec = reconstructed clutter-free image

kov & Erden, 2007; Tretyakov & Erden, 2008; Erden, 2017). The electromagnetic properties of buried target also have effect on the performance. The higher conductivity of the buried material causes the attenuation of the GPR signal and the detection probability of the non-metal targets decreases. Another important factor is the depth of the buried target. The penetration of the electromagnetic wave attenuates with depth and the targets close to the surface of the ground are mixed with the clutter (ground-bounce

Figure 6. MDBF decomposition of GPR image

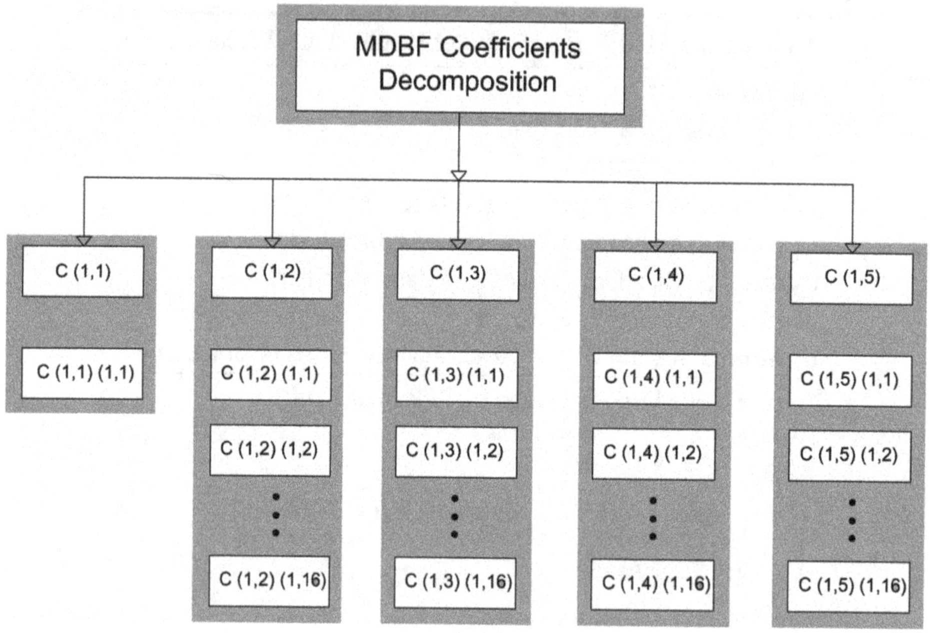

and direct wave) (Leucci, 2008). Thus, it can be observed that these parameters have a significant influence on the GPR performance. So, these parameters are systematically changed to observe their influence on the detection performance. The dielectric constants and electrical conductivities of the soil and target materials used in the simulated dataset are presented in Table 2. These values are obtained from (Bostanudin, 2013; Deiana & Anitori, 2010).

Table 2 presents the electromagnetic properties of the two objects and the 6 different soil types. The 7th soil type is a heterogeneous soil that includes water puddles. The properties of the heterogeneous soil are defined as clay fraction 0.5, sand fraction 0.5, bulk density 2 g/cm3, sand particle density of 2.66 g/cm3, and a volumetric water fraction range of 0.001 - 0.25. The experimental setup of the simulated dataset is presented in Figure 7. As shown in the figure, cylindrical targets are buried at 2 cm to simulate objects that buried close to the surface of the ground such as landmines. Only a single object is buried in each simulation, and multiple object detection is out of focus in this simulated dataset. The radius and the height of the cylinder are defined as 2.8 cm and 4 cm to match the dimensions of the antipersonnel landmine. The discretization of the model is selected as $\Delta x = \Delta y = \Delta z = 1$ mm and the simulation domain is defined as 1000x120x400 mm. Antennas are placed at 5 cm above the surface and moved 1 cm in each simulation to obtain the A-scan along x-direction and in each scenario, antennas are moved 80 times to obtain the B-scan or GPR image. Therefore, the scanned distance is 80 cm (it is also called down-track or antenna position) and it starts from 10 cm to 90 cm (Temlioglu et al., 2017).

A-scan is a one-dimensional signal produced by gprMax software in one simulation and B-scan is two-dimensional signal resulting from a collection of adjacent A-scan signals as shown in Figure 8(a) and (b). Simulated dataset is designed for 4 different scenarios.

Table 1. Systematic construction of the simulated dataset

Scenarios				Soil Type	Target Material	Burial Depth	Image No
Scenario 1: Flat surface	**Scenario 2:** Flat surface + grass	**Scenario 3:** Rough surface	**Scenario 4:** Rough surface + water puddles	Heterogeneous	Plastic	2 cm	GPR image 1
						3 cm	GPR image 2
					Aluminium	2 cm	GPR image 3
						3 cm	GPR image 4
				Dry sand	Plastic	2 cm	GPR image 5
						3 cm	GPR image 6
					Aluminium	2 cm	GPR image 7
						3 cm	GPR image 8
				Damp sand	Plastic	2 cm	GPR image 9
						3 cm	GPR image 10
					Aluminium	2 cm	GPR image 11
						3 cm	GPR image 12
				Wet sand	Plastic	2 cm	GPR image 13
						3 cm	GPR image 14
					Aluminium	2 cm	GPR image 15
						3 cm	GPR image 16
				Dry clay soil	Plastic	2 cm	GPR image 17
						3 cm	GPR image 18
					Aluminium	2 cm	GPR image 19
						3 cm	GPR image 20
				Wet clay soil	Plastic	2 cm	GPR image 21
						3 cm	GPR image 22
					Aluminium	2 cm	GPR image 23
						3 cm	GPR image 24
				Dry loam soil	Plastic	2 cm	GPR image 25
						3 cm	GPR image 26
					Aluminium	2 cm	GPR image 27
						3 cm	GPR image 28

In the first scenario, it is assumed that the ground surface is flat. This is the simplest scenario among all the scenarios. In the second scenario, 100000 numbers of blades of grass are applied to the flat surface area. The height of the root of the grass is selected as 2 cm and blade of the grass is 2 cm. The influence of the vegetation in the detection performance is tested for this scenario. In the third scenario, a rough ground surface is applied. Most of the false alarms stem from the surface roughness. This scenario is crucial for the testing of the robustness and the accuracy of the algorithm. The amount of fluctuation on the ground surface is nearly 2 cm. In the last scenario the surface is rough with water puddles. The rough surface is defined similar to the third scenario. The depth of the water is defined as 0.5 cm. The raw images obtained from gprMax simulation tool for four different scenarios are presented in Figure

Table 2. Electromagnetic properties of the materials

Material	Dielectric Constant (F/m)	Electrical Conductivity (S/m)
Dry sand	3.0	0.001
Damp sand	8.0	0.01
Wet sand	20.0	0.1
Dry clay soil	10.0	0.01
Wet clay soil	12.0	0.01
Dry loam soil	10.0	0.001
Aluminum	3.1	2.3×10^{-7}
Plastic	3.0	0.01

Figure 7. The setup of the scenario for simulated dataset

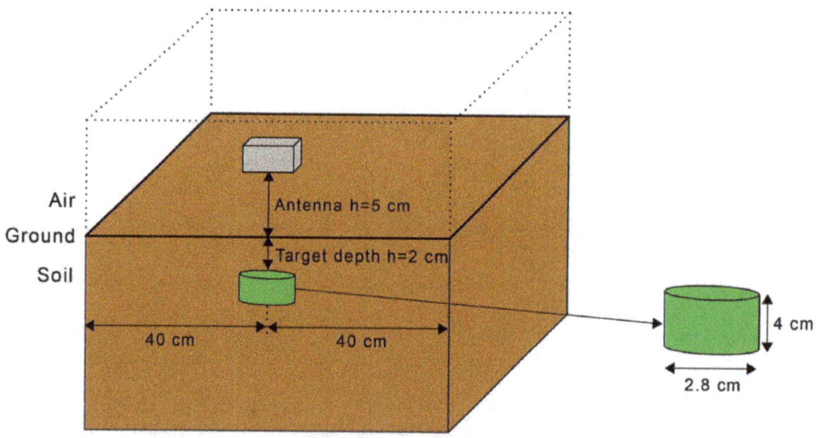

Figure 8. (a) A-scan signal, (b) B-scan signal constructed from concatenation of A-scans

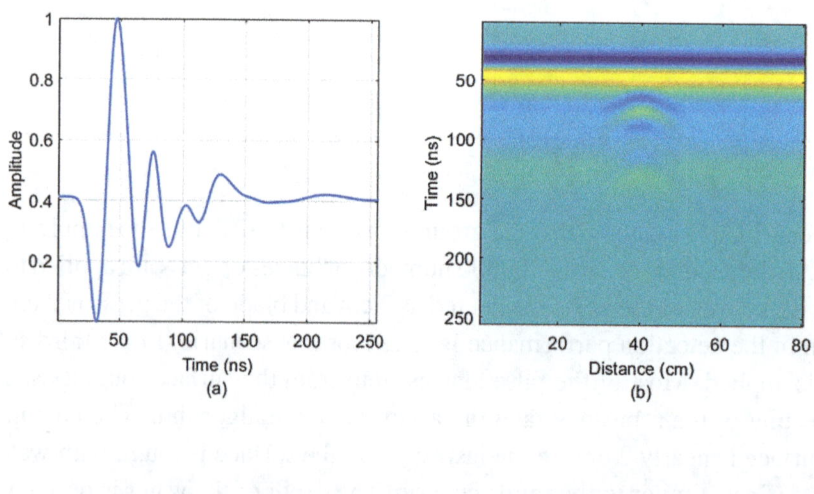

9(a)-(d). Figure 9(a) presents the scenario 1 that is the simple scenario with flat surface. In the Figure 9(a), the soil type is heterogeneous, the target material is aluminum and the burial depth of the target is 2 cm that is close to surface ground (GPR image 3 in Table 1). As shown in the figure, the target can be observed visually, since the scenario is simple and the target is aluminum which means more conductive. Figure 9(b) presents the scenario 2 that is the flat ground surface with grass. As shown in the figure, the effect of vegetation can be observed. The type of the soil is dry sand that is a perfect environment for landmine detection since both the dielectric constant, and the electrical conductivity is less. The target material is aluminum and it is buried to 3 cm, therefore it can also be observed visually (GPR image 8 in Table 1). Figure 9(c) presents the scenario 3 that is the rough surface. As seen in the figure, the surface roughness is distinctive compared to scenario 1 and if the target is buried very close to surface for this scenario, it cannot be even detected with human eyes. The soil type is wet sand that attenuates the electromagnetic propagation of GPR. The target material is selected as plastic and it is buried at 2 cm (GPR image 12 in Table 1). Plastic is also less conductive than aluminum. As a result of these selected parameters, the target can be barely detected visually. Figure 9(d) presents the scenario 4 that is the rough surface with water puddles. The effect of the water puddles can be observed in the surface. The soil type is wet clay soil that also attenuates the electromagnetic wave propagation. The target material is aluminum and it is buried to 3 cm (GPR image 24 in Table 1). Since the target material is aluminum and the soil type is wet clay soil, it is more visible compared to the scenario 3 in Figure 9(c). The visual effect of the selected parameters can be observed clearly in Figure 9(a)-(d).

Figure 9. GPR images for four different scenarios (a) Flat ground surface, (b) Flat ground surface with grass, (c) Rough ground surface, (d) Rough ground surface contains water puddles

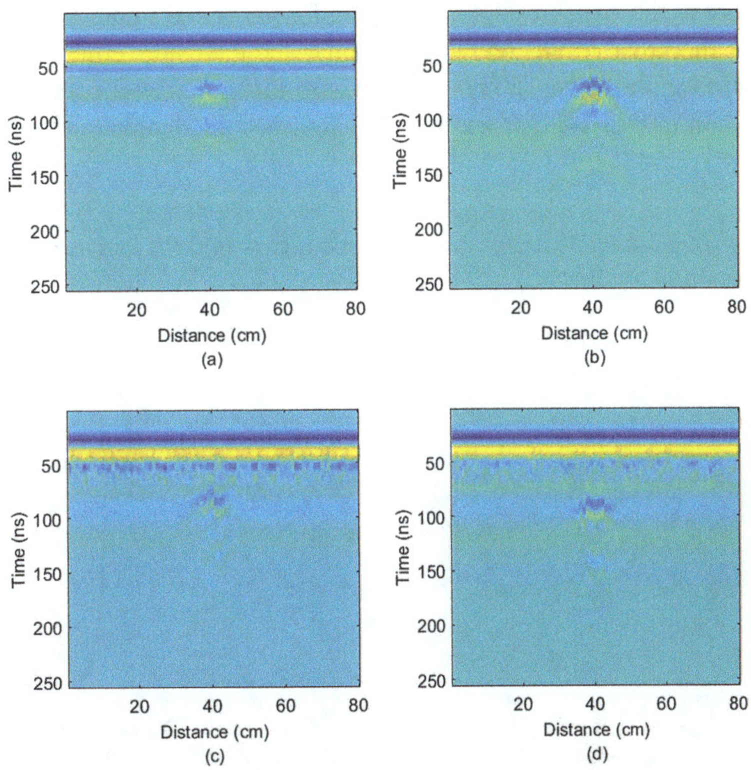

RESULTS AND DISCUSSION

Clutter removal techniques explained in the section above are applied to simulated datasets and the clutter-free target images are compared based on the ROC curves to report the quantitative analysis results about the overall detection performance for all the soil types, objects and burial depths. Probability of Detection (PD) versus False Alarm Rate (FAR) is used to calculate the ROC curves by means of the energy function of the obtained clutter-free target image. PD is equal to the number of correct alarms divided by the total number of B-scans or GPR image that is 112 for this work. FAR is equal to the false alarms divided by the number of maximum possible false alarms.

For this calculation, a detection region is determined using the exact down-track location and the dimension of the buried object. Then, remaining areas are defined to be the false alarm regions. A detection alarm is given if there are 5 consecutive samples in the inspection area with detection values greater than the threshold, which corresponds to a target width of about 2 cm. The number of maximum possible false alarms is the ceiling number of false alarm positions divided by 5, which is the alarm size. Figure 10(a) shows the center of the buried landmine (green line) and detection start and stop regions (red lines) which are between $(25-55)$ and the outside of this area $(0-25$ and $55-80)$ is false alarm region. Figure 10(b) shows the energy of the given clutter-free B-scan image, the black line is the normalized energy and blue line is the smoothed energy via low pass filter. The images are smoothed to remove the short-term fluctuations and to obtain the long-term trend of the energy function.

The clutter-free visual results are shown in Figure 11(a)-(f) for the PCA, SVD, ICA, MCA, CT, and MDBF methods. The original GPR image is selected from scenario 1 that is the simple scenario; the soil type is selected as heterogeneous, the target material is aluminum and depth of the target is 2 cm (GPR image 3 in Table 1). Since the scenario is simple and appropriate soil type and target material are chosen, the quality of the visual results is satisfying for the clutter removal algorithms. As seen in Figure 11(a) and (c), the visual results seem quite similar and there are vertical lines below the target compared to results in Figure 11(b). The results in Figure 11(b) are better compared to others except Figure 11(d). The result in Figure 11(d) seems satisfying since the dictionaries are optimized for this

Figure 10. (a) Clutter-free landmine image and detection region in GPR image, (b) Energy and filtered energy plot of the detection algorithm

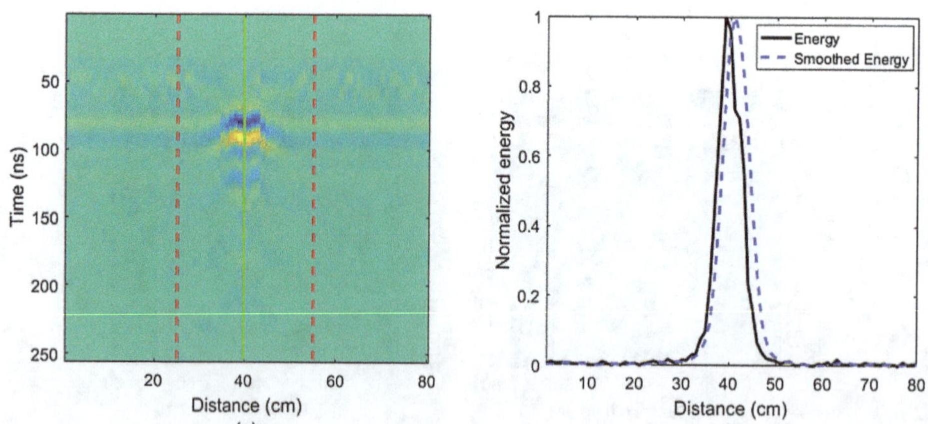

scenario however for the other scenarios the visual results are not as good as expected. The visual quality of the results in Figure 11(e) seems poorer compared to subspace-based methods since it is using different approach for clutter removal. Therefore, there is a distortion in the target and there is also some remaining clutter on the surface. However, the target location and amplitude can be seen by visually. The result in Figure 11(f) has the same methodology as in Figure 11(e), however, it has a better visual result compared to Figure 11(e).

The ROC curve results in Figure 12 show the performance of the methods for different scenarios. Figure 12(a) is the results for scenario 1 and this scenario has a flat ground surface. Thus it is the simplest scenario. As seen in the figure, all the methods except PCA and ICA show excellent performances and their detection probabilities are almost equal to *1*.

Figure 12(b) is the results for scenario 2 and this scenario has a flat ground surface with grass. In this scenario, PCA and ICA also have the worst performance and CT has a poorer performance compared to the others and MDBF has the best performance. Figure 12(c) shows the results for scenario 3 and this scenario has a rough ground surface. As seen in Figure 12(c), performances of all the algorithms are degraded and CT has the best performance. Figure 12(d) is the results for scenario 4 and this scenario has a rough ground surface with water puddles. As seen in Figure 12(d) MDBF has the best performance and MCA takes the second place. Briefly, performance of the algorithms gets poorer when the scenarios get challenging. The results show that multiresolution-based methods show a better performance compared to the others when the scenarios get challenging and MCA needs separate dictionaries for each scenario in order to obtain good results. The performance of PCA, SVD and ICA get poorer with respect to the difficulty of the scenarios.

Figure 11. Clutter-free GPR image obtained by different methods for simulated dataset (a) PCA (b) SVD (c) ICA (d) MCA (e) CT (f) MDBF

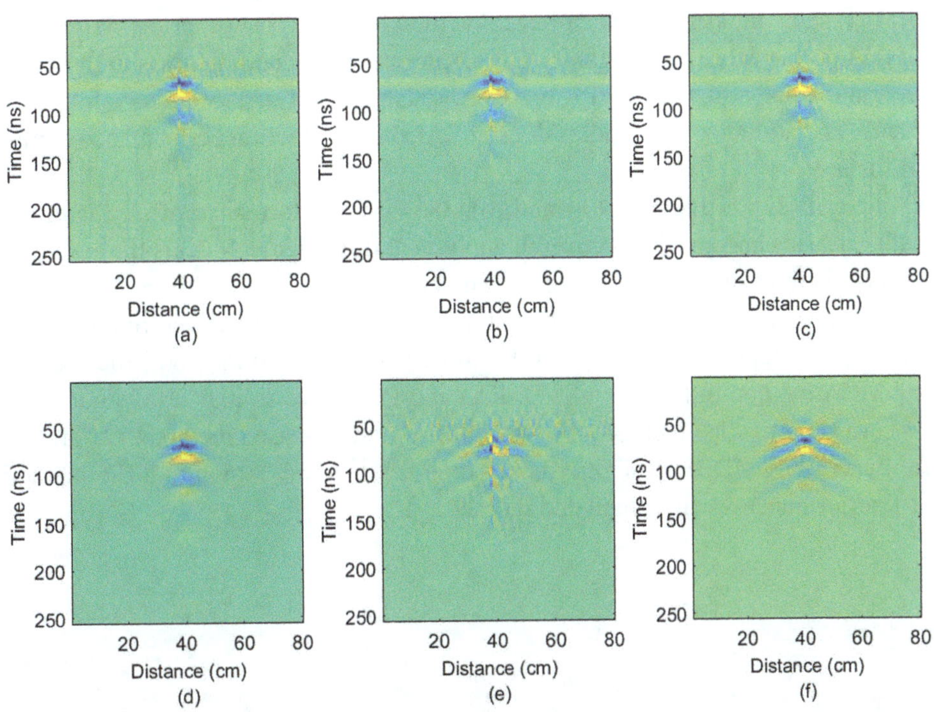

Figure 12. Performance evaluation of methods for different scenarios in terms of ROC curves (a) Scenario 1, (b) Scenario 2, (c) Scenario 3, (d) Scenario 4

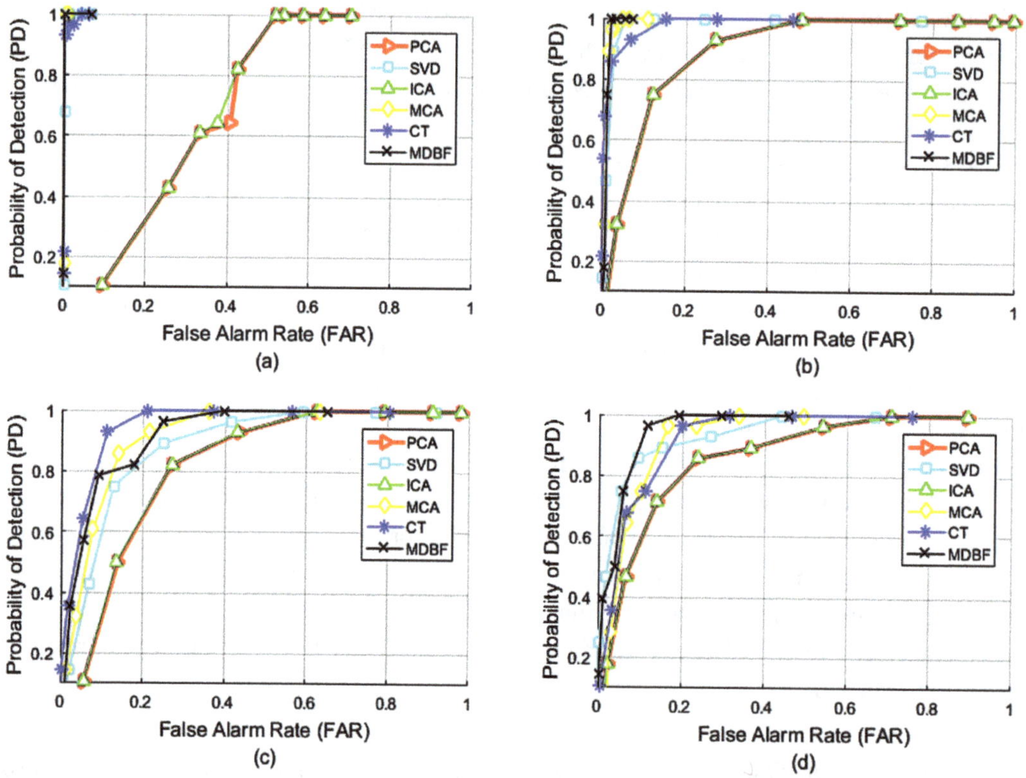

Figure 13(a)-(f) show the performance of each method for different scenarios. As shown in the figure, PCA and ICA performances are instable. The figure shows that adding grass to the ground surface has little effect on the performance of GPR however, rough ground surface affects the performance of GPR significantly and adding water puddles does not affect the detection performance compared to the rough ground surface.

The ROC curve results in Figure 14(a) and (b), show the performance of all the clutter removal methods for all scenarios and provide an overall analysis ability. Figure 14(b) is the zoomed version of Figure 14(a) since there is slight difference between performances of algorithms. It can be seen in the figure that subspace-based methods and multiresolution-based methods show a similar performance for the simulated dataset except for PCA and ICA. For the subspace-based methods, the best performance is achieved by MCA method followed by SVD. ICA and PCA have similar results but they have poor performances compared to other methods. For the multiresolution-based methods, CT and MDBF have similar performances but MDBF is slightly better than CT method. Among all the methods, MDBF shows the best performance for this simulated dataset.

Figure 13. Performance evaluation of each method for different scenarios in terms of ROC curves (a) PCA, (b) SVD, (c) ICA, (d) MCA, (e) CT, (f) MDBF

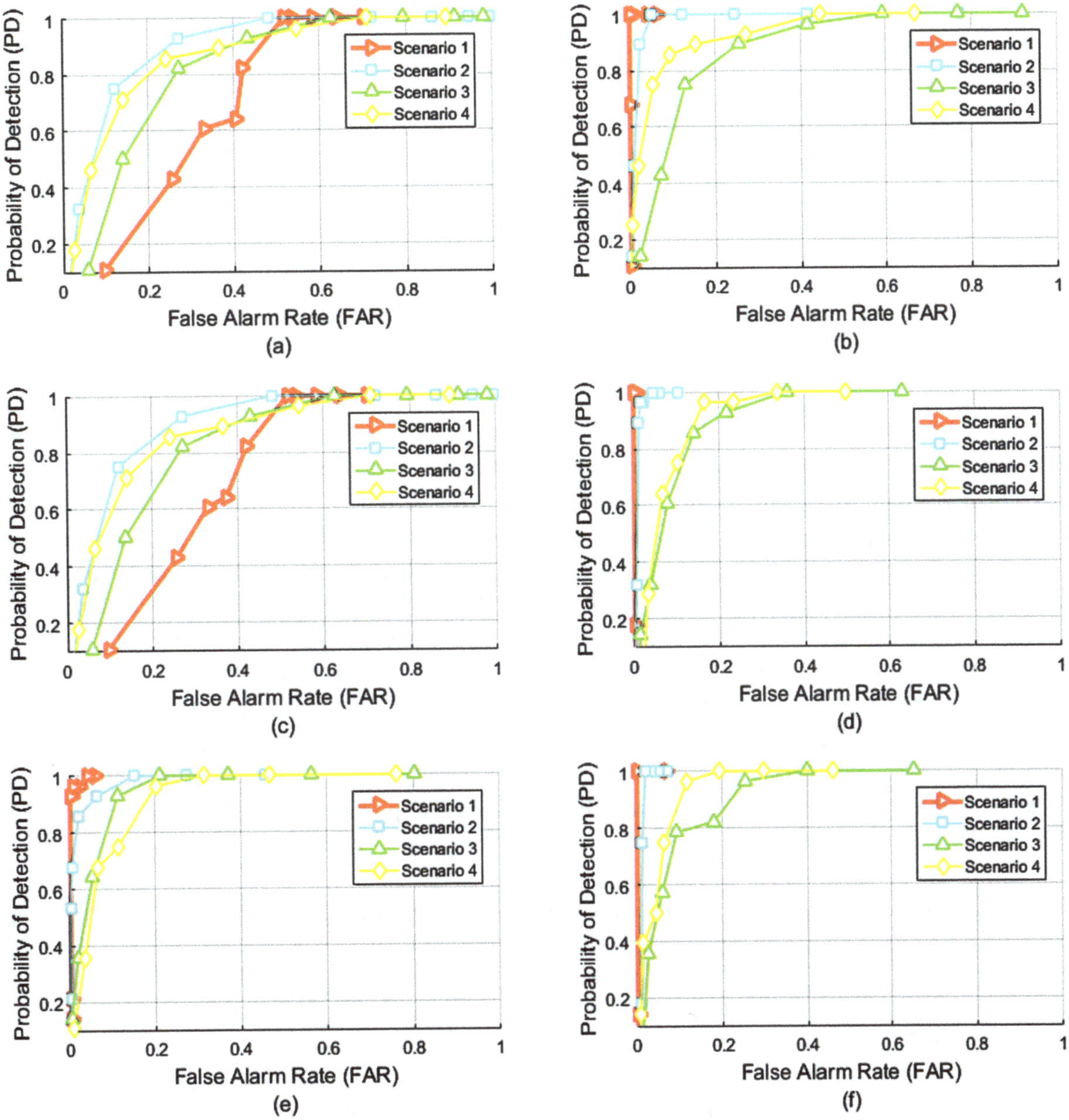

CONCLUSION

The GPR clutter removal methods can be grouped into two categories: subspace-based and multiresolution-multidirection based methods. In the subspace-based methods, MCA shows a superior performance compared to the widely used SVD, PCA and ICA based methods. It uses a different methodology and assumes that the GPR image is composed of clutter and target components which are sparsely modeled

Figure 14. Performance evaluation of all methods (a) ROC curves, (b) Zoomed version of ROC curves

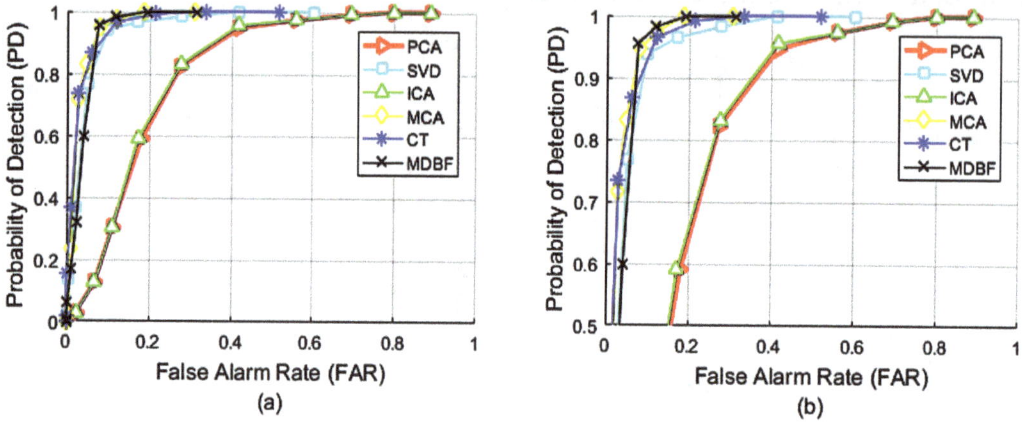

with appropriate dictionaries. Thus the performance heavily depends on the dictionary choice and can be further enhanced if learned dictionaries are used instead of the fixed ones. In the second group the best performance is achieved by the recently proposed MDBF. Multiresolution-multidirection based methods have two consecutive steps. Firstly, the GPR image is decomposed into different scale subbands and then each scale subband is decomposed into directional subbands. Therefore, the performance of the directional subbands depends on the performance of scale subbands. In MDFB, bilateral filter, a member of edge-preserving filters family available in the literature, is used for the first step. Since bilateral filters' parameters can be optimized for a given application, the overall performance will improve.

Another contribution of this work is the introduction of a challenging dataset created by the new version of the gprMax electromagnetic simulation software used for GPR studies. The new dataset, unlike the older ones, contains rough surfaces, vegetation and water puddles, thus it is more realistic. The different clutter removal methods described in this chapter are implemented and their performances are compared in terms of ROC curves and visual analysis. The results also show that some methods present different performances according to the complexity of the scenarios. In order to test the robustness of the algorithms, more challenging scenarios can be added to this simulated dataset. We hope that such datasets will serve as a source for fair comparison of the future work in this area.

REFERENCES

Abujarad, F., Jostingmeier, A., & Omar, A. S. (2004, June). Clutter removal for landmine using different signal processing techniques. *Tenth International Conference on Ground Penetrating Radar*, Delft, The Netherlands.

Abujarad, F., Nadim, G., & Omar, A. (2005, May). Clutter reduction and detection of landmine objects in ground penetrating radar data using singular value decomposition (SVD). *3rd International Workshop on Advanced Ground Penetrating Radar*, Delft, The Netherlands. 10.1109/AGPR.2005.1487840

Abujarad, F., & Omar, A. (2006, April). *GPR data processing using the component-separation methods PCA and ICA. IEEE International Workshop on Imagining Systems and Techniques*, Minori, Italy. 10.1109/IST.2006.1650776

Bao, Q. Z., Li, Q. C., & Chen, W. C. (2014). GPR data noise attenuation on the curvelet transform. *Applied Geophysics*, *11*(3), 301–310. doi:10.100711770-014-0444-2

Bobin, J., Moudden, Y., Starck, J. L., & Elad, M. (2006). Morphological diversity and source separation. *IEEE Signal Processing Letters*, *13*(7), 409–412. doi:10.1109/LSP.2006.873141

Bostanudin, N. (2013). *Computational methods for processing ground penetrating radar data* (Doctoral dissertation). University of Portsmouth, Portsmouth, UK.

Candes, E., Demanet, L., Donoho, D., & Ying, L. (2006). Fast discrete curvelet transforms. *Multiscale Modeling & Simulation*, *5*(3), 861–899. doi:10.1137/05064182X

Da Cunha, A. L., Zhou, J., & Do, M. N. (2006). The nonsubsampled contourlet transform: Theory, design, and applications. *IEEE Transactions on Image Processing*, *15*(10), 3089–3101. doi:10.1109/TIP.2006.877507 PMID:17022272

Daniels, D. J. (2005). *Ground penetrating radar*. John Wiley & Sons, Inc. doi:10.1002/0471654507.eme152

Daniels, D. J. (2009). Ground penetrating radar for buried landmine and IED detection. *Unexploded Ordnance Detection and Mitigation*, 89-111.

Deiana, D., & Anitori, L. (2010, June). Detection and classification of landmines using AR modeling of GPR data. *13th International Conference on Ground Penetrating Radar*, Lecce, Italy. 10.1109/ICGPR.2010.5550141

Erden, F. (2017). Evolutionary approach to solve a novel time-domain cavity problem. *IEEE Transactions on Antennas and Propagation*, *65*(11), 5918–5931. doi:10.1109/TAP.2017.2752240

Fattal, R., Agrawala, M., & Rusinkiewicz, S. (2007). Multiscale shape and detail enhancement from multi-light image collections. *ACM Transactions on Graphics*, *26*(3), 51. doi:10.1145/1276377.1276441

Furuta, K., & Ishikawa, J. (2009). *Anti-personnel landmine detection for humanitarian demining*. Springer. doi:10.1007/978-1-84882-346-4

Golub, G. H., & Reinsch, C. (1970). Singular value decomposition and least squares solutions. *Numerische Mathematik*, *14*(5), 403–420. doi:10.1007/BF02163027

Hu, J., & Li, S. (2012). The multiscale directional bilateral filter and its application to multisensor image fusion. *Information Fusion*, *13*(3), 196–206. doi:10.1016/j.inffus.2011.01.002

Hyvarinen, A. (1999). Fast and robust fixed-point algorithms for independent component analysis. *IEEE Transactions on Neural Networks*, *10*(3), 626–634. doi:10.1109/72.761722 PMID:18252563

Hyvarinen, A., Karhunen, J., & Oja, E. (2004). *Independent component analysis* (Vol. 46). John Wiley & Sons.

Kaplan, N. H., & Erer, I. (2012, July). Bilateral pyramid based pansharpening of multispectral satellite images. In *Geoscience and Remote Sensing Symposium (IGARSS)*, Munich, Germany. 10.1109/IGARSS.2012.6351017

Kaplan, N. H., & Erer, I. (2014). Bilateral filtering-based enhanced pansharpening of multispectral satellite images. *IEEE Geoscience and Remote Sensing Letters*, *11*(11), 1941–1945. doi:10.1109/LGRS.2014.2314389

Karlsen, B., Larsen, J., Sorensen, H. B. D., & Jakobsen, K. B. (2001, August). Comparison of PCA and ICA based clutter reduction in GPR systems for anti-personal landmine detection. *11th IEEE Signal Processing Workshop on Statistical Signal Processing*. 10.1109/SSP.2001.955243

Kaynak, S., Kumlu, D., & Erer, I. (2017, December). Multiscale directional bilateral filter based fusion of satellite images. *International Conference on Electrical and Electronics Engineering (ELECO)*, Bursa, Turkey.

Kumlu, D., & Erer, I. (2017a, July). Multiscale directional bilateral filter based clutter removal in GPR image analysis. In *Geoscience and Remote Sensing Symposium (IGARSS)*, Fort Worth, TX. 10.1109/IGARSS.2017.8127461

Kumlu, D., & Erer, I. (2017b, April). A comparative study on clutter reduction techniques in GPR images. *International Conference on Electrical and Electronic Engineering (ICEEE)*, Ankara, Turkey. 10.1109/ICEEE2.2017.7935843

Leucci, G. (2008). The electromagnetic signal attenuation and maximum penetration depth. *Scholarly Research Exchange*.

Naik, G. R., & Wang, W. (2014). *Blind source separation*. Berlin: Springer. doi:10.1007/978-3-642-55016-4

Riaz, M. M., & Ghafoor, A. (2012). Information theoretic criterion based clutter reduction for ground penetrating radar. *Progress In Electromagnetics Research B*, *45*, 147–164. doi:10.2528/PIERB12080802

Robledo, L., Carrasco, M., & Mery, D. (2009). A survey of land mine detection technology. *International Journal of Remote Sensing*, *30*(9), 2399–2410. doi:10.1080/01431160802549435

Solimene, R., Cuccaro, A., Dell'Aversano, A., Catapano, I., & Soldovieri, F. (2014). Ground Clutter Removal in GPR surveys. *IEEE Journal of Selected Topics in Applied Earth Observations and Remote Sensing*, *7*(3), 792–798. doi:10.1109/JSTARS.2013.2287016

Starck, J. L., Candes, E. J., & Donoho, D. L. (2002). The curvelet transform for image denoising. *IEEE Transactions on Image Processing*, *11*(6), 670–684. doi:10.1109/TIP.2002.1014998 PMID:18244665

Starck, J. L., Moudden, Y., Bobin, J., Elad, M., & Donoho, D. L. (2005, July). Morphological component analysis. *Proceedings of the Society for Photo-Instrumentation Engineers*, *5914*, 1–15. doi:10.1117/12.615237

Temlioglu, E., & Erer, I. (2016). Clutter removal in ground-penetrating radar images using morphological component analysis. *IEEE Geoscience and Remote Sensing Letters*, *13*(12), 1802–1806. doi:10.1109/LGRS.2016.2612582

Temlioglu, E., Erer, I., & Kumlu, D. (2017, July). A least mean square approach to buried object detection for ground penetrating radar. In *Geoscience and Remote Sensing Symposium (IGARSS)*, Fort Worth, TX. 10.1109/IGARSS.2017.8128084

Terrasse, G., Nicolas, J. M., Trouvé, E., & Drouet, É. (2015, July). Application of the curvelet transform for pipe detection in GPR images. In *Geoscience and Remote Sensing Symposium (IGARSS)*, Milan, Italy. 10.1109/IGARSS.2015.7326779

Tomasi, C., & Manduchi, R. (1998, January). Bilateral filtering for gray and color images. In *Computer Vision, 1998. Sixth International Conference on* (pp. 839-846). IEEE. 10.1109/ICCV.1998.710815

Tretyakov, O. A., & Erden, F. (2007). Separation of the instantaneous and dynamic polarizations in studies of dispersive dielectrics. *The Sixth Int. Kharkov Symp. on Phys. and Eng. of Microwaves, Millimeter and Submillimeter Waves*, 42-48.

Tretyakov, O. A., & Erden, F. (2008). Temporal cavity oscillations caused by a wide-band waveform. *Progress in Electromag. Res. B*, *6*, 183.

Tzanis, A. (2015). The Curvelet Transform in the analysis of 2-D GPR data: Signal enhancement and extraction of orientation-and-scale-dependent information. *Journal of Applied Geophysics*, *115*, 145–170. doi:10.1016/j.jappgeo.2015.02.015

Van Der Merwe, A., & Gupta, I. J. (2000). A novel signal processing technique for clutter reduction in GPR measurements of small, shallow land mines. *IEEE Transactions on Geoscience and Remote Sensing*, *38*(6), 2627–2637. doi:10.1109/36.885209

Verma, P. K., Gaikwad, A. N., Singh, D., & Nigam, M. J. (2009). Analysis of clutter reduction techniques for through wall imaging in UWB range. *Progress In Electromagnetics Research B*, *17*, 29–48. doi:10.2528/PIERB09060903

Walsh, N. E., & Walsh, W. S. (2003). Rehabilitation of landmine victims: The ultimate challenge. *Bulletin of the World Health Organization*, *81*(9), 665–670. PMID:14710508

Warren, C., Giannopoulos, A., & Giannakis, I. (2016). gprMax: Open source software to simulate electromagnetic wave propagation for ground penetrating radar. *Computer Physics Communications*, *209*, 163–170. doi:10.1016/j.cpc.2016.08.020

Wold, S., Esbensen, K., & Geladi, P. (1987). Principal component analysis. *Chemometrics and Intelligent Laboratory Systems*, *2*(1-3), 37–52. doi:10.1016/0169-7439(87)80084-9

Compilation of References

Abi-Zeid, I., & Frost, J. R. (2005). SARPlan: A decision support system for Canadian Search and Rescue Operations. *European Journal of Operational Research*, *162*(3), 630–653. doi:10.1016/j.ejor.2003.10.029

Aboueljinane, L., Jemai, Z., & Sahin, E. (2012, December). Reducing ambulance response time using simulation: The case of Val-de-Marne department Emergency Medical Service. *Simulation Conference (WSC) Proceedings of the*, *2012*(Winter), 1–12.

Aboueljinane, L., Sahin, E., & Jemai, Z. (2013). A review on simulation models applied to emergency medical service operations. *Computers & Industrial Engineering*, *66*(4), 734–750.

Aboueljinane, L., Sahin, E., Jemai, Z., & Marty, J. (2014). A simulation study to improve the performance of an emergency medical service: Application to the French Val-de-Marne department. *Simulation Modelling Practice and Theory*, *47*, 46–59.

Abujarad, F., Jostingmeier, A., & Omar, A. S. (2004, June). Clutter removal for landmine using different signal processing techniques. *Tenth International Conference on Ground Penetrating Radar*, Delft, The Netherlands.

Abujarad, F., Nadim, G., & Omar, A. (2005, May). Clutter reduction and detection of landmine objects in ground penetrating radar data using singular value decomposition (SVD). *3rd International Workshop on Advanced Ground Penetrating Radar*, Delft, The Netherlands. 10.1109/AGPR.2005.1487840

Abujarad, F., & Omar, A. (2006, April). *GPR data processing using the component-separation methods PCA and ICA. IEEE International Workshop on Imagining Systems and Techniques*, Minori, Italy. 10.1109/IST.2006.1650776

Acar, M. (2015). *Optimization of Turkish Air Force SAR Units Forward Deployment Points for a Central Based SAR Force Structure (No. AFIT-ENS-MS-15-M-148)*. Air Force Institute of Technology Wright-Patterson AFB OH Graduate School Of Engineering And Management.

Ackoff, R. L. (n.d.). *Resource Allocation*. Retrieved August 2017, from https://www.britannica.com/topic/operations-research/Resource-allocation

Adams, E. S., & Mesterton-Gibbons, M. (2003). Lanchester's attrition models and fights among social animals. *Behavioral Ecology*, *14*(5), 719–723. doi:10.1093/beheco/arg061

Afshartous, D., Guan, Y., & Mehrotra, A. (2009). US Coast Guard air station location with respect to distress calls: A spatial statistics and optimization based methodology. *European Journal of Operational Research*, *196*(3), 1086–1096. doi:10.1016/j.ejor.2008.04.010

Ahmadi-Javid, A., Seyedi, P., & Syam, S. S. (2017). A survey of healthcare facility location. *Computers & Operations Research*, *79*, 223–263. doi:10.1016/j.cor.2016.05.018

Ahn, J., de Weck, O., Geng, Y., & Klabjan, D. (2012). Column generation based heuristics for a generalized location routing problem with profits arising in space exploration. *European Journal of Operational Research, 223*(1), 47–59. doi:10.1016/j.ejor.2012.06.018

Ahn, J., De Weck, O., & Hoffman, J. (2008). An optimization framework for global planetary surface exploration campaigns. *JBIS. Journal of the British Interplanetary Society, 61*(12), 487.

Aidala, V. (1979). Kalman Filter Behavior in Bearings-Only Tracking Applications. *IEEE Transactions on Aerospace and Electronic Systems, AES-15*(1), 29–39. doi:10.1109/TAES.1979.308793

Ai, Y. F., Lu, J., & Zhang, L. L. (2015). The Optimization Model for the Location of Maritime Emergency Supplies Reserve Bases and the Configuration of Salvage Vessels. *Transportation Research Part E, Logistics and Transportation Review, 83*, 170–188. doi:10.1016/j.tre.2015.09.006

Akbari, A., Eiselt, H. A., & Pelot, R. (2017a). A maritime search and rescue location analysis considering multiple criteria, with simulated demand. *INFOR*, 1–23.

Akbari, A., Pelot, R., & Eiselt, H. A. (2017b). A modular capacitated multi-objective model for locating maritime search and rescue vessels. *Annals of Operations Research*, 1–26.

Akca, Z., Berger, R. T., & Ralphs, T. K. (2009). A branch-and-price algorithm for combined location and routing problems under capacity restrictions. In *Operations research and cyber-infrastructure* (pp. 309–330). Boston, MA: Springer. doi:10.1007/978-0-387-88843-9_16

Albadvi, A., Chaharsooghi, S. K., & Esfahanipour, A. (2007). Decision Making in Stock Trading: An Application of PROMETHEE. *European Journal of Operational Research, 177*(2), 673–683. doi:10.1016/j.ejor.2005.11.022

Albrecht, T., & Muller, M. (2009). Dynamic Time Warping (DTW). *Information Retreival for Music and Motion*, 70-83.

Alkanat, O. (2008). *Determining the Surface-to-Air Missile Requirement for Western and Southern Part of the Turkish Air Defense System. AFIT- Air Force Institute of Technology, AFIT/GOR/ENS/08-01*. Dayton, OH: Wright Patterson AFB.

Álvarez, J. F. (2007). *A Lagrangian relaxation approach to the vessel planning problem*. Working paper, Pompeu Fabra University, Barcelona.

Álvarez, J. F. (2006). A heuristic for vessel planning in a reach stacker terminal. *Journal for Maritime Research, 3*(1), 3–16.

Ambrosino, D., & Sciomachen, A. (1998). A constraint satisfaction approach for master bay plans. *Water Stud. Ser.*, 175–184.

Ambrosino, D., Sciomachen, A., & Tanfani, E. (2004). Stowing a containership: The master bay plan problem. *Transport Res. A, 38*, 81–99.

Ambrosino, D., Sciomachen, A., & Tanfani, E. (2006). A decomposition heuristics for the container ship stowage problem. *Journal of Heuristics, 12*(3), 211–233. doi:10.100710732-006-5905-1

Amer, M., & Daim, T. U. (2011). Selection of renewable energy technologies for a developing county: A case of Pakistan. *Energy for Sustainable Development, 15*(4), 420–435.

Amiri, M. P. (2010). Project selection for oil-fields development by using the AHP and fuzzy TOPSIS methods. *Expert Systems with Applications, 37*(9), 6218–6224.

Amouzegar, M. A., Tripp, R. S., & Galway, L. A. (2004b). Integrated logistics planning for the air and space expeditionary force. *The Journal of the Operational Research Society, 55*(4), 422–430. doi:10.1057/palgrave.jors.2601704

Amouzegar, M. A., Tripp, R. S., McGarvey, R. G., Chan, E. W., & Roll, C. R. Jr. (2004a). *Supporting air and space expeditionary forces: analysis of combat support basing options*. RAND Project Air Force.

Anagnostopoulos, K., Doukas, H., & Psarras, J. (2008). A Linguistic Multicriteria Analysis System Combining Fuzzy Sets Theory, Ideal and Anti-Ideal Points for Location Site Selection. *Expert Systems with Applications*, *35*(4), 2041–2048. doi:10.1016/j.eswa.2007.08.074

Apte, A. U., Rendon, R. G., & Salmeron, J. (2011). An optimization approach to strategic sourcing: A case study of the United States Air Force. *Journal of Purchasing and Supply Management*, *17*(4), 222–230. doi:10.1016/j.pursup.2011.03.002

Araz, C., Ozfırat, P. M., & Ozkarahan, I. (2007). An Integrated Multicriteria Decision-Making Methodology for Outsourcing Management. *Computers & Operations Research*, *34*(12), 3738–3756. doi:10.1016/j.cor.2006.01.014

Archetti, C., Hertz, A., & Speranza, M. G. (2007). Metaheuristics for the team orienteering problem. *Journal of Heuristics*, *13*(1), 49–76. doi:10.100710732-006-9004-0

Arslan, O. (2009). *Developing a Tool for the Location Optimization of the Alert Aircraft with Changing Threat Anticipation (No. AFIT/GOR/ENS/09-02)*. Air Force Inst of Tech Wright-Patterson AFB OH Graduate School of Engineering and Management.

Ashari, H., & Parsaei, M. (2014). Application of the multi-criteria decision method ELECTRE III for the weapon selection. *Decision Science Letters*, *3*(4), 511–522. doi:10.5267/j.dsl.2014.6.001

Athanasopoulos, G., Riba, C. R., & Athanasopoulou, C. (2009). A Decision Support System for Coating Selection Based on Fuzzy Logic and Multi-Criteria Decision Making. *Expert Systems with Applications*, *36*(8), 10848–10853. doi:10.1016/j.eswa.2009.01.016

Avellar, G. S., Pereira, G. A., Pimenta, L. C., & Iscold, P. (2015). Multi-uav routing for area coverage and remote sensing with minimum time. *Sensors (Basel)*, *15*(11), 27783–27803. doi:10.3390151127783 PMID:26540055

Avriel, M., & Penn, M. (1993). Exact and approximate solutions of the container ship stowage problem. *Computers & Industrial Engineering*, *25*(1-4), 271–274. doi:10.1016/0360-8352(93)90273-Z

Avriel, M., Penn, M., & Shpirer, N. (2000). Container ship Stowage Problem: Complexity and Connection to the Colouring of Circle Graphs. *Discrete Applied Mathematics*, *103*(1-3), 271–279. doi:10.1016/S0166-218X(99)00245-0

Avriel, M., Penn, M., Shpirer, N., & Witteboon, S. (1998). Stowage planning for container ships to reduce the number of shifts. *Annals of Operations Research*, *76*, 55–71. doi:10.1023/A:1018956823693

Awasthi, A., Chauhan, S. S., & Omrani, H. (2011). Application of fuzzy TOPSIS in evaluating sustainable transportation systems. *Expert Systems with Applications*, *38*(10), 12270–12280. doi:10.1016/j.eswa.2011.04.005

Ayöperken, E., & Ermiş, M. (2011). Modeling and optimizing the bases of unmanned aerial vehicles as a set-coverage problem. *Havacilik ve Uzay Teknolojileri Dergisi*, *5*(1), 61–71.

Azizi, A., Aikhuele, D. O., & Souleman, F. S. (2015). A Fuzzy TOPSIS Model to Rank Automotive Suppliers. *Procedia Manufacturing*, *2*, 159–164. doi:10.1016/j.promfg.2015.07.028

Azofra, M., Pérez-Labajos, C. A., Blanco, B., & Achutegui, J. J. (2007). Optimum placement of sea rescue resources. *Safety Science*, *45*(9), 941–951. doi:10.1016/j.ssci.2006.09.002

Back, Fogel, & Michalewicz. (2000). *Evolutionary Computation 1 Basic Algorithms and Operators*. Institute of Physics Publishing.

Badri, M. A., Mortagy, A. K., & Alsayed, C. A. (1998). A multi-objective model for locating fire stations. *European Journal of Operational Research*, *110*(2), 243–260. doi:10.1016/S0377-2217(97)00247-6

Bahri, S. (2016, April). 0-1 integer linear programming model for location selection of fire station: A case study in Indonesia. In AIP Conference Proceedings (Vol. 1723, No. 1, p. 030004). AIP Publishing.

Baker, D., Bridges, D., Hunter, R., Johnson, G., Krupa, J., & Murphy, J. (2001, December). Retrieved January 02, 2012, from Guidebook to Decision-Making Methods, WSRC-IM-2002-00002, Department of Energy, USA: http://www.dss.dpem.tuc.gr/pdf/Decision%20Making%20Guidebook_2002.pdf

Baker, A. (2013). Configurable Obsolescence Mitigation Methodologies. *Procedia CIRP*, *11*, 352–356. doi:10.1016/j.procir.2013.07.013

Balci, O. (1994). Principles of Simulation Model Validation, Verification, and Testing. Technical Report. Virginia Polytechnic Institute & State University.

Balcik, B., & Beamon, B. M. (2008). Facility location in humanitarian relief. *International Journal of Logistics*, *11*(2), 101–121. doi:10.1080/13675560701561789

Baldacci, R., Mingozzi, A., & Wolfler Calvo, R. (2011). An exact method for the capacitated location-routing problem. *Operations Research*, *59*(5), 1284–1296. doi:10.1287/opre.1110.0989

Bandyopadhyay, S., & Saha, S. (2013). Some Single- and Multiobjective Optimization. In *Unsupervised Classification* (pp. 17–58). Springer. doi:10.1007/978-3-642-32451-2_2

Bandyopadhyay, S., & Saha, S. (2013). *Unsupervised Classification* (1st ed.). Springer-Verlag Berlin Heidelberg. doi:10.1007/978-3-642-32451-2

Banks, J. (1999). Introduction to simulation. In *Simulation Conference Proceedings*, 1999 Winter (Vol. 1, pp. 7-13). IEEE.

Bao, Q. Z., Li, Q. C., & Chen, W. C. (2014). GPR data noise attenuation on the curvelet transform. *Applied Geophysics*, *11*(3), 301–310. doi:10.100711770-014-0444-2

Barreto, S., Ferreira, C., Paixao, J., & Santos, B. S. (2007). Using clustering analysis in a capacitated location-routing problem. *European Journal of Operational Research*, *179*(3), 968–977. doi:10.1016/j.ejor.2005.06.074

Barsocchi, P., Celandroni, N., Davoli, F., Ferro, E., Giambene, G., Castano, F. J. G., ... Todorova, P. (2005). Radio resource management across multiple protocol layers in satellite networks: A tutorial overview. *International Journal of Satellite Communications and Networking*, *23*(5), 265–305. doi:10.1002at.820

Barton, D. L., & Chawla, P. (2003). Designing Systems for Future Obsolescence. *Digital Avionics Systems Conference, 2003. DASC '03. The 22nd*, *2*, 1–11.

Başar, A., Kabak, Ö., Topçu, Y. İ., & Bozkaya, B. (2015). Location analysis in banking: a new methodology and application for a Turkish bank. In *Applications of Location Analysis* (pp. 25–54). Springer International Publishing. doi:10.1007/978-3-319-20282-2_2

Basdemir, M. M. (2000). *Locating search and rescue stations in the Aegean and Western Mediterranean regions of Turkey (No. AFIT/GOR/ENS/00M-03). AFIT- Air Force Institute of Technology*. Dayton, OH: Wright Patterson AFB.

Batchelor, T. P., & Briffa, M. (2011). Fight tactics in wood ants: individuals in smaller groups fight harder but die faster. *Proceedings of the Royal Society of London B: Biological Sciences, 278*(1722), 3243-3250. 10.1098/rspb.2011.0062

Bates, A., Bell, Z., Mountford, A., & Evangelista, P. (2015). Military Resource Allocation as a Set Covering Problem. *Industrial and Systems Engineering Review*, *3*(1), 1–6.

Beasley, J. E., & Chu, P. C. (1996). A genetic algorithm for the set covering problem. *European Journal of Operational Research, 94*(2), 392–404. doi:10.1016/0377-2217(95)00159-X

Beasley, J. E., & Jörnsten, K. (1992). Enhancing an algorithm for set covering problems. *European Journal of Operational Research, 58*(2), 293–300. doi:10.1016/0377-2217(92)90215-U

Behzadian, M., Kazemzadeh, R. B., Albadvi, A., & Aghdasi, M. (2010). PROMETHEE: A Comprehensive Literature Review on Methodologies and Applications. *European Journal of Operational Research, 200*(1), 198–215. doi:10.1016/j.ejor.2009.01.021

Belenguer, J. M., Benavent, E., Prins, C., Prodhon, C., & Calvo, R. W. (2011). A branch-and-cut method for the capacitated location-routing problem. *Computers & Operations Research, 38*(6), 931–941. doi:10.1016/j.cor.2010.09.019

Bell, J. E. (2003). *A simulated annealing approach for the composite facility location and resource allocation problem: a study of strategic positioning of US Air Force munitions (No. C102-927)*. Auburn Univ.

Bell, J. E., Griffis, S. E., Cunningham, W. A. III, & Eberlan, J. A. (2011). Location optimization of strategic alert sites for homeland defense. *Omega, 39*(2), 151–158. doi:10.1016/j.omega.2010.05.004

Berman, O., & Krass, D. (2002). The generalized maximal covering location problem. *Computers & Operations Research, 29*(6), 563–581. doi:10.1016/S0305-0548(01)00079-X

Bernstein, P. L. (1998). *Against the Gods: The Remarkable Story of Risk*. New York: John Wiley & Sons, Inc.

Bjorklund, P., Varbrand, P., & Yuan, D. (2003, March). Resource optimization of spatial TDMA in ad hoc radio networks: A column generation approach. In *INFOCOM 2003. Twenty-Second Annual Joint Conference of the IEEE Computer and Communications. IEEE Societies* (Vol. 2, pp. 818-824). IEEE. 10.1109/INFCOM.2003.1208919

Bobin, J., Moudden, Y., Starck, J. L., & Elad, M. (2006). Morphological diversity and source separation. *IEEE Signal Processing Letters, 13*(7), 409–412. doi:10.1109/LSP.2006.873141

Boginski, V., Pasiliao, E. L., & Shen, S. (2015). Special issue on optimization in military applications. *Optimization Letters, 9*(8), 1475-1476.

Bonder, S. (1981). Mathematical modeling of military conflict situations. In Proceedings of Symposia in Applied Mathematics (Vol. 25, pp. 1-51). American Mathematical Society.

Bonder, S., & Farrell, R. (1970). *Development of models for defense systems planning (No. SRL-2147-TR-70-2)*. Michigan Univ Ann Arbor Systems Research Lab.

Borden, K. D. (2001). *Optimizing the number and employment of combat logistics force shuttle ships, with a case study of the T-AKE ship* (Master's thesis). Naval Postgraduate School, Monterey, CA.

Borges, J. L. (1998). *The library of Babel: Collected fictions*. Academic Press.

Bostanudin, N. (2013). *Computational methods for processing ground penetrating radar data* (Doctoral dissertation). University of Portsmouth, Portsmouth, UK.

Boudahri, F., Aggoune-Mtalaa, W., Bennekrouf, M., & Sari, Z. (2013). Application of a clustering based location-routing model to a real agri-food supply chain redesign. In *Advanced methods for computational collective intelligence* (pp. 323–331). Berlin, Heidelberg: Springer. doi:10.1007/978-3-642-34300-1_31

Bouly, H., Dang, D. C., & Moukrim, A. (2010). *A memetic algorithm for the team orienteering problem*. 4OR: A Quarterly Journal of Operations Research, 8(1), 49-70.

Boussier, S., Feillet, D., & Gendreau, M. (2007). *An exact algorithm for team orienteering problems.* 4OR: A Quarterly Journal of Operations Research, 5(3), 211-230.

Bracken, J., Kress, M., & Rosenthal, R. E. (1995). *Special Issue on Air-Land-Naval Warfare Models* (Vol. 42). Naval Research Logistics.

Brandsma, M.G., Vliet, A.J., & Griffioen-Young, H.J. (2006). *Analysis Methods and Models for Small Unit Operations.* TNO-DV 2006 A228, TNO Report.

Brooks, A., Bankes, S., & Bennett, B. (1997). *Weapon Mix and Exploratory Analysis, A Case Study. Project Air Force.* RAND.

Brown, G. G., & Carlyle, W. M. (2008). Optimizing the U.S. Navy's combat logistics force. *Naval Research Logistics, 55*(8), 800–810.

Brown, G. G., Carlyle, W. M., Kelton, D., Kline, J., & Salmerón, J. (2009). Operational planning tools for U.S. Navy Maritime Commanders. In *International Conference on Harbor, Maritime and Multimodal Logistics Modeling and Simulation*. Universidad de La Laguna.

Brown, G. G., Carlyle, W. M., & Burson, P. (2010). *Replenishment at Sea Planner (RASP) model.* Monterey, CA: Operations Research Department, Naval Postgraduate School.

Buchanan, L., & O'Connell, A. (2006). A Brief History of Decision Making. Harvard Business School Publishing.

Buede, D. M., & Bresnick, T. A. (1992). Applications of Decision Analysis to the Military Systems Acquisition Process. *Interfaces, 22*(6), 110–125. doi:10.1287/inte.22.6.110

Burke, J. F., Love, R. J., Macal, C. M., Howard, D. L., & Jackson, J. (2000). *Modeling force deployment from army installations using the transportation system capability (TRANSCAP) model.* Argonne National Laboratory.

Burrell, R. S. (2011). *The Ghosts of Iwo Jima* (Vol. 102). Texas A&M University Press.

Butt, S. E., & Ryan, D. M. (1999). An optimal solution procedure for the multiple tour maximum collection problem using column generation. *Computers & Operations Research, 26*(4), 427–441. doi:10.1016/S0305-0548(98)00071-9

Büyüközkan, G., Feyzioğlu, O., & Nebol, E. (2008). Selection of the Strategic Alliance Partner in Logistics Value Chain. *International Journal of Production Economics, 113*(1), 148–158. doi:10.1016/j.ijpe.2007.01.016

Byrne, P. J., & Heavey, C. (2006). The impact of information sharing and forecasting in capacitated industrial supply chains: A case study. *International Journal of Production Economics, 103*(1), 420–437.

Caldwell, B., Hartman, J., Parry, S., Washburn, A., & Yungren, M. (2000). *Aggregated Combat Models*. Academic Press.

Cambell, E. (2014, June 10). Spratly Islands: Foreign correspondent visits remote reef flash point where Filipino marines hold out against Chinese navy. *ABC News*. Retrieved from http://www.abc.net.au/news/2014-05-20/oil-key-in-spratlyislands- dispute-south-china sea/5463080

Can, D., & Saraclar, M. (2011). Lattice indexing for spoken term detection. *IEEE Transactions on Audio, Speech, and Language Processing, 19*(8), 2338–2347. doi:10.1109/TASL.2011.2134087

Candes, E., Demanet, L., Donoho, D., & Ying, L. (2006). Fast discrete curvelet transforms. *Multiscale Modeling & Simulation, 5*(3), 861–899. doi:10.1137/05064182X

Caprara, A., Toth, P., & Fischetti, M. (2000). Algorithms for the set covering problem. *Annals of Operations Research, 98*(1-4), 353–371. doi:10.1023/A:1019225027893

Cavallaro, F. (2010). Fuzzy TOPSIS approach for assessing thermal-energy storage in concentrated solar power (CSP) systems. *Applied Energy*, *87*(2), 496–503. doi:10.1016/j.apenergy.2009.07.009

Cebi, S., & Kahraman, C. (2010). Developing a Group Decision Support System Based on Fuzzy Information Axiom. *Knowledge-Based Systems*, *23*(1), 3–16. doi:10.1016/j.knosys.2009.07.005

Çelik, M., Cebi, S., Kahraman, C., & Er, I. D. (2009). Application of Axiomatic Design and TOPSIS Methodologies Under Fuzzy Environment for Proposing Competitive Strategies on Turkish Container Ports in Maritime Transportation Network. *Expert Systems with Applications*, *36*(3), 4541–4557. doi:10.1016/j.eswa.2008.05.033

Chandrashekar, K., Dekhordi, M. R., & Baras, J. S. (2004, October). Providing full connectivity in large ad-hoc networks by dynamic placement of aerial platforms. In *Military Communications Conference, 2004. MILCOM 2004. 2004 IEEE* (Vol. 3, pp. 1429-1436). IEEE. 10.1109/MILCOM.2004.1495151

Chan, Y. (2001). *Location theory and decision analysis*. South-Western College Pub.

Chan, Y., Mahan, J. M., Chrissis, J. W., Drake, D. A., & Wang, D. (2008). Hierarchical maximal-coverage location–allocation: Case of generalized search-and-rescue. *Computers & Operations Research*, *35*(6), 1886–1904. doi:10.1016/j.cor.2006.09.018

Chao, I. M., Golden, B. L., & Wasil, E. A. (1996). A fast and effective heuristic for the orienteering problem. *European Journal of Operational Research*, *88*(3), 475–489. doi:10.1016/0377-2217(95)00035-6

Chen, G., Yilmaz, O., Trmal, J., Povey, D., & Khudanpur, S. (2013, December). Using proxies for OOV keywords in the keyword search task. In *Automatic Speech Recognition and Understanding (ASRU), 2013 IEEE Workshop on* (pp. 416-421). IEEE. 10.1109/ASRU.2013.6707766

Chen, X. W., Wang, X. Z., Ma, L., & Kong, L. (2013). Equipment Maintenance Position Location Uncertain Decision Method in Wartime. *Fire Control & Command Control*, *4*, 21.

Chen, X., Jiang, N., Jing, Y., Stojanovski, G., & Dimirovski, G. M. (2011). Differential game model and its solutions for force resource complementary via Lanchester square law equation. *IFAC Proceedings Volumes*, *44*(1), 14229-14233. 10.3182/20110828-6-IT-1002.01042

Chen, C.-T., Lin, C.-T., & Huang, S.-F. (2006). A Fuzzy Approach for Supplier Evaluation and Selection in Supply Chain Management. *International Journal of Production Economics*, *102*(2), 289–301. doi:10.1016/j.ijpe.2005.03.009

Cheng, C. H. (1997). Evaluating naval tactical missile systems by fuzzy AHP based on the grade value of membership function. *European Journal of Operational Research*, *96*(2), 343–350. doi:10.1016/S0377-2217(96)00026-4

Cheng, C. H., & Lin, Y. (2002). Evaluating the best main battle tank using fuzzy decision theory with linguistic criteria evaluation. *European Journal of Operational Research*, *142*(1), 174–186. doi:10.1016/S0377-2217(01)00280-6

Cheng, C. H., Yang, K. L., & Hwang, C. L. (1999). Evaluating attack helicopters by AHP based on linguistic variable weight. *European Journal of Operational Research*, *116*(2), 423–435. doi:10.1016/S0377-2217(98)00156-8

Chen, J., Wang, B., & Liu, W. (2015). Constructing perimeter barrier coverage with bistatic radar sensors. *Journal of Network and Computer Applications*, *57*, 129–141. doi:10.1016/j.jnca.2015.07.015

Chen, M.-F., & Tzeng, G.-H. (2004). Combining Grey Relation and TOPSIS Concepts for Selecting an Expatriate Host Country. *Mathematical and Computer Modelling*, *40*(13), 1473–1490. doi:10.1016/j.mcm.2005.01.006

Chen, R. S., & Shyu, J. Z. (2006). Selecting a weapon system using zero-one goal programming and analytic network process. *Journal of Information and Optimization Sciences*, *27*(2), 379–399. doi:10.1080/02522667.2006.10699697

Chen, S. M. (1996). Evaluating weapon systems using fuzzy arithmetic operations. *Fuzzy Sets and Systems*, *77*(3), 265–276. doi:10.1016/0165-0114(95)00096-8

Chief of Naval Operations (CNO). (2007). *Sustainment at Sea* (Navy Warfare Publication 4-01.2). Washington, DC: United States Navy. Retrieved from https://ndls.nwdc.navy.mil

Chircop, J., & Buckingham, C. D. (2014). A multiple pheromone ant clustering algorithm. In Nature Inspired Cooperative Strategies for Optimization (NICSO 2013) (pp. 13-27). Springer International Publishing. doi:10.1007/978-3-319-01692-4_2

Cho, B., Kim, D. Y., Kim, S. H., & Youn, C. (2007). Real-Time Distributed Simulation Environment for Air Defense System Using A Software Framework. *Journal of Defense Modeling and Simulation.*, *4*(2), 64–79. doi:10.1177/154851290700400201

Church, R. L., & Murray, A. T. (2009). *Business site section, location analysis and GIS*. New York: Wiley.

Church, R. L., & Murray, A. T. (2009). *Business site selection, location analysis and GIS*. New York: Wiley.

Church, R. L., & Scaparra, M. P. (2007). Protecting critical assets: The *r*-interdiction median problem with fortification. *Geographical Analysis*, *39*(2), 129–146. doi:10.1111/j.1538-4632.2007.00698.x

Church, R. L., Scaparra, M. P., & Middleton, R. (2004). The r-interdiction median problem and the r-interdiction covering problem. *Annals of the Association of American Geographers*, *94*, 491–502. doi:10.1111/j.1467-8306.2004.00410.x

Church, R., & ReVelle, C. (1974). The maximal covering location problem. *Papers in Regional Science*, *32*(1), 101–118. doi:10.1007/BF01942293

Cioppa, T. M., & Lucas, T. W. (2007). Efficient nearly orthogonal and space-filling Latin hypercubes. *Technometrics*, *49*(1), 45–55. doi:10.1198/004017006000000453

Claxton, J. D., Cavoli, C., & Johnson, C. (2005). *Test and evaluation management guide*. Defense Acquisition Univ Ft Belvoir VA.

Coelho, B. N., Coelho, V. N., Coelho, I. M., Ochi, L. S., Zuidema, D., Lima, M. S., & da Costa, A. R. (2017). A multi-objective green UAV routing problem. *Computers & Operations Research*, *88*, 306–315. doi:10.1016/j.cor.2017.04.011

Colegrave, R. K., & Hyde, J. M. (1993). The Lanchester square-law model extended to a (2,2) conflict. *IMA Journal of Applied Mathematics*, *51*(2), 95–109. doi:10.1093/imamat/51.2.95

Collins, G., & Erickson, A. (2010, December 26). China deploys first long-range, landbased carrier killer: DF-21D Anti-Ship Ballistic Missile (ASBM) reaches "Initial Operation Capability (IOC). *China Signpost.* Retrieved from http://www.andrewerickson.com/2010/12/china-deploys-world%E2%80%99sfirst-

Commander in Chief, Atlantic Fleet (CINCLANTFLT) (2001), *Fleet analysis of the combat logistics force*. Commander in Chief, Atlantic Fleet. This document is FOR OFFICIAL USE ONLY

Commander Strike Fighter Wing Pacific. (2016). *NAS Lemoore inflight guide*. Lemoore, CA: Author.

Contardo, C., Cordeau, J. F., & Gendron, B. (2011). *A branch-and-cut-and-price algorithm for the capacitated location-routing problem*. CIRRELT.

Coulson, S. G. (2018). Lanchester modelling of intelligence in combat. *IMA Journal of Management Mathematics*. doi:10.1093/imaman/dpx014

Council, N. R. (1998). *Statistics, Testing, and Defense Acquisition: New Approaches and Methodological Improvements* (M. L. Cohen, J. E. Rolph, & D. L. Steffey, Eds.). Washington, DC: The National Academies Press.

Craparo, E. M., & Karatas, M. (2018). *A Method for Placing Sources in Multistatic Sonar Networks.* Technical Report NPS-OR-18-001. Naval Postgraduate School.

Craparo, E. M., Fügenschuh, A., Hof, C., & Karatas, M. (2018). Optimizing Source and Receiver Placement in Multistatic Sonar Networks to Monitor Fixed Targets. *European Journal of Operational Research, 2018.* doi:10.1016/j.ejor.2018.02.006

Craparo, E. M., Karatas, M., & Kuhn, T. U. (2017). Sensor placement in active multistatic sonar networks. *Naval Research Logistics, 64*(4), 287–304. doi:10.1002/nav.21746

Cruz, J. B., Simaan, M. A., Gacic, A., Jiang, H., Leitellier, B., Li, M., & Liu, Y. (2001, October). Game-theoretic modeling and control of a military air operation. *IEEE Transactions on Aerospace and Electronic Systems, 37*(4), 1393–1405. doi:10.1109/7.976974

Cuculoski, V. (2013). Obsolescence Management of Electronic and Control Systems. *Australian Control Conference,* (7), 342–47. 10.1109/AUCC.2013.6697296

Cura, T. (2014). An artificial bee colony algorithm approach for the team orienteering problem with time windows. *Computers & Industrial Engineering, 74,* 270–290. doi:10.1016/j.cie.2014.06.004

Current, J., Min, H., & Schilling, D. (1990). Multiobjective analysis of facility location decisions. *European Journal of Operational Research, 49*(3), 295–307. doi:10.1016/0377-2217(90)90401-V

Da Cunha, A. L., Zhou, J., & Do, M. N. (2006). The nonsubsampled contourlet transform: Theory, design, and applications. *IEEE Transactions on Image Processing, 15*(10), 3089–3101. doi:10.1109/TIP.2006.877507 PMID:17022272

Dağdeviren, M., Yavuz, S., & Kılınç, N. (2009). Weapon selection using the AHP and TOPSIS methods under fuzzy environment. *Expert Systems with Applications, 36*(4), 8143–8151.

Dagdeviren, M., Yavuz, S., & Kılınç, N. (2009). Weapon Selection Using the AHP and TOPSIS Methods Under Fuzzy Environment. *Expert Systems with Applications, 36*(4), 8143–8151. doi:10.1016/j.eswa.2008.10.016

Dahmann, J. S., & Baldwin, K. J. (2008). Understanding the Current State of US Defense Systems of Systems and the Implications for Systems Engineering. *2008 2nd Annual IEEE Systems Conference,* 1-7.

Dambreville, F. (2006, July). Optimal area covering by sensors for planning a Tracks Collection. In *Information Fusion, 2006 9th International Conference on* (pp. 1-7). IEEE. 10.1109/ICIF.2006.301821

Dang, D. C., El-Hajj, R., & Moukrim, A. (2013a, May). A branch-and-cut algorithm for solving the team orienteering problem. In *International Conference on AI and OR Techniques in Constriant Programming for Combinatorial Optimization Problems* (pp. 332-339). Springer, Berlin, Heidelberg. 10.1007/978-3-642-38171-3_23

Dang, D. C., Guibadj, R. N., & Moukrim, A. (2011, April). *A pso-based memetic algorithm for the team orienteering problem.* In *European Conference on the Applications of Evolutionary Computation* (pp. 471-480). Springer, Berlin, Heidelberg. 10.1007/978-3-642-20520-0_48

Dang, D. C., Guibadj, R. N., & Moukrim, A. (2013b). An effective PSO-inspired algorithm for the team orienteering problem. *European Journal of Operational Research, 229*(2), 332–344. doi:10.1016/j.ejor.2013.02.049

Daniels, D. J. (2009). Ground penetrating radar for buried landmine and IED detection. *Unexploded Ordnance Detection and Mitigation,* 89-111.

Daniels, D. J. (2005). *Ground penetrating radar.* John Wiley & Sons, Inc. doi:10.1002/0471654507.eme152

Danskin, J. M. (1967). *The Theory of Max-Min and its Application to Weapons Allocation Problems.* Berlin: Springer-Verlag. doi:10.1007/978-3-642-46092-0

Dantzig, G. B., & Ramser, J. H. (1959). The truck dispatching problem. *Management Science*, *6*(1), 80–91. doi:10.1287/mnsc.6.1.80

Darwin, C. (1929). *The Origin of Species by Means of Natural Selection or the Preservation of Favored Races in the Struggle for Life*. New York: Random House.

Daskin, M. (1997). Network and Discrete Location: Models, Algorithms and Applications. *The Journal of the Operational Research Society*, *48*(7), 763–763. doi:10.1057/palgrave.jors.2600828

Daskin, M. S. (1983). A maximum expected covering location model: Formulation, properties and heuristic solution. *Transportation Science*, *17*(1), 48–70. doi:10.1287/trsc.17.1.48

Daskin, M. S. (1995). *Network and discrete location: Models, algorithms and applications*. New York: John Wiley and Sons. doi:10.1002/9781118032343

Daskin, M. S. (2008). What you should know about location modelling. *Naval Research Logistics*, *55*(4), 283–294. doi:10.1002/nav.20284

Daskin, M. S., & Stern, E. H. (1981). A hierarchical objective set covering model for emergency medical service vehicle deployment. *Transportation Science*, *15*(2), 137–152. doi:10.1287/trsc.15.2.137

Dawson, M. C., Bell, J. E., & Weir, J. D. (2007). A Hybrid Location Method for Missile Security Team Positioning. *Journal of Business and Management*, *13*, 5–17.

Defence, The Secretary of. (1994). *Specifications & Standards - A New Way of Doing Business*. Retrieved from http://sw-eng.falls-church.va.us/perry94.html

Defense Acquisition University. (2015). *Glossary of defense acquisition acronyms and terms*. Retrieved from https://dap.dau.mil/glossary/Pages/Default.aspx

Deiana, D., & Anitori, L. (2010, June). Detection and classification of landmines using AR modeling of GPR data. *13th International Conference on Ground Penetrating Radar*, Lecce, Italy. 10.1109/ICGPR.2010.5550141

Deitchman, S. J. (1962). A Lanchester model of guerrilla warfare. *Operations Research*, *10*(6), 818–827. doi:10.1287/opre.10.6.818

Dell, R. F., Ewing, P. L., & Tarantino, W. J. (2008). Optimally stationing army forces. *Interfaces*, *38*(6), 421–435. doi:10.1287/inte.1080.0401

Department of Defense, *Unmanned Systems Integrated Roadmap FY2011-2036* (Washington, DC: Government Printing Office, October 2010), 89.

Doğançay, K. (2005). On the efficiency of a bearings-only instrumental variable estimator for target motion analysis. *Signal Processing*, *85*(3), 481–490. doi:10.1016/j.sigpro.2004.10.014

Dong, L., Xing-hua, K., & Xiang-tao, Y. (2009, October). A model for supply chain critical facility protection planning based on time satisfaction. In *Intelligent Computation Technology and Automation, ICICTA'09. Second International Conference on* (Vol. 3, pp. 903-906). IEEE.

Dorigo, M. (1992). *Optimization, learning and natural algorithms*. Ph. D. Thesis, Politecnico di Milano, Italy.

Dorigo, M., & Stützle, T. (2010). Ant colony optimization: overview and recent advances. In Handbook of metaheuristics (pp. 227-263). Springer US. doi:10.1007/978-1-4419-1665-5_8

Doyle, D. E. (2006). *Evaluation of fleet ownership versus global allocation of ships in the combat logistics force* (Master's thesis). Naval Postgraduate School, Monterey, CA.

Drexl, M., & Schneider, M. (2015). A survey of variants and extensions of the location-routing problem. *European Journal of Operational Research, 241*(2), 283–308. doi:10.1016/j.ejor.2014.08.030

Drezner, Z. (Ed.). (1995). *Facility location: A survey of applications and methods*. Berlin, Germany: Springer Verlag. doi:10.1007/978-1-4612-5355-6

Drezner, Z., & Hamacher, H. W. (Eds.). (2002). *Facility location: Applications and theory*. Berlin, Germany: Springer-Verlag. doi:10.1007/978-3-642-56082-8

Drummond, H. (2001). *The Art of Decision Making Mirrors of Imagination, Masks of Fate*. West Sussex, UK: John Wiley & Sons ltd.

Dubrovsky, O., Levitin, O. G., & Penn, M. (2002). A genetic algorithm with a compact solution encoding for the container ship stowage problem. *Journal of Heuristics, 8*(6), 585–599. doi:10.1023/A:1020373709350

Duhamel, C., Lacomme, P., Prins, C., & Prodhon, C. (2010). A GRASP× ELS approach for the capacitated location-routing problem. *Computers & Operations Research, 37*(11), 1912–1923. doi:10.1016/j.cor.2009.07.004

Eberlan, J. A. (2004). *Location optimization of continental United States strip alert sites supporting homeland defense (No. AFIT/GLM/ENS/04-02)*. Air Force Inst of Tech Wright-Patterson AFB OH Dept of Operational Sciences.

Ehrgott, M., & Gandibleux, X. (2002). Multiple Criteria Optimization, State of the Art Annotated Bibliographic Surveys. Springer US.

Eiselt, H. A., & Marianov, V. (Eds.). (2015). *Applications of Location Analysis*. Springer International Publishing Switzerland; doi:10.1007/978-3-319-20282-2

Ellis, D. R. (2013). *A distributed logistics concept for flotilla surface combatant support, joint campaign analysis*. Monterey, CA: Naval Postgraduate School.

Engel, J. H. (1954). A verification of Lanchester's law. *Journal of the Operations Research Society of America, 2*(2), 163–171. doi:10.1287/opre.2.2.163

Enright, J. J., Frazzoli, E., Pavone, M., & Savla, K. (2015). UAV routing and coordination in stochastic, dynamic environments. In *Handbook of Unmanned Aerial Vehicles* (pp. 2079–2109). Springer Netherlands. doi:10.1007/978-90-481-9707-1_28

Erden, F. (2017). Evolutionary approach to solve a novel time-domain cavity problem. *IEEE Transactions on Antennas and Propagation, 65*(11), 5918–5931. doi:10.1109/TAP.2017.2752240

Erickson, G. M. (1992). Empirical analysis of closed-loop duopoly advertising strategies. *Management Science, 38*(12), 1732–1749. doi:10.1287/mnsc.38.12.1732

Ertuğrul, İ., & Karakaşoğlu, N. (2009). Performance Evaluation of Turkish Cement Firms with Fuzzy Analytic Hierarchy Process and TOPSIS Methods. *Expert Systems with Applications, 36*(1), 702–715. doi:10.1016/j.eswa.2007.10.014

Eshlaghy, A. T., & Homayonfar, M. (2011). MCDM Methodologies and Applications: A Literature Review from 1999 to 2009. *Research Journal of International Studies, 21*, 86–137.

Esnaf, Ş., & Küçükdeniz, T. (2009). A fuzzy clustering-based hybrid method for a multi-facility location problem. *Journal of Intelligent Manufacturing, 20*(2), 259–265.

Estay-Ossandon, C., Mena-Nieto, A., & Harsch, N. (2017). Using a fuzzy TOPSIS-based scenario analysis to improve municipal solid waste planning and forecasting: A case study of Canary archipelago (1999–2030). *Journal of Cleaner Production.*

Farahani, R. Z., & Asgari, N. (2007). Combination of MCDM and covering techniques in a hierarchical model for facility location: A case study. *European Journal of Operational Research, 176*(3), 1839–1858. doi:10.1016/j.ejor.2005.10.039

Farahani, R. Z., Asgari, N., Heidari, N., Hosseininia, M., & Goh, M. (2012). Covering problems in facility location: A review. *Computers & Industrial Engineering, 62*(1), 368–407. doi:10.1016/j.cie.2011.08.020

Farahani, R. Z., & Hekmatfar, M. (Eds.). (2009). *Facility location: Concepts, models, algorithms and case studies.* Heidelberg, Germany: Physica Verlag. doi:10.1007/978-3-7908-2151-2

Farahani, R. Z., Hekmatfar, M., Fahimnia, B., & Kazemzadeh, N. (2014). Hierarchical facility location problem: Models, classifications, techniques, and applications. *Computers & Industrial Engineering, 68*, 104–117. doi:10.1016/j.cie.2013.12.005

Farahani, R. Z., Steadie Seifi, M., & Asgari, N. (2010). Multiple criteria facility location problems: A survey. *Applied Mathematical Modelling, 34*(7), 1689–1709. doi:10.1016/j.apm.2009.10.005

Farina, A. (1999). Target tracking with bearings – Only measurements. *Signal Processing, 78*(1), 61–78. doi:10.1016/S0165-1684(99)00047-X

Fattal, R., Agrawala, M., & Rusinkiewicz, S. (2007). Multiscale shape and detail enhancement from multi-light image collections. *ACM Transactions on Graphics, 26*(3), 51. doi:10.1145/1276377.1276441

Feldman, K., & Sandborn, P. (2007). Integrating Technology Obsolescence Considerations into Product Design Planning. *Information in Engineering Conference*, 1–8. 10.1115/DETC2007-35881

Feng, D. (2007). *Optimizing Lifetime Buy Quantities to Minimize Lifecycle Cost.* Academic Press.

Feng, D., Singh, P., & Sandborn, P. (2007). Lifetime Buy Optimization to Minimize Lifecycle Cost. *Proceedings of the 2007 Aging.* Retrieved from http://www.enme.umd.edu/ESCML/Papers/AgingAircraft07-LTB.pdf

Ferrer, G. (2010). Open architecture, inventory pooling and maintenance modules. *International Journal of Production Economics, 128*(1), 393–403. doi:10.1016/j.ijpe.2010.07.034

Fiscus, J. G., Ajot, J., Garofolo, J. S., & Doddingtion, G. (2007). Results of the 2006 spoken term detection evaluation. In Proc. sigir (Vol. 7, pp. 51-57). Academic Press.

Fox, B., Boito, M., Graser, J. C., & Younossi, O. (2004). Test and evaluation trends and costs for aircraft and guided weapons (No. RAND/MG-109-AF). RAND Corp.

Fox, W. P. (2017). *The Battle of Trafalgar: An Example of Modeling with Technology.* Academic Press.

Francis, R. L., Lowe, T. J., Rayco, M. B., & Tamir, A. (2009). Aggregation error for location models: Survey and analysis. *Annals of Operations Research, 167*(1), 171–208.

Francis, R. L., McGinnis, L. F., & White, J. A. (1992). *Facility layout and location: An analytical approach* (2nd ed.). Englewood Cliffs, NJ: Prentice-Hall.

Francis, R. L., & White, J. A. (1974). *Facility layout and location an analytical approach* (1st ed.). Englewood Cliffs, NJ: Prentice-Hall.

Franks, N. R., & Partridge, L. W. (1993). Lanchester battles and the evolution of combat in ants. *Animal Behaviour, 45*(1), 197–199. doi:10.1006/anbe.1993.1021

Fredrikson, A., North, C., Plaisant, C., & Shneiderman, B. (1999, November). Temporal, geographical and categorical aggregations viewed through coordinated displays: a case study with highway incident data. In *Proceedings of the 1999 workshop on new paradigms in information visualization and manipulation in conjunction with the eighth ACM international conference on Information and knowledge management* (pp. 26-34). ACM.

Froggett, S., & Liu, S. (2001). System Performance Modeling in C4ISR/Weapon System Design and Development. *Naval Engineers Journal, 113*(2), 55-64. DOI:10.1111/j.1559-3584.2001.tb00034.x

Fu, M. C. (1994). Optimization via simulation: A review. *Annals of Operations Research, 53*(1), 199–247.

Furuta, K., & Ishikawa, J. (2009). *Anti-personnel landmine detection for humanitarian demining*. Springer. doi:10.1007/978-1-84882-346-4

Gamberini, R., Grassi, A., & Rimini, B. (2006). A New Multi-Objective Heuristic Algorithm for Solving the Stochastic Assembly Line Re-Balancing Problem. *International Journal of Production Economics, 102*(2), 226–243. doi:10.1016/j.ijpe.2005.02.013

GAMS. (2015). *General Algebraic Modeling System Development Corporation, version 24.4.2 r51415 (64 bit)*. Retrieved from https://www.gams.com

Gandhe, A., Qin, L., Metze, F., Rudnicky, A., Lane, I., & Eck, M. (2013, December). Using web text to improve keyword spotting in speech. In *Automatic Speech Recognition and Understanding (ASRU), 2013 IEEE Workshop on* (pp. 428-433). IEEE. 10.1109/ASRU.2013.6707768

Garcia, A., & Gish, H. (2006, May). Keyword spotting of arbitrary words using minimal speech resources. In *Acoustics, Speech and Signal Processing, 2006. ICASSP 2006 Proceedings. 2006 IEEE International Conference on* (Vol. 1, pp. I-I). IEEE. 10.1109/ICASSP.2006.1660179

Gardiner, S. R. (Ed.). (1891). *A School Atlas of English History*. Longmans, Green.

Geiser, M. T. (2012). *Improving aircraft refueling procedures at NAS Oceana* (Master's thesis). Retrieved from Calhoun: http://calhoun.nps.edu/handle/10945/7346

Gencer, C., Aydogan, E. K., & Soydemır, A. (2008). Chemical agent detector placement methodology. *Applied Mathematics and Computation, 195*(2), 542–557. doi:10.1016/j.amc.2007.05.033

Genc, H. M. (2010). A new solution approach for the bearing only target tracking problem. *4th International Workshop on Soft Computing Applications*. 10.1109/SOFA.2010.5565618

Genc, T. (2015). Application of ELECTRE III and PROMETHEE II in evaluating the military tanks. *International Journal of Procurement Management, 8*(4), 457–475. doi:10.1504/IJPM.2015.070743

Georgopoulou, E., Sarafidis, Y., Mirasgedis, S., Zaimi, S., & Lalas, D. P. (2003). A Multiple Criteria Decision-Aid Approach in Defining National Priorities for Greenhouse Gases Emissions Reduction in the Energy Sector. *European Journal of Operational Research, 146*(1), 199–215. doi:10.1016/S0377-2217(02)00250-3

Gerber, C. A., & Clark, J. A. (2013). *More fight–less fuel: Reducing fuel burn through ground process improvement* (Master's thesis). Retrieved from Calhoun: http://calhoun.nps.edu/handle/10945/34667

Ghanmi, A. (2011). Canadian forces global reach support hubs: Facility location and aircraft routing models. *The Journal of the Operational Research Society, 62*(4), 638–650. doi:10.1057/jors.2010.22

Ghose, D., Krichman, M., Speyer, J. L., & Shamma, J. S. (2002, May). Modeling and analysis of air campaign resource allocation: A spatio-temporal decomposition approach. *IEEE Transactions on Systems, Man and Cybernetics. Part A*, *32*(3), 403–418.

Ghose, D., Speyer, J. L., & Shamma, J. S. (2002). A game-theoretical multiple resource interaction approach to temporal resource allocation in an air campaign. *Annals of Operations Research*, *109*(1/4), 15–40. doi:10.1023/A:1016335731847

Gibbs, B. P. (2011). *Advanced Kalman filtering, least-squares and modeling: a practical handbook*. Hoboken, NJ: A John Wiley & Sons.

Giuliani, J. (2015, April 14). *The Challenge of Illegal Immigration in the Mediterranean*. Retrieved from http://www.robert-schuman.eu/en/doc/questions-d-europe/qe-352-en.pdf

Givens, R. D. (2002). *A comparison of operational potential and capability of two combat logistics force alternatives* (Master's thesis). Naval Postgraduate School, Monterey, CA.

Glade, D. (2000). *Unmanned aerial vehicles: Implications for military operations*. Air Univ Press Maxwell AFB Al.

Godzdanker, R., Rutherford, M. J., & Valavanis, K. P. (2011, July). ISLANDS: a self-leveling landing platform for autonomous miniature UAVs. In *IEEE/ASME International Conference on Advanced Intelligent Mechatronics (AIM)*, 2011 (pp. 170-175). IEEE. 10.1109/AIM.2011.6027085

Goerlandt, F., Torabihaghighi, F., & Kujala, P. (2013). A model for evaluating performance and reliability of the voluntary maritime rescue system in the Gulf of Finland. In *11th International Probabilistic Safety Assessment and Management Annual European Safety and Reliability Conference* (pp. 1-6). Academic Press.

Golabi, M., Shavarani, S. M., & Izbirak, G. (2017). An edge-based stochastic facility location problem in UAV-supported humanitarian relief logistics: A case study of Tehran earthquake. *Natural Hazards*, 1–21.

Golden, B. L., Levy, L., & Vohra, R. (1987). The orienteering problem. *Naval Research Logistics*, *34*(3), 307–318. doi:10.1002/1520-6750(198706)34:3<307::AID-NAV3220340302>3.0.CO;2-D

Goldsman, D., & Goldsman, P. (2015). *Modeling and Simulation in the Systems Engineering Life Cycle*. London: Springer.

Göleç, A., Gürbüz, F., & Şenyiğit, E. (2016). Determination of Best Military Cargo Aircraft with Multi-Criteria Decision-Making Techniques. *MANAS Journal of Social Studies*, *5*(5), 87–101.

Golub, G. H., & Reinsch, C. (1970). Singular value decomposition and least squares solutions. *Numerische Mathematik*, *14*(5), 403–420. doi:10.1007/BF02163027

Gong, X., Zhang, J., Cochran, D., & Xing, K. (2016). Optimal placement for barrier coverage in bistatic radar sensor networks. *IEEE/ACM Transactions on Networking*, *24*(1), 259–271. doi:10.1109/TNET.2014.2360849

González, E., & Villena, M. (2011). Spatial Lanchester models. *European Journal of Operational Research*, *210*(3), 706–715. doi:10.1016/j.ejor.2010.11.009

Gorry, G. A., & Morton, M. S. (1971). *A Framework for Management Information Systems. Alfred P.* Cambridge, MA: Sloan School of Management.

Green, L. V., & Kolesar, P. J. (2004). Anniversary article: Improving emergency responsiveness with management science. *Management Science*, *50*(8), 1001–1014.

Greiner, M. A., Fowler, J. W., Shunk, D. L., Carlyle, W. M., & McNutt, R. T. (2003, May). A hybrid approach using the analytic hierarchy process and integer programming to screen weapon systems projects. *IEEE Transactions on Engineering Management*, *50*(2), 192–203. doi:10.1109/TEM.2003.810827

Grewal, M. S., & Andrews, A. P. (1993). *Kalman filtering: theory and practice*. Englewood Cliffs, NJ: Prentice Hall.

Grodzevich, O., & Romanko, O. (2006). Normalization and other topics in multi-objective optimization. In *Proceedings of the Fields–MITACS Industrial Problems Workshop* (pp.89-101.), Toronto, Canada: Academic Press.

Gue, K. R. (2003). A dynamic distribution model for combat logistics. *Computers & Operations Research, 30*(3), 367–381. doi:10.1016/S0305-0548(01)00104-6

Gumus, M., Kaminsky, P., Tiemroth, E., & Ayik, M. (2008). A multi-stage decomposition heuristic for the container stowage problem. *Proceedings of the 2008 MSOM Conference*.

Günal, M. M., & Pidd, M. (2010). Discrete event simulation for performance modelling in health care: A review of the literature. *Journal of Simulation, 4*(1), 42–51.

Gunasekera, N. V. (2010). Collaboration of the Left and the Right Brain in Decision Making. *Chartered Institute of Management Accountants, 2*(2).

Gunawan, A., Lau, H. C., & Vansteenwegen, P. (2016). Orienteering problem: *A survey of recent variants, solution approaches and applications. European Journal of Operational Research, 255*(2), 315–332. doi:10.1016/j.ejor.2016.04.059

Gündogdu, B., Sarı, L., Çetinkaya, G., & Saraçlar, M. (2016). Template-Based Keyword Search with Pseudo Posteriorgrams. In *2016 24th Signal Processing and Communications Applications Conference (SIU)* (pp. 973-976). Academic Press. 10.1109/SIU.2016.7495904

Gündoğdu, B., & Saraçlar, M. (2017a). Distance Metric Learning for Posteriorgram Based Keyword Search. *International Conference on Acoustics, Speech and Signal Processing (ICASSP)*. 10.1109/ICASSP.2017.7953240

Gündoğdu, B., & Saraçlar, M. (2017b). Similarity Learning Based Query Modeling for Keyword Search. *Proc. Interspeech, 2017*, 3617–3621. doi:10.21437/Interspeech.2017-1273

Gündoğdu, B., & Saraçlar, M. (2017c). Novel score normalization methods for keyword search. In *Signal Processing and Communications Applications Conference (SIU), 2017 25th* (pp. 1-4). IEEE. 10.1109/SIU.2017.7960262

Guo, S., & Zhao, H. (2015). Optimal site selection of electric vehicle charging station by using fuzzy TOPSIS based on sustainability perspective. *Applied Energy, 158*, 390–402. doi:10.1016/j.apenergy.2015.08.082

Haidt, J. (2006). *The happiness hypothesis: Finding modern truth in ancient wisdom*. Basic Books.

Hajkowicz, S., & Higgins, A. (2008). A Comparison of Multiple Criteria Analysis Techniques for Water Resource Management. *European Journal of Operational Research, 184*(1), 255–265. doi:10.1016/j.ejor.2006.10.045

Hakimi, S. L. (1964). Optimum locations of switching centers and the absolute centers and medians of a graph. *Operations Research, 12*(3), 450–459. doi:10.1287/opre.12.3.450

Hakimi, S. L. (1965). Optimum distribution of switching centers in a communication network and some related graph theoretic problems. *Operations Research, 13*(3), 462–475. doi:10.1287/opre.13.3.462

Hall, N. G., & Hochbaum, D. S. (1992). The multicovering problem. *European Journal of Operational Research, 62*(3), 323–339. doi:10.1016/0377-2217(92)90122-P

Han, Y., Wang, J., & Liao, X. (2013, July). Optimization of disposition for terminal air defence system based on set-covering model. In *Control Conference (CCC), 2013 32nd Chinese* (pp. 8461-8464). IEEE.

Handler, G. Y., & Mirchandani, P. B. (1979). *Location on networks: Theory and algorithms*. Cambridge, MA: MIT Press.

Harper, M. (2014). *IARPA babel program*. Retrieved from https://www. iarpa. gov/index. php/research-programs/babel

Hatami-Marbini, A., & Kangi, F. (2017). An extension of fuzzy TOPSIS for a group decision making with an application to tehran stock exchange. *Applied Soft Computing*, *52*, 1084–1097. doi:10.1016/j.asoc.2016.09.021

Helmbold, R. L. (1965). Letter to the Editor—A Modification of Lanchester's Equations. *Operations Research*, *13*(5), 857–859. doi:10.1287/opre.13.5.857

Herva, M., & Roca, E. (2013). Ranking municipal solid waste treatment alternatives based on ecological footprint and multi-criteria analysis. *Ecological Indicators*, *25*, 77–84.

Hillier, F. S., & Lieberman, G. J. (2004). *Introduction to Operations Research*. McGraw-Hill Science/ Engineering/Math.

Hinton, G., Deng, L., Yu, D., Dahl, G. E., Mohamed, A. R., Jaitly, N., & Kingsbury, B. (2012). Deep neural networks for acoustic modeling in speech recognition: The shared views of four research groups. *IEEE Signal Processing Magazine*, *29*(6), 82–97. doi:10.1109/MSP.2012.2205597

Ho, W. (2008). Integrated analytic hierarchy process and its applications–A literature review. *European Journal of Operational Research*, *186*(1), 211–228.

Howland, D., & Bonder, S. (1963). *The Tank Weapon System. Report No. RF 573 AR 63-1 (U). Systems Research Group*. Columbus, OH: Ohio State Univ.

http://www.au.af.mil/au/awc/awcgate/army/fm101-5_mdmp.pdf; (21/11/2017)

Huang, X., Acero, A., & Hon, H. W. (2001). *Spoken language processing: A guide to theory, algorithm, and system development*. Prentice Hall PTR.

Huang, D. K., Chiu, H. N., Yeh, R. H., & Chang, J. H. (2009). A Fuzzy Multi-Criteria Decision-Making Approach for Solving a Bi-Objective Personnel Assignment Problem. *Computers & Industrial Engineering*, *56*(1), 1–10. doi:10.1016/j.cie.2008.03.007

Hughes, W. P. Jr. (2009). *The new navy fighting machine: A study of the connections between contemporary policy, strategy, sea power, naval operations, and the composition of the United States fleet*. Monterey, CA: Naval Postgraduate School.

Hu, J., & Li, S. (2012). The multiscale directional bilateral filter and its application to multisensor image fusion. *Information Fusion*, *13*(3), 196–206. doi:10.1016/j.inffus.2011.01.002

Hundt, C., Schmidt, B., & Schömer, E. (2014, September). Cuda-accelerated alignment of subsequences in streamed time series data. In *Parallel Processing (ICPP), 2014 43rd International Conference on* (pp. 10-19). IEEE. 10.1109/ICPP.2014.10

Hu, Q., & Lim, A. (2014). An iterative three-component heuristic for the team orienteering problem with time windows. *European Journal of Operational Research*, *232*(2), 276–286. doi:10.1016/j.ejor.2013.06.011

Hyvarinen, A. (1999). Fast and robust fixed-point algorithms for independent component analysis. *IEEE Transactions on Neural Networks*, *10*(3), 626–634. doi:10.1109/72.761722 PMID:18252563

Hyvarinen, A., Karhunen, J., & Oja, E. (2004). *Independent component analysis* (Vol. 46). John Wiley & Sons.

İç, Y. T. (2012). Development of a credit limit allocation model for banks using an integrated Fuzzy TOPSIS and linear programming. *Expert Systems with Applications*, *39*(5), 5309–5316. doi:10.1016/j.eswa.2011.11.005

İç, Y. T., & Yurdakul, M. (2009). Development of a Decision Support System for Machining Center Selection. *Expert Systems with Applications*, *36*(2), 3505–3513. doi:10.1016/j.eswa.2008.02.022

İç, Y. T., & Yurdakul, M. (2010). Development of a quick credibility scoring decision support system using fuzzy TOPSIS. *Expert Systems with Applications*, *37*(1), 567–574. doi:10.1016/j.eswa.2009.05.038

Ilachinski, A. (2004). *Artificial war: Multiagent-based simulation of combat.* World Scientific. doi:10.1142/5531

Imai, A., Nishimura, E., & Papadimitriou, S. (2005a). The dynamic berth allocation problem for a container port. *Transport Res B*, *39*(3), 197. doi:10.1016/j.trb.2004.03.004

Imai, A., Sasaki, K., Nishimura, E., & Papadimitriou, S. (2006b). Multi-objective simultaneous stowage and load planning for a container ship with container rehandle in yard stacks. *European Journal of Operational Research*, *171*(2), 373–389. doi:10.1016/j.ejor.2004.07.066

Ince, L., Sezen, B., Saridogan, E., & Ince, H. (2009). An evolutionary computing approach for the target motion analysis (TMA) problem for underwater tracks. *Expert Systems with Applications*, *36*(2), 3866–3879. doi:10.1016/j.eswa.2008.02.046

Ingolfsson, A., Erkut, E., & Budge, S. (2003). Simulation of single start station for Edmonton EMS. *The Journal of the Operational Research Society*, *54*(7), 736–746.

Ishizaka, A., Balkenborg, D., & Kaplan, T. (2011). Does AHP help us make a choice? An experimental evaluation. *The Journal of the Operational Research Society*, *62*(10), 1801–1812.

Işıklar, G., & Büyüközkan, G. (2007). Using a multi-criteria decision making approach to evaluate mobile phone alternatives. *Computer Standards & Interfaces*, *29*(2), 265–274.

Jahangirian, M., Eldabi, T., Naseer, A., Stergioulas, L. K., & Young, T. (2010). Simulation in manufacturing and business: A review. *European Journal of Operational Research*, *203*(1), 1–13.

Jain, A. K. (2010). Data clustering: 50 years beyond K-means. *Pattern Recognition Letters*, *31*(8), 651–666.

Jain, A. K., & Dubes, R. C. (1988). *Algorithms for clustering data.* Prentice-Hall, Inc.

Jaiswal, N. K. (1997). Optimization Problems. In Military Operations Research: Quantitative Decision Making (pp. 129-167). Springer Science+Business Media, LLC.

Jane's Fighting Ships. (2015, March 26). *Henry J Kaiser class (Oilers) (AOH).* Retrieved from https://janes.ihs.com.libproxy.nps.edu/CustomPages/Janes/DisplayPage.aspx?DocType=Reference&ItemId=+++1355951&Pubabbrev=JFS_

Jane's Fighting Ships. (2015, March 26). *Lewis and Clark class (Dry Cargo/Ammunition Ships) (AKEH).* Retrieved from https://janes.ihs.com.libproxy.nps.edu/CustomPages/Janes/DisplayPage.aspx?DocType=Reference&ItemId=+++1355951&Pubabbrev=JFS_

Jane's Strategic Weapons Systems. (2015, April 3). *DF-21 (CSS-5).* Retrieved from https://janes.ihs.com/CustomPages/Janes/DisplayPage.aspx?DocType=Reference&ItemId=+++1316682&Pubabbrev=JSWS

Jia, J., Chen, J., Chang, G., Li, J., & Jia, Y. (2007, March). Coverage optimization based on improved NSGA-II in wireless sensor network. In *Integration Technology, 2007. ICIT'07. IEEE International Conference on* (pp. 614-618). IEEE. 10.1109/ICITECHNOLOGY.2007.4290391

Jing, Y., Zhang, S., & Dimirovski, G. M. (2017). Complexity of warfare command, communication and control systems simplified: Optimal resource partitioning via Lanchester equations. In *Smart Technologies, IEEE EUROCON 2017-17th International Conference on* (pp. 611-617). IEEE.

Jisong, F., Hui, R., & Chengxiang, T. (2017). An Analysis of Wargame Rules Simulation Based on Stochastic Lanchester Models. In *Proceedings of the 2017 VI International Conference on Network, Communication and Computing* (pp. 135-139). ACM. 10.1145/3171592.3171612

Johnson, D. D., & MacKay, N. J. (2015). Fight the power: Lanchester's laws of combat in human evolution. *Evolution and Human Behavior*, *36*(2), 152–163. doi:10.1016/j.evolhumbehav.2014.11.001

Johnstone, D. P. (2002). *Modeling the pre-positioning of air force precision guided munitions (No. AFIT/GOR/ENS/02-09)*. Air Force Inst of Tech Wright-Patterson AFB OH Dept of Operational Science.

Josias, Terpenny, & McLean. (2004). Component Obsolescence Risk Assessment. *Proceedings of the 2004 industrial engineering research conference (IERC), 15*, 2–7. Retrieved from http://www.enme.umd.edu/ESCML/ObsSGER/IERC2004-Josias-Terpenny-McLean-376.pdf

Jourdan, D. B., & de Weck, O. L. (2004). Multi-objective genetic algorithm for the automated planning of a wireless sensor network to monitor a critical facility. In *Defense and Security* (pp. 565–575). International Society for Optics and Photonics. doi:10.1117/12.541685

Jun, M., & D'Andrea, R. (2003). Path planning for unmanned aerial vehicles in uncertain and adversarial environments. In Cooperative control: models, applications and algorithms (pp. 95-110). Springer US. doi:10.1007/978-1-4757-3758-5_6

Kahraman, C., Ateş, N., Çevik, S., Gülbay, M., & Erdoğan, S. (2007). Hierarchical Fuzzy TOPSIS Model for Selection Among Logistics Information Technologies. *Journal of Enterprise Information Management*, *20*(2), 143–168. doi:10.1108/17410390710725742

Kannan, D., Jabbour, A. B. S. L., & Jabbour, C. J. C. (2014). Selecting green suppliers based on GSCM practices: Using fuzzy TOPSIS applied to a Brazilian electronics company. *European Journal of Operational Research*, *233*(2), 432–447. doi:10.1016/j.ejor.2013.07.023

Kannan, G., Pokharel, S., & Kumar, P. S. (2009). A hybrid approach using ISM and fuzzy TOPSIS for the selection of reverse logistics provider. *Resources, Conservation and Recycling*, *54*(1), 28–36. doi:10.1016/j.resconrec.2009.06.004

Kaplan, N. H., & Erer, I. (2012, July). Bilateral pyramid based pansharpening of multispectral satellite images. In *Geoscience and Remote Sensing Symposium (IGARSS)*, Munich, Germany. 10.1109/IGARSS.2012.6351017

Kaplan, N. H., & Erer, I. (2014). Bilateral filtering-based enhanced pansharpening of multispectral satellite images. *IEEE Geoscience and Remote Sensing Letters*, *11*(11), 1941–1945. doi:10.1109/LGRS.2014.2314389

Karacan, İ. (2015). *A New Hybrid Decision Support Tool and an Application to Health Technology Selection* (Doctoral dissertation). Naval Science and Engineering Institute, Department of Operations Research, Turkish Naval Academy.

Karacan, İ., Karatas, M., Tozan, H., & Sulukan, E. (2015). Evaluating the Energy Management Performance in Country Scale by Using a Multi Criteria Decision Making Methodology. *5th International 100% Renewable Energy Conference and Exhibition*.

Karacan, İ., Tozan, H., & Karatas, M. (2016). Multi Criteria Decision Methods in Health Technology Assessment: A Brief Literature Review. *Eurasian Journal of Health Technology Assessment*, *1*(1), 12–19.

Karakaya, M., & Sevinç, E. (2017). An Efficient Genetic Algorithm for Routing Multiple UAVs under Flight Range and Service Time Window Constraints. *International Journal of InformaticsTechnologies*, *10*(1).

Karakitsiou, A. (2015). *Modeling discrete competitive facility location*. Springer International Publishing. doi:10.1007/978-3-319-21341-5

Karakos, D., & Schwartz, R. (2014). Subword and phonetic search for detecting out-of-vocabulary keywords. *Fifteenth Annual Conference of the International Speech Communication Association*.

Karasakal, O., & Karasakal, E. K. (2004). A maximal covering location model in the presence of partial coverage. *Computers & Operations Research, 31*(9), 1515–1526. doi:10.1016/S0305-0548(03)00105-9

Karataş, M., & Gecili, H. (2012). The role of decision support systems in steel industry. *Engineering Science & Technology, an International Journal, 15*(1).

Karatas, M. (2013). A multi foci closed curve: Cassini oval, its properties and applications. *Dogus University Journal, 14*(2), 233–250.

Karatas, M. (2017). A multi-objective facility location problem in the presence of variable gradual coverage performance and cooperative cover. *European Journal of Operational Research, 262*(3), 1040–1051. doi:10.1016/j.ejor.2017.04.001

Karatas, M. (2017b). Multiattribute Decision Making Using Multiperiod Probabilistic Weighted Fuzzy Axiomatic Design. *Systems Engineering, 20*(4), 318–334. doi:10.1002ys.21397

Karatas, M. (2018). Optimal Deployment of Heterogeneous Sensor Networks for a Hybrid Point and Barrier Coverage Application. *Computer Networks, 132*(26), 129–144. doi:10.1016/j.comnet.2018.01.001

Karatas, M., & Akman, G. (2014). An Extension to Multi-Attribute Decision Making Method: Dynamic Fuzzy Axiomatic Design Approach. *Joint International Symposium on CIE44 and IMSS'14 Proceedings.*

Karataş, M., & Akman, G. (2015). Bistatic sonobuoy deployment configuration for stationary targets. *Journal of Naval Science and Engineering, 11*(2), 1–10.

Karatas, M., & Craparo, E. (2015). Evaluating the direct blast effect in multistatic sonar networks using Monte Carlo simulation. In *Proceedings of the 2015 Winter Simulation Conference* (pp. 1184-1194). IEEE Press. 10.1109/WSC.2015.7408244

Karatas, M., Gunal, M. M., & Craparo, E. M. (2016b). Performance Evaluation of Mobile Multistatic Search Operations via Simulation. In *Proceedings of the 2016 Spring Simulation Conference*. IEEE Press.

Karatas, M., & Onggo, B. S. (2016). Validating an Integer Non-Linear Program Optimization Model of a Wireless Sensor Network Using Agent-Based Simulation. In *Proceedings of the 2016 Winter Simulation Conference*. IEEE Press. 10.1109/WSC.2016.7822188

Karatas, M., Razi, N., & Gunal, M. M. (2017b). An ILP and simulation model to optimize search and rescue helicopter operations. *The Journal of the Operational Research Society, 68*(11), 1335–1351. doi:10.105741274-016-0154-7

Karatas, M., Razi, N., & Tozan, H. (2016a). A Comparison of p-median and Maximal Coverage Location Models with Q–coverage Requirement. *Procedia Engineering, 149*, 169–176. doi:10.1016/j.proeng.2016.06.652

Karatas, M., Razı, N., & Tozan, H. (2017a). A multi-criteria assessment of the p-median, maximal coverage and p-center location models. *Tehnicki Vjesnik-Technical Gazette, 24*(Suppl. 2), 399–407. doi:10.17559/TV-20160508131905

Karatas, M., Razi, N., & Tozan, H. (2017a). A multi-criteria assessment of the p-median, maximal coverage and p-centre location models. *Technical Gazette, 24*(2), 399–407.

Karatas, M., & Yakıcı, E. (2018). An iterative solution approach to a multi-objective facility location problem. *Applied Soft Computing, 62*, 272–287. doi:10.1016/j.asoc.2017.10.035

Karlsen, B., Larsen, J., Sorensen, H. B. D., & Jakobsen, K. B. (2001, August). Comparison of PCA and ICA based clutter reduction in GPR systems for anti-personal landmine detection. *11th IEEE Signal Processing Workshop on Statistical Signal Processing*. 10.1109/SSP.2001.955243

Kass, R. A. (2006). The logic of warfighting experiments. Assistant Secretary of Defense (C3I/Command Control Research Program). doi:10.21236/ADA457164

Kaszkurewicz, E., & Bhaya, A. (2018). *Modeling Market Share Dynamics Under Advertising Effort and Word-of-Mouth Interactions Between Customers. IEEE Transactions on Computational Social Systems.*

Kaya, T., & Kahraman, C. (2011). Multicriteria Decision Making in Energy Planning Using a Modified Fuzzy TOPSIS Methodology. *Expert Systems with Applications*, *38*(6), 6577–6585. doi:10.1016/j.eswa.2010.11.081

Kaynak, S., Kumlu, D., & Erer, I. (2017, December). Multiscale directional bilateral filter based fusion of satellite images. *International Conference on Electrical and Electronics Engineering (ELECO)*, Bursa, Turkey.

Ke, L., Archetti, C., & Feng, Z. (2008). Ants can solve the team orienteering problem. *Computers & Industrial Engineering*, *54*(3), 648–665. doi:10.1016/j.cie.2007.10.001

Kelemenis, A., Ergazakis, K., & Askounis, D. (2011). Support managers' selection using an extension of fuzzy TOPSIS. *Expert Systems with Applications*, *38*(3), 2774–2782. doi:10.1016/j.eswa.2010.08.068

Kerdprasop, K., Kerdprasop, N., & Sattayatham, P. (2005). Weighted k-means for density-biased clustering. *Data Warehousing and Knowledge Discovery*, 488-497.

Keshtkaran, M., Ziarati, K., Bettinelli, A., & Vigo, D. (2016). Enhanced exact solution methods for the team orienteering problem. *International Journal of Production Research*, *54*(2), 591–601. doi:10.1080/00207543.2015.1058982

Khalil, W. A.-S., Shanableh, A., Rigby, P., & Kokot, S. (2005). Selection of Hydrothermal Pre-Treatment Conditions of Waste Sludge Destruction Using Multicriteria Decision Making. *Journal of Environmental Management*, *75*(1), 53–64. doi:10.1016/j.jenvman.2004.11.007 PMID:15748803

Kierstead, D. P., & DelBalzo, D. R. (2003). A genetic algorithm applied to planning search paths in complicated environments. *Military Operations Research*, *8*(2), 45–59. doi:10.5711/morj.8.2.45

Kimball, G. E. (1957). Some industrial applications of military operations research methods. *Operations Research*, *5*(2), 201–204. doi:10.1287/opre.5.2.201

Kim, K. H., & Bae, J. W. (1999). A dispatching method for automated guided vehicles to minimize delay of containership operations. *International Journal of Management Sciences*, *5*, 1–25.

Kim, K. H., & Bae, J. W. (2004). A look-ahead dispatching method for automated guided vehicles in automated port container terminals. *Transportation Science*, *38*(2), 224–234. doi:10.1287/trsc.1030.0082

Kim, Y., Chung, E.-S., Jun, S.-M., & Kim, S. U. (2013). Prioritizing the best sites for treated wastewater instream use in an urban watershed using fuzzy TOPSIS. *Resources, Conservation and Recycling*, *73*, 23–32. doi:10.1016/j.resconrec.2012.12.009

Kladis, G. P., Economou, J. T., Tsourdos, A., White, B. A., & Knowles, K. (2008, June). An emergency refuelling problem over a dynamically changing environment in the context of Unmanned Aerial Vehicles. In *Control and Automation, 2008 16th Mediterranean Conference on* (pp. 703-708). IEEE. 10.1109/MED.2008.4602182

Kodinariya, T. M., & Makwana, P. R. (2013). Review on determining number of Cluster in K-Means Clustering. *International Journal (Toronto, Ont.)*, *1*(6).

Konak, A., Kulturel-Konak, S., & Snyder, L. (2017). A Multi-Objective Approach to the Competitive Facility Location Problem. *Procedia Computer Science*, *108*, 1434–1442. doi:10.1016/j.procs.2017.05.035

Konoza, A. J. (2012, April). An Evaluation of End of Maintenance Dates and Lifetime Buy Estimations for Electronic Systems Facing Obsolescence. *ProQuest Dissertations and Theses*, 165. Retrieved from http://search.proquest.com/docview/1346194418?accountid=14511%5Cnhttp://sfx.ucl.ac.uk/sfx_local?url_ver=Z39.88-2004&rft_val_fmt=info:ofi/fmt:kev:mtx:dissertation&genre=dissertations+%26+theses&sid=ProQ:ProQuest+Dissertations+%26+Theses+Global&atitle=&titl

Konoza, A., & Sandborn, P. (2012). An Evaluation of End of Maintenance Dates for Electronic Assemblies. In *Aircraft Airworthiness*. Sustainment Conference.

Konoza, S. (2013). Evaluating the End of Maintenance Dates for Electronic Assemblies Composed of Obsolete Parts. *Journal of Mechanical Design*, *c*, 1–35.

Konur, D., Farhangi, H., & Dagli, C. H. (2016). A multi-objective military system of systems architecting problem with inflexible and flexible systems: Formulation and solution methods. *OR-Spektrum*, *38*(4), 967–1006. doi:10.100700291-016-0434-2

Korpela, J., & Tuominen, M. (1996). A Decision Aid in Warehouse Site Selection. *International Journal of Production Economics*, *45*(1), 169–180.

Kou, G., Shi, Y., & Wang, S. (2011). Multiple Criteria Decision Making and Decision Support Systems. *Decision Support Systems*, *51*(2), 247–249. doi:10.1016/j.dss.2010.11.027

Kowalski, K., Stagl, S., Madlener, R., & Omann, I. (2009). Sustainable Energy Futures: Methodological Challenges in Combining Scenarios and Participatory Multi-Criteria Analysis. *European Journal of Operational Research*, *197*(3), 1063–1074. doi:10.1016/j.ejor.2007.12.049

Kress, M., Caulkins, J. P., Feichtinger, G., Grass, D., & Seidl, A. (2018). Lanchester model for three-way combat. *European Journal of Operational Research*, *264*(1), 46–54. doi:10.1016/j.ejor.2017.07.026

Kress, M., Lin, K., & MacKay, N. J. (2017). The Attrition Dynamics of Multilateral War. *Operations Research*.

Kress, M., & Royset, J. O. (2008). Aerial search optimization model (ASOM) for UAVs in special operations. *Military Operations Research*, *13*(1), 23–33. doi:10.5711/morj.13.1.23

Krichman, M., Ghose, D., Speyer, J. L., & Shamma, J. S. (2001). Theater level campaign resource allocation. *Proceedings of the American Control Conference*, 4716-4721. 10.1109/ACC.2001.945727

Krohling, R. A., & Campanharo, V. C. (2011). Fuzzy TOPSIS for Group Decision Making: A Case Study for Accidents with Oil Spill in the Sea. *Expert Systems with Applications*, *38*(4), 4190–4197. doi:10.1016/j.eswa.2010.09.081

Kuhn, H. W., & Tucker, A. W. (1951). Nonlinear Programming. In *Proceeding of Second Berkeley Symposium on Mathematical Statistics and Probability*. Berkeley, CA: University of California Press.

Kumar & Saranga. (2008). *Optimal Selection of Obsolescence Mitigation Strategi ~ Using a Class of Bandit Models*. Academic Press.

Kumlu, D., & Erer, I. (2017a, July). Multiscale directional bilateral filter based clutter removal in GPR image analysis. In *Geoscience and Remote Sensing Symposium (IGARSS)*, Fort Worth, TX. 10.1109/IGARSS.2017.8127461

Kumlu, D., & Erer, I. (2017b, April). A comparative study on clutter reduction techniques in GPR images. *International Conference on Electrical and Electronic Engineering (ICEEE)*, Ankara, Turkey. 10.1109/ICEEE2.2017.7935843

Kurban, Ö. F., & Tuncay, C. A. N. (2016). Allocation of mini unmanned aerial vehicles for urgent intelligence, surveillance and reconnaissance request. *Journal of Marmara University Social Sciences Institute/Öneri, 12*(45).

Labadie, N., Mansini, R., Melechovský, J., & Calvo, R. W. (2012). The team orienteering problem with time windows: An lp-based granular variable neighborhood search. *European Journal of Operational Research, 220*(1), 15–27. doi:10.1016/j.ejor.2012.01.030

Labadie, N., Melechovský, J., & Calvo, R. W. (2011). Hybridized evolutionary local search algorithm for the team orienteering problem with time windows. *Journal of Heuristics, 17*(6), 729–753. doi:10.100710732-010-9153-z

Lanchester, F. W. (1916). *Aircraft in warfare: The dawn of the fourth arm.* Constable limited.

Lanchester, F. W. (1956). Mathematics in warfare. *The World of Mathematics, 4*, 2138-2157.

Lang, T. E., & McGarvey, R. G. (2016). Determining Reliable Networks of Prepositioning Materiel Warehouses for Public-Sector Rapid Response Supplies. *Advances in Operations Research.*

Laporte, G., Nickel, S., & da Gama, F. S. (2015). Location science. In The p-Median Problem (p. 644). Berlin: Springer.

Lazić, Lj., & Velašević, D. (2004, December). Applying simulation and design of experiments to the embedded software testing process. Journal of Software: Testing. *Verification and Reliability., 4*(14), 257–282. doi:10.1002tvr.299

Le, N. T., Saha, N., Mondal, R. K., Chae, S., & Jang, Y. M. (2014, February). Balanced energy and coverage guaranteed protocol for wireless sensor networks. In *Electrical Information and Communication Technology (EICT), 2013 International Conference on* (pp. 1-6). IEEE. 10.1109/EICT.2014.6777881

Lee, S. W., Tanaka, K., & Itoh, Y. (2005, March). Combining multiple subword representations for open-vocabulary spoken document retrieval. In *Acoustics, Speech, and Signal Processing, 2005. Proceedings. (ICASSP'05). IEEE International Conference on* (Vol. 1, pp. I-505). IEEE.

Lee, H., Lee, S., & Park, T. (2009). Selection of Technology Acquisition Mode Using the Analytic Network Process. *Mathematical and Computer Modelling, 49*(5-6), 1274–1282. doi:10.1016/j.mcm.2008.08.010

Lee, J., Kang, S. H., Rosenberger, J., & Kim, S. B. (2010). A hybrid approach of goal programming for weapon systems selection. *Computers & Industrial Engineering, 58*(3), 521–527. doi:10.1016/j.cie.2009.11.013

Lee, L. S., Glass, J., Lee, H. Y., & Chan, C. A. (2015). Spoken content retrieval—beyond cascading speech recognition with text retrieval. *IEEE/ACM Transactions on Audio, Speech, and Language Processing, 23*(9), 1389–1420.

Leucci, G. (2008). The electromagnetic signal attenuation and maximum penetration depth. *Scholarly Research Exchange.*

Liao, K., & Guo, D. (2008). A Clustering-Based Approach to the Capacitated Facility Location Problem. *Transactions in GIS, 12*(3), 323–339.

Liberatore, F., Scaparra, M. P., & Daskin, M. S. (2011). Analysis of facility protection strategies against an uncertain number of attacks: The stochastic R-interdiction median problem with fortification. *Computers & Operations Research, 38*(1), 357–366. doi:10.1016/j.cor.2010.06.002

Li, J., Prins, C., & Chu, F. (2012). A Scatter Search for A Multi-Type Transshipment Point Location Problem with Multicommodity Flow. *Journal of Intelligent Manufacturing, 23*(4), 1103–1117.

Li, K., Wen, Z., & Li, S. (2015, November). Coverage Optimization for Wireless Sensor Networks by Evolutionary Algorithm. In *International Symposium on Intelligence Computation and Applications* (pp. 52-63). Springer.

Lima, F. R. Junior, Osiro, L., & Ribeiro Carpinetti, L. C. (2014). A comparison between Fuzzy AHP and Fuzzy TOPSIS methods to supplier selection. *Applied Soft Computing, 21*, 194–209. doi:10.1016/j.asoc.2014.03.014

Lim, G. J., Kim, S., Cho, J., Gong, Y., & Khodaei, A. (2016). Multi-UAV Pre-positioning and Routing for Power Network Damage Assessment. *IEEE Transactions on Smart Grid*, 1. doi:10.1109/TSG.2016.2637408

Lin, C.-M., & Gen, M. (2008). Multi-Criteria Human Resource Allocation for Solving Multistage Combinatorial Optimization Problems Using Multiobjective Hybrid Genetic Algorithm. *Expert Systems with Applications*, *34*(4), 2480–2490. doi:10.1016/j.eswa.2007.04.016

Lin, C.-T., & Tsai, M.-C. (2009). Development of an Expert Selection System to Choose Ideal Cities for Medical Service Ventures. *Expert Systems with Applications*, *36*(2), 2266–2274. doi:10.1016/j.eswa.2007.12.056

Lin, K. Y., & MacKay, N. J. (2014). The optimal policy for the one-against-many heterogeneous Lanchester model. *Operations Research Letters*, *42*(6-7), 473–477. doi:10.1016/j.orl.2014.08.008

Lin, S. W. (2013). Solving the team orienteering problem using effective multi-start simulated annealing. *Applied Soft Computing*, *13*(2), 1064–1073. doi:10.1016/j.asoc.2012.09.022

Lirn, T. C., Thanopoulou, H. A., Beynon, M. J., & Beresford, A. K. C. (2004). An application of AHP on transhipment port selection: A global perspective. *Maritime Economics & Logistics*, *6*(1), 70–91.

Liu, J. X., Zhao, D. M., & Wang, F. (2013). Networks Attack-Defense model based on the improved Lanchester equation. In *Machine Learning and Cybernetics (ICMLC), 2013 International Conference on* (Vol. 3, pp. 1083-1086). IEEE.

Liu, X. F., Gao, L. M., Guang, Z. W., & Song, Y. Q. (2013). A UAV allocation method for traffic surveillance in sparse road network. [English Edition]. *Journal of Highway and Transportation Research and Development*, *7*(2), 81–87.

Liu, X. F., Guan, Z. W., Song, Y. Q., & Chen, D. S. (2014). An optimization model of UAV route planning for road segment surveillance. *Journal of Central South University*, *21*(6), 2501–2510. doi:10.100711771-014-2205-z

Liu, X. F., Peng, Z. R., Chang, Y. T., & Zhang, L. Y. (2012). Multi-objective evolutionary approach for UAV cruise route planning to collect traffic information. *Journal of Central South University*, *19*(12), 3614–3621. doi:10.100711771-012-1449-8

Love, R., Morris, J., & Wesolowsky, G. O. (1988). *Facility location: Models and methods*. Amsterdam: North-Holland.

Luo, Z., Liu, Z., & Shi, J. (2017). A Two-Echelon Cooperated Routing Problem for a Ground Vehicle and Its Carried Unmanned Aerial Vehicle. *Sensors (Basel)*, *17*(5), 1144. doi:10.339017051144 PMID:28513552

Luthra, S., Kumar, S., Garg, D., & Haleem, A. (2015). Barriers to renewable/sustainable energy technologies adoption: Indian perspective. *Renewable & Sustainable Energy Reviews*, *41*, 762–776.

MacCalman, A. D. (2012). *DesignCreatorv2 spreadsheet*. Retrieved from http://harvest.nps.edu

Mahdavi, I., Mahdavi-Amiri, N., Heidarzade, A., & Nourifer, R. (2008). Designing a model of fuzzy TOPSIS in multiple criteria decision making. *Applied Mathematics and Computation*, *206*(2), 607–617. doi:10.1016/j.amc.2008.05.047

Mahdevari, S., Shahriar, K., & Esfahanipour, K. (2014). Human health and safety risks management in underground coal mines using fuzzy TOPSIS. *The Science of the Total Environment*, *488–489*, 85–99. doi:10.1016/j.scitotenv.2014.04.076 PMID:24815558

Marianov, V., & Revelle, C. (1994). The queuing probabilistic location set covering problem and some extensions. *Socio-Economic Planning Sciences*, *28*(3), 167–178. doi:10.1016/0038-0121(94)90003-5

Marshall, M. M., & Lambert, K. R. (2008). Insights into Supporting Complex Systems under Conditions of Obsolescence. PICMET: Portland International Center for Management of Engineering and Technology, Proceedings, 1918–23. doi:10.1109/PICMET.2008.4599811

Mathew, T., & Webb, D. W. (2005). Generalized p Values and Confidence Intervals for Variance Components: Applications to Army Test and Evaluation. *Technometrics*, *47*(3), 312–322. doi:10.1198/004017005000000265

Mavi, R. K., Goh, M., & Mavi, N. K. (2016). Supplier selection with Shannon entropy and fuzzy TOPSIS in the context of supply chain risk management. *Procedia: Social and Behavioral Sciences*, *235*, 216–225. doi:10.1016/j.sbspro.2016.11.017

Mavrotas, G., Diakoulaki, D., & Caloghirou, Y. (2006). Project Prioritization Under Policy Restrictions. A Combination of MCDA with 0-1 Programming. *European Journal of Operational Research*, *171*(1), 296–308. doi:10.1016/j.ejor.2004.07.069

Maybeck, P. S. (1979). *Stochastic Models, Estimation, and Control: Vol.: 1*. New York: Academic Press.

McShea, B. (2010). Test and Evaluation of Aircraft Avionics and Weapons Systems. *Sciences et Techniques (Paris)*.

Meilunas, Ng, Rajagopal, & Chaudhury. (2007). *Overview of the Simulation, Test and Evaluation Process Network (STEPNET) Program*. American Institute of Aeronautics and Astronautics (AIAA) Atmospheric Flight Mechanics Conference and Exhibit, Hilton Head, SC. doi:10.2514/6.2007-6572

Meng, X., Thornberg, B., & Olsson, L. (2014). Strategic Proactive Obsolescence Management Model. *IEEE Transactions on Components, Packaging, and Manufacturing Technology*, *4*(6), 1099–1108. doi:10.1109/TCPMT.2014.2316212

Mersheeva, V., & Friedrich, G. (2015, April). Multi-UAV Monitoring with Priorities and Limited Energy Resources. In ICAPS (pp. 347-356).

Michalewicz, Z. (1996). *Genetic Algorithms + Data Structures = Evolution Programs* (3rd ed.). Springer-Verlag. doi:10.1007/978-3-662-03315-9

Microsoft. (2009). *Microsoft Windows 7 Professional (64 bit)*. Retrieved from https://www.nps.edu/Technology/SoftwareLib

Microsoft. (2013). *Microsoft Office Professional Plus 2013: Microsoft Excel version 15.0.4885.1000 (64 bit)*. Retrieved from https://www.nps.edu/Technology/SoftwareLib

Mirchandani, P. B., & Francis, R. L. (Eds.). (1990). *Discrete location theory*. New York: Wiley-Interscience-Interscience.

Mohri, M., Pereira, F., & Riley, M. (2002). Weighted finite-state transducers in speech recognition. *Computer Speech & Language*, *16*(1), 69–88. doi:10.1006/csla.2001.0184

Mon, D. L., Cheng, C. H., & Lin, J. C. (1994). Evaluating weapon system using fuzzy analytic hierarchy process based on entropy weight. *Fuzzy Sets and Systems*, *62*(2), 127–134. doi:10.1016/0165-0114(94)90052-3

Montazer, G. A., Saremi, Q. H., & Ramezani, M. (2009). Design a New Mixed Expert Decision Aiding System Using Fuzzy ELECTRE III Method for Vendor Selection. *Expert Systems with Applications*, *36*(8), 10837–10847. doi:10.1016/j.eswa.2009.01.019

Montemanni, R., & Gambardella, L. M. (2009). An ant colony system for team orienteering problems with time windows. *Foundation Of Computing And Decision Sciences*, *34*(4), 287.

Morse, T. C. (2008). *Optimization of combat logistics force required to support major combat operations* (Master's thesis). Naval Postgraduate School, Monterey, CA.

Morse, P., & Kimball, G. (1950). *Method of Operations Research*. New York: John Wiley & Sons.

Morteza, Z., Reza, F. M., Seddiq, M. M., Sharareh, P., & Jamal, G. (2016). Selection of the optimal tourism site using the ANP and fuzzy TOPSIS in the framework of Integrated Coastal Zone Management: A case of Qeshm Island. *Ocean and Coastal Management*, *130*, 179–187. doi:10.1016/j.ocecoaman.2016.06.012

Morton, D. P., Salmeron, J., & Wood, R. K. (2009). A stochastic program for optimizing military sealift subject to attack. *Military Operations Research, 14*(2), 19–39.

Mozaffari, M., Saad, W., Bennis, M., & Debbah, M. (2016, May). Optimal transport theory for power-efficient deployment of unmanned aerial vehicles. In *Communications (ICC), 2016 IEEE International Conference on* (pp. 1-6). IEEE.

Mufalli, F., Batta, R., & Nagi, R. (2012). Simultaneous sensor selection and routing of unmanned aerial vehicles for complex mission plans. *Computers & Operations Research, 39*(11), 2787–2799. doi:10.1016/j.cor.2012.02.010

Murthy, P. R. (2005). Operations Research (2nd ed.). New Delhi, India: New Age International (P) Limited, Publishers.

Murty, K. G., & Djang, P. A. (1999). The US Army national guard's mobile training simulators location and routing problem. *Operations Research, 47*(2), 175–182. doi:10.1287/opre.47.2.175

Muthuswamy, S., & Lam, S. S. (2011). Discrete particle swarm optimization for the team orienteering problem. *Memetic Computing, 3*(4), 287–303. doi:10.100712293-011-0071-x

Naik, G. R., & Wang, W. (2014). *Blind source separation.* Berlin: Springer. doi:10.1007/978-3-642-55016-4

Nardone, S., & Aidala, V. (1981). Observability Criteria for Bearings-Only Target Motion Analysis. *IEEE Transactions on Aerospace and Electronic Systems, AES-17*(2), 162–166. doi:10.1109/TAES.1981.309141

Naval Air Systems Command. (2002). *Aircraft refueling manual (NAVAIR 00-80T-109).* Washington, DC: Department of the Navy.

Nelson III & Sandborn. (2013). Strategic Management of Component Obsolescence Using Constraint-Driven Design Refresh Planning. *International Journal of Product Lifecycle Management, 6*(2), 99–120. Retrieved from http://search.proquest.com/docview/1328233150?accountid=26636%5Cnhttp://link.periodicos.capes.gov.br/sfxlcl41?url_ver=Z39.88-2004&rft_val_fmt=info:ofi/fmt:kev:mtx:journal&genre=article&sid=ProQ:ProQ:pqdieng&atitle=Strategic+management+of+component+obsoles

Nelson, C., Boros, E., Roberts, F., Rubio-Herrero, J., Kantor, P., McGinity, C., . . . Hanson, K. (2014). ACCAM Global Optimization Model for the USCG Aviation Air Stations. In *Proceedings of IIE Annual Conference*. Institute of Industrial Engineers-Publisher.

Nemes, M. I., Carvalho, H. B., & Souza, M. F. (2004). Antiretroviral therapy adherence in Brazil. *AIDS (London, England), 18*, S15–S20. PMID:15322479

Ngenkaew, W., Ono, S., & Nakayama, S. (2008). The deposition of multiple pheromones in ant-based clustering. *International Journal of Innovative Computing, Information, & Control, 4*(7), 1349–4198.

Nickel, S., & Puerto, J. (2005). *Location theory: A unified approach.* Berlin, Germany: Springer Verlag.

Nikolic, D., Jovanovic, I., Mihajlovic, I., & Zivkovic, Z. (2009). Multi-Criteria Ranking of Copper Concentrates According to Their Quality - An Element of Environmental Management in the Vicinity of Copper - Smelting Complex in Bor, Serbia. *Journal of Environmental Management, 91*(2), 509–515. doi:10.1016/j.jenvman.2009.09.019 PMID:19833429

Nogueira, L. C., Pinto, L. R., & Silva, P. M. S. (2016). Reducing Emergency Medical Service response time via the reallocation of ambulance bases. *Health Care Management Science, 19*(1), 31–42. PMID:24744263

Onggo, B. S., & Karatas, M. (2015, December). Agent-based model of maritime search operations: A validation using test-driven simulation modelling. In *Winter Simulation Conference (WSC)*, 2015 (pp. 254-265). IEEE.

Onggo, B. S., & Karatas, M. (2016). Test-driven simulation modelling: A case study using agent-based maritime search-operation simulation. *European Journal of Operational Research, 254*(2), 517–531.

Önüt, S., & Soner, S. (2008). Transshipment site selection using the AHP and TOPSIS approaches under fuzzy environment. *Waste Management (New York, N.Y.)*, *28*(9), 1552–1559. doi:10.1016/j.wasman.2007.05.019 PMID:17768038

Overholts, D. L. II, Bell, J. E., & Arostegui, M. A. (2009). A location analysis approach for military maintenance scheduling with geographically dispersed service areas. *Omega*, *37*(4), 838–852. doi:10.1016/j.omega.2008.05.003

Owen, S. H., & Daskin, M. S. (1998). Strategic facility location: A review. *European Journal of Operational Research*, *111*(3), 423–447. doi:10.1016/S0377-2217(98)00186-6

Ozturk, N., & Tozan, H. (2015). A fuzzy-based decision support model for selecting the best dialyser flux in haemodialysis. *Journal of Healthcare Engineering, Vol*, *6*(:3), 303–324. doi:10.1260/2040-2295.6.3.303 PMID:26753437

Paksoy, T., Pehlivan, N. Y., & Kahraman, C. (2012). Organizational Strategy Development in Distribution Channel Management Using Fuzzy AHP and Hirarchical Fuzzy TOPSIS. *Expert Systems with Applications*, *39*(3), 2822–2841. doi:10.1016/j.eswa.2011.08.142

Panton, D., John, M., Lucas, S., & Mason, A. (2003). Flight test data cycle map optimisation. *European Journal of Operational Research*, *146*(3), 486–497. doi:10.1016/S0377-2217(02)00232-1

Papadopoulos, A., & Karagiannidis, A. (2008). Application of the Multi-Criteria Analysis Method ELECTRE III for the Optimisation of Decentralised Energy Systems. *Omega*, *36*(5), 766–776. doi:10.1016/j.omega.2006.01.004

Parsons, D. J., & Krause, L. C. (1999). Tactical logistics and distribution systems (TLoaDS) simulation. In *Simulation Conference Proceedings*, 1999 Winter (Vol. 2, pp. 1174-1178). IEEE.

Pecht & Humphrey. (2008). Addressing Obsolescence - The Uprating Option. *IEEE Transactions on Components and Packaging Technologies*, *31*(3), 741–45.

Pelot, R., Akbari, A., & Li, L. (2015). Vessel Location Modeling for Maritime Search and Rescue. In H. A. Eiselt & V. Marianov (Eds.), *Applications of Location Analysis* (pp. 369–402). Springer International Publishing. doi:10.1007/978-3-319-20282-2_16

Peterson, R. H. (1967). Letter to the Editor—On the "Logarithmic Law" of Attrition and its Application to Tank Combat. *Operations Research*, *15*(3), 557–558. doi:10.1287/opre.15.3.557

Pettit, L. I., Wiper, M. P., & Young, K. D. S. (2003). Bayesian inference for some Lanchester combat laws. *European Journal of Operational Research*, *148*(1), 152–165. doi:10.1016/S0377-2217(02)00462-9

Pillon, D., Perez-Pignol, A., & Jauffret, C. (2016). Observability: Range-only vs. bearings-only target motion analysis for a leg-by-leg observers trajectory. *IEEE Transactions on Aerospace and Electronic Systems*, *52*(4), 1667–1678. doi:10.1109/TAES.2016.150016

Pingle, P. (2015). *Selection of Obsolescence Resolution Strategy Based on a Multi Criteria Decision Model*. Iowa State University.

Plunkett & Verma. (2003). *Technology Refreshment and Obsolescence Avoidance Strategy and Plan*. Academic Press.

Poggi, M., Viana, H., & Uchoa, E. (2010). The team orienteering problem: Formulations and branch-cut and price. In *OASIcs-OpenAccess Series in Informatics* (Vol. 14). Schloss Dagstuhl-Leibniz-Zentrum fuer Informatik.

Porter, Z. (1998). An Economic Method for Evaluating Electronic Component Obsolescence Solutions. *Boeing Information Space & Defence Systems*, 1–9. Retrieved from http://scholar.google.com/scholar?hl=en&btnG=Search&q=intitle:An+Economic+Method+for+Evaluating+Electronic+Component+Obsolescence+Solutions#0

Prabhakar, V. J., Allison, H., Sandborn, P., & Eriksson, B. (2013). Optimizing Part Sourcing Strategies for Low-Volume, Long Life Cycle Products Using Second Sourcing Ad Part Hoarding. International Design Engineering Technical Conferences.

Prabhakar, V. J., & Sandborn, P. (2013). A Model for Comparing Sourcing Strategies for Parts in Long Life Cycle Products Subject to Long-Term Supply Chain Disruptions. *International Journal of Product Lifecycle Management*, *6*(3), 228–249. doi:10.1504/IJPLM.2013.055880

Prins, C., Prodhon, C., & Calvo, R. W. (2006). *Solving the capacitated location-routing problem by a GRASP complemented by a learning process and a path relinking*. FOR: A Quarterly Journal of Operations Research, 4(3), 221-238.

Prodhon, C., & Prins, C. (2014). A survey of recent research on location-routing problems. *European Journal of Operational Research*, *238*(1), 1–17. doi:10.1016/j.ejor.2014.01.005

Qin, X. S., Huang, G. H., Chakma, A., Nie, X. H., & Lin, Q. G. (2008). A MCDM-Based Expert System for Climate-Change Impact Assessment and Adaptation Planning - A Case Study for the Georgia Basin, Canada. *Expert Systems with Applications*, *34*(3), 2164–2179. doi:10.1016/j.eswa.2007.02.024

Qiu, J., & Sharkey, T. C. (2013). Integrated dynamic single-facility location and inventory planning problems. *IIE Transactions*, *45*(8), 883–895. doi:10.1080/0740817X.2013.770184

Rabiner, L. R. (1989). A tutorial on hidden Markov models and selected applications in speech recognition. *Proceedings of the IEEE*, *77*(2), 257–286. doi:10.1109/5.18626

Radovilsky, Z., & Koermer, T. (2007). Allocation of US Coast Guard boats utilizing integer programming. *Journal of the Academy of Business and Economics*, *7*(2), 130–135.

Radovilsky, Z., & Wagner, M. R. (2014). Optimal Allocation of Resources at US Coast Guard Boat Stations. *Journal of Supply Chain and Operations Management*, *12*(1), 50.

Raghavan, S., & Chowdhury, B. (2015). State Diagram-Based Life Cycle Management Plans for Power Plant Components. *IEEE Transactions on Smart Grid*, *6*(2), 965–972. doi:10.1109/TSG.2015.2390232

Raghavan, S., & Member, S. (2012). Developing Life Cycle Management Plans For Power Plant Components. *North American Power Symposium (NAPS)*. 10.1109/NAPS.2012.6336327

Rai, R., & Terpenny, J. (2008). Product Obsolescence. *IEEE Transactions on Components and Packaging Technologies*, *31*(4), 880–889. doi:10.1109/TCAPT.2008.2005115

Raissi-Dehkordi, M., Chandrashekar, K., & Baras, J. S. (2004). *UAV placement for enhanced connectivity in wireless ad-hoc networks*. Technical Research Report.

Raju, K. L., Rao, S. K., Das, R. P., Santhosh, M. N., & Murthy, A. S. (2015). Passive Target Tracking using Unscented Kalman Filter based on Monte Carlo Simulation. *Indian Journal of Science and Technology*, *8*(29). doi:10.17485/ijst/2015/v8i1/76981

Rappold, J. A., & Van Roo, B. D. (2009). Designing multi-echelon service parts networks with finite repair capacity. *European Journal of Operational Research*, *199*(3), 781–792. doi:10.1016/j.ejor.2008.08.006

Rathinam, S., & Sengupta, R. (2007). Algorithms for routing problems involving uavs. In *Innovations in Intelligent Machines-1* (pp. 147–172). Springer Berlin Heidelberg. doi:10.1007/978-3-540-72696-8_6

Razi, N., & Karatas, M. (2016). A multi-objective model for locating search and rescue boats. *European Journal of Operational Research*, *254*(1), 279–293. doi:10.1016/j.ejor.2016.03.026

Razi, N., Karatas, M., & Gunal, M. M. (2016). A combined optimization and simulation based methodology for locating search and rescue helicopters. In *Proceedings of the 2016 Spring Simulation Conference*. IEEE Press.

Razi, N., Karatas, M., & Gunal, M. M. (2016). A Combined Optimization and Simulation Based Methodology for Locating Search and Rescue Helicopters. In *Proceedings of the 2016 Spring Simulation Conference*. IEEE Press.

Razi, N., Karatas, M., & Gunal, M. M. (2016). A combined optimization and simulation based methodology for locating search and rescue helicopters. In *Proceedings of the 49th Annual Simulation Symposium* (p. 5). Society for Computer Simulation International.

Ren, J., & Sovacool, B. K. (2014). Enhancing China's energy security: Determining influential factors and effective strategic measures. *Energy Conversion and Management*, *88*, 589–597.

Rettke, A. J., Robbins, M. J., & Lunday, B. J. (2016). Approximate dynamic programming for the dispatch of military medical evacuation assets. *European Journal of Operational Research*, *254*(3), 824–839. doi:10.1016/j.ejor.2016.04.017

ReVelle, C. S., & Eiselt, H. A.ReVelle. (2005). Location Analysis: A Synthesis and Survey. *European Journal of Operational Research*, *165*(1), 1–19. doi:10.1016/j.ejor.2003.11.032

ReVelle, C. S., & Rosing, K. E. (2000). Defendens imperium romanum: A classical problem in military strategy. *The American Mathematical Monthly*, *107*(7), 585–594. doi:10.1080/00029890.2000.12005243

Revelle, C., & Hogan, K. (1989). The maximum reliability location problem and α-reliablep-centre problem: Derivatives of the probabilistic location set covering problem. *Annals of Operations Research*, *18*(1), 155–173. doi:10.1007/BF02097801

Riaz, M. M., & Ghafoor, A. (2012). Information theoretic criterion based clutter reduction for ground penetrating radar. *Progress In Electromagnetics Research B*, *45*, 147–164. doi:10.2528/PIERB12080802

Robledo, L., Carrasco, M., & Mery, D. (2009). A survey of land mine detection technology. *International Journal of Remote Sensing*, *30*(9), 2399–2410. doi:10.1080/01431160802549435

Rodriguez-Fuentes, L. J., & Penagarikano, M. (2013). *MediaEval 2013 spoken web search task: system performance measures. n. TR-2013-1*. Department of Electricity and Electronics, University of the Basque Country.

Rojo, F. J. R. (2012). A Cost Estimating Framework for Electronic, Electrical and Electromechanical (EEE) Components Obsolescence within the Use-Oriented Product-Service Systems Contracts. *Proceedings of the Institution of Mechanical Engineers. Part B, Journal of Engineering Manufacture*, *226*(1), 154–166. doi:10.1177/0954405411406774

Ross, D. B., & Harmon, J. A. (2012). *New navy fighting machine in the South China Sea* (Master's thesis). Naval Postgraduate School, Monterey, CA.

Ross, S. M. (2014). *Introduction to probability models*. Academic Press.

Rouhani, S., Ghazanfari, M., & Jafari, M. (2012). Evaluation model of business intelligence for enterprise systems using fuzzy TOPSIS. *Expert Systems with Applications*, *39*(3), 3764–3771. doi:10.1016/j.eswa.2011.09.074

Rowden, T., Gumataotao, P., & Fanta, P. (2015). Distributed Lethality. *Proceedings*, *141*(1), 18–23.

Roy, B. (1991). The Outranking Approach and the Foundations of ELECTRE Methods. *Theory and Decision*, *31*(1), 49–73. doi:10.1007/BF00134132

Royset, J. O., Carlyle, W. M., & Wood, R. K. (2009). Routing Military Aircraft With A Constrained Shortest Path Algorithm. *Military Operations Research*, *14*(3), 31–52. doi:10.5711/morj.14.3.31

Rubin, J. (1973). A technique for the solution of massive set covering problems, with application to airline crew scheduling. *Transportation Science*, *7*(1), 34–48. doi:10.1287/trsc.7.1.34

Ryan, J. L., Bailey, T. G., Moore, J. T., & Carlton, W. B. (1998, December). Reactive tabu search in unmanned aerial reconnaissance simulations. In *Proceedings of the 30th conference on Winter simulation* (pp. 873-880). IEEE Computer Society Press. 10.1109/WSC.1998.745084

Sadi-Nezhad, S., & Damghani, K. K. (2010). Application of a Fuzzy TOPSIS Method Based on Modified Preference Ratio and Fuzzy Distance Measurement in Assessment of Traffic Police Centers Performance. *Applied Soft Computing*, *10*(4), 1028–1039. doi:10.1016/j.asoc.2009.08.036

Saginaw, D. J., & Perakis, A. N. (1989). A decision support system for containership stowage planning. *Materials Technology*, *26*, 47–61.

Sahraeian, R., & Kazemi, M. S. (2011, December). A fuzzy set covering-clustering algorithm for facility location problem. In *Industrial Engineering and Engineering Management (IEEM), 2011 IEEE International Conference on* (pp. 1098-1102). IEEE.

Sakoe, H., & Chiba, S. (1978). Dynamic programming algorithm optimization for spoken word recognition. *IEEE Transactions on Acoustics, Speech, and Signal Processing*, *26*(1), 43–49. doi:10.1109/TASSP.1978.1163055

Salhi, S., & Rand, G. K. (1989). The effect of ignoring routes when locating depots. *European Journal of Operational Research*, *39*(2), 150–156. doi:10.1016/0377-2217(89)90188-4

Sandborn, P. (2008, June). Strategic Management of DMSMS in Systems. *DSP Journal*, 24–30.

Sandborn, P., Prabhakar, V., & Feng, D. (2007). DMSMS Lifetime Buy Characterization Via Data Mining of Historical Buys. *DMSS Conference*. Retrieved from http://www.enme.umd.edu/ESCML/obsolescence.htm

Sandborn, Terpenny, & Zheng. (2011). Modeling Constraints in Design Refresh Planning. *ASME 2011 International Design Engineering Technical Conferences*, 1–11.

Sandborn. (2014). *A New "Availability-Payment" Model for Pricing Performance-Based Logistics Contracts*. Academic Press.

Sandborn, P. (2013). Design for Obsolescence Risk Management. *Procedia CIRP*, *11*, 15–22. doi:10.1016/j.procir.2013.07.073

Sandborn, P. A., Herald, T. E., Houston, J., & Singh, P. (2003). Optimum Technology Insertion into Systems Based on the Assessment of Viability. *IEEE Transactions on Components and Packaging Technologies*, *26*(4), 734–738. doi:10.1109/TCAPT.2003.820984

Sandborn, P. A., Mauro, F., & Knox, R. (2007). A Data Mining Based Approach to Electronic Part Obsolescence Forecasting. *IEEE Transactions on Components and Packaging Technologies*, *30*(3), 397–401. doi:10.1109/TCAPT.2007.900058

Sandborn, P., & Goudrazi, N. (2015). A Real Options Approach to Quantity and Cost Optimization for Lifetime and Bridge Buys of Parts. *Acquisition Research Symposium.*

Sandborn, P., Prabhakar, V., & Ahmad, O. (2011). Forecasting Electronic Part Procurement Lifetimes to Enable the Management of DMSMS Obsolescence. *Microelectronics and Reliability*, *51*(2), 392–399. doi:10.1016/j.microrel.2010.08.005

Sandborn, P., Singh, P., Geiser, T., & Lorenson, D. (2002). Electronic Part Obsolescence Driven Product Redesign Optimization. *International Journal of Advanced Manufacturing Systems*, *7*(2), 1–13. http://www.enme.umd.edu/ESCML/Papers/AgingAircraft.pdf

Sang, X., Liu, X., & Qin, J. (2015). An analytical solution to fuzzy TOPSIS and its application in personnelselection for knowledge-intensive enterprise. *Applied Soft Computing*, *30*, 190–204. doi:10.1016/j.asoc.2015.01.002

Santhosh, M. N., Rao, S. K., Das, R. P., & Raju, K. L. (2015). Underwater Target Tracking using Unscented Kalman Filter. *Indian Journal of Science and Technology*, *8*(31). doi:10.17485/ijst/2015/v8i31/77054

Saraclar, M., & Sproat, R. (2004). Lattice-based search for spoken utterance retrieval. *Urbana (Caracas, Venezuela)*, *51*, 61801.

Saremi, M., Mousavi, S. F., & Sanayei, A. (2009). TQM consultant selection in SMEs with TOPSIS under fuzzy environment. *Expert Systems with Applications*, *36*(2), 2742–2749. doi:10.1016/j.eswa.2008.01.034

Sarıçiçek, İ., & Akkuş, Y. (2015). Unmanned Aerial Vehicle hub-location and routing for monitoring geographic borders. *Applied Mathematical Modelling*, *39*(14), 3939–3953. doi:10.1016/j.apm.2014.12.010

Sarikaya, N. (2009). *Determining the Orbit Locations of Turkish Airborne Early Warning and Control Aircraft over the Turkish Airspace. AFIT-Air Force Institute of Technology, AFIT/GOR/ENS/09-14*. Dayton, OH: Wright Patterson AFB.

Sarı, L., Gündoğdu, B., & Saraçlar, M. (2015). Fusion of LVCSR and Posteriorgram Based Keyword Search. *INTERSPEECH*, *2015*, 824–828.

Savuran, H., & Karakaya, M. (2015). Route optimization method for unmanned air vehicle launched from a carrier. *Lecture Notes on Software Engineering*, *3*(4), 279–284. doi:10.7763/LNSE.2015.V3.204

Scaparra, M. P., & Church, R. L. (2008). A bilevel mixed-integer program for critical infrastructure protection planning. *Computers & Operations Research*, *35*(6), 1905–1923. doi:10.1016/j.cor.2006.09.019

Schick, W. G. (1992). *Location an Imaging Radar in Canada for Identifying Spaceborne Objects*. Air Force Inst. of Technology, Wright Patterson AFB. Retrieved from http://handle.dtic.mil/100.2/ADA258969

Sciomachen, A., & Tanfani, E. (2003). The master bay plan problem: A solution method based on its connection to the three-dimensional bin packing problem. *IMA Journal of Management Mathematics*, *14*(3), 251–269. doi:10.1093/imaman/14.3.251

Sciomachen, A., & Tanfani, E. (2006). A 3D-BPP approach for optimising stowage plans and terminal productivity. *European Journal of Operational Research*. doi:10.1016/j.ejor

Seçme, N. Y., Bayrakdaroğlu, A., & Kahraman, C. (2009). Fuzzy Performance Evaluation in Turkish Banking Sector Using Analytic Hierarchy Process and TOPSIS. *Expert Systems with Applications*, *36*(9), 11699–11709. doi:10.1016/j.eswa.2009.03.013

Segall, R. S. (2000). Some quantitative methods for determining capacities and locations of military emergency medical facilities. *Applied Mathematical Modelling*, *24*(5), 365–389. doi:10.1016/S0307-904X(99)00045-1

Şengül, Ü., Eren, M., Shiraz, S. E., Gezder, V., & Şengül, A. B. (2015). Fuzzy TOPSIS method for ranking renewable energy supply systems in Turkey. *Renewable Energy*, *75*, 617–625. doi:10.1016/j.renene.2014.10.045

Shanian, A., & Savadogo, O. (2009). A Methodological Concept for Material Selection of Highly Sensitive Components Based on Multiple Criteria Decision Analysis. *Expert Systems with Applications*, *36*(2), 1362–1370. doi:10.1016/j.eswa.2007.11.052

Sharma, V., Sabatini, R., & Ramasamy, S. (2016). UAVs assisted delay optimization in heterogeneous wireless networks. *IEEE Communications Letters*, *20*(12), 2526–2529. doi:10.1109/LCOMM.2016.2609900

Shawki, K. M., Kilani, K., & Gomaa, M. A. (2015). Analysis of earth-moving systems using discrete-event simulation. *Alexandria Engineering Journal, 54*(3), 533–540.

Sheeba, P. S., & Ghose, D. Optimal Resource Partitioning in a Military Conflict based on Lanchester Attrition Models. In *Proceedings of the 44th IEEE Conference on Decision and Control, and the European Control Conference 2005*. Seville, Spain: IEEE. 10.1109/CDC.2005.1583098

Shen, W., White, C. M., & Hazen, T. J. (2009). *A comparison of query-by-example methods for spoken term detection*. Massachusetts Inst of Tech Lexington Lincoln LAB.

Shen, Y., & Willems, S. P. (2014). Modeling Sourcing Strategies to Mitigate Part Obsolescence. *European Journal of Operational Research, 236*(2), 522–533. doi:10.1016/j.ejor.2014.01.025

Shephard, R. W. (1963). War gaming as a technique in the study of operational research problems. *The Journal of the Operational Research Society, 14*(2), 119–130. doi:10.1057/jors.1963.20

Shetty, V. K., Sudit, M., & Nagi, R. (2008). Priority-based assignment and routing of a fleet of unmanned combat aerial vehicles. *Computers & Operations Research, 35*(6), 1813–1828. doi:10.1016/j.cor.2006.09.013

Sheu, J.-B. (2008). A Hybrid Neuro-Fuzzy Analytic Approach to Mode Choice of Global Logistics Management. *European Journal of Operational Research, 189*(3), 971–986. doi:10.1016/j.ejor.2006.06.082

Shields, J. J. (1984). Container-ship stowage: A computer-aided preplanning system. *Materials Technology, 21*, 370–383.

Shi, X., & Xue, B. (2016). Deriving a minimum set of viewpoints for maximum coverage over any given digital elevation model data. *International Journal of Digital Earth, 9*(12), 1153–1167. doi:10.1080/17538947.2016.1207718

Shubik, M. (Ed.). (1983). *Mathematics of Conflict*. New York: North-Holand.

Shyur, H.-J., & Shih, H.-S. (2006). A Hybrid MCDM Model for Strategic Vendor Selection. *Mathematical and Computer Modelling, 44*(7-8), 749–761. doi:10.1016/j.mcm.2005.04.018

Silver, M. S. (1991). *Systems that Support Decision Makers: Description and Analysis*. West Sussex, UK: John Wiley & Sons.

Singh, Sandborn, & Singh. (2004). Forecasting Technology Insertion Concurrent with Design Refresh Planning for COTS-Based Electronics Systems. *Logistics Spectrum, 38*(2), 349–54. Retrieved from http://www.enme.umd.edu/ESCML/Papers/solemocapaper.pdf%5Cnhttp://ieeexplore.ieee.org/lpdocs/epic03/wrapper.htm?arnumber=1408387

Singh, P., & Sandborn, P. (2006). Obsolescence Driven Design Refresh Planning for Sustainment-Dominated Systems. *The Engineering Economist, 51*(2), 115–139. doi:10.1080/00137910600695643

Singh, P., Sandborn, P., Lorenson, D., & Geiser, T. (2002). Determining Optimum Redesign Plans for Avionics Based on Electronic Part Obsolescence Forecasts. *World Avioation Congress and Exposition*. 10.4271/2002-01-3012

Singh, R. K., & Benyoucef, L. (2011). A fuzzy TOPSIS based approach for e-sourcing. *Engineering Applications of Artificial Intelligence, 24*(3), 437–448. doi:10.1016/j.engappai.2010.09.006

Sivilevičius, H., & Maskeliūnaite, L. (2010). The criteria for identifying the quality of passengers' transportation by railway and their ranking using AHP method. *Transport, 25*(4), 368–381.

Smith, J. S. (2003). Survey on the use of simulation for manufacturing system design and operation. *Journal of Manufacturing Systems, 22*(2), 157–171.

Solimene, R., Cuccaro, A., Dell'Aversano, A., Catapano, I., & Soldovieri, F. (2014). Ground Clutter Removal in GPR surveys. *IEEE Journal of Selected Topics in Applied Earth Observations and Remote Sensing*, *7*(3), 792–798. doi:10.1109/JSTARS.2013.2287016

Solomon, R., Sandborn, P. A., & Pecht, M. G. (2000). Electronic Part Life Cycle Concepts and Obsolescence Forecasting. *IEEE Transactions on Components and Packaging Technologies*, *23*(4), 707–717. doi:10.1109/6144.888857

Sonmez, H. H., Hocaoglu, A. K., & Genc, H. M. (2017). A new solution method for the passive target motion analysis problem using evolutionary strategies. *2017 25th Signal Processing and Communications Applications Conference (SIU)*. doi:10.1109iu.2017.7960401

Souffriau, W., Vansteenwegen, P., Vanden Berghe, G., & Van Oudheusden, D. (2013). The multiconstraint team orienteering problem with multiple time windows. *Transportation Science*, *47*(1), 53–63. doi:10.1287/trsc.1110.0377

Spitz, G. (2007). *Mission resources allocation in the Gulf of Guinea* (Master's thesis). Naval Postgraduate School, Monterey, CA.

Stahlbock, R., & Voß, S. (2007). Operations research at container terminals: A literature update. *OR-Spektrum*, *30*(1), 1–52. doi:10.100700291-007-0100-9

Stanescu, M., Barriga, N., & Buro, M. (2015). Using Lanchester attrition laws for combat prediction in StarCraft. *Eleventh Annual AAAI Conference on Artificial Intelligence and Interactive Digital Entertainment (AIIDE)*, 86-92.

Starck, J. L., Candes, E. J., & Donoho, D. L. (2002). The curvelet transform for image denoising. *IEEE Transactions on Image Processing*, *11*(6), 670–684. doi:10.1109/TIP.2002.1014998 PMID:18244665

Starck, J. L., Moudden, Y., Bobin, J., Elad, M., & Donoho, D. L. (2005, July). Morphological component analysis. *Proceedings of the Society for Photo-Instrumentation Engineers*, *5914*, 1–15. doi:10.1117/12.615237

Status of the Navy. (n.d.). Retrieved April 26, 2015, from http://www.navy.mil/navydata/nav_legacy.asp?id=146

Steenken, D., Voß, S., & Stahlbock, R. (2004). Container terminal operation and operations research – a classification and literature review. *OR-Spektrum*, *26*(1), 3–49. doi:10.100700291-003-0157-z

Stein, E. W. (2013). A comprehensive multi-criteria model to rank electric energy production technologies. *Renewable & Sustainable Energy Reviews*, *22*, 640–654.

Stewart, K. A. (2013). *NPS-Developed replenishment at sea program could save millions*. Office of Naval Postgraduate School Public Affairs. Retrieved from http://www.navy.mil/submit/display.asp?story_id=72111

Stützle, T., & Hoos, H. (1997). MAX-MIN ant system and local search for the traveling salesman problem. In *IEEE International Conference on Evolutionary Computation (ICEC'97)*. 10.1109/ICEC.1997.592327

Stützle, T., & Hoos, H. H. (2000). MAX–MIN ant system. *Future Generation Computer Systems*, *16*(8), 889–914. doi:10.1016/S0167-739X(00)00043-1

Sun, C.-C. (2010). A performance evaluation model by integrating fuzzy AHP and fuzzy TOPSIS methods. *Expert Systems with Applications*, *37*(12), 7745–7754. doi:10.1016/j.eswa.2010.04.066

Sun, C.-C., & Lin, G. T. (2009). Using Fuzzy TOPSIS Method for Evaluating the Competitive Advantages of Shopping Websites. *Expert Systems with Applications*, *36*(9), 11764–11771. doi:10.1016/j.eswa.2009.04.017

Sundar, K., & Rathinam, S. (2014). Algorithms for routing an unmanned aerial vehicle in the presence of refueling depots. *IEEE Transactions on Automation Science and Engineering*, *11*(1), 287–294. doi:10.1109/TASE.2013.2279544

Sylvester, J. J. (1857). A question in the geometry of situation. *Quarterly Journal of Pure and Applied Mathematics*, 1.

Talbi. (2009). *Metaheuristics: From Design to Implementation*. Academic Press.

Tang, H., & Miller-Hooks, E. (2005). A tabu search heuristic for the team orienteering problem. *Computers & Operations Research*, *32*(6), 1379–1407. doi:10.1016/j.cor.2003.11.008

Tansel, B. C., Francis, R. L., & Lowe, T. J. (1983). State of the art—location on networks: a survey. Part I: the p-center and p-median problems. *Management Science*, *29*(4), 482–497. doi:10.1287/mnsc.29.4.482

Taylan, O., Bafail, A. O., Abdulaal, R. M. S., & Kabli, M. R. (2014). Construction projects selection and risk assessment by fuzzy AHP and fuzzy TOPSIS methodologies. *Applied Soft Computing*, *17*, 105–116. doi:10.1016/j.asoc.2014.01.003

Taylor, J. G. (1980). *Lanchester-Type Models of Warfare* (Vol. 1). Naval Postgraduate School Monterey CA.

Temlioglu, E., & Erer, I. (2016). Clutter removal in ground-penetrating radar images using morphological component analysis. *IEEE Geoscience and Remote Sensing Letters*, *13*(12), 1802–1806. doi:10.1109/LGRS.2016.2612582

Temlioglu, E., Erer, I., & Kumlu, D. (2017, July). A least mean square approach to buried object detection for ground penetrating radar. In *Geoscience and Remote Sensing Symposium (IGARSS)*, Fort Worth, TX. 10.1109/IGARSS.2017.8128084

Temuçin, T., & Tozan, H. (2016). A fuzzy based decision support system for air conditioner selection and an application to Turkish construction sector. *Technical Gazette-TEHNIČKI VJESNIK, 23*(:4), 1117–1122.

Temuçin, T., Tozan, H., Valíček, J., & Harničárova, M. (2013). A fuzzy based decision support model for nontraditional machining process selection. *Technical Gazette-TEHNIČKI VJESNIK, 20*(Iss:3), 683–687.

Temuçin, T., Tozan, H., Vayvay, O., Valíček, J., & Harničárova, M. (2014). A fuzzy based decision model for nontraditional machining process selection. *International Journal of Advanced Manufacturing Technology, Vol, 70*(9-12), 2275–2282. doi:10.100700170-013-5474-z

Terrasse, G., Nicolas, J. M., Trouvé, E., & Drouet, É. (2015, July). Application of the curvelet transform for pipe detection in GPR images. In *Geoscience and Remote Sensing Symposium (IGARSS)*, Milan, Italy. 10.1109/IGARSS.2015.7326779

Teunter, R. H., & Fortuin, L. (1999). End-of-Life Service. *International Journal of Production Economics*, *59*(1-3), 487–497. doi:10.1016/S0925-5273(98)00112-1

Tian, J., Shen, L., & Zheng, Y. (2006). Genetic algorithm based approach for multi-UAV cooperative reconnaissance mission planning problem. *Lecture Notes in Computer Science*, *4203*, 101–110. doi:10.1007/11875604_13

Tian, J., Yu, D., & Ma, S. (2013). A fuzzy TOPSIS model via chi-square test for information source selection. *Knowledge-Based Systems*, *37*, 515–527. doi:10.1016/j.knosys.2012.09.010

Tibshirani, R., Walther, G., & Hastie, T. (2001). Estimating the number of clusters in a data set via the gap statistic. *Journal of the Royal Statistical Society. Series B, Statistical Methodology*, *63*(2), 411–423.

Tiernan, T. A., Richardson, N., Azeredo, P., Najm, W. G., & Lochrane, T. (2017). *Test and Evaluation of Vehicle Platooning Proof-of-Concept Based on Cooperative Adaptive Cruise Control. Research Report. U.S. Department of Transportation John A.* Volpe National Transportation Systems Center Advanced Vehicle Technology Division.

Ting, C. J., & Chen, C. H. (2013). A multiple ant colony optimization algorithm for the capacitated location routing problem. *International Journal of Production Economics*, *141*(1), 34–44. doi:10.1016/j.ijpe.2012.06.011

Todd, D. S., & Sen, P. (1997). A multiple criteria genetic algorithm for containership loading. *Proceedings of the Seventh International Conference on Genetic Algorithms*, 674–681.

Toker, D., Tozan, H., & Vayvay, O. (2013). A decision model for pharmaceutical marketing and a case study in Turkey. *Ekonomska istraživanja-Economic Research, 26*(1), 101-114.

Tomasi, C., & Manduchi, R. (1998, January). Bilateral filtering for gray and color images. In *Computer Vision, 1998. Sixth International Conference on* (pp. 839-846). IEEE. 10.1109/ICCV.1998.710815

Toregas, C., Swain, R., ReVelle, C., & Bergman, L. (1971). The Location of Emergency Service Facilities. *Operations Research, 19*(6), 1363–1373. doi:10.1287/opre.19.6.1363

Torlak, G., Sevkli, M., Sanal, M., & Zaim, S. (2011). Analyzing business competition by using fuzzy TOPSIS method: An example of Turkish domestic airline industry. *Expert Systems with Applications, 38*(4), 3396–3406. doi:10.1016/j.eswa.2010.08.125

Tosun, K. (1992). *İşletme Yönetimi (Genel Esaslar) (6th ed.)*. Ankara: Savaş Yayınları.

Toyoglu, H., Karasan, O. E., & Kara, B. Y. (2011). Distribution network design on the battlefield. *Naval Research Logistics, 58*(3), 188–209. doi:10.1002/nav.20402

Tozan, H. (2011). Fuzzy AHP based decision support system for technology selection in abrasive water jet cutting processes. *Technical Gazette-TEHNIČKI VJESNIK, Vol, 18*(:2), 187–191.

Tretyakov, O. A., & Erden, F. (2007). Separation of the instantaneous and dynamic polarizations in studies of dispersive dielectrics. *The Sixth Int. Kharkov Symp. on Phys. and Eng. of Microwaves, Millimeter and Submillimeter Waves*, 42-48.

Tretyakov, O. A., & Erden, F. (2008). Temporal cavity oscillations caused by a wide-band waveform. *Progress in Electromag. Res. B, 6*, 183.

Trickey, W. R., & Grenwald, M. L. (2014). Navy Logistics Resiliency In A WESTPAC A2AD Scenario. CNA Corporation. (This document is classified SECRET)

Trickey, W. R. (2014). *Navy Logistics Resiliency Model Description*. Arlington, VA: CNA Corporation.

Tricoire, F., Romauch, M., Doerner, K. F., & Hartl, R. F. (2010). Heuristics for the multi-period orienteering problem with multiple time windows. *Computers & Operations Research, 37*(2), 351–367. doi:10.1016/j.cor.2009.05.012

Tsoukias, A. (2008). From Decision Theory to Decision Aiding Methodology. *European Journal of Operational Research, 187*(1), 138–161. doi:10.1016/j.ejor.2007.02.039

Tsoutsos, T., Drandaki, M., Frantzeskaki, N., Losifidis, E., & Kiosses, L. (2009). Sustainable Energy Planning by Using Multi-Criteria Analysis Application in the Island of Crete. *Energy Policy, 37*(5), 1587–1600. doi:10.1016/j.enpol.2008.12.011

Turban, E., & Aronson, J. E. (2001). Decision Support Systems and Intelligent Systems (6th ed.). Prentice-Hall, Inc.

Tzanis, A. (2015). The Curvelet Transform in the analysis of 2-D GPR data: Signal enhancement and extraction of orientation-and-scale-dependent information. *Journal of Applied Geophysics, 115*, 145–170. doi:10.1016/j.jappgeo.2015.02.015

UNHCR. (n.d.). *Global Appeal 2015 Update*. Retrieved from http://www.unhcr.org/5461e5f80.html

US Department of Defense. (1997). *Simulation, Test, and Evaluation Process Step Guidelines*. Retrieved from http://www.dtic.mil/get-tr-doc/pdf?AD=ADA341111

Vaidya, O. S., & Kumar, S. (2006). Analytic hierarchy process: An overview of applications. *European Journal of Operational Research, 169*(1), 1–29.

Vaillancourt, K., & Waaub, J.-P. (2004). Equity in International Greenhouse Gases Abatement Scenarios: A Multicriteria Approach. *European Journal of Operational Research, 153*(2), 489–505. doi:10.1016/S0377-2217(03)00170-X

Valavanis, K. P., & Vachtsevanos, G. J. (2014). *Handbook of unmanned aerial vehicles*. Springer Publishing Company, Incorporated.

Van Der Merwe, A., & Gupta, I. J. (2000). A novel signal processing technique for clutter reduction in GPR measurements of small, shallow land mines. *IEEE Transactions on Geoscience and Remote Sensing, 38*(6), 2627–2637. doi:10.1109/36.885209

Van Kooten, J. P. J., & Tan, T. (2007). The Final Order Problem for Repairable Spare Parts under Condemnation. *Operations Research Society, 53*(9), 1689–1699.

Van Thong, J. M., Moreno, P. J., Logan, B., Fidler, B., Maffey, K., & Moores, M. (2002). Speechbot: An experimental speech-based search engine for multimedia content on the web. *IEEE Transactions on Multimedia, 4*(1), 88–96. doi:10.1109/6046.985557

Vansteenwegen, P., Souffriau, W., Berghe, G. V., & Van Oudheusden, D. (2009a). A guided local search metaheuristic for the team orienteering problem. *European Journal of Operational Research, 196*(1), 118–127. doi:10.1016/j.ejor.2008.02.037

Vansteenwegen, P., Souffriau, W., Berghe, G. V., & Van Oudheusden, D. (2009b). Metaheuristics for tourist trip planning. In *Metaheuristics in the service industry* (pp. 15–31). Springer Berlin Heidelberg. doi:10.1007/978-3-642-00939-6_2

Vansteenwegen, P., Souffriau, W., Berghe, G. V., & Van Oudheusden, D. (2009c). Iterated local search for the team orienteering problem with time windows. *Computers & Operations Research, 36*(12), 3281–3290. doi:10.1016/j.cor.2009.03.008

Vansteenwegen, P., Souffriau, W., & Van Oudheusden, D. (2011). The orienteering problem: A survey. *European Journal of Operational Research, 209*(1), 1–10. doi:10.1016/j.ejor.2010.03.045

Vargas, R. A. (2012). *Unmanned Systems: Operational Considerations for the 21st Century Joint Task Force Commander and Staff*. ARMY COMMAND AND GENERAL STAFF COLL FORT LEAVENWORTH KS.

Varol, A. E., & Gunal, M. M. (2015). Simulating prevention operations at sea against maritime piracy. *The Journal of the Operational Research Society, 66*(12), 2037–2049.

Vemuganti, R. R. (1998). Applications of set covering, set packing and set partitioning models: A survey. In Handbook of combinatorial optimization (pp. 573-746). Springer US. doi:10.1007/978-1-4613-0303-9_9

Verderber, R. F., Verderber, K. S., & Sellnow, D. D. (2009). Communicate! (13th ed.). Boston: Wadsworth Publishing.

Verma, P. K., Gaikwad, A. N., Singh, D., & Nigam, M. J. (2009). Analysis of clutter reduction techniques for through wall imaging in UWB range. *Progress In Electromagnetics Research B, 17*, 29–48. doi:10.2528/PIERB09060903

Wagner, M. R., & Radovilsky, Z. (2012). Optimizing Boat Resources at the US Coast Guard: Deterministic and Stochastic Models. *Operations Research, 60*(5), 1035–1049. doi:10.1287/opre.1120.1085

Walczak, D., & Rutkowska, A. (2017). Project rankings for participatory budget based on the fuzzy TOPSIS method. *European Journal of Operational Research, 260*(2), 706–714. doi:10.1016/j.ejor.2016.12.044

Walsh, N. E., & Walsh, W. S. (2003). Rehabilitation of landmine victims: The ultimate challenge. *Bulletin of the World Health Organization, 81*(9), 665–670. PMID:14710508

Wang, D., Frankel, J., Tejedor, J., & King, S. (2008, March). A comparison of phone and grapheme-based spoken term detection. In *Acoustics, Speech and Signal Processing, 2008. ICASSP 2008. IEEE International Conference on* (pp. 4969-4972). IEEE. 10.1109/ICASSP.2008.4518773

Wang, W. F., Liu, X. L., & Guo, B. (2008). Research on the multi-criteria integrated military support facility location-allocation method. *Systems Engineering-Theory & Practice*, *5*, 20.

Wang, B., Chen, J., Liu, W., & Yang, L. T. (2016). Minimum cost placement of bistatic radar sensors for belt barrier coverage. *IEEE Transactions on Computers*, *65*(2), 577–588. doi:10.1109/TC.2015.2423679

Wang, F., & Lim, A. (2007). A Stochastic Beam Search for the Berth Allocation Problem. *Decision Support Systems*, *42*(4), 2186–2196. doi:10.1016/j.dss.2006.06.008

Wang, H., Huo, D., & Alidaee, B. (2014). Position unmanned aerial vehicles in the mobile ad hoc network. *Journal of Intelligent & Robotic Systems*, *74*(1-2), 455–464. doi:10.100710846-013-9939-y

Wang, J.-J., & Yang, D.-L. (2007). Using a Hybrid Multi-Criteria Decision Aid Method for Information Systems Outsourcing. *Computers & Operations Research*, *34*(12), 3691–3700. doi:10.1016/j.cor.2006.01.017

Wang, P., Meng, P., & Song, B. (2014). Response surface method using grey relational analysis for decision making in weapon system selection. *Journal of Systems Engineering and Electronics*, *25*(2), 265–272. doi:10.1109/JSEE.2014.00030

Wang, T. C., & Lee, H. D. (2009). Developing A Fuzzy TOPSIS Approach Based on Subjective Weights and Objective Weights. *Expert Systems with Applications*, *36*(5), 8980–8985. doi:10.1016/j.eswa.2008.11.035

Wang, T.-C., & Chang, T.-H. (2007). Application of TOPSIS in Evaluating Initial Training Aircraft Under a Fuzzy Environment. *Expert Systems with Applications*, *33*(4), 870–880. doi:10.1016/j.eswa.2006.07.003

Wang, X., & Triantaphyllou, E. (2008). Ranking Irregularities When Evaluating Alternatives by Using Some ELECTRE Methods. *Omega*, *36*(1), 45–63. doi:10.1016/j.omega.2005.12.003

Wang, Y.-J. (2008). Appliying FMCDM to Evaluate Financial Performance of Domestic Airlines in Taiwan. *Expert Systems with Applications*, *34*(3), 1837–1845. doi:10.1016/j.eswa.2007.02.029

Wang, Y.-J. (2014). The evaluation of financial performance for Taiwan containershipping companies by fuzzy TOPSIS. *Applied Soft Computing*, *22*, 28–35. doi:10.1016/j.asoc.2014.03.021

Wang, Y.-J., & Lee, H.-S. (2007). Generalizing TOPSIS for fuzzy multiple-criteria group decision-making. *Computers & Mathematics with Applications (Oxford, England)*, *53*(11), 1762–1772. doi:10.1016/j.camwa.2006.08.037

Wang, Y.-M., & Elhag, T. M. (2006). Fuzzy TOPSIS Method Based on Alpha Level Sets with An Application to Bridge Risk Assessment. *Expert Systems with Applications*, *31*(2), 309–319. doi:10.1016/j.eswa.2005.09.040

Want China Times. (2013, January 23). PLA 'sinks' US carrier in DF-21D missile test in Gobi. *Want China Times*. Retrieved from http://www.wantchinatimes.com/newssubclass-cnt.aspx?id=20130123000112&cid=1101

Ward, P. W. (2016). *NAS Lemoore modeling project. Presented at OA4611 Joint Logistics Models*. Monterey, CA: Naval Postgraduate School.

Warren, C., Giannopoulos, A., & Giannakis, I. (2016). gprMax: Open source software to simulate electromagnetic wave propagation for ground penetrating radar. *Computer Physics Communications*, *209*, 163–170. doi:10.1016/j.cpc.2016.08.020

Washburn, A. R. (2010). A Multistatic Sonobuoy Theory. Naval Postgraduate School report NPS-OR-10-005. doi:10.21236/ADA526686

Washburn, A. (2000). *Lanchester systems*. Naval Postgraduate School Monterey CA. doi:10.21236/ADA383409

Washburn, A., & Karatas, M. (2015). Multistatic search theory. *Military Operations Research*, *20*(1), 21–38.

Weber, A. (1929). *Theory of the Location of Industries* (C. J. Friedrich, Trans.). Academic Press.

Wei Lam, S. S., Zhang, Z. C., Oh, H. C., Ng, Y. Y., Wah, W., & Hock Ong, M. E. (2014). Reducing ambulance response times using discrete event simulation. *Prehospital Emergency Care*, *18*(2), 207–216. PMID:24134647

Wei, R., & Murray, A. T. (2014). A multi-objective evolutionary algorithm for facility dispersion under conditions of spatial uncertainty. *The Journal of the Operational Research Society*, *65*(7), 1133–1142. doi:10.1057/jors.2013.58

White, J., & Case, K. (1974). On covering problems and the central facility location problem. *Geographical Analysis*, *281*, 1974.

Williams, E. J., & Çelik, H. (1998, December). Analysis of conveyor systems within automotive final assembly. In *Proceedings of the 30th conference on Winter simulation* (pp. 915-920). IEEE Computer Society Press.

Wilson, I.D., & Roach, P.A. (1999). Principles of combinatorial optimization applied to container-ship stowage planning. *Journal of Heuristics*, *5*(4), 403–418.

Wilson, I. D., & Roach, P. A. (2000). Container stowage planning: A methodology for generating computerized solutions. *The Journal of the Operational Research Society*, *51*(11), 1248–1255. doi:10.1057/palgrave.jors.2601022

Wold, S., Esbensen, K., & Geladi, P. (1987). Principal component analysis. *Chemometrics and Intelligent Laboratory Systems*, *2*(1-3), 37–52. doi:10.1016/0169-7439(87)80084-9

Wu, T.T., Powell, W.B., & Whisman, A. (2008, September). The Optimizing-Simulator: An Illustration using the Military Airlift Problem. *ACM Transactions on Computational Logic*.

Wu, C. H., & Hwang, K. P. (2009). Using a Discrete-event Simulation to Balance Ambulance Availability and Demand in Static Deployment Systems. *Academic Emergency Medicine*, *16*(12), 1359–1366. PMID:20053259

Wu, H.-Y., Tzeng, G.-H., & Chen, Y.-H. (2009). A Fuzzy MCDM Approach for Evaluating Banking Performance Based on Balanced Scorecard. *Expert Systems with Applications*, *36*(6), 10135–10147. doi:10.1016/j.eswa.2009.01.005

Xiaofeng, L. I. U., Zhongren, P. E. N. G., Zhang, L., & Li, L. I. (2012). Unmanned aerial vehicle route planning for traffic information collection. *Journal of Transportation Systems Engineering and Information Technology*, *12*(1), 91–97. doi:10.1016/S1570-6672(11)60186-4

Xidonas, P., Mavrotas, G., & Psarras, J. (2009). A Multicriteria Methodology for Equity Selection Using Financial Analysis. *Computers & Operations Research*, 36.

Yakıcı, E. (2016). Solving location and routing problem for UAVs. *Computers & Industrial Engineering*, *102*, 294–301. doi:10.1016/j.cie.2016.10.029

Yakıcı, E. (2016b). Generalization of a UAV location and routing problem by time windows. *Journal of Naval Science and Engineering*, *12*(2), 67–78.

Yanar, L. (2011). *A Fuzzy Hybrid Decision Support Software Development for Multi-Criteria Decision Making Problems and an Application to Abrasive Water Jet*. İstanbul: Turkish Naval Academy.

Yanar, L., & Tozan, H. (2012). A fuzzy hybrid decision support system for interceptor baywatch boat propulsion system selection. *Technical Gazette-TEHNIČKI VJESNIK*, *19*(2), 407–413.

Yang, M., Cao, J., Song, J., & Guo, J. (2008, December). A Postorder Traversal Location Model Taking Security on Routes into Consideration. In *Intelligent Information Technology Application Workshops, IITAW'08. International Symposium on* (pp. 535-538). IEEE.

Yang, T., Wen, Y.-F., & Wang, F.-F. (2011). EValuation of Robustness of Supply Chain Information - Sharing Strategies Using a Hybrid Taguchi and Multiple Criteria Decision-Making Method. *International Journal of Production Economics, 134*(2), 458–466. doi:10.1016/j.ijpe.2009.11.018

Yildirim, M. F. (2016). *Optimization of hub locations of UAVs for border monitoring as coverage problem* (Doctoral dissertation). State University of New York at Binghamton.

Yıldırım, U. Z., Tansel, B. Ç., & Sabuncuoğlu, İ. (2009). A multi-modal discrete-event simulation model for military deployment. *Simulation Modelling Practice and Theory, 17*(4), 597–611.

Yurdakul, M., & İç, Y. T. (2009). Analysis of the benefit generated by using fuzzy numbers in a TOPSIS model developed for machine tool selection problems. *Journal of Materials Processing Technology, 209*(1), 310–317. doi:10.1016/j.jmatprotec.2008.02.006

Yu, V. F., & Hu, K.-J. (2010). An Integrated Fuzzy Multi-Criteria Approach for the Performance Evaluation of Multiple Manufacturing Plants. *Computers & Industrial Engineering, 58*(2), 269–277. doi:10.1016/j.cie.2009.10.005

Zammori, F. A., Braglia, M., & Frosolini, M. (2009). A Fuzzy Multi-Criteria Approach for Critical Path Definition. *International Journal of Project Management, 27*(3), 278–291. doi:10.1016/j.ijproman.2008.03.006

Zhang, J., Jia, L., Niu, S., Zhang, F., Tong, L., & Zhou, X. (2015). A space-time network-based modeling framework for dynamic unmanned aerial vehicle routing in traffic incident monitoring applications. *Sensors (Basel), 15*(6), 13874–13898. doi:10.3390150613874 PMID:26076404

Zheng, L. (2011). Knowledge Representation and Decision Support for Managing Product Obsolescence. *ProQuest Dissertations and Theses*, 174. Retrieved from http://search.proquest.com/docview/1020897067?accountid=14511%5Cnhttp://sfx.ucl.ac.uk/sfx_local?url_ver=Z39.88-2004&rft_val_fmt=info:ofi/fmt:kev:mtx:dissertation&genre=dissertations+&+theses&sid=ProQ:ProQuest+Dissertations+&+Theses+Global&atitle=&title=Kn

Zheng, L., Terpenny, J., & Sandborn, P. (2015). Design Refresh Planning Models for Managing Obsolescence. *IIE Transactions, 47*(12), 1407–1423. doi:10.1080/0740817X.2014.999898

Zheng, Nelson, & Sandborn, & Terpenny. (2013). Ontology-Based Knowledge Representation for Obsolescence Forecasting. *Journal of Computing and Information Science in Engineering*, 13.

Zhenhua, W., Weiguo, Z., Jingping, S., & Ying, H. (2008, September). *UAV route planning using multiobjective ant colony system.* In Cybernetics and Intelligent Systems, 2008 IEEE Conference on (pp. 797-800). IEEE.

Zhen, L., Wang, K., Hu, H., & Chang, D. (2014). A simulation optimization framework for ambulance deployment and relocation problems. *Computers & Industrial Engineering, 72*, 12–23.

Zhu, N., Liu, Y., Ma, S., & He, Z. (2014). Mobile traffic sensor routing in dynamic transportation systems. *IEEE Transactions on Intelligent Transportation Systems, 15*(5), 2273–2285. doi:10.1109/TITS.2014.2314732

Zitzler, E. (2002). *Evolutionary Algorithms for Multiobjective*. Academic Press.

Zitzler, E., & Thiele, L. (1999). Multiobjective evolutionary algorithms: A comparative case study and the strength Pareto approach. *IEEE Transactions on Evolutionary Computation, 3*(4), 257–271.

Zouggari, A., & Benyoucef, L. (2012). Simulation based fuzzy TOPSIS approach for group multi-criteria supplier selection problem. *Engineering Applications of Artificial Intelligence, 25*(3), 507–519. doi:10.1016/j.engappai.2011.10.012

About the Contributors

Hakan Tozan is currently working as Professor and the Head of the Department of Industrial Engineering in the School of Engineering and Natural Sciences of Istanbul Medipol University. He is also the founder and director of the Healthcare Systems Engineering program and Healthcare Modeling and Simulation Research Group in the same institution. His research interests include decision analysis, operations research, healthcare systems optimization and health technology assessment. He acts as an adviser for some projects of Health Technologies Assessment Department of the Turkish Ministry of Health. During his academic career in the Turkish Naval Academy, he acted as the head of the Department of Industrial Engineering, as the director of Military Operations Research Program and the director of Naval Sciences and Engineering Institute.

Mumtaz Karatas graduated from the Turkish Naval Academy in 2001. He received his MS degree in Industrial and Operations Engineering from the University of Michigan and his PhD in Industrial Engineering from Kocaeli University, Turkey. He spent two years at the Naval Postgraduate School as a visiting researcher and postdoctoral fellow between 2011 and 2013. He is currently an Associate Professor in the Industrial Engineering Department at the Turkish Naval Academy, National Defense University. His current research areas include optimization and mathematical modelling.

* * *

Alper Aytun is currently a Ph.D. student at the Industrial Engineering Department of Marmara University, Turkey. He graduated from Turkish Naval Academy in 2003 with a B.Sc. Degree in Industrial Engineering. He received his M.Sc. in Electronic Warfare Systems Engineering degree from US Naval Postgraduate School in 2010. His research interests include Operations Research, Evolutionary Algorithms, Signal Processing, Radar and Electronic Warfare Systems.

Mevlüt Savaş Bilican is currently a Ph.D. student at the Industrial Engineering Department of İstanbul Technical University, Turkey. He graduated from Turkish Naval Academy in 1997 with a B.Sc. Degree in Mechanical Engineering. He received his M.Sc. in Operations Research from Rutgers University in 2002. His research interests include Optimization algorithms, Data Mining, Big Data, and Stochastic Processes.

Serol Bulkan graduated from Management Engineering Department of Istanbul Technical University in 1986. He got his first MSc degree from the same department in 1989, a second MSc degree in Operations Research from Florida Institute of Technology in 1992 and a PhD degree in Industrial Engineering from Cleveland State University in 1999. Dr. Bulkan is currently at Industrial Engineering Department of Marmara University, Turkey. His teaching research areas are Operations Research, Scheduling Theory, Network Theory, Production Planning.

Brian Colburn was commissioned via the 2-Year Naval ROTC Program in December, 2001. He earned a Bachelor of Science degree in Civil Engineering from San Diego State University in 2001 and completed a Master of Science degree with an Energy Certificate in Operations Logistics from the Navy Postgraduate School, Monterey in March, 2015. LCDR Colburn is a qualified Surface Warfare Supply Corps Officer, Naval Aviation Supply Corps Officer, and Aviator.

Emily Craparo is an Assistant Professor of Operations Research at the Naval Postgraduate School. She obtained her Ph.D., S.M., and S.B. in Aeronautics and Astronautics at the Massachusetts Institute of Technology. Prior to joining the Operations Research faculty at the Naval Postgraduate School, she completed a National Research Council Postdoctoral Fellowship in the same department. Her research areas include mathematical modeling, optimization, and military applications.

Irem Duzdar Argun is an Assistant Professor Ph.D. in Department of Industrial Engineering instructor at Düzce University. She received her PhD in Engineering Management at Marmara University in Turkey. Dr. Irem DUZDAR ARGUN is currently working as Assistant Professor and the Head of the Department of Industrial Engineering in the Faculty of Engineering of the Duzce University.

Yeliz Ekinci is an Assistant Professor in the Business Informatics Department at Istanbul Bilgi University, Turkey. Her work focuses on decision making, statistics, and operations research topics. Her refereed articles have appeared in a variety of journals and in the proceedings of major national and international conferences.

Işın Erer has received B.Sc., M.Sc. and Ph.D. degrees from Istanbul Technical University in 1991, 1993 and in 2001 respectively, all in Electrical and Electronics Engineering. She has been a research assistant and also worked as a research associate in some of the SAR processing projects of ITU Remote Sensing Center between 1996-2001 Since 2002, she has been an associate professor at the department of Electronics and Communications, Istanbul Technical University where she continues to teach undergraduate/graduate level courses on digital signal processing, statistical signal processing, image processing and radar signal processing. Her research interests include remote sensing, image processing, SAR/ISAR imaging and ground penetrating radar. Dr. Erer is a member of IEEE Geoscience Society. and has been the author of several journal papers and conference proceedings in her research areas.

Levent Erişkin graduated from Turkish Naval Academy in 1998. He received his MSc. degree from US Naval Postgraduate School in Operations Research in 2003 and Phd. from Middle East Technical University in Industrial Engineering in 2015. He works as Logistics Analysis Division Head in Turkish Navy Headquarters, Ankara. His research interests include Multi Criteria Decision Making (MCDM), statistical learning and military operations research.

Murat Gunal is an expert in discreet event simulation. After he gained his BSc at the Turkish Naval Academy, he went on to gain an MsC in Operational Research and later his PhD in Management Science at Lancaster University. He's worked on a number of projects including modelling hospitals for performance management, developing a generic warehouse simulation for a large UK company, contact centre modelling for a UK police force, and air defence simulation modelling for the Turkish Navy. In addition, he worked in NHS hospitals in the UK, to map their healthcare processes. He also developed simulation models for pharmaceutical companies for health technology assessment. His view of simulation helped clients understand and structure problems, create solutions and implement these solutions successfully.

Batuhan Gundogdu received his B.S. degree from Electrical and Electronics Engineering Department, National Defense University, Naval Academy Istanbul, Turkey in 2006. He received his M.S. degree from the Ming Hsieh Department of Electrical Engineering, Viterbi School of Engineering, University of Southern California in 2010 where he studied with the special fellowship granted from the Turkish Naval Forces. He obtained his Ph.D. degree in the Electrical and Electronic Engineering Department, Bo˘gazic¸i University, Istanbul, Turkey in 2017. He is currently serving as an instructor at Electrical and Electronics Engineering Department, National Defense University, Naval Academy, Istanbul, Turkey. His research interests include machine learning, speech retrieval and statistical signal processing.

Ilknur Karacan graduated from Electrical & Electronics Engineering Department of Turkish Naval Academy in 2004. She achieved her MSc degree in Naval Operations Research from Naval Sciences & Engineering Institute of Turkish Naval Academy in 2015. She is a PhD candidate at Industrial Engineering Department of Marmara University since 2015. Her research areas are Operations Research, Multi Criteria Decision Making, and Scheduling Theory.

Ismet Karacan graduated from Industrial Engineering Department of Turkish Naval Academy in 2003. He achieved his MSc degree in Naval Operations Research from Naval Sciences & Engineering Institute of Turkish Naval Academy in 2014. He is a PhD candidate at Industrial Engineering Department of Marmara University since 2014. His research areas are Operations Research, Data Mining, and Scheduling Theory.

Melis Almula Karadayi is an Assistant Professor in the Industrial Engineering Department at Istanbul Medipol University, Turkey. Her areas of interest include healthcare management, waste management, multiple criteria decision making, data envelopment analysis, fuzzy optimization, and applications of mathematical programming and fuzzy set theory.

Deniz Kumlu received the B.S. degree in Electrical and Electronics Engineering Department fom Turkish Naval Academy, Istanbul, Turkey in 2007 and the M.S. degree in Electrical Engineering Department from University of Southern California, Los Angeles, U.S.A. in 2012. He is currently working as an instructor in the Electrical and Electronics Engineering Department at National Defense University, Naval Academy, Istanbul, Turkey and continue his Ph.D. degree in Istanbul Technical University, Department of Electronics and Communications Engineering, Istanbul, Turkey. His research interests include, clutter removal algorithms in ground penetrating radar, signal processing and pattern recognition.

Mehmet Gokhan Metin graduated from Computer Engineering Department of Turkish Naval Academy in 1997. He got his M.Sc. degree from Computer Engineering Department at Middle East Technical University in 2006. Mehmet Gökhan is currently a PhD student in Industrial Engineering Department at Marmara University. His research areas are Operations Research, Operations Management, and Scheduling Theory.

Barış Egemen Özkan graduated from Electrical Engineering/Control Systems department of Turkish Naval Academy in 1998. After serving 4 years on Turkish Navy as branch officer, he has received his first MSc degree in Computer Science from Naval Postgraduate School/Ca USA in 2004 with distinguished degree. Following his master degree, he started as an engineer at Turkish Navy Research Center Command. He received his second degree in MBA from Gebze High Tech Institute in 2006. While he was working at various roles in project development and management for various military systems, he had quite extensive experience on military system development and the cost impacts of using COTS parts in military systems over total product life cycle. He is currently pursuing his PhD in Marmara University/Istanbul in Industrial Engineering Department in Operations Research, researching Obsolescence Management for Military Systems using predominantly COTS parts.

Nasuh Razi graduated from the Turkish Naval Academy in 2009. He completed his MS study in Naval Operation Research at the Naval Sciences Institute at the Turkish Naval Academy in 2016. His research areas include optimization, simulation and military applications.

Manuel Rosas is a native of Bucaramanga, Colombia. He graduated in 1994 from Universidad Industrial de Santander in Colombia where he received his Bachelor's degree in "Ingeniería de Sistemas". He is also a graduate of the Universidad Santo Tomas in Colombia, holding a Master's degree in Business Administration, and the Excelsior College, NY with a Bachelor's degree in Liberal Arts. LCDR Rosas earned a Master's degree of Science in Operations Research at the Naval Postgraduate School in Monterey, CA in 2017. Lieutenant Commander Rosas enlisted in the US Navy in 2003 as Ship's Serviceman and received his commission in 2006 through the Officer Candidate School. He is a qualified Surface Warfare Supply Corps Officer, and he currently serves as Operations Research Analyst at NAVSUP WSS.

Murat Saraçlar received his B.S degree in 1994 from the Electrical and Electronics Engineering Department at Bilkent University, M.S.E. degree in 1997 and Ph.D. degree in 2001 from the Electrical and Computer Engineering Department at the Johns Hopkins University. From 2000 to 2005, he was with the multimedia services department at AT&T Labs Research. In 2005, he joined the Boğaziçi University Electrical and Electronic Engineering Department where he is currently a full professor. He was a visiting research scientist at Google Inc, New York (2011-2012) and an academic visitor at IBM T.J. Watson Research Center (2012-2013). Dr. Saraçlar was awarded the AT&T Labs Research Excellence Award in 2002, the Turkish Academy of Sciences Young Scientist (TUBA-GEBIP) Award in 2009, and the IBM Faculty Award in 2010. He has more than 100 publications in journals and conference proceedings. He served as an associate editor for IEEE Signal Processing Letters (2009-2012) and IEEE

Transactions on Audio, Speech, and Language Processing (2012-2016), was on the editorial board of Language Resources and Evaluation journal (2012- 2016) and is currently serving on the editorial board of Computer Speech and Language. He is also a member of the IEEE Signal Processing Society Speech and Language Technical Committee (2007–2009, 2015-2018).

Tolga Temucin graduated from Industrial Engineering Department of Turkish Naval Academy in 2004. He got his MSc degree from the same department in 2012. Currently, Tolga Temuçin is a PhD student in Industrial Engineering Department of Marmara University. From 2012 to present, he has been studying as a force structure analyst at Scientific Decision Support division within Turkish Naval Headquarters, Ankara, Turkey. His research areas include primarily related to Operations Research, Multi Criteria Decision Making, Production Planning, and Forecasting Techniques/Time Series. He has published papers in several international journals.

Ertan Yakıcı graduated from the Turkish Naval Academy, Istanbul, Turkey, in 1999, He received his M.S. degree in Industrial Engineering from the Georgia Institute of Technology, Atlanta, USA, in 2004 and his Ph.D. degree in Industrial Engineering from the Middle East Technical University, Ankara, Turkey, in 2013. From 2004 to 2014, he served as an analyst in the Department of Decision Support in the Navy. In 2014 he joined the Department of Industrial Engineering in Turkish Naval Academy of National Defense University, Istanbul, Turkey. He had been in the Naval Postgraduate School, USA as a visiting professor between 2015 and 2016. His current research interests include Military Operations Research Problems, Routing Problems and Location-Routing Problems.

Oktay Yılmaz graduated from the Turkish Naval Academy, Istanbul, Turkey, in 2010. He is a graduate student pursuing M.S. degree in Industrial Engineering in Barbaros Naval Sciences and Engineering Institute of National Defense University.

Index

I

K

L

M

N

O

P

Q

R

S

T

U

W

www.ingramcontent.com/pod-product-compliance
Lightning Source LLC
LaVergne TN
LVHW080533300926
849162LV00007B/1596